Food Identity Preservation and Traceability

Safer Grains

Food Identity Preservation and Traceability

Safer Grains

Gregory S. Bennet

CRC Press
Taylor & Francis Group
Boca Raton London New York

CRC Press is an imprint of the
Taylor & Francis Group, an **informa** business

CRC Press
Taylor & Francis Group
6000 Broken Sound Parkway NW, Suite 300
Boca Raton, FL 33487-2742

First issued in paperback 2017

ISBN 13: 978-1-138-11776-1 (pbk)
ISBN 13: 978-1-4398-0486-5 (hbk)

Library of Congress Cataloging-in-Publication Data

Bennet, Gregory S.
    Food identity preservation and traceability : safer grains / Gregory S. Bennet.
       p. ; cm.
    Includes bibliographical references and index.
    ISBN 978-1-4398-0486-5 (hardcover : alk. paper)
    1. Food--Safety measures. 2. Grain. I. Title.
    [DNLM: 1. Food Contamination--prevention & control. 2. Food Industry--standards.
WA 701 B469f 2010]

    TX531.B46 2010
    363.19'26--dc22
                                                    2009030740

Visit the Taylor & Francis Web site at
http://www.taylorandfrancis.com

and the CRC Press Web site at
http://www.crcpress.com

To my wife, Cynthia D. Bennet, who has been tremendously supportive in my research, and for her endless encouragement and patience. Words cannot express what she means to me—I am so fortunate.

Contents

SECTION I General Introduction

SECTION II   Programs and Standards

SECTION III   Auditors and Laboratories

SECTION IV   Consultative and Service Contributors

SECTION V Research Instruments

List of Figures

List of Tables

Preface

Melamine in milk, *Salmonella* in peanut products, tainted ingredients, *Aflatoxin* in pet food, counterfeit ingredients, and *Escherichia coli* found on peppers are becoming common headlines on the evening news. These occurrences will be increasingly more common as global trade of agricultural products expands. Hand in hand with this will be the nearly epidemic numbers of people becoming sick and dying from food-related safety issues. The world population is expected to reach 10 billion by 2050. Production agriculture and the food supply chain are already facing challenges because of environmental degradation and climate change, which further reduce the land available for production. Scarcity of food and greed for profit push some producers to cut safety corners or to outright substitution of inferior or counterfeit ingredients. In researching the root causes and common elements of the expanding food safety issue, I discovered how diverse, broad, and fragmented the food system was. Books on how the food safety system works with regard to how identity preservation and traceability affects food safety are few and focus on selected, more narrowly focused aspects of the food supply chain. Regulations, articles, and software packages abound regarding particular issues; yet, there is little offered that pulls the entire food safety system together. Thus, this book was written.

This book explains, in a working person's terms, the reasons why and how identity preservation and traceability helps improve food safety. The hope is that the reader may gain a better understanding of the complexity of identity preservation and traceability systems, rules that it functions under, how identity preservation and traceability is shaped and modified, and primary support and ancillary components that govern it.

The approach used to read this book is very much dependent upon the reader's needs and desires. The table of contents is key for navigation and provides the best map of the book. For those who only wish for a basic understanding of identity preservation and traceability, Part I provides the basics, history, design, and components. Part II offers abridged versions of the programs and available standards through which identity preservation and traceability navigates. Auditing and laboratory testing are becoming increasingly important and required. Part III provides an introduction and summary of the more popular auditing programs and laboratory testing services. Especially important are the outside forces that can shape identity preservation and traceability rules, regulations, and efficiency. Part IV provides the reader with a sampling of organizations that influence the direction of identity preservation and traceability, food safety policy, software systems, process facilitators, food recalls, and insurance. Part V includes research instruments for evaluating compliance, cost-benefit analyses, and farmer surveys. This work is generally encyclopedic in nature rather than narrative and is intended as a reference work. Every effort has been made to provide the identity preservation and traceability story in sequential or hierarchal order; unfortunately this is not always possible because the identity preservation and traceability story is diverse and fragmented.

Walking through a bookstore or browsing online, a reader finds numerous books and articles that highlight genetically modified organisms and food recalls. Some books and articles have titles such as from "Farm to Fork" or "From Dirt to Dinner Plate." This book goes into greater detail about the components and structure of the food supply system than previous works and, yet, without getting into too much detail, provides summaries and glimpses of rules, regulations, systems, companies, etc., that impact and contribute to food identity preservation and traceability. We have seen how the commercial market has jumped in with solutions that range from radio frequency identification technology and deoxyribonucleic acid laboratory testing to third-party auditors and network providers. None of these examples truly encompasses the enormity of identity preservation and traceability within the food supply chain. This book is the first to provide the complete story of identity preservation and traceability as it applies to food safety. Any chapter could be expanded upon and developed into a more finely focused subject. However, what makes identity preservation and traceability unique, as with this book, is in its approach—it necessitates interactions between businesses, computer systems, and policy organizations. The strength of this book is that it addresses and pulls together the various independent yet interactive parts of the food supply chain to develop a holistic view of how identity preservation and traceability contributes to food safety.

This book, although not exhaustive, is illustrative of trends. Identity preservation and traceability are not new concepts; however, the growth of public and business interest and concerns regarding identity preservation and traceability has grown tremendously during the past decade because of many food safety events. This has resulted in these concepts joining together (identity preservation and traceability) within a single concept. This book, while attempting to be thorough, will highlight selected major systems of identity preservation and traceability from a U.S. grain perspective. It is understood that fundamental identity preservation and traceability concepts follow similar systems structures regardless of product (i.e., livestock, fruits and vegetables, fish, etc.).

Most recently, identity preservation and traceability of food has fallen under regulatory requirements to protect against bioterrorism. The system of documentation is moving from paper-based to computer-or electronic-based for storage and information sharing. However, many of our more modern technology systems (computer hardware and software) are still very fragmented, discrete, and uncomplimentary in regard to integrating individual identity preservation and traceability systems with one another along the supply chain. In addition, training of management on identity preservation and traceability processes needs to be more thorough, transparent, linked, and standardized to improve interactions and performance. The cost of diverse government regulations, differing proprietary services, and incompatible commercial solutions to the consumers, companies, and the global supply chain calls for defining identity preservation and traceability as a business process. This book helps bring together not only the many primary and secondary participants but also the ancillary groups, which help shape our food supply system.

NOTE ON THE LITERATURE

This book contains information obtained from a wide variety of highly regarded references and sources. Numerous resources provided information regarding facets of identity preservation and traceability. Of special note are the works from the U.S. Department of Agriculture written by Elise Golan and others. The most comprehensive works about the state of the art of identity preservation and traceability were by Dennis Strayer, *Identity-Preserved Systems: A Reference Handbook* (2002) and *Improving Traceability in Food Processing and Distribution*, edited by Ian Smith and Anthony Furness (2006), for which I wish to thank them for their contribution to this field of research.

DISCLAIMER

The information within this book is derived from official websites and published literature as cited. Excerpts from these sources have been used and condensed for brevity and all efforts have been made to credit these sources. It is the intent of the author to not change the meaning or intent of the original work or publication. Any omissions or errors are solely the responsibility of the author; however, reasonable efforts have been made to publish reliable data and information. In addition, the author is not responsible for claims made by individuals or organizations as to being true. The use of product or service names does not imply endorsement by the author. This overview is intended to assist the reader to better understand the range and scope of identity preservation and traceability as it applies to local to global food chains.

Acknowledgments

I wish to thank Iowa State University's faculty and staff for their support and guidance, and especially Dr. Charles Hurburgh for the resources and opportunities provided in the development of the foundational materials for this book. I also wish to thank Iowa State's Grain Quality Laboratory and the Iowa Grain Quality Initiative for their support.

I also want to thank the faculty of Iowa State University's history department for their guidance in preparing me for my research and writing, and for their enthusiastic support of this book.

Finally, the team of experts that embody the Taylor and Francis publishing family (at CRC Press) have my heartfelt appreciation, especially Steve Zollo (Senior Editor, Food Science & Technology), Andrea Grant (Editorial Assistant), Amber Donley (Project Coordinator, Editorial Project Development), Glenon Butler (Project Editor), Jennifer Spotto (Senior Marketing Manager), Katherine Jones (Copywriter, Creative Services), James Miller (Art & Design), Mindy Rosenkrantz (Permissions Coordinator), and the Cadmus Communications staff of Marc Johnston (Senior Project Manager) and Chris Schoedel (Assistant Project Manager) for their patient guidance and help during this process.

While I had the help and support of many people and organizations, any omissions or errors in this publication are solely my own responsibility. It is my goal to fully credit and retain the original intent of others' work that I have included in my publication. Every effort has been made to publish reliable data and information.

Author

Gregory S. Bennet had a successful military career in the U.S. Army and U.S. Coast Guard before retiring and pursuing his PhD degree. A helicopter rescue pilot, Greg was involved in the weather phenomenon now known as "The Perfect Storm." He achieved an MBA and a Master's degree in history, focusing on agricultural business. His PhD is in agricultural and biosystems engineering with a focus on sustainable agriculture. Greg and his wife Cynthia live in Ames, Iowa.

List of Abbreviations

AI—application identifier
ALOP—appropriate level of protection
AMS—Agricultural Marketing Service
APEC—Asia-Pacific Economic Cooperation
ASEAN—Association of Southeast Asian Nations
ASN—Advance Shipment Notice
BEM—biological exposure monitoring
BOL—bill of lading
BT Bt—*Bacillus thuringiensis*
CBD—Convention on Biological Diversity
CCIA—Canadian Cattle Identification Agency
CCP—critical control point
DNA—deoxyribonNucleic acid
DOA—Department of Agriculture
DOT—Department of Transportation
ECCC—Electronic Commerce Council of Canada
EDI—electronic data interchange
EIA—enzyme immunoassay
ELISA—enzyme-linked immunoabsorbent assay
EPA—Environmental Protection Agency
EPC—Electronic Product Code
FAO—Food and Agriculture Organization of the United Nations
FDA—Food and Drug Administration
FSMS—Food Safety Management System (or Standard)
GAP—Good Agricultural Practices
GATT—General Agreement on Tariffs and Trade
GC—gas chromatography
GDPs—Good Distribution Practices
GE—genetically engineered
GEPIR—Global EAN Party Information Register
GFSI—Global Food Safety Initiative

GIAI—Global Individual Asset Identifier GLN—Global Location Number
GMO—genetically modified organism
GMPs—Global or Good Manufacturing Practices
GRAI—Global Returnable Asset Identifier
GTIN—Global Trade Item Number
HACCP—Hazard Analysis and Critical Control Point
HPLC—high-performance liquid chromatography
IAEA—International Atomic Energy Agency
INFOSAN—International Food Safety Authorities Network
IP—identity preservation
IP—intellectual property
IPM—integrated pest management
IPM—integrated product management
IPPC—International Plant Protection Convention
IPT—identity preservation and traceability
ISO—International Organization for Standardization
JECFA—Joint FAO/WHO Expert Committee on Food Additives
JEMRA—Joint FAO/WHO Expert Meetings on Microbiological Risk Assessment
JMPR—Joint FAO/WHO Meetings on Pesticide Residues
MRL—maximum residue limit
NAFTA—North American Free Trade Agreement
NGO—non-governmental organization
NOP—National Organic Program
NOSB—National Organic Standards Board

OECD—Organisation for Economic Co-operation and Development
OIE—World Organization for Animal Health
OSHA—Occupational Safety and Health Administration
PCR—polymerase chain reaction
PO—purchase order
POS—point of sale
PPB or ppb—parts per billion
PPM or ppm—parts per million
RFID—radio frequency identification
RSS—reduced space symbology
SCC—shipping container code
SCC—Standards Council of Canada
SQFI—Safe Quality Food Institute

SOP—state organic program
SSCC—serial shipping container code
SSM—supportive safety measure
TBT—technical barriers to trade
TCC—Transaction Certificate of Compliance
UCC—Uniform Code Council
U.N.—United Nations
UNECE—U.N. Economic Commission for Europe
UPC—Universal Product Code
UREC—unavoidable residual environmental contamination
WHO—World Health Organization
WTO—World Trade Organization
XML—Extensible Markup Language

Section I

General Introduction

INTRODUCTION, HISTORY, AND THEORY; DESIGN; AND COMPONENTS OF IDENTITY PRESERVATION AND TRACEABLILITY

Part I of this work provides the reader with an overall introduction to identity preservation and traceability (IPT). The idea of identity preservation (IP—tracking from origin to customers), traceability (tracing from the customer to origin), and their incorporated systems and programs has become increasingly important to customers from local food markets to global traders. The first three chapters bring together the story of IPT. The first chapter provides an introduction (the fundamentals of IPT), the second chapter provides an overall historical view of how it came into being, and the third chapter covers IPT theory, design, components, an interpretation, analytical techniques, and introduction to batch processing challenges.

Although the story's origins appear fragmented and disconnected, the resultant systems and programs come together as organizations, and various entities bring forth solutions to sometimes abstract questions or demands that society asks of its food supply system.

The subsequent parts include Part II, covering IPT programs and standards; Part III, which concerns auditors and laboratories; Part IV discusses consultative and service contributors; and Part V covers scorecard matrices, spreadsheets, and questionnaires. At the very end is the interpretation and conclusions.

A reminder to the limitations of this work: this section is not designed to be interpretive or judgmental. The goal of the main body of the state of the science is to provide an introduction to and summary of IPT systems and programs available and develop a conceptual model of IPT at the farmer level. Interpretations regarding the IPT are at the end of this work.

1 Introduction to Identity Preservation and Traceability

1.1 INTRODUCTION TO IDENTITY PRESERVATION AND TRACEABILITY

This work attempts to describe the who, what, where, when, and why of identity preservation and traceability (IPT) as it applies to the food chain up until early 2007. The perspective is primarily of a U.S. grain production viewpoint. However, many other views are included.

The information obtained is derived from official websites and published literature. Excerpts from these sources have been used and condensed for brevity, and all efforts have been made to credit these sources. It is the intent of the author to not change the meaning or intent of the original publication. Any omissions or errors are solely the author's. However, the author is not responsible for claims made by individuals or organizations as to being true. This work is a compilation of many diverse entities that go into an IPT system. This overview is intended to assist and better understand the range and scope of IPT.

So what is IPT? First, we must explain each of these terms.

1.2 IPT SYSTEM

IPT is considered a market solution system (singular) that answers two market needs. The first, identity preservation (IP), holds the notion that any given product has a value (which is desirable to maintain for various consumers), from less valuable commodity grains—U.S. Department of Agriculture (USDA) inspected—to more valuable specialty crops (e.g., organic certified). To accomplish this, businesses implement systems to preserve particular trait(s) and credence attribute(s). The second market need, traceability, is needed for business logistics purposes and many times is required by food safety regulations. For business, this represents inventory control and a method to recall defective products; for food safety, this represents the mechanism during an outbreak of disease to remove affected products and locate the source of contamination. For food chain participants, the tracking (from seed to plate) and tracing (from outbreak to source) often entails using one and the same paper and/or electronic documentation procedures, tests, certifiers, etc. IPT represents a system or program in which industry can meet the traceability requirements that society demands and also profit, with overlapping systems, by providing increased identity-preserved product for lower costs.

IP envelops the idea that specific traits and/or credence attributes are important for various customers to maintain or realize. Often the term "value-added" is used, especially to connote an economic aspect of a trait for the farmer, processor, or society. For soybean or corn farmers, the traits of interest when they purchase their seeds may include oil and protein content of harvested crop, harvest yields, drought tolerance, Roundup Ready, etc. Farmers hope to gain increased profits from greater yields or less use of pesticides. For grain elevators, the traits of interest may be in accepting yellow versus white corn, genetically modified (GM) grain versus non-GM grain, etc. The difference in quality and content may affect income from contracts. For processors, the traits of interest may be starch content, or how well certain varieties process or extend shelf life.[*]

In addition to physical traits of interest, there are "credence attributes" of interest. Crop or product innovations may involve credence attributes, characteristics that consumers cannot discern even after consuming the product. Credence attributes can describe content or process characteristics of the product. ***Content attributes*** affect the physical properties of a product, although they may be difficult for consumers to perceive. For example, consumers are unable to determine the amount of isoflavones in a glass of soymilk, or otherwise distinguish between conventional corn oil and oil made from genetically engineered (GE) corn. ***Process attributes*** do not affect final product content but refer to characteristics of the production process. Process attributes include country of origin, organic, free-range, animal welfare, dolphin-safe, shade-grown, earth-friendly, wage and fair-trade, etc. In general, neither consumers nor specialized laboratory testing equipment can detect process attributes. Governments may also be interested in the origins of the food or origins of a particular process, thus providing a form of brand or regional name of value and labeling regulations. All of these traits, many others not mentioned, and some yet to be determined, are traits and credence attributes that comprise identity preserved products[†] (Golan, Krissoff, and Kuchler 2004).

Third-party verification may be used to ensure credence attributes or content attributes that are difficult or costly to measure. The only way to verify the existence of these attributes is through recordkeeping that establishes their creation and preservation. Government may also require that firms producing foods with credence attributes substantiate their claims through mandatory traceability systems; for example, some governments require that firms producing organic foods verify their claims. If firms are not required to prove that credence attributes exist, some may try to gain price premiums by passing off standard products as products with credence attributes (Golan, Krissoff, and Kuchler 2004).

---

[*] These are terms used throughout this paper. Sometimes genetically modified organism (GMO) and GM are used interchangeably. I attempted to standardize their use but reverted to using the same description as the organization uses the term. Therefore, if there is a noted difference throughout the book, it is because of the organization's use of a term.

[†] Functional Food—New concepts like functional food, nutraceuticals, fortified foods, and dietary supplements are created by the industry trying to open new market segments.

1.3 THE DILEMMA: WHAT CONSUMERS WANT AND ARE WILLING TO PROVIDE FOR

For IP to be credible, it must have a tracking mechanism and be profitable. The term "IP-*T*" is used here to represent IP and its tracking mechanism. Generally, IP-*T* works from the food origins, includes many processes and events, and continues up until the final purchase. It is the way that IP products retain their value-added qualities or credence attributes. On the other hand, traceability (sometimes referred to as "back-tracing") works in the opposite direction—from consumer or store shelf backward to the food or ingredient's origin or source. Traditionally this has been used for business logistics to know when a product was sold or ingredient consumed. As in recent food scares, traceability has been used for food recalls, mislabeling, etc., and has been used as a tool to more quickly remove selected products from the market.

1.4 THE SECOND PART OF IPT IS TRACEABILITY

Traceability has existed for years, although it has and does go by other names such as logistics control, inventory management and, on the food safety side, involves product recalls. Historically, when a defect or mislabeling occurred the firm recalled the defective product. When the defect or contamination was found, the organization would attempt to "back trace" to locate the source of deficiency. Lots, batches, pallets, and production lines would be involved and checked. Traceability uses informal (industry) and formal (national) rules and regulations. Traceability mechanisms were traditionally regulated primarily by industry; however, because of recent food security issues, the guiding force behind mandatory traceability has been government.

Many organizations are developing or including a system, be it under quality control, safety, etc., that utilizes both an IP tracking IP-*T* system (this may include documentation, audits, and laboratory testing) and traceability (back-tracing) system (this too may include, but to a lesser degree, documentation, audits, and laboratory testing).

As unique and different as IPT systems are from each other, they both utilize many of the same concepts and processes, documentation, third-party audits, laboratory tests. Each of these concepts may start from opposite ends of the food chain from one another; however, each system will incorporate the functions of documentation, auditing, and tests to insure IP or traceability (see Figure 1.1 for illustration).

Terminology Review

Identity Preservation (IP): Trait(s) and/or credence attribute(s) of interest.

Identity Preservation-Tracking (IP-*T*): Mechanisms that track product or ingredient from origin to customer.

Traceability (T): Mechanisms that trace product or ingredient from shelf backward to origin or source; for example, from a consumer or point of food safety event back through the various processes and players, then back to the source of defect.

Identity Preservation & Traceability (IPT): Includes the mechanisms that enable both way tracking and tracing by paper and/or electronic trails, and may include third-party audits and laboratory tests. The mechanisms that track forward or trace backward need not be exclusive.

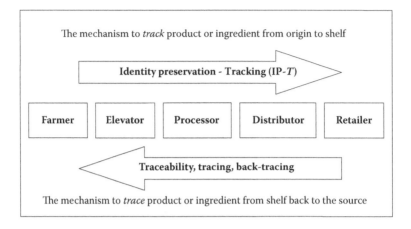

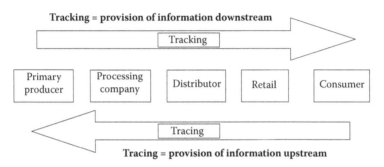

FIGURE 1.1 Terminology review. (Reprinted from Schwägele, F, *Meat Science* 71, 164–173, 2005. With permission.)

1.5 AN IMPORTANT NOTE

Many published works use the terms "track" or "trace" interchangeability when pursuing forward or backward information of a product or ingredient within the food supply chain. Thus it is important to understand for what purpose(s) the tracking or tracing is being used. For this book, *tracking* will always be regarded as the mechanism used to follow a product or ingredient; for example, from seed through various processes and entities on until the product is purchased. *Tracing* will always be regarded as the mechanism used to follow a product or ingredient from the point of sale or concern (e.g., mislabeled product on shelf) backward through the various entities, processes, and players, on until the source of defect or event origins.

Another good pictorial example of how both tracking and tracing systems work together is from John Deere FoodOrigins' illustration (Figure 1.2). Although they use this diagram to promote their own IPT software program, the illustration does graphically point to the connectivity of the food supply chain and how easy it would be, in the case of recall, to recall nontargeted product or ingredient, which was not involved with recall, merely because it was a similar or like ingredient. This was the

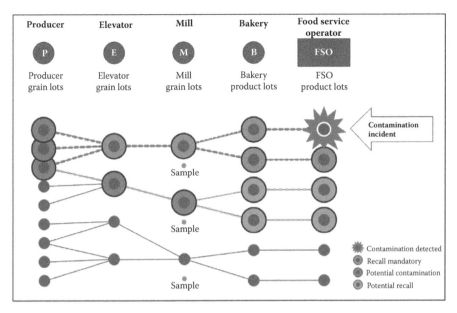

FIGURE 1.2 FoodOrigins' tracking and tracing. (From Bantham, A. "Rapid Response—Identification and Isolation." In *International Food and Agribusiness Management Association Deere & Company USA*, Slide 11, 2004.)

case with the 2006 spinach recall in the United States, in which nearly all spinach was recalled because of weak traceability programs by that industry.

1.6 WHY IPT CAME INTO EXISTENCE

Food security has long been an issue for society. For the most part, local governments through laws and codes tailored food safety to meet local needs that dealt with growing, cooking, labeling, and packaging of food. Usually this was enough, or at worst kept pace to meet situations such as regional disease outbreaks, mislabeled products, or production hygiene issues. However, most recently two major events affecting two different large regions have affected local, national, regional, and international consumers. (Chapter 6 provides greater detail of how the various standards are implemented.)

In Europe, the crisis that damaged the public's confidence in their food safety was the outbreak of bovine spongiform encephalopathy (BSE), or "mad cow disease", which overwhelmed authorities. As perceived by the populace, government could not handle the outbreak, was ill prepared, and fell short of expectations in protecting consumers. The drive for strict standards imposed by the grocery industry across Europe was furthered by a series of food safety crises, including diesel fuel in palm oil, sewage waste in feed, *Listeria* in cheese, *Salmonella* and antibiotics in poultry, and *Escherichia coli* in animal meat, which undermined consumer confidence in their food supply (Moe, 1998). Shortly after these food crises, activists and environmentalists pointed toward the next possible threat to the food supply and the environment—genetically modified organisms (GMOs). These groups' concern was well justified, as far as governments' ability to conduct proper oversight and protect its people because governments fell short of expectations. Governments and the food industry let the customer or consumer down. These groups gained a greater voice in heralding the dangers of GMOs and pressured European producers to restrict perceived unsafe, untested food products. A decade of food safety scares, and well-organized "green" and consumer movements in Europe revolving around food crises, have had greater results than pressures put on American producers. Thus, European agriculture moved more aggressively to institutionalize changes than the United States. Europe's approach to food safety is in its mandatory government mandate of rules and laws, which involve documentation, testing, tolerances, and labeling. To protect its food system the European Union (EU) uses the "precautionary principal" to guide it in its determination of whether or not a food is safe (Glassheim, Nagel, and Roele 2005).

The United States has had its share of food safety incidents, although none reached the near panic level of concern that was felt in Europe toward food safety. However, for the United States, the attacks on the World Trade Center's twin towers and Pentagon on September 11, 2001 heightened the government's and public's concern over food safety issues because of terrorism, and more specifically, bioterrorism. The notion that terrorists could contaminate crops and livestock along any part of the food supply chain scared authorities. The solution was not much different than that of European authorities' traceability and the ability to trace food products backward. The major difference in philosophy between the United States and Europe is that in Europe the rules of how to accomplish compliance are most often determined or directed by government. In the United States, the government determines the requirements or criteria and lets the market (i.e., industry, producers, etc.) determine the best course of action or how to meet government mandates. Regarding GMOs, for Americans, the risks and threat from GMOs are minimal. Americans have had years of GMO use and consumption, so aside from government approval for various crops used for human or animal consumption, labeling of GMO content is not required. The notion of "substantially equivalent" is how the U.S. government views approved GMO products.

For the Australian food industry, additional examples of recent food safety incidents illustrate the need for greater emphasis on food safety, including: (1) the 1995 *Garibaldi* incident, in which one person died and 23 people were hospitalized; (2) the 1996 salmonellosis scare in peanut butter; and (3) the endosulfan scare, in which endosulfan was detected in meat for export. Other food safety incidents raised public awareness of these issues as well, and thus all manufacturers are extremely conscious of what the implications are if such a food safety incident should occur in their industry (Smith 1998) (see Chapter 6 on the SQF Institute).

Since the occurrences of many of the above incidents, follow-up issues of tracking food shipments to reduce the risk of tampering, and on traceability systems to detail country of origin, animal welfare, and genetic composition have become paramount. In addition, tracing particular risks identified in the areas of chemical hazards (chemical residues, weed seed toxins) and microbiological hazards (mycotoxins and *Salmonella*) have been included in many countries' new and improved food safety regulations. Heightened awareness of food-related safety issues among today's consumers, coupled with a more educated public, is driving the demand for more information about food's vertically integrated supply chain. Recent animal health and food-borne illness scares in all parts of the globe are creating a demand for source verification, food safety, and supply chain identification of food products.

There are several factors driving food safety. They include the following:

- Increased consumer awareness
- Tighter government regulations
- Increased scientific knowledge and more accurate methods of testing
- Increased publicity given to food safety incidents in recent years
- Increased number of value-added products on the market

1.7 TOWARD CONSUMER SOLUTIONS— SAFETY AND CONSUMER CHOICE

The increasing implementation of Good Manufacturing Practice (GMP) and International Organization for Standardization (ISO) 9000 quality management in food manufacturing have resulted in traceability systems becoming more advanced and involving increased amounts of information and more steps in the production chain. However, the BSE crisis and debates about transgenic crops have drawn new attention to chain traceability (Moe 1998). Increased awareness of food safety issues among consumers, along with a more educated and informed public, is driving the demand for more information about the food supply chain. Recent animal health and food-borne illness scares from all corners of the world are creating increased demand for source verification, food safety, and supply chain identification of food products. Although most industries and governments have established processes and systems to ensure food quality and safety (i.e., Hazard Analysis and Critical Control Points, or HACCP), these systems are often applied independently at various points in the food continuum. Traceability systems assist by making the necessary linkages between a specific product and the application of these food safety and quality assurance systems at various points along the food continuum (Can-Trace website).

From a public health perspective, improving the speed and accuracy of tracking and tracing food items can help limit the risk associated with a failure in the system. Rapid and effective traceability can also minimize the unnecessary expenditure of private and public resources and reduce consumer concerns. Furthermore, tracing food items may help public health services and industry operators in determining potential causes of a problem, thereby providing data to identify and minimize food-borne public health hazards (Can-Trace, http://www.can-trace.org).

1.8 TRACEABILITY BENEFITS FOR BUSINESS

Traceability provides many potential benefits for business, including

- Meeting regulatory requirements
- Recall and risk management: perception related to reduced risks

- Process improvements—efficiency and quality: improved customer service/ response time
- Addressing customer and market needs

The last bullet highlights where businesses can benefit from government-required traceability mandates by businesses being able use their traceability infrastructure to focus on customer and market needs, which attest to prescribed traits and credence attributes of interest. This notion of providing what can be considered value-added often results in a new profit center for the company and additional benefit to consumers, the environment, animals, the region, etc.

Thus, from a business perspective, the requirements of government to enforce traceability regulations and resultant corporate infrastructure to support this mandate help facilitate the aspects of IP of traits and credence attributes of interest. A business or corporation that effectively combines both traceability and IP is said to have a bidirectional IPT program or system.

1.9 WHAT HAS BEEN ESTABLISHED IN RESPONSE TO FOOD CRISES—WHAT IS OUT THERE?

Globally, many changes have occurred regarding traceability during the past decade. Still, nations and regions around the world have reacted in different ways. Of particular clarity is Guillaume P. Gruère's work (2006) titled *An Analysis of Trade Related International Regulations of Genetically Modified Food and their Effects on Developing Countries*, which provides an excellent overview of traceability country by country. According to Gruère, because of consumer, environmental, ethical or political reasons, many countries have adopted stringent regulation regarding the approval and marketing of food and feed products, especially those derived from GM origins.

International regulations of GM food vary widely among developed countries. In particular, the EU and the United States have adopted different approaches on the marketing of GM food. EU regulations follow an approach based on the "precautionary principle" and consumers' "right to know," with stringent approval, labeling, and traceability standards on any food produced from or derived from GM ingredients. By contrast, the U.S. regulatory approach is based on differences in end-product characteristics and includes a voluntary safety consultation and voluntary labeling guidelines for GM food.* Most other developed countries, including Japan, Canada, or Australia, have introduced intermediary regulations that fall between mandatory and voluntary systems (Gruère 2006).

In the developing world, some of the large agricultural traders (such as Brazil) have developed biosafety and marketing regulations of GM food, but at the same time many other developing countries have not adopted any specific regulation of GM food because they lack the capacity to do so, or perhaps they have adopted a position of "wait and see" (Gruère 2006).

---

* Regarding labeling, nonsubstantially equivalent GM foods have to display the difference with conventional products, but there is no labeling requirement related to the fact that they were produced with genetic engineering.

As of 2005, 10 years after the introduction of the first GM crop, from nation to nation there is large variation in the regulation of GM food. At a macro level, countries can be divided as follows into three groups according to the status or type of their regulations: (1) countries with a comprehensive and stringent regulatory framework applied to GM food, including mandatory safety approval and mandatory labeling; (2) countries that have adopted a more pragmatic regulatory approach on the basis of the notion of substantial equivalence with voluntary labeling instead of mandatory labeling for GM food; and (3) many countries either without regulations or pending toward adopting certain regulations on GM food approval and marketing. Currently, developed countries are in the first and second group, whereas most developing countries are in the third group, with a few notable exceptions. The distinction between voluntary and mandatory labeling is important, because it drives several necessary regulatory requirements. *Mandatory* labeling requirement affects the whole agrofood channel from the retailers to the producers, requiring them to acquire and transmit information about the presence or origin for each food product, whereas *voluntary* labeling is driven by private incentives and the presence of market niches for non-GM food* (Gruère 2006).

Among the countries with regulations, there are two main groups of countries: those that rely on a test of substantial equivalence (substantial equivalent products are exempt from specific requirements) and those that generally do not and whose regulatory procedure depends on the production process (which means that any food produced with or derived from transgenic crops is subject to GM food regulations). Each country has also adopted its set of safety approval and labeling policies with specific characteristics. More stringent regulations will generally require more costly procedures on behalf of exporters and more comprehensive policies may have a more important trade effect (Gruère 2006).

The large producers and exporters of GM crops have well-defined regulations, but most of them are in Group 4 (Canada, United States, Argentina, South Africa), with pragmatic regulations of GM food, whereas the last two are in Group 2 (Brazil and China), with stringent regulations.

National regulations reveal that there is a large variation in regulations among countries, first in terms of development stages of regulatory framework, and second between countries with well-defined regulations. Developed countries differ in their general approach of regulations, with most GM producers and exporters in groups with pragmatic regulations whereas importers tend to have more stringent marketing regulations for GM food and GM derived products. Developing countries tend to have fewer regulations in place (Gruère 2006).

1.10 STANDARDS—REACTIONS TO FOOD SAFETY CRISES

For expanded information on standards see Chapter 6 on standards, which highlights and reviews various standards (i.e., United States, Canada, EU, international, organic, and regional, and religious).

---

* For more information on an economic comparison between voluntary and mandatory labeling, see Runge and Jackson (2003) and Carter and Gruère (2003).

ISO has referred to traceability in such a manner that others have borrowed from them.* ISO, which develops voluntary international standards for products and services, defines traceability as the "ability to trace the history, application, or location of that which is under consideration." This definition is quite broad. It does not specify a standard measurement for "that which is under consideration" (a grain of wheat or a truckload), a standard location size (field, farm, or county), a list of processes that must be identified (pesticide applications or animal welfare), or a standard identification technology (pen and paper or computer). It does not specify that a hamburger be traceable to the cow or that the wheat in a loaf of bread be traceable to the field. It does not specify which type of system is necessary for preserving the identity of tofu-quality soybeans, controlling the quality of grain used in a particular cereal, or guaranteeing correct payments to farmers for different grades of apples. This leaves much to be determined by producers, governments, and consumers (Golan, Krissoff, and Kuchler 2004).

According to Jenkins (2003), overall governmental traceability programs have focused on bioterrorism, GMOs, country-of-origin labeling (COOL), biofarming, overall food safety, and legislation to monitor the industry, which has created a more informed consumer base and that contributes to a shift in global food supply networks. Furthermore, consumers exert pressure on farmers, food processors, and manufacturers because of concerns about overall safety and genetic heritage of the groceries they purchase. Food producers differentiate products over a wide variety of quality attributes (taste, texture, nutritional content, origin); consumers can easily detect some attributes (color, etc.) but other innovations involve credence attributes (i.e., characteristics that consumers cannot discern even after consuming the product). Identification and traceability are essential for marketing food products, and, if food products are being differentiated via content and/or process credence attributes, record keeping, auditing and validation are essential elements of verification for IP and authenticity management (Smith et al. 2005).

1.10.1   EU Perspective

On the European continent, the general public has demanded increased food safety because of several food crises. Governments' response has been to establish traceability systems that provide information on origin, processing, retailing, and final destination of foodstuffs. Such systems enhance consumer confidence in food and enable the regulatory authorities to identify and withdraw health hazards from the market. Animal feeds are an element in this "food-to-farm" approach to public health. Such feedstuffs are preliminary elements of some foods for human consumption and hence are an inherent element of the food chain. A harmonized EU food traceability protocol greatly assists authorities in detecting fraud and dangerous substances. The food chain comprises a range of sequential and parallel stages,

---

* Some believe that ISO, as an international organization consisting of individual countries with equal voting rights or equal rights to voice concerns and participate, is a more democratic venue and less biased in their judgments.

bridging the full spectrum from agricultural production to the consumable food-stuffs by consumers[*] (Schwägele 2005).

The General Food Law (i.e., Regulation (EC) 178 (2002)) of the European Parliament and the Council outlines the general principles and requirements of food law, establishes the European Food Safety Authority (EFSA), and provides procedures in matters of food safety (i.e., among other things, the implementation of traceability systems in the food and feed supply chains in Europe) (Schwägele 2005).

The EU traceability legislation consists of four major points (excerpts and condensed from Schwägele 2005):

1. The *traceability* of food, feed, food-producing animals, and any other substance intended to be, or expected to be, incorporated into a food or feed shall be *established at all stages* of production, processing, and distribution.
2. Food and feed business operators *shall be able to identify any person* from whom they have been supplied with a food, a feed, a food-producing animal, or any substance intended to be, or expected to be, incorporated into a food or feed. To this end, such operators shall have in place systems and procedures that allow for this information to be made available to the competent authorities on demand.
3. Food and feed business operators *shall have in place systems and procedures to identify the other businesses* to which their products have been supplied. This information shall be made available to the competent authorities on demand.
4. Food or feed that is placed on the market or is likely to be placed on the market in the community *shall be adequately labeled or identified to facilitate its traceability*, through relevant documentation or information in accordance with the relevant requirements of more specific provisions.

1.10.1.1 Traceability along the Full Supply Chain

To be able to trace products and retrieve related information, producers must collect information and keep track of products during all stages of production (primary production, processing, distribution, retailing, and consumer). Therefore, traceability can be divided into two key functions: tracking and tracing.[†] Tracking can be defined as the ability to follow the path of an item as it moves forward through the supply chain from its origin to the shelf. Tracing is just the opposite and incorporates the ability to identify the origin of an item or group of items, through records, backward through the supply chain[‡] (Schwägele 2005).

Aside from mandated traceability, and depending upon the IP trait(s) or credence attribute(s) of interest, other process verifications and tests may need to be used. For

---

[*] Reprinted from Schwägele, F. "Traceability from a European Perspective." *Meat Science* 71, 2005: 164–173. Copyright 2005 with permission from Elsevier.

[†] This is the case where the term "traceability" incorporates the notion of tracking (origin to shelf) and tracing (shelf to origin).

[‡] The General Food Law covers the entire supply chain [Regulation (EC) 178 (2002), Article 18, paragraph 1].

example, within the EU, both farm and environmental sustainability have become hot topics. However, the meaning of sustainability differs from country to country.* In short, the idea of sustainability, in its broadest sense, should include elements of environmental health, societal and rural development, animal welfare, food quality and safety, and human health issues, which may require protocols, tests, and audits outside of normal food safety mandates (Glassheim et al. 2005).

1.10.1.2 EU Social Agenda

Within the EU, some see non-food safety issues or credence attributes as conflict of interests for society. This pressure is increased because of the tension between the expectations of the "citizen" and the "consumer" as two sides of mankind. For example, the citizen expects animal welfare, care for the environment, a nice landscape, and if possible, an organic agriculture. On the other hand, the consumer is not always prepared to pay an adequate price for these demands. Many producers find themselves caught between these two expectations, often mentioning that foreign competitors can sell food at lower prices because they have fewer environmental rules. Sometimes this is true, sometimes not. Psychologically, many farmers feel trapped between the supermarket (as a representative of the consumer) and the government (as a representative of the citizen). Within the EU, as in many other countries and regions, credence attributes of a social nature take on greater importance, especially when they have to do with local communities benefiting from brand naming their prized local product to the area or brand naming a process, plant, animal, or quality to that area or region. This represents an economic force that governments are dealing with (Glassheim et al. 2005).

1.10.2 U.S. Perspective

The events of September 11, 2001 in the United States caused Congress to recognize that safety of the nation's food supply could be compromised easily by a bioterrorist attack. In response, the U.S. Congress passed into law the Public Health Security and Bioterrorism Preparedness and Response Act on June 12, 2002. Under that law, the U.S. Food and Drug Administration (FDA) has authority to order the detention of any food if, as determined during an inspection, examination, or investigation, there exists "credible evidence or information" indicating that the article "presents a threat of serious adverse health consequences or death to humans or animals" (Smith et al. 2005).

Golan et al. (2004) concluded that U.S. private-sector food firms are developing, implementing, and maintaining substantial traceability systems designed to (1) improve food supply management, (2) facilitate trace-back for food safety and quality, and differentiate and market foods with subtle or undetectable quality attributes. Despite this, and although the United States has typically set the operating

---

* In northern Europe, talking about sustainability is commonly considered a discussion about environmental affairs. In southern Europe, more attention is paid to social issues. The discussion about sustainability can also be divided into two "mental maps": one group of people is in search of concrete, consistent, and scientific definitions of what sustainability is. The other group considers sustainability more as a process, even a political or societal process.

standard for international food handling, the U.S. food industry has been lagging in regards to food traceability. There is currently no standard process that identifies a traceable product, nor brand or social equity product.*

Studies within the United States have shown that (1) traceability is an objective-specific concept, (2) that the private sector in the United States has developed a significant capacity to trace, and (3) industry/product characteristics lead to systematic variation in traceability systems. Golan et al. (2004) found that efficient traceability systems vary across industries and over time as firms balance costs and benefits to determine the efficient breadth, depth, and precision of their traceability systems.

Government may consider mandating traceability to increase food safety, but this may impose inefficiencies on already efficient private traceability systems. The widespread voluntary adoption of traceability complicates the application of a centralized system because firms have developed so many different approaches and systems of tracking. If mandatory systems do not allow for variations in traceability systems, they will likely end up forcing firms to make adjustments to already efficient systems or creating parallel systems (Golan et al. 2004).

Not unlike the EU, fines become the tool of government to modify business behavior. Policy aimed at increasing the cost of distributing unsafe foods (e.g., fines or plant closures) or policies that increase the probability of catching unsafe food producers (e.g., increased safety testing or food-borne illness surveillance) also provide firms with incentives to strengthen their traceability systems. When the cost of distributing unsafe food goes up, so too do the benefits of traceability systems (Golan et al. 2004).

Although governments may define regulations, these are but tools for achieving several different objectives while dealing with a complex problem. As a result, no traceability system is complete. Even a hypothetical system for tracking beef, in which consumers scan their packet of beef at the checkout counter and access the animal's date and location of birth, lineage, vaccination records, and use of mammalian protein supplements, is incomplete. This system does not provide traceability with respect to bacterial control in the barn, use of genetically engineered feed, or animal welfare attributes like hours at pasture. This form of traceability is based on fulfilling regulatory requirements, which are generally broad and provide minimal hurtles, but to the contrary, IPT systems are usually tailored to customers' wants and their ability or willingness to pay (Golan et al. 2004).

A key notion with U.S. traceability is flexibility. A single system for tracking every input and process to satisfy every objective would be enormous and very costly. Consequently, firms across the U.S. food supply system have developed varying amounts and kinds of traceability. Firms determine the necessary breadth, depth, and precision of their traceability systems depending on characteristics of their production process and their traceability objectives. For example, an important aspect of developing regulations is appropriate focus. One difficulty with mandatory (EU) traceability is that they often fail to differentiate between valuable quality attributes (those for which verification is needed) and less valuable attributes for which no

---

* G. Smith et al., 2005 cites the Sparks 2002 publication, *Food Traceability: Standards and Systems for Tracing and Tracking Food and Agri-Products.* Memphis, TN: Sparks Companies, Inc.

verification is needed. This can be very costly for business and hurt trade or provide an unfair advantage to competitors* (Golan et al. 2004).

Within the United States, firms build traceability systems, aside from fulfilling rules and regulations toward food safety, to also improve supply-side management and construct lower-cost distribution systems. But simply knowing where a product is in the supply chain does not improve supply management unless the traceability system is paired with a real-time delivery system or inventory-control system (Golan, Krissoff, and Kuchler 2004 and Golan et al. 2004). A vital element of any supply management strategy is the collection of information on each product from production to delivery or point of sale; the idea is "to have an information trail that follows the product's physical trail." Throughout the food industry, companies are adopting new electronic traceability systems to track production, purchases, inventory, and sales to provide a basis for good supply management, allowing them to more efficiently manage resources (Smith et al. 2005).

1.10.2.1   U.S. Industry Efforts to Encourage Differentiation

Third-party entities provide objective validation of quality attributes and traceability systems. They reassure input buyers and final consumers that the product's attributes are as advertised. Third-party verification of credence attributes can be provided by a wide variety of entities, including consumer groups, producer associations, private third-party entities, and international organizations. For example, Food Alliance and Veri-Pure, private for-profit entities, provide independent verification of food products that are grown in accordance with the principles of sustainable agriculture. Third-party entities certify attributes as wide ranging as kosher, free-range, location of production, and "slow food." Governments can also provide voluntary third-party verification services. For example, to facilitate marketing, producers may voluntarily abide by commodity grading systems established and monitored by the government (Golan et al. 2004).

In some cases (e.g., branded pork, beef for export), verification is required. "To verify" is defined as "to prove the truth or accuracy of, or to substantiate, by the presentation of evidence or testimony." Source verification requires substantiation of the origin (e.g., breed, strain, geographic area) of the livestock, poultry, or meat.

---

* According to Smith et al. (2005), traceability of a food consists of development of "an information trail that follows the food product's physical trail," which may include process changes of importance to the customer and/or government regulations. Traceability (for livestock, poultry, and meat) in its broadest context, can, could, or will eventually be used to (1) to ascertain origin and ownership and to deter theft and misrepresentation of animals and meat; (2) for surveillance, control, and eradication of foreign animal diseases; (3) for biosecurity protection of the national livestock population; (4) for compliance with requirements of international customers; (5) for compliance with country-of-origin labeling requirements; (6) for improvement of supply-side management, distribution/delivery systems, and inventory controls; (7) to facilitate value-based marketing; (8) to facilitate value-added marketing; (9) to isolate the source and extent of quality-control and food-safety problems; and (10) to minimize product recalls and make crisis management protocols more effective. Domestically and internationally, it has now become essential that producers, packers, processors, wholesalers, exporters, and retailers ensure that livestock, poultry, and meat are identified; that recordkeeping ensures traceability through all or parts of the complete life cycle; and residuals that, in some cases, the source, the production-practices, and/or the process of generating final products, can be verified.

Production practice verification involves authentication of things done (e.g., grass-fed, free-range, raised/handled humanely) or things not done (e.g., no antibiotics, no hormonal growth promoters, not fed animal byproducts) during rearing of the animals. The USDA Process Verification Program (PVP) provides suppliers of agricultural products the opportunity to assure customers of their ability to provide consistent quality products. The PVP is accomplished by having documented manufacturing processes verified through independent, third-party audits, and it enables suppliers to make marketing claims (e.g., breed, feeding practices, or other raising and processing claims) and market themselves as "USDA Process Verified." Beef export verification is based on substantiation of conditions required by an importing company, of the exporting country, as verified by the USDA Quality System Assessment (QSA) program (e.g., beef export verification, Japan) (Smith et al. 2005).

1.11 HOW IP-*T*, TRACEABILITY, AND IPT SYSTEM PROGRAMS WORK: THE FUNDAMENTALS

1.11.1 IP-*T*

The global agricultural commodity system is being revolutionized as an increasing number of crops and livestock are being differentiated to ensure that their value or uniqueness is captured and maintained throughout the supply chain (Smyth and Phillips 2002). Again, IP refers to the trait(s) or credence attribute(s) of interest, whereas IP-*T* refers to the mechanism of software, documentation, tests, and audits that are used to insure that the IP trait(s) or attribute(s) are within tolerance or meet regulatory compliance.

The first product differentiation system is IP-*T* (in some literature it is called identity-preserved production and marketing, or IPPM), which has evolved over time in the grain and oilseed industry. Purchasers of raw products became more demanding about the quality and purity of the product they were purchasing, so the grain handling system gradually developed distinct channels to market the differing grades of grains and oilseeds. All grains and oilseeds are purchased by a grading system in today's marketplace; this grading system has premiums that rise as one moves from low to high grades. The relationship of premiums to differing grades for private market incentives is the defining feature of an IP system (Smyth and Phillips 2002).

IP-*T* systems have been initiated by the grain and oilseed industry to extract premiums from a marketplace that has expressed willingness to pay for an identifiable and marketable product trait or feature. An IP-*T* system is generally a closed-loop channel that facilitates the production and delivery of an assured quality by allowing identification of a commodity from the germplasm or breeding stock to the processed product on a retail shelf. Grain and oilseed IP-*T* systems are predominantly voluntary, private firm-based initiatives that range between systems that are loosely structured (e.g., malting barley) with high tolerance levels and those with rigid structures (e.g., non-GMO EU markets) with minimal tolerance levels. Firms operating in minimal tolerance systems achieve this by developing and adhering to strict protocols that specify production standards, provide for sampling, and ensure appropriate documentation to audit the flow of product* (Smyth and Phillips 2002).

Numerous IP-*T* systems operate around the world. Some extend only between the breeders and the wholesale market or processor, while others extend right up to the retailer. Their structure depends on the attribute being preserved. For instance, some novel oils, such as low linolenic oils that are more stable in fryers, only have value at the processing level, while others, such as high oleic oils, have health attributes that can be marketed to consumers. IP-*T* systems are important for providing information to consumers about the origin of a product, as those attributes are not visible or detectable in the product itself† (Smyth and Phillips 2002).

1.11.2   IP-*T* Segregation

The second product differentiation system, segregation, has frequently been applied incorrectly to the grading of different classes of grains and oilseeds to receive a higher price for the commodity than if it were allowed to be commingled. Segregation is a step between commodity processing (low value) and identity preserved (high value). It represents both a middle value and mid-level involvement of management to ensure its quality. Segregation systems have a formal structure and, in fact, can act as regulatory standards. Segregation differs from IP-*T* in that the focus of the system is not on capturing premiums, but rather on ensuring that potentially hazardous crops are prevented from entering supply chains that have products destined for human consumption. Segregation can be viewed as a regulatory tool that is required for variety approval and commercial release of grain and oilseed varieties that could enter the supply chain and create the potential for serious health hazards. Segregation systems can be developed as part of a variety registration process, in which government regulators use contract registration to ensure that certain novel varieties will not enter the handling system of like varieties. The private firm seeking registration of the novel variety has to demonstrate that there is a segregation system developed to ensure the containment of the variety‡ (Smyth and Phillips 2002).

---

* A survey of the literature on IP shows that although there is growing discussion about IP systems, there are very few working definitions. It has been suggested that an IP system is a more stringent (and expensive) handling process and requires that strict separation, typically involving containerized shipping, is maintained at all times. IP lessens the need for additional testing as control of the commodity changes hands, and it lowers liability and risk of biotech and non-biotech commingling for growers and handlers (Smyth and Phillips 2002).

† The body of literature pertaining to aspects of IP is limited but is growing. Many of the works relate to IP systems relating to theoretical and operational uses of IP systems. Bullock, Desquilbet, and Nitsi (2000) and Bullock and Desquilbet (2001) discuss differentiation between GM and non-GM products, and Herrman, Boland, and Heishman (1999) examine the feasibility of wheat IP. Bender (2003) has released a series of papers on handling specialty corn and soybean crops, with costs being the focus, not the defining of the system used to handle the specialty crop. Additionally, Miranowski et al. (1999) offer some perspectives on the economics of IP, and Kalaitzandonakes, Maltsbarger, and Barnes (2001) provide a solid theoretical model for examining the cost of IP.

‡ The distinction between "IP" and "segregation" is often blurred, and a "strict segregation" system may be more precise than a loose IP system. The level of precision of the traceability system may also influence recordkeeping costs (Golan et al. 2004).

Segregation is focused on ensuring that the integrity of the special trait is not allowed to "adventitiously commingle" with other products destined for the food and feed supply chain. Production contracts are used by the private firms to ensure that the entire commodity being segregated is collected and that the producer retains no amount of seed* (Smyth and Phillips 2002).

1.11.2.1   Important Issue: Internal versus External Traceability

The Food Standards Agency of the European Community recognizes two levels of IP-*T* within the food industry. The first level, called "internal tracking," takes place within one link of the chain (Moe 1998). Considerable internal tracking already exists within the food industry, providing individual firms the ability to follow product logistics through their internal operations; however, only very limited information actually follows the product to the next step (Golan et al. 2004 and Pape 2006). In addition, the real difficulty in designing and implementing IP-*T* lies within the complexity of the second level, called "external (chain) tracking" (Moe 1998). Chain tracking, which provides information paths between individual entities throughout the entire food chain, cannot be achieved without considerable knowledge-based vertical integration and may entail any number of entities in the seafood industry including fishers, buyers, processor, wholesalers, transporters, and retailers [†] (Moe 1998 and Pape 2006).

When looking at IP-*T* systems it is important to distinguish between internal tracking and external (chain) tracking. Internal tracking is within a company or location that is under consideration. In terms of a product, it relates to the origin of materials, the processing history, and the distribution of the product after delivery. Chain or external tracking is, on the other hand, focused on the maintenance of product information from one link in the chain to the next. It describes which data are transmitted and received and how. Chain tracking is between companies and countries and depends on the presence of internal traceability in each link. In some literature the terms internal or external traceability are used instead of internal or external tracking (Moe 1998 and Pape 2006).

Most IP-*T* regulations focus primarily on external or chain tracking. Legislation demands that each producer has control over input ingredients and is able to identify from whom they bought the raw material and to whom they delivered the finished products. This is a major gap within the notion of food and ingredient accountability.

---

* Buffer zones are required for segregation systems as a preventative measure for reducing crosspollination. Producers may also have restrictions placed on what crop varieties are allowed to be grown the following year on fields that produced segregated crops. Premiums are available in the short and long term to ensure that product supply is maintained.

† Two important motives for the formation and coordination of information in vertical supply chains are to manage liability associated with adulteration or contamination and to identify and preserve quality traits. Traceability systems can be defined by these motives. Segregation systems attempt to separate batches of food and ingredients from each other during processing, whereas IP-*T* systems identify the source and nature of each batch, requiring considerable information to guarantee that the traits and qualities of the product are maintained throughout the supply chain. The type of system to be used will depend on what the producers want to accomplish and how much information they want to make available to other firms in the supply chain. Information on products and production practices must remain in the control of the entity responsible for these processes (Golan et al. 2004).

For example, a processor should be able to document all of the different input ingredients as they arrive on their loading dock for use. Many loads of flour may arrive from different sources and be poured into one of several bins. Over time, as one bin empties, the flour from another bin (from still other sources) will be introduced into the process. Regarding rules that processors must follow, the processor should also be able to document, as product leaves its loading dock for its next destination, what ingredients are in the product. Unfortunately, internal tracking is often lacking in accounting for mixing of in-house bins (because bins are constantly being filled, and because product is continuously used in production). This has been the main focus during the past decade and today there exist several standards and/or solutions that will solve the internal traceability issues (Moe 1998 and Pape 2006).

1.11.2.2 Internal Tracking (In-House, Processor)

Many advantages can accrue from having internal tracking. A minimum of internal tracking, being able to track the raw material that went into a product, is in the interest of most food manufacturers. Establishing internal tracking may be easy enough for individual batch processing; however, for continuous or semicontinuous processing it can be very difficult. Under such conditions the ideal traceable resource unit (TRU) can be very small and therefore many food processors do not have tracking down to the ideal TRU. Instead they have a sort of "sufficient" tracking, in which products processed within a period of time are known to come from a certain raw material batch, with some mixing at both ends. However, only an internal tracking system coming close to tracing the ideal TRU can be used as a grid for combining data from process control, quality management, and other management systems (Moe 1998 and Pape 2006).

Achieving external or chain tracking requires comprehensive planning during the initial stages of development, particularly when addressing the three issues most crucial to the success of any traceability system. These include compatibility, meaning it must be possible to track products from one entity to another;* data standardization, or compatible data transmission protocols and computer applications to integrate knowledge-based operations, which may include product handling and processes, including transformation, value addition, packaging, transport, and storage; and the definition of a TRU. Defining a TRU may be one of the most difficult steps involved in the design of a traceability system† (see Appendix A regarding IP-*T* systems at seed production, processing, and retail stages).

---

* This requires that all entities within the chain are able to communicate and transmit data efficiently. Having the ability to transmit and receive data does not, in itself, ensure traceability; it only provides a means. Rapid advances in information technology and increased compatibility between available operating systems have provided the necessary tools to improve knowledge-based vertical integration.

† A TRU is simply defined as a unit of trade, such as a whole fish or a batch of fish at the initial stage. However, this will invariably change during processing because new TRUs are assigned at each step within the food chain. The initial TRU must follow each fish or lot, through all steps of processing, distribution, and retail. This process can become very complicated, especially during processing, and it may be difficult to keep from mixing fish from several batches, especially when processing may include portioning, additional ingredients, processes, storage, and transportation. Mixing of batches can occur between resource units, which may cause problems in identifying individual batches. Each

1.11.2.3   Quality Management

IP-*T* is also an essential subsystem of quality management. The development of advanced internal IP-*T* systems can, however, also be spurred by the search for improving the efficiency of data collection, plant control, and quality assurance. That search has resulted in an increasing interest in coupling data from more than one control or management system, which in turn, requires that a traceability system with a high degree of detail be established. Traceability is also a system in itself and its establishment should be given proper attention and suited to actual needs using a systematic approach. To do this well requires awareness of the various features of traceability that are addressed in this book (Moe 1998).

1.11.3   IP-*T* Systems

According to Golan, Krissoff, and Kuchler (2004),[*] an IP-*T* system can be split into two elements; namely, the routes of the product and the extent of tracking desired or willing to be paid for. Routes describe the path along which, and the means by which, products can be identified throughout the manufacturing, distribution, and retail system. Extent defines the scope of tracking. This is elaborated below. The descriptors *depth*, *breadth*, and *precision* highlighted in Golan works will be used to describe overall IPT concepts.

- *Depth* is how far back or forward the system tracks the relevant information. For example, an IPT system for decaffeinated coffee would extend back only to the processing stage. An IPT system for fair-trade coffee would only extend to information on price and terms of trade between coffee growers and processors. An IPT system for fair wages would extend to harvest; for shade grown, to cultivation; and for non-genetically engineered, to the bean or seed. For food safety, the depth of the traceability system depends on where hazards and remedies can enter the food production chain. For some health hazards such as BSE, ensuring food safety requires establishing safety measures at the farm. For other health hazards, such as food-borne pathogens, firms may need to establish several critical control points along the entire production and distribution chain. The key here is to know what traits/attributes are desired and/or what safety level is needed for who or what (e.g., for labors, processors, consumers, environment, animals, etc.) (Golan, Krissoff, and Kuchler 2004).

(Continued)

firm must develop a system of assigning new TRUs during processing, distribution, and retail (Moe 1998 and Pape 2006).

[*] Golan, Krissoff, and Kuchler have written extensively on food issues such IPT systems. Her work with others in *Traceability in the U.S. Food Supply: Economic Theory and Industry Studies* and *Food Traceability One Ingredient in a Safe and Efficient Food Supply* is the basis for greater comprehension of how IPT regulations and pragmatic realities of how these regulations are used serve as a standard in clarity of understanding these topics. I wish to express my appreciation for their work and how it has added to this research paper. The portion of IP-*T* that addresses breadth, depth, and precision, is borrowed, shortened, and modified from her works.

- *Breadth* describes the amount of information collected. A recordkeeping system cataloging all of a food's attributes would be enormous, unnecessary, and expensive. Take, for example, a cup of coffee. The beans could come from any number of countries, be grown with numerous pesticides or just a few, be grown on huge corporate organic farms or small family-run conventional farms, be harvested by children or by machines, be stored in hygienic or pest-infested facilities, and be decaffeinated using a chemical solvent or hot water. Few, if any, producers or consumers would be interested in all this information. The breadth of most IPT systems would exclude some of these attributes (Golan, Krissoff, and Kuchler 2004).
- *Precision* reflects the degree of assurance with which the IPT system can pinpoint a particular food product's movement or characteristics.* In some cases, the objectives of the system will dictate a precise system, whereas for other objectives a less precise system will suffice. For more traditional systems, such as in bulk grain markets, a less precise system of traceability from the elevator back to a handful of farms is usually sufficient because the elevator serves as a key quality control point for the grain supply chain. Elevators clean and sort deliveries by variety and quality (e.g., protein level). Elevators then blend shipments to achieve a homogeneous quality and to meet sanitation and quality standards. Once blended, only the new grading information is relevant; there is no need to track the grain back to the farm to control for quality problems. Strict tracking and segregation by farm would prevent the ability of elevators to mix shipments for homogeneous product[†] (Golan, Krissoff, and Kuchler 2004).

1.11.4 What Does an IPT Chain Do?

Firms have three primary objectives in using IPT systems: (1) improve supply management, (2) facilitate trace-back for food safety and quality, and (3) differentiate and market foods with subtle or undetectable quality attributes. With regards to business, the benefits associated with these objectives include lower cost distribution systems, reduced recall expenses, and expanded sales of products with attributes that are difficult to discern. In every case, the benefits of IPT translate into larger net revenues for the firm. These benefits are driving the widespread development of traceability systems across the U.S. food supply chain (Golan, Krissoff, and Kuchler 2004).

---

* Precision in trace-back to the farm declines the further one goes down the production chain. As grain is funneled from a wider geographic area, it is more difficult to pinpoint from where and from whom the commodities came. Traceability at the port elevator level typically extends only back to the country or subterminal elevator (Golan, Krissoff, and Kuchler 2004).

† When farmers deliver their crops to local elevators, they are given receipts that indicate the commodity sold, its weight, price received, time of purchase, and any premiums or discounts for quality factors such as extra moisture, damage, pests, or dockage (easily removable foreign material). Country elevators keep this information, thus establishing a recordkeeping link from the product in an elevator at a point in time to the farmers who supplied the product. An elevator operator knows the farmers that delivered grain and oilseeds at that location and the geographic area from which they came. This is the minimum level of IPT that is required by the USDA (Golan, Krissoff, and Kuchler 2004).

1.11.4.1   Third Parties: Options to Enhance IPT

In cases in which markets do not supply enough traceability for product differentiation, individual firms and industry groups have developed systems for policing and advertising the authenticity of credence claims. Third-party safety/quality auditors are at the heart of these efforts. These auditors provide consumers with verification that traceability systems exist to substantiate credence claims. For example, auditors from Food Alliance, a nonprofit organization, certify foods grown with a specific set of sustainable agricultural practices. Many buyers, including many restaurants and some grocery stores, now require their suppliers to establish IPT systems and to verify, often through third-party certification, that such systems are in compliance. The growth of third-party standards and certifying agencies is helping push the whole food industry, not just those firms that employ third-party auditors, toward documented, verifiable traceability systems (Golan, Krissoff, and Kuchler 2004).

For some crops, farmers may be asked to submit their shipments for testing. For example, the oil content of corn and the protein level in wheat are routinely tested. Tests may be performed by the elevator or by independent third-party verifiers. Elevators usually keep records of test results, including the identity of the farms that sold the commodities to them. For some specialty crops, buyers may simply require farmers to certify that the crops are as specified. This was the case early in the development of differentiated markets for non-genetically engineered crops (Golan et al. 2004).

Most, if not all, third-party food-safety/quality certifiers such as the Swiss-based Société Générale de Surveillance (SGS) and the American Institute of Baking (AIB) recognize traceability as the centerpiece of a firm's safety management system. AIB's standard food safety audit specifies several very specific activities[*] (American Institute of Baking 2003 and Golan et al. 2004).

According to Golan, Krissoff, and Kuchler (2004), electronic systems for tracking inventory, purchases, production, and sales have become an integral part of doing business in the United States. A few big retailers such as Wal-Mart and Target have even created proprietary supply-chain information systems that they require their suppliers to adopt. In addition to private systems, U.S. firms may also use industry-standard coding systems such as UPC codes. These systems are not confined to packaged products. The food industry has developed several complex coding systems to track the flow of raw agricultural inputs to the products on grocery store shelves. These systems help to create a supply management system stretching from the farm to the retailer.

---

[*] Third-party standards and certifying agencies are used across the food industry. In 2002, AIB audited 5,954 food facilities in the United States and was slated to audit 6,697 in 2003; SGS expected to perform over 1,000 U.S. food safety audits in 2003; and ISO management standards are implemented by more than 430,000 organizations in 158 countries (ISO website). Food sectors using third-party verifiers cover the spectrum from spices and seasoning to fruit and vegetables to meat and seafood to bakery products and dough. The growth of third-party standards and certifying agencies is helping to push the whole food industry, not just those firms that use third-party auditors, toward documented, verifiable traceability systems (Golan, Krissoff, and Kuchler 2004).

1.11.4.2   Risk Accumulates

The benefits of precise tracking and tracing for food safety and quality control are greater with the increased likelihood and cost of safety or quality failures. Where the likelihood and cost of failure are high, manufacturers have large financial incentives to reduce the size of the standard recall lot and to adopt a more precise traceability system. The benefits of traceability are also likely to be high if other options for safety control are few (Golan et al. 2004).

1.11.4.3   Traceability Chain (Back-tracing)

Another benefit of IPT systems is that they may help firms establish the extent of their liability in cases of food safety failure and potentially shift liability to others in the supply chain (see Chapter 12 regarding recalls and liability issues) If a firm can produce documentation to establish that the safety failure did not occur in its plant, then it may be able to protect itself from liability or other negative consequences (Golan et al. 2004).

Despite the important role safety plays within traceability systems, it is only one element of a firm's overall safety/quality control system. In and of themselves, traceability systems do not produce safer or high-quality products or determine liability. Traceability systems provide information, looking backward, about whether control points in the production or supply chain were operating correctly or not. In cases where markets do not supply enough traceability for food safety trace-back, several industry groups have developed food safety and trace-back standards. For example, the California cantaloupe industry has incorporated traceability requirements in their marketing order to monitor food safety practices. In addition, buyers in every sector are increasingly relying on contracting, vertical integration, or associations to improve product traceability and facilitate the verification of safety and quality attributes. Many hog operations are now integrated by ownership or contractually connected to slaughtering firms. As a result, identification by herd or batch is much easier today than 50 years ago (Golan, Krissoff, and Kuchler 2004).

1.11.5   Traceability

Traceability (back-tracing) can also be considered another product differentiation system commonly used in the food industry. Retail products found with unacceptable bacteria levels or intolerable levels of pesticide or chemical residues need to be quickly and completely removed from store shelves. Traceability systems allow for retailers and the supply chain to identify the source of contamination and thereby initiate procedures to remedy the situation. The key focus of traceability is on food safety. Additionally, the focus for developing traceability systems for new sectors of the marketplace has shifted to include extracting premiums from products that possess traits of value. Extracting market premiums could never be the driver for developing a traceability system. In and of themselves, traceability systems do not motivate quality, they simply trace it (Smyth and Phillips 2002).

Various traceability systems have been established in Europe, North America, and elsewhere. In Canada, traceability was developed in conjunction with a quality assurance (QA) system to reassure export markets about the quality of Canadian beef products.* In a similar QA effort, the Canadian grain and oilseed industries conducted a 2-year pilot project in 2002 and 2003 to evaluate the costs and benefits of an on-farm HACCP-based traceability system (Smyth and Phillips 2002).

Traceability (or retrospective analysis) is required to recall what has already occurred and, in use, traceability works backward. This means that the recordings concerning the TRU must be designed from the viewpoint that they will be retrospectively interrogated. Furthermore, a stable, accessible record system is essential (Moe 1998).

1.11.5.1   Advantages of Traceability

Traceability offers the following advantages:

- Establishes the basis for efficient recall procedures to minimize losses.
- Information about the raw material can be used for better quality and process control.
- Avoids unnecessary repetition of measurements in two or more successive steps.
- Improves incentive for maintaining inherent quality of raw materials.
- Makes possible the marketing of special raw material or product features.
- Meets current and possible future requirements (e.g., confirming country of origin).

Most food processing companies establish end-product traceability to secure efficient product recall procedures. Product recall systems only require traceability in part of the chain from the production step to the consumer. However, if the problem stems from the supply of raw material, traceability back to the supplier improves the possibility of either correcting faults, avoiding reoccurrence, or placing the responsibility there. Recall systems can be established on a minimum of traceability information (e.g., production date); however, the more subdescriptors that are included (e.g., production time, batch number, production conditions) the more focused the product recall can be, thereby minimizing loss of money and reputation (Moe 1998).

1.11.5.2   Combining Forward-Tracking Systems
with Back-Tracing Systems—IPT

IPT can be used in four distinct contexts, each with a different implied sense.

1. Product: IPT may relate materials, their origin, processing history, and their distribution and location after delivery.
2. Data: IPT relates calculations and data generated throughout the quality loop, sometimes back to the requirements for quality.

---

* In this case however, it should be noted that this system has been met with great resistance at the farm level, because producers do not want to allow government regulators onto their farms or provide regulators with any sensitive farm information.

3. Calibration: IPT relates measuring equipment to national or international standards, primary standards, basic physical constants or properties, or reference materials.
4. Software and programming: IPT relates design and implementation back to the requirements for a system.

Product and data cover the fundamental concepts included in independent traceability and tracking systems relating to products and their processing (Moe 1998). These important issues are somewhat neglected in the literature on food processing and are therefore the subject of this book. Calibrating measuring equipment using standards that are trackable and traceable to national or international standards is essential to all food business to provide a common base for assessment of product quality and performance in accordance with specification. This is well discussed in the literature and Chapter 8. The IPT system is explained in greater detail in Chapter 10.

1.11.5.3 Overall Supply Chain IPT Management

In addition to traceability as a food safety mechanism, it is also crucial for providing access to new categories of products. Many markets have demanded documentation regarding product composition before allowing market access. Consumer information is fundamental for traceability systems, because they are designed to increase information regarding food safety to consumers. Information is also provided back up the supply chain to regulators and processors. Labeling is important to traceability to ensure high-quality standards and allow consumers to identify with this feature. In this way, market premiums may be available for products that show evidence of continuous traceability.

1.12 SUMMARY OF CHAPTERS

Part I consists of Chapter 1, "Introduction to Identity Preservation and Traceability;" Chapter 2, "IPT History;" and Chapter 3, "Overview of IPT System Components, Theory, and Design." Part II (Chapters 4–6) provides examples of official seed agencies, industry programs, and standards including U.S., Canadian, EU, international, organic, regional, and religious standards. Part III includes Chapters 7 and 8, a sampling and explanation of auditors and laboratories. Part IV (Chapters 9–12), reviews domestic and foreign policy and advisory organizations, software providers, IPT process facilitators, and information about food recalls and insurance. Part V (Chapters 13 and 14) provides examples of a spreadsheet and questionnaire. The last portion of this work contains the conclusion; appendices; related products, services, and organizations; glossary of terms; directory of resources; and bibliography.

REFERENCES

Agriculture and Agri-Food Canada. "Can-Trace," http://www.can-trace.org (accessed 2009).
Glassheim, Eliot, Jerry Nagel, and Cees D. Roele. "The New Marketplace in European Agriculture: Environmental and Social Values within the Food Chain." *Plains Speaking* 8, no. 1, 2005: 5–12.

Golan, Elise, Barry Krissoff, and Fred Kuchler. "Food Traceability One Ingredient in a Safe and Efficient Food Supply." *Amber Waves* 2, no. 2, 2004: 14–21.

Golan, Elise, Barry Krissoff, Fred Kuchler, Linda Clavin, Kenneth Nelson, and Gregory Price. *Traceability in the U.S. Food Supply: Economic Theory and Industry Studies.* Agricultural Report Number 830. Washington, DC: U.S. Department of Agriculture Economic Research Service, 2004.

Gruère, Guillaume P. *An Analysis of Trade Related International Regulations of Genetically Modified Food and their Effects on Developing Countries.* Environment and Production Technology Discussion Paper 147, Washington, DC: International Food Policy Research Institute, 2006.

Jenkins, C. "HACCP further up the Food Chain." *Food Quality* 10, no. 3, 2003: 55–56.

Moe, Tina. "Perspectives on Traceability in Food Manufacture." *Trends in Food Science & Technology* 9, no. 5, 1998: 211–214.

Pape, Will. "Connecting the Dots for Food Recalls." *Food Traceability Report* 1, 2006:12–13.

Schwägele, Fredi. "Traceability from a European Perspective." *Meat Science* 71, 2005: 164–173.

Smith, Gary C., J. Daryl Tatum, Keith E. Belk, John A. Scanga, Temple Grandin, and John N. Sofos. "Traceability from a US Perspective." *Meat Science* 71, 2005: 174–193.

Smyth, Stuart, and Peter W.B. Phillips. "Product Differentiation Alternatives: Identity Preservation, Segregation, and Traceability." *AgBioForum* 5, no. 2, 2002: 30–42.

2 History of Identity Preservation and Traceability

2.1 INTRODUCTION

This chapter provides a short overview of identity preservation and traceability (IPT) history, primarily from a European Union (EU) and U.S. perspective, which includes a blending of eras of events and legislative initiatives. This section will not be completely fluid or chronologically continuous. IPT history is a blending of *re-actions* from food scares and *pro-actions* to help mitigate future food problems, by the private sector and governments. It also has taken on a perspective of credence attributes not associated with food safety, such as animal and labor welfare, food source origins, etc.* For the United States, the events of the World Trade Center bombings motivated the most recent wave of change. For Europeans food safety issues really came to the forefront with the discovery of bovine spongioform enceph-alopathy, (BSE, or mad cow disease) and have been amplified by concerns of geneti-cally modified organisms (GMOs). The results have given rise to strict government regulations put forth by public demand, incited by government failures and activists. From this we can see how the U.S. and EU perspectives have started at near polar opposites but are working slowly and more closely together to help resolve important issues faced by differing cultures and governments. Although not a complete his-tory of IPT, the goal of this chapter is to bring the reader up to speed as to why and how different paths are being taken toward answering the challenges regarding food issues. Woven through this chapter are the more important U.S. and EU legislations that affect IPT programs.

This chapter will highlight historical aspects of IPT, U.S. rules history, the Green Revolution, the Gene Revolution, country-of-origin labeling (COOL), EU rules his-tory, EU labeling/segregation/IPT, and concerns on the horizon.

2.2 THE EU AND THE UNITED STATES

Traceability has become the focus of a major trade dispute between the EU and the United States. It has also sparked debate in the Codex Alimentarius,† the interna-tional body co-sponsored by the U.N. Food and Agriculture Organization (FAO)

---

* See Chapter 6 regarding how many public standards work for more specific details.
† See Chapter 6 on international standards for more information as to specific details.

and World Trade Organization (WTO). Codex sets international food standards and guidelines that are referenced by the WTO in trade disputes.*

The current debate over traceability is globally more of a clash of differing regulatory cultures. "Traceability" is a term few in the United States had heard of before 2000. U.S. regulators prefer the terms "product tracing" or "traceback," which have a history of use in illness outbreak investigations and food product recalls for public health purposes through imposed industry standard systems. In the United States, the notion is that product tracing should only be required, if at all, for food safety purposes. A traceback system should be able to trace one level forward and one level back and not require excessive documentation. Furthermore, the system would be industry imposed, monitored, and policed. If private groups or industries desire traceability for identity preservation, or for organic or kosher labeling, then those groups or industries would be most efficient and cost-effective in designing standards. But governments should inherently not be in the business of requiring traceability for reasons other than public health and safety (Clapp 2002).

The EU, on the other hand, has depended upon on traceability and labeling as solutions to low consumer confidence in the safety of its food supply. Europe has been overwhelmed by one food safety crisis after another: mad cow disease, dioxin in chicken feed, foot-and-mouth disease, fear of genetically modified foods, and, residues of a banned herbicide in organic chicken feed. Food safety scandals have toppled European governments, caused cabinet ministers to resign, and forced a major overhaul of the European Commission (EC), the EU's executive branch (Clapp 2002).

In the EU, because of vocal advocacy groups and sensationalistic reports by the media, biotech food products have been especially focused upon and lumped together with these other food safety concerns. EC officials acknowledge that biotech foods are no less safe than conventional foods, and may even offer important advantages to developing countries. However, they argue that consumer confidence in Europe can only be restored if biotech products are clearly labeled, and the ingredients can be traced backward to the source and tracked forward to the customer (Clapp 2002).

GMO labeling policy for foods is under intense development. Countries are choosing mandatory labeling or adherence to voluntary labeling. Challenges to mandatory labeling are unlikely to be successful under current WTO rules. Marketers and trade negotiators recognize this and are moving toward living with a variety of labeling policies (Caswell 2000).

Traceability became a transatlantic political fight after the turn of the century. The EC approved proposals requiring traceability and labeling for biotech foods. Traceability and labeling were seen as critical pieces of the dilemma that would enable the EU to end an informal, yet real moratorium on approving new biotech products. EC officials acknowledge that this moratorium was illegal. The commission ended the moratorium, but for political reasons it does not dare to overrule the member states that embrace the moratorium (Clapp 2002).

---

* Excerpts with permission from "Traceability: a Global Perspective." *Food Traceability Report* 1: 16–19, by Stephen Clapp.

Consumers are at yet another important crossroad on the path that will determine the market acceptance of foods produced with the use of biotechnology via the use of traceability. Individual governments are managing a range of policies that affect biotechnology and credence attributes, including those on research and development, intellectual property rights, regulatory approval (safety assessment), and labeling requirements. They are taking divergent policy paths that make for market uncertainty. At the same time, companies are announcing their intentions regarding the use or non-use of GMOs in their products. For companies, these intentions make the market less uncertain for sales but raise the stakes in predicting the choices that other companies make (Caswell 2000).

The EU views its traceability and labeling proposals as the solution to its political problems with food safety and biotechnology. Europe already has a strict labeling law that has for practical purposes cleared supermarket shelves of products bearing any biotech stigma. The EU's new rules require labeling of products in which no altered deoxyribonucleic acid (DNA) or proteins can be detected. Product categories covered by the proposals include highly refined corn oil, soybean oil, canola oil, glucose syrup produced from cornstarch, and animal feed made from corn gluten or soybean meal. The new traceability rules require records of genetic transformation events to be kept throughout the production process. The proposal includes a 1% threshold for the adventitious presence of transgenic materials in non-biotech commodities. Producers must be able to show that the traces were "technically unavoidable," and the EU must have approved the transgenic material. This is where full accountability of a product or traceability is critical.[*] (Clapp 2002)

As we will see, legislative proposals are usually implemented to solve problems. So too are IPT programs' attempts to answer not only the biotech question, but also developing questions regarding the public's demand on other attributes such as organic, food origins, animal and labor welfare, etc. Once again, the U.S. and EU visions of traceability are in conflict. The U.S. government would like to only restrict mandatory traceability systems to food safety, whereas the EU stresses not only the importance of safety but also non-safety aspects, such as labeling and identity preservation for social welfare, the environment, etc.[†]

Another example of regulatory culture clash is that Codex has defined traceability as the "ability to trace the history, application or location of an entity by means of recorded identifications." Traceability is closely linked to product identity, but it can also relate to the origin of materials and parts, product processing history, and the

---

[*] In addition, European consumer groups would like to label meat, milk, and eggs from animals fed biotech feed, but the EC has resisted this idea. The United States is currently able to sell biotech feed grains to Europe despite consumer resistance.

[†] Recent legislation in the EU and the United States imposes increasingly stringent information requirements on food supply chain participants for the purpose of ensuring food safety and food security. Agri-food industry players are working to achieve compliance, through the implementation of auditable and verifiable traceability systems that integrate information across the supply chain, ensuring credibility of origin and brand claims, delivering rapid response and improving recordkeeping. Traceability systems have proven that they can connect information across the food supply chain to simultaneously support food safety requirements/rapid response to food security issues and improve business performance, offsetting the cost of regulatory compliance by creating new value through productivity gains (Bantham and Duval 2004).

distribution and location of the product after delivery. After length, Codex came up with a compromising language; for example, the text lists among risk management tools "the tracing of products for the purpose of facilitating withdrawal from the market when a risk to human health has been identified, or to support post-market monitoring" in specified circumstances on a case-by-case basis. A footnote acknowledges that other applications of traceability (e.g., labeling and identity preservation) are currently under consideration elsewhere in Codex (Clapp 2002).

2.3 HISTORICAL CONTEXT

As the demand for food and fiber has grown during the past 300 years because of expanding population and rising per capita incomes, society has met this demand first by increasing the land area under cultivation and later by improving crops so that their yields became higher. Before 1900, land was abundant almost everywhere, and in the United States new lands were brought into production as the frontier moved across the country between 1700 and 1900. In addition, the great crop exchange between different continents permitted high-yielding crops like potatoes (*Solanum tuberosum*) to be grown in Europe and rice (*Oryza sativa*) in the United States. Improvement was by selection of the fittest that flourished in its new environment. By 1900, the frontier was closed in the United States, and this increased the urgency of finding new methods for increasing crop yields (Huffman 2004).

The evolution of commercial food production and products, although influenced by several economic, regulatory, and environmental factors, was most influenced by the expectations of society, which initiated the major changes and acted as the true driver throughout the twentieth century. As these expectations have evolved, dominant themes have emerged in such a way that they can be seen to define distinct eras throughout the twentieth century (Jones and Rich 2004).

2.3.1 UNITED STATES

The U.S. Government has a long, albeit limited, history of mandating programs that contain traceability requirements. Government regulations have a diverse set of objectives. Often, they take into consideration numerous views, ensuring a level of food safety, preventing and limiting animal diseases, or facilitating market transactions. Some of these regulations entail establishing traceability systems for select attributes in particular food subsectors, whereas other regulations have broader objectives but, in effect, require firms to develop tracing capacity. Whether the intent of the regulation is to address food safety or animal disease concerns or other issues, government-imposed demands for traceability usually require information about the sellers and buyers (name, address, phone, etc.) and product-related information. The demands on recordkeeping are usually one-up, one-back traceability. Less frequently required, but becoming more in demand by the public, are traceability systems for regional source, process, social, and quality credence attributes, which have become more prevalent, although there are exceptions, such as the U.S. national organic food (NOP) standard. The following briefly highlights some important regulations that require traceability systems, their relevant legislation, objectives of the regulations, product coverage, and recordkeeping requirement(s). The list is not intended to be encyclopedic, but, instead, illustrative of important and recent legislation that affects tracing by food suppliers (Golan et al. 2004 and Golan, Krissoff, and Kuchler 2004).

In the United States, the food manufacturing industry during the first half of the twentieth century was focused on production. During this time, the United States and others experienced world wars, the Depression, and other natural and man-made events that influenced society and agriculture. In this era, a large proportion of the workforce was involved in unskilled/physical labor. The consumers' expectation of food was to provide sufficient rations to meet energy requirements and for its satisfaction or fullness value. Society's concerns with the environment were almost non-existent, exhibiting an attitude that the natural environment provided an unlimited supply of raw materials and storehouse for waste. Although the community held narrow expectations in terms of food functionality, it did require foods to be safe. Industry responded with continuous increases in production output and processed foods (Jones and Rich 2004).

In the United States, one of the early legislations was the **Meat, Poultry, and Egg Inspection Acts**. Legislation was passed in 1906 for meats, 1957 for poultry, and 1970 for eggs. The Wholesome Meat and Poultry Acts of 1967 and 1968 substantially amended the initial legislation. The Meat, Poultry, and Egg Inspection Acts have the primary goals of preventing adulterated or misbranded livestock, meat, poultry, shell eggs, and egg products from being sold as food and to ensure that meat and meat products are slaughtered and processed under sanitary conditions. The Food Safety and Inspection Service (FSIS) of the U.S. Department of Agriculture (USDA) is responsible for ensuring that these products are safe and accurately labeled. The acts call for complete and accurate recordkeeping and disclosure of all transactions in conducting commerce in livestock, meat, poultry, and eggs. For example, packers, renderers, animal food manufacturers, or other businesses slaughtering, preparing, freezing, packaging, or labeling any carcasses must keep records of their transactions. Businesses only need to maintain one-up, one-back records (Golan et al. 2004 and Golan, Krissoff, and Kuchler 2004).

For imported meat, poultry, and egg products, importers must satisfy requirements of the U.S. Customs Service and two USDA agencies—FSIS and the Animal and Plant Health Inspection Service (APHIS). Imported meat and poultry must be certified, not only by country, but by individual establishment within a country. Certificates are issued by the government of the exporting country and are required to accompany imported meat, poultry, and egg products to identify products by country and plants of origin, destination, shipping marks, and amount. FSIS demands that the country of origin provide a health certificate indicating the product was inspected and passed by the country's inspection service and is eligible for export to the United States (Golan et al. 2004 and Golan, Krissoff, and Kuchler 2004).

In 1930, the **Perishable Agricultural Commodities Act (PACA)** was enacted. PACA is intended to promote fair trading practices in the fruit and vegetable industry. The objective of the recordkeeping is to help facilitate the marketing of fruit and vegetables, to verify claims, and to minimize any misrepresentation of the condition of the item, particularly when long distances separate the traders. PACA calls for complete and accurate recordkeeping and disclosure for shippers, brokers, and other first handlers of produce selling on behalf of growers. PACA has extensive record-keeping requirements on who buyers and sellers are, what quantities and kinds of produce are transacted, and when and how the transaction takes place. PACA regulations recognize that the varied fruit and vegetable industries will have different types of recordkeeping needs, and the regulations allow for this variance. Records need to be kept for 2 years from the closing date of the transaction (Golan et al. 2004 and Golan, Krissoff, and Kuchler 2004).

In the United States, plant breeding for almost all crops was undertaken first in the public sector by the USDA and the state agricultural experiment stations, and then, wherever large markets for seed existed and genetic improvements could be protected, the private sector emerged as a major source of crop improvement. In self-pollinated crops like small grains and soybeans (*Glycine max*), protection of crop improvements largely did not exist before the early 1970s when plant variety protection legislation was enacted. In the case of cross-pollinated crops such as corn (*Zea mays*) and sorghum (*Sorghum bicolor*), hybridization discovered early in the twentieth century proved a type of natural protection to developers/discoverers of genetic improvement because hybrids cannot reproduce themselves (Huffman 2004).

Hybrid corn was not a commercial success in the United States until after the first commercial double cross was developed in 1920. More than another decade was required before superior double-cross varieties were generally available to farmers in the Midwest. Starting in the 1930s, hybrid corn varieties jointly developed by the public and private sectors rapidly replaced open-pollinated corn varieties. Farmers in the center of the U.S. Corn Belt were the first to have superior hybrids made available to them because that region promised the greatest profits to the seed companies. Despite the additional cost, farmers rapidly adopted the new hybrids because they were profitable. Outside of the Corn Belt, superior hybrids were made available later, but farmers less rapidly adopted them. Thirty-five years later, single crosses largely replaced double crosses, and in the Midwest the private sector hybrid corn companies (e.g., Pioneer, DeKalb, Pfister, Funk Seeds) soon took control of the development of corn hybrids, commercial reproduction, and commercial distribution. In contrast, for small grains, soybeans, legumes, and grasses, the public sector remained an important developer of new varieties. This was due in large part to hybrid corn's offspring being sterile, and its inability to be used the following year. It was not until the Plant Variety Protection Act (PVPA), signed into law in 1970 and amended in 1994, that greater gains in genetic research were observed for soybeans.[*] In other developed countries (Europe, Japan, Australia, etc.), the public sector was also the main developer of improved crop varieties (Huffman 2004).

The *Food Assistance Programs* also have IPT qualities. The National School Lunch Act was enacted in 1946, after World War II. The purpose of the program was to guarantee that foods (flour, grains, oils and shortenings, dairy, red meat, fish,

---

[*] Breeders had known for sometime that hybrid soybeans and their offspring were not sterile and that seeds once bought were less likely to be purchased again because of the replanting of progeny. It was not until PVPA that private seed companies could benefit from their expensive research by farmer contracts, which were recognized by law and established that farmers could not replant part of their harvest for the next year's crop.

poultry, egg, fruits, vegetables, and peanut products) are strictly American; producers who win USDA contracts must provide documentation establishing the origin of each ingredient in a food product. The producer pays USDA inspectors to review the traceability documents and certify the origin of each food. Starting with the "code" or lot number on a processed product, inspectors use producer-supplied documentation to trace product origins all the way back to a grower's name and address (Golan et al. 2004 and Golan, Krissoff, and Kuchler 2004).

2.3.2   CONSUMERISM ERA (1950–1980)

The so-called world food crisis of 1972–1974 triggered new interest in the global availability of food in what became known as "food security." This era saw the following milestones: the World Food Congress in 1974, the establishment of the International Food Policy Research Institute in 1975, and the first meetings of the World Food Council. Although broader environmental issues had not yet captured the global community's interest, during this era pollution had become an issue in the communities of developed nations. The subsequent global landmarks indicated the mood of the era: publication of Rachel Carson's *Silent Spring* (1962) exposing the hazards of dichloro-diphenyl-trichloroethane (DDT); the first U.N. international conference on the environment (held in Paris, 1968); and the U.N. international conference on the environment (held in Stockholm, 1972), with the recommendation for the creation of the U.N. Environment Programme (UNEP) (Jones and Rich 2004).

2.3.3   GREEN REVOLUTION

Just as the United States was establishing rules and programs during the early to mid-twentieth century, for developing countries the production of their modern crop varieties started in earnest in the 1950s. Notably, in the mid-1960s scientists such as Norman Borlaug (1970 Nobel Peace Prize recipient) developed modern varieties of rice and wheat that were subsequently released to farmers in Latin America and Asia. The success of these modern varieties has been coined the "Green Revolution." The new rice and wheat varieties were rapidly adopted in tropical and subtropical regions with good irrigation systems or reliable rainfall. These modern varieties were associated with the first two major international agricultural research centers; the International Center for Wheat and Maize Improvement (CIMMYT)[*] in Mexico and the International Rice Research Institute (IRRI)[†] in the Philippines (Huffman 2004).

Over the period from 1960 to 2000, many of the international agricultural research centers, applying largely traditional breeding techniques in collaboration with national research programs but with negligible private sector input, contributed to the development of modern varieties for many crops. These varieties contributed to large increases in crop production in Asia and Latin America. These Green Revolution productivity gains were applauded; however, yield gains were uneven

---

[*] See http://www.cimmyt.org.
[†] See http://www.irri.org.

across crops (larger in rice and wheat than other crops) and across regions (largest in Asia and Latin America and very small in Africa). Consumers in developing countries generally benefited from declines in food prices relative to other household purchases (declines in food prices have averaged about 1% per year since 1960), and farmers in developing countries benefited only when cost reductions exceeded price reductions. One striking feature is that gains from modern varieties were larger in the 1980s and 1990s than in the preceding 2 decades, despite popular perceptions that the Green Revolution was effectively over by the 1980s. Overall, the productivity data suggest that the Green Revolution is best understood not as a one-time jump in yields, occurring in the late 1960s, but rather as a long-term increase in the trend growth rate of productivity. This occurred because successive generations of modern varieties were developed, each contributing gains over previous generations (Huffman 2004).

The next major influence came with more advanced genetic crop improvement. Genetic crop improvement, or plant breeding, is most notably a late twentieth century phenomenon. Traditional gene exchange occurs only in sexually compatible species. Most of the genetic variation is created through crossing. Selection is conducted and determined by measuring plant characteristics such as grain yields, and the genes that underlie these characteristics were unknown. Traditional or conventional breeding also does not require knowledge at the DNA level (Huffman 2004).

2.3.4 GENE REVOLUTION

The 1990s brought us the "Gene Revolution" in crop improvement. Genetic modification in this era is a relatively new and complex process that involves insertion of a gene, often from a different species (transgenic), into a plant or animal. (It is this procedure that opponents to GMOs cite as being unsafe for the environment and food safety.) The process is sometimes referred to as genetic engineering and genetic modification, and the crops are referred to as genetically modified (GM) organisms (GMOs), or just GM crops. Since the beginning of farming, farmers and others have been genetically modifying plants to enhance the quantity of desirable attributes. However, since the early 1990s, the term "genetic modification" has been associated with a much narrower set of techniques that use recombinant DNA or gene splicing technology to facilitate the transfer of genes across species. Foods made using this type of GM material have become commonly known as GM foods[*] (Huffman 2004).

Major GM crop varieties became available to U.S. farmers starting in the mid-1990s with insect-resistant [e.g. *Bacillus thuringiensis* (Bt)] cotton (*Gossypium hirsutum*), and herbicide-tolerant [e.g., "Round-Up Ready" (RR)] cotton, soybean, and corn. Later, insect-resistant (e.g., Bt) corn became available. Insect-resistant technology uses Bt, which encodes proteins that are toxic to plant-feeding insects, and RR

---

[*] In 1973, Cohen and Boyer discovered the basic technique for recombinant DNA, which launched a new field of genetic engineering. The Cohen-Boyer patent on gene-splicing technology was awarded in 1980 to Stanford University and the University of California (Office of Technology Assessment 1989). They built on the 1953 discovery by Watson and Crick of the structure of DNA and of the suggestion about how it replicates.

technology uses plants that have been encoded with a protein, the enzyme mEPSPS, which makes the plant tolerant to glyphosate, the active ingredient in Round-Up herbicide. When Round-Up is applied to a RR crop variety every plant is killed except for the RR plants. Newer varieties are being developed and used such as YieldGard Plus by Heartland Hybrids.[*] In corn, the triple-stack seed traits mean that genetic technology was used to introduce three types of focused protections for the plant for in-plant protection against: (1) European and Southwestern Corn Borers, (2) Western and Northern Corn Rootworms, and (3) RR for herbicide protection[†] (Huffman 2004).

2.3.5   GM CONTROVERSY

The application of GM technology to crop production has been hailed by some (e.g., the biotech industry, including the Council for Biotechnology Education) as the greatest invention since the beginning of farming, but international environmental groups such as Greenpeace, Friends of the Earth, and Action Aid counter that GM technology has not been proven safe for humans or the environment, that it benefits only big business and not the consumers, and that it creates "Frankenfoods." The growing controversy over GM food products and consumers' attempts to make improved or better food purchasing decisions have stimulated interest in food labeling, identity preservation, and new sources of information. For example, two international non-governmental organizations (NGOs)—Greenpeace and Friends of the Earth—believe that GM labeling would benefit consumers, and these groups promote the use of labels on GM foods to give consumers the right to choose whether or not to consume GM foods. In fact, they have demanded mandatory labeling, which they believe would benefit consumers. However, microbial contamination of foods is and has been a much greater food safety concern (even in developed countries, let alone developing countries) than GM content, but in the case of GM foods, the international NGOs have made GM food their number one issue. This is but one more example of where IPT programs can assist both consumers and industry for better understanding of the food supply chain (Huffman 2004).

Just as the Gene Revolution was picking up speed and attention, more of the public became focused on the environment and health. This era is distinguished by the influence of a series of international issues of historic significance: The discovery of the AIDS virus in 1981, the global warming alert, the discovery of a "hole" in

---

[*] See Heartland's website for more information at http://www.heartlandhybrids.com.
[†] In addition, Bt technology has been effective in reducing insecticide application rates dramatically in cotton in the southern United States and in India. It replaced chemical insecticides that are quite toxic to the environment and humans. RR soybeans brought more effective weed control into the management toolkit of the poorest farm managers, although some extension agricultural economists indicate very little difference in the cost of production for RR soybean varieties relative to traditional soybean varieties (they fail to count the value of reduced risk of effective weed control due to weather or other delays using conventional practices) and bean yields would be expected to be higher. However, farmers find the technology to be easy to apply, not timing-critical, and effective in a 2-year crop rotation. These are undoubtedly the reasons why RR soybeans and cotton have been such large commercial successes in the United States.

the Earth's ozone layer (1985), Surgeon General's Report on Nutrition and Health (1980s), and National Research Council's Report - *Diet and Health: Implications for Reducing Chronic Disease Risk* (1980s). Societal concerns embraced environmental issues such as pollution and its subsequent costs (financial and natural) and saw the progress from waste treatment to loss monitoring and waste minimization and the beginnings of cleaner production processes. Community expectations of food moved to "clean and green," in which the produce was free of chemicals and the environmental damage was limited through the restricted use of herbicides and pesticides to the emergence of "organic" foods (Jones and Rich 2004).

The next major legislative law that greatly emphasizes identity preservation and traceability systems is the ***Organic Foods Production Act*** of 1990. This act, in many ways, is the modern-day template used by many in designing product, industry, and food chain IPT programs. The act was subsequently amended and rules went into effect October 2002. The objective was to establish national standards governing the marketing of certain agricultural products as organically produced products to assure consumers that organically produced products meet national production, handling, and labeling standards and to facilitate commerce in fresh and processed foods that are organically produced. Organic food certifiers work with growers and handlers to develop an individualized recordkeeping system to ensure traceability of food products grown, marketed, and distributed in accordance with national organic standards. Records can be adapted to the particular business as long as they fully disclose all activities and transactions in sufficient detail to be readily understood, have an audit trail sufficient to prove that they are in compliance with the act, and are maintained for at least 5 years. Many different types of records are acceptable. For example, documents supporting an organic system may include field, storage, breeding, animal purchase, and health records, as well as sales invoices, general ledgers, and financial statements. For the attribute "organic" to be preserved, growers and handlers must maintain traceability from receiving point to point of sale and ensure that only organic or approved materials are used throughout the supply chain. Thus, for a traceability system for organic products to be viable, it must confer depth (Golan et al. 2004 and Golan, Krissoff, and Kuchler 2004).

COOL has taken on greater meaning with increased food scares, GMOs, and bioterrorism. The legislation amends the Agricultural Marketing Act of 1946 by incorporating COOL in the Farm Security and Rural Investment Act of 2002 (Public Law 107-171). Specific guidelines for voluntary labeling were issued in 2002 and are currently in effect. The objective is to provide consumers with more information regarding the country where covered commodities originate. The legislation affects the labeling of beef, pork, lamb, fish, shellfish, fresh fruit, vegetables, and peanuts. COOL is not required if these foods are ingredients in processed food items or are a combination of substantive food components. Examples include bacon, orange juice, peanut butter, bagged salad, seafood medleys, and mixed nuts (Golan et al. 2004 and Golan, Krissoff, and Kuchler 2004).

Food service establishments, such as restaurants, food stands, and similar facilities including those within retail stores (e.g., delicatessens and salad bars) are exempt from the requirements. Moreover, grocery stores that have an annual invoice value of less than $230,000 of fruits and vegetables are exempt from COOL requirements.

As a result, retail food outlets (e.g., butcher shops and fish markets that do not sell fruit and vegetables) are not included under COOL requirements. Retailers may use a label, stamp, mark, placard, or other clear and visible sign on the covered commodity, or on the package, display, holding unit, or bin containing the commodity at the final point of sale (Golan et al. 2004 and Golan, Krissoff, and Kuchler 2004).

The acts and rules, again, reflect increased IPT attributes, such as having stringent requirements on the *depth of recordkeeping*. First, the supplier responsible for initiating the country-of-origin declaration must establish and maintain records that substantiate the claim. If a firm already possesses records, then it is not necessary to create and maintain additional information. As a *vertical supply chain*, there must be a verifiable audit trail to ensure the integrity of the traceability system; that is, firms must assure the transfer of information of the country-of-origin claim. As a consequence, firms along the supply chain must maintain records to establish and identify the immediate previous source and the immediate subsequent recipient of the transaction. For an imported product, the traceability system must extend back to at least the port of entry into the United States. Firms have flexibility in the types of records that need to be maintained and systems that transfer information. Records need to be kept for 2 years (Golan et al. 2004 and Golan, Krissoff, and Kuchler 2004).

The *Public Health Security and Bioterrorism Preparedness and Response Act* of 2002 provides new authority to the U.S. Federal Drug Administration (FDA).[*] The objective is to protect the nation's food supply against the threat of serious adverse health consequences to human and animal health from intentional contamination. All foods are subject to the legislation except meat, poultry, and eggs (which are under USDA's jurisdiction) (Golan et al. 2004 and Golan, Krissoff, and Kuchler 2004).

In response to concerns about terrorist contamination of the food supply following the events of September 11, 2001, Congress passed the U.S. Public Health Security and Bioterrorism Preparedness Act of 2002. The act enables the FDA to prevent and respond to intentional and unintentional food-borne illness outbreaks by granting it the authority to require facilities that manufacture, process, pack, distribute, receive, hold, or import food to register with the FDA, submit notice to the FDA before importing any food into the United States, and maintain records (for up to 2 years) sufficient to allow the FDA to identify the immediate previous sources and the immediate subsequent recipients of food and its packaging (Bantham and Duval 2004).

The act has changed the way domestic and foreign food and feed facilities are required to operate since December 12, 2003, when the registration and prior notice interim final rules went into effect. Also on December 12, 2003, the FDA published the recordkeeping interim final rule.[†] The FDA used the pharmaceutical industry as the precedent for the 4-hour response standard. There is a scientific basis for tracking quickly, supported by studies on BSE in the United Kingdom indicating that it is necessary to know where a contamination event occurs within 24 hours in order to contain it. The act permits conducting "trace-back" and "trace-forward"

---

[*] See Chapter 6 on U.S. standards for more information as to specific details.
[†] A deadline that was first extended to March 2004 and subsequently extended to May 2004.

investigations in the event that the FDA has reasonable belief that an article of food is adulterated and poses a threat of serious adverse health consequences or death to humans or animals, which suggests that food supply chain management will become increasingly necessary (Bantham and Duval 2004).

The act requires both domestic and foreign facilities to register with the FDA. Facilities subject to these provisions are those that manufacture, process, pack, transport, distribute, receive, hold, or import food. The act exempts farms, restaurants, other retail food establishments, nonprofit food establishments in which food is prepared for or served directly to the consumer, and fishing vessels from the requirement to register. Also, foreign facilities subject to the registration requirement are limited to those that manufacture, process, pack, or hold food only if food from such a facility is exported to the United States without further processing or packaging outside of the United States (Golan et al. 2004 and Golan, Krissoff, and Kuchler 2004).

The act requires the creation and maintenance of records needed to determine the *immediate previous sources* and the *immediate subsequent recipients* of food (i.e., one-up, one-down). For imported food, the rules also require prior notice of shipment and a description of the article including code identifiers; the name, address, telephone, fax, and e-mail of the manufacturer, shipper, and the grower (if known); the country of origin; the country from which the article is shipped; and anticipated arrival information. Records are required to be retained for 2 years except for perishable products and animal foods (e.g., pet foods), for which 1 year of recordkeeping is allowed. Records may be stored offsite (Golan et al. 2004 and Golan, Krissoff, and Kuchler 2004).

2.3.6   EU—Identity Preservation and Traceability Initiatives and their Basis[*]

EU traceability history initiatives before Regulation (EC) No. 178/2002 of 2002 included Regulation (EC) No. 820/1997, in which, following the BSE crises in 1997, the EU decided to set up the identification, recording, and labeling of beef meat. Regulation (EC) No. 1760/2000 followed it on July 17, 2000, seeking to establish a system of identification and recording of beef and labeling of beef, veal, and bovine meat products and specifying that the relationship between the identification of meat and the animal or animals concerned must be guaranteed (Article 1), a correlation between the arrivals and the departures must be assured (Article 1), and the size of

---

[*] Additional notes: Some coin this period the Sustainability and Functionality Era (1990 to present). During this era, the warnings of the scientific community raised earlier on the state of the environment, which had largely been played down by the governments of developed nations, were gaining a foothold in the community consciousness, driven in part by the earth summits in Rio de Janeiro in 1992, Kyoto (Rio plus 5) in 1997 and Johannesburg (Rio plus 10) in 2002. In addition to these summits, this era saw the following significant milestones: National Food Authority (NFA) proposed the introduction of HACCP-based food safety plans (1994), the Garibaldi smallgoods incident and subsequent death of a four-year-old girl from food poisoning (1995; see Chapter 6 regarding SQF development in Australia), the British government confirmed a link between BSE and Variant Creutzfeldt-Jakob Disease (CJD), and the International Organization for Standardization (ISO) finalized the guide for environmental management systems ISO 14001 (1996).

a batch can not exceed 1 day of production (Article 4) (Bantham and Duval 2004, Golan et al. 2004 and Golan, Krissoff, and Kuchler 2004).

In May 2001, the French Ministry of Agriculture [Direction Générale de l'Alimentation (DGAL)] spearheaded operations in the beef chain to test and compare computerized systems for the management of traceability. The objectives of these pilot operations were to develop effective means of traceability, innovate and adapt to the needs of operators, and set them up in real conditions within an identified chain to allow for the control of sanitary risks. Work was conducted over a 15-month duration beginning June 2002, quality was controlled by a steering committee whose secretariat was ensured by the Bureau of Quality, and pilot results were presented to industry players in the beef supply chain during a conference in Paris in March 2004 (Bantham and Duval 2004).

The EU, through the Council Decision of September 30, 2002, adopted a specific program for research, technological development, and demonstration, called "Integrating and Strengthening the European Research Area (2002–2006)," focusing on traceability processes all along the production chain. The objective was to strengthen the scientific and technological basis for ensuring complete traceability (e.g., of GMOs), including those based on recent biotechnology developments from raw material origin to purchased food products, thereby increasing consumer confidence in the food supply (Bantham and Duval 2004).

The key EU IPT regulations include Regulation (EC) No. 178/2002, which established the European Food Safety Authority (EFSA), from which Regulation (EC) No. 1829/2003 (concerning GM food and feed), and Regulation (EC) No. 1830/2003 (concerning the traceability and labeling of GMOs and the traceability of food and feed products produced from GMOs) and its amending Directive 2001/18/EC.

2.3.7 Labeling, Segregation, and Identity Preservation

In the United States, truthful labeling has been used historically to provide consumers with information on calories, nutrients, and food ingredients under regulatory guidelines. But the federal government only requires explicit labeling of GM food if it has distinctive characteristics relative to the non-GM version. In contrast, the EC adopted GM food labels in 1997. The EC requires each member country to enact a law requiring labeling of all new products containing substances derived from GM organisms. Japan, Australia, and many other countries have also passed laws requiring GM labels for major foods. The international environmental lobby has frequently argued that "consumers have the right to know whether their food is GM or not." However, labeling involves real costs, especially the costs of testing for the presence of GM, segregating the crops, variable costs of monitoring for truthfulness of labeling and enforcement of the regulations that exist, and risk premiums for being out of contract (Huffman 2004).

2.3.8 IPT

An effective labeling policy also requires effective segregation, or an identity preservation system. To the extent that there is a market for non-GM crops, buyers of crops would be expected to specify in their purchase contracts some limit on GM content and/or precise prescriptions regarding production, marketing, and/or handling processes. One can envision a marketplace of buyers with differentiated demand according to their aversion to GM content. To make this differentiation effective, new costs and risks are incurred. Additional testing involves costs of conducting the tests for which there are several technologies of varying accuracy. The risk is that GM and non-GM varieties will be commingled and detected in customers' shipments under contract limits on GM content. This is a serious economic problem as agents seek to determine the optimal strategy for testing and other risk mitigation strategies (Huffman 2004).

"Tolerances" are an important issue in identity preservation and segregation. Tolerance refers to the maximum impurity level for GM content that is tolerated in a product that still carries the non-GM label. There are two levels in which tolerances apply: one is defined by regulatory agencies such as the U.S. Food and Drug Administration, and the other is commercial tolerances. Individual firms can and seem likely to adopt different tolerances, subject to any regulation. Moreover, different countries are likely to have different tolerance levels and this increases the risks and costs of identity preservation (Huffman 2004).

Dual market channels could develop privately without regulated tolerance levels. This system would require growers to declare GM content at the point of first delivery and be subject to their own uncertainty about GM content. This is commonly referred to as "GM declaration" and has been an important element of the evolution of markets for GM grains. At the delivery point, a grain elevator could segregate within its own facilities, or each elevator could specialize in handling only GM versus non-GM grain. Or, it could be a vertically integrated firm with some delivery points specializing in GM and others in non-GM commodities or different GM commodities (Huffman 2004).

Major risks arise in segregation and identity preservation. Growers face three sources of risk: (1) "volunteer or feral plants" in subsequent crops, (2) pollen drift, and (3) on-farm adventitious commingling. The volunteer plant rate is highest during the first year after planting a crop and decreases as subsequent years pass. At some cost to farmers, this population can be reduced through mechanical weeding or selective application of chemical herbicides. Pollen drift is modest in self-pollinated crops (e.g., wheat, rice, and soybeans) but very high in open-pollinated crops (e.g., corn and sorghum). Even in self-pollinated crops, out-crossing occurs at a nonzero rate for most plants. Farmers can reduce the likelihood of pollen drift in the crops by establishing physical barriers (buffer strips) and physiological barriers (staggering pollination dates). On-farm adventitious commingling can be expected to occur at a significant rate on farms producing GM and/or non-GM crops and other GM crops. This problem would decrease as a farmer becomes more specialized in one non-GM crop, but if this resulted in more monocultures, then it would increase costs from pests that thrive on monocultures, soil erosion, and higher commercial fertilizer rates (Huffman 2004).

Although private sector handlers routinely segregate and blend grains as a primary function of their business, new risks arise when handling GM grains because of the added risk of adventitious commingling. Currently in the United States, this risk may be about 4% at the elevator level. Farmer-processor contracting of specialty crops could reduce this margin by specializing in the product being delivered. Another source of risks is testing because no test is 100% accurate. This risk varies with the technology, tolerance, and variety of products handled and seems likely to be falling over time as the technology of testing advances (Huffman 2004).

A recent study showed that with current GM technology, standard-labeled and non-GM-labeled products would sell at a premium. For this reason, growers and handlers of non-GM grains have a private incentive to "signal" their "superior quality." This signaling is costly; that is, it involves segregation and identity preservation. Because GM grains would currently sell at a discount, GM growers and handlers do not have any incentive to undertake costly identifying and segregating non-GM from GM grains. In fact, because non-GM would sell for more, they have an incentive for adventitious commingling of GM and non-GM products. Hence, only products destined to be non-GM would need to be tested. Furthermore, setting of tolerance levels must take into consideration that the science of detection of impurity is steadily rising, so "a zero-tolerance level" is very costly. Studies have shown that consumers would pay a significant amount for what they perceived as a zero-tolerance level in vegetable oil, tortilla chips, and russet potatoes, but they were indifferent between a 1 and 5% tolerance level; that is, indifferent between a non-GM labeled product with a 1 and 5% GM impurity rate (Huffman 2004).

In a marketing system with identity preservation or segregation, end-users and buyers would need to express their needs and aversions to GM in contracts with tolerances. Ultimately, it is important for buyers, who want to limit GM content in non-GM shipments, to specify limits/restrictions in their purchase contracts. Those who are not opposed to GM would not have to do anything special. Grower declarations on grain shipped is a critical first step in this process. Therefore, it is important that growers know the purity of the varieties they plant or at least have the capability of knowing. This provides a wealth of information that needs to be conveyed to the marketing system. To the extent that farmers do not have perfect control of their production process (e.g., use purchased seed that may not be 100% non-GM; grow crops in the open air where windblown contamination can occur rather than in greenhouses; and produce GM and non-GM crops, which leads to adventitious commingling) they may be reluctant to declare that their delivery of grain is GM-free (Huffman 2004).

2.4 NEW SAFETY CONCERNS ON THE HORIZON— HOW WILL IPT MEET THESE CHALLENGES?

Below are three short selections of challenges that will be faced by the food industry and consumers. Following this is a concern regarding traceability of food *after* it is purchased.

1. "Send in the Clones: FDA Set to Approve Food from Cloned Animals." The FDA recently released a preliminary safety assessment that clears the way for marketing of meat and dairy products from cloned animals for human consumption. According to consumer groups, such as the Center for Food Safety, the assessment and the agency's expected endorsement of cloned food comes despite widespread concern among scientists and food safety

advocates over the safety of such products. The move to market cloned milk and meat also ignores dairy and food industry concerns and recent consumer opinion polls showing that most Americans do not want these experimental foods (*The Organic & Non-GMO Report* 2007).

2. "Germans Find Italian Organic Standards Wanting." A German report claims that 17% of "organic" food products imported from Italy in 2005 were not actually organic, *AgriHolland* reports. By comparison, only 2.5% of Germans claimed falsely to be organic* (*Agra Europe Weekly* 2006).

3. "From GMO to Nano: A Familiar Debate over a New Technology." Scientists have developed a new technology they claim will revolutionize food production and create healthier foods. Critics have raised concerns that the technology poses great risks to human health and the environment. Government agencies have had difficulty regulating the technology. The new technology is not genetic engineering, but nanotechnology. The theory behind nanotechnology is that by manipulating and assembling molecules and atoms—the so-called building blocks of matter—in certain configurations scientists can create almost anything (*The Non-GMO Report* 2006).

One area where industry has no incentive to create traceability systems is for tracking food once it has been sold and consumed. No firm has an incentive to monitor the health of the nation's consumers to speed the detection of unsafe product. Government-supplied systems for monitoring the incidence of food-borne illness, such as FoodNet and PulseNet, are one option for helping close this gap in the food system's traceability network. Food-borne illness surveillance systems increase the capability of the entire food supply chain to respond to food safety problems before they grow and affect more consumers (Golan, Krissoff, and Kuchler 2004).

2.5 RESULTS OF IPT—IMPROVED FOOD SAFETY AND CONSUMER CONFIDENCE

The results from businesses that combine identity preservation tracking systems and traceability systems, which results in a comprehensive IPT program, have been a tremendous improvement in food safety and confidence building toward public and private food chain participants. In practice, the challenge of food safety and preserving a food product's identity is complicated by the many times that ingredients and products change hands between the seed supplier and the food manufacturer. For example, a medium-sized food company has more than 1,000 suppliers of over 8,000 ingredients that go through more than 30 processing plants and end up in some 6,000 different finished products (Anonymous 2005).

As we have seen, reactions to food crises have brought about increased legislation and rules around the globe. Players in the food chain include principal components such as parent seed companies, farmers, elevators/cooperatives, transportation, storage, processors, and retailers. But the food chain also includes a direct support cadre

---

* Report from *Ökomonitoring* 2005, published by CVUA Stuttgart, July 2006. Available online at http://www.untersuchungsaemter-bw.de/pdf/oekomonitoring2005.pdf (accessed April 17, 2009).

that includes software developers, auditors, labs, training personnel, etc. Policy-makers have motivated producers to develop new tools and approaches toward IPT. From rules and programs implemented by industry and academia through studies, leverage points and ways to improve IPT are being discovered.

Changes in the local to global supply chain have not been easy, nor without cost. There are many factors that affect costs and benefits of IPT systems. These include the following:

Factors affecting costs

- The wider the breadth of traceability, the more information to record and the higher the costs of traceability.
- The greater the depth and the number of transactions, the higher the costs of traceability.
- The greater the precision, the smaller and more exacting the tracking units, and the higher the costs of traceability.
- The greater the degree of product transformation, the more complex the traceability system, and the higher the costs of traceability.
- The larger the number of new segregation or identity preservation activities, the higher the costs of traceability.
- The larger the number of new accounting systems and procedures, the more expensive the start-up costs of traceability.
- The greater the technological difficulties of tracking, the higher the cost of traceability.

Factors affecting benefits

- The higher the value of coordination along the supply chain, the larger the benefits of traceability for supply-side management.
- The larger the market, the larger the benefits of traceability for supply-side management, safety and quality control, and credence attribute marketing.
- The higher the value of the food product, the larger the benefits of traceability for safety and quality control.
- The higher the likelihood of safety or quality failures, the larger the benefits of reducing the extent of failure with traceability systems for safety and quality control.
- The higher the penalty for safety or quality failures, in which penalties include loss of market, legal expenses, or government-mandated fines, the greater the benefits of reducing the extent of safety or quality failures with traceability.
- The higher the expected premiums, the larger the benefits of traceability for credence attribute marketing.

REFERENCES

Anonymous. "Consumers Drive Changes in the Food System." *Feed & Grain* 44, no. 3, 2005: 39–40.

Bantham, Amy, and Jean-Louis Duval. *Connecting Food Chain Information for Food Safety/ Security and New Value—Area III: Food Safety and Food Security.* Submitted to IAMA Business Office, April 30, 2004.

Caswell, Julie A. "Labeling Policy for GMOs: To Each His Own?" *AgBioForum* 3, no. 1, 2000: 53–57.

Clapp, Stephen. "Traceability: A Global Perspective." *Food Traceability Report* 1, 2002: 16–19.

Golan, Elise, Barry Krissoff, Fred Kuchler, Linda Clavin, Kenneth Nelson, and Gregory Price. *Traceability in the U.S. Food Supply: Economic Theory and Industry Studies.* Agricultural Report Number 830. Washington, DC: U.S. Department of Agriculture Economic Research Service, 2004.

Golan, Elise, Barry Krissoff, and Fred Kuchler. "Food Traceability One Ingredient in a Safe and Efficient Food Supply." *Amber Waves* 2, no. 2, 2004: 14–21.

Huffman, Wallace E. "Production, Identity Preservation, and Labeling in a Marketplace with Genetically Modified and Non-Genetically Modified Foods." *Plant Physiology* 134, 2004: 3–10.

Jones, Michael K., and Bruce R. Rich. "The Food Industry from Feeling Full to Fulfillment." Paper presented to the Social Change in the 21st Century Conference. Centre for Social Change Research, Queensland University of Technology, 2004.

The 2006 Non-GMO Sourcebook. Edited by Ken Roseboro. Fairfield, IA: Writing Solutions, Inc., 2006. "Send in the Clones: FDA Set to Approve Food from Cloned Animals." *The Organic & Non-GMO Report*, 7, no. 2, 2007: 1, 3.

"Germans Find Italian Organic Standards Wanting." *Agra Europe Weekly* no. 2218:N/62006.

3 Identity Preservation and Traceability Theory, Design, Components, and Interpretation

3.1 INTRODUCTION

The key to this chapter is to understand that all components of an identity preservation and traceability (IPT) program must work together. Each must be able to not only function on its own, but also be passed along to the next component in the process. This chapter primarily covers IPT process theory, system design, and system components. At the end of the chapter there will also be an introduction to types of laboratory analyses and challenges to IPT through batch processing.

Traditionally the food supply system was made up of independent farmers selling their product to elevators or cooperatives. Elevators and cooperatives attempted to meet minimum commodity standards and hoped to prevent spoilage or infestation. Some farmers may have been on contract and sold directly to processors. Transportation providers were loosely governed and regulated, in the same accordance as storage facilities aimed to meet the minimum standards. Processors received truck- or trainloads of commodities that would be added to bins of like commodities. Mixing and processing occurred nearly continuously, with batch production becoming more common over time. Product and resultant mixtures would in turn be packaged, stored, shipped, and used as ingredients in final-use products. Again, this final product was made up of many ingredients and numerous processes and would be packaged, stored, shipped, warehoused, and at some point put on shelves for sale to a customer or end consumer. The chain was typically fragmented in regard to food safety, accountability standards, etc. The goal was efficient food production (commoditized) and an abundance of food inexpensively provided to consumers and customers.

The advent of recent domestic and global events has given cause to increase food safety and consumer choice. IPT offers a solution. Throughout the IPT food chain of events, numerous parties are directly and indirectly involved with IPT practices such as management, documentation, processes, verification, certification, analysis, procedures, etc. To better understand the parties that are involved with IPT, it is important to understand how IPT fits into each party's area of responsibility. To do this it is important to know what IPT is, which is discussed next. After that, the ideas involved in designing an IPT program are discussed. Following this, the major components of IPT will be divided into four groups. The first group will be that of rules and regulations that govern the IPT program. The second group includes the primary parties that are directly involved with the food chain (i.e., its farming, transportation, etc.). The third group is comprised of support parties that facilitate IPT (e.g., software providers, trainers, auditors, laboratories, equipment, and chemical manufactures, etc.). The fourth and last group is comprised of ancillary parties such as advisory policy groups, lobbyists, and insurance organizations. A laboratory analysis section will highlight its growing importance within product conformity, quality control of traits, etc. Finally, of increasing difficulty and complexity is the monitoring of batch production processes for IPT compliance. This section will provide an overview of how difficult it is to provide an accurate accounting of ingredients of (1) inbound bulk commodity products that arrive at the loading dock for processing to (2) outbound products that have been processed.

3.2　THEORY—WHAT IS ASSUMED WITH IPT?

What is IPT? There are many definitions and ways to describe this term. In essence, identity preservation (IP) is a term used in the food chain that helps describe a level of tolerance, and/or processes, and/or other attributes that customers along the food chain may desire or demand. Tolerance may include percentage of genetically modified organism (GMO) material in the crop or percentage of protein in soybeans. Processing may include knowing what chemicals were applied to the crop or if the process was in accordance with religious rules. Other attributes may include animal and labor welfare or geolocation of food origin. IP is generally viewed as starting from the seed and soil and following the crop, process, etc., through the food chain until it arrives on retail shelves. Many organizations characterize this as from "dirt to dinner plate" or "farm to fork."

Traceability, on the other hand, moves in just the opposite direction: backward from dinner plate to the crop's origins or back to a specific event. Traceability can be viewed more as an accounting-type function based on paper and electronic records kept by participants and members of the food chain. Often the traceability or traceback is tied directly to an IP system or program used in crop production, transportation, processing, etc. In this way IP is usually more extensive than traceability, but no less important.

A premier work on traceability is Golan's U.S. Department of Agriculture (USDA) research publication on traceability, which is groundbreaking in its clarity. In her work titled "Food Traceability One Ingredient in a Safe and Efficient Food Supply," she emphasizes that "[f]irms determine the necessary *breadth*, *depth*, and *precision* of their traceability systems depending on characteristics of their production process and their traceability objectives."* Although her work coined the terms "breadth,"

---

* See pages 24–25 for greater details of these terms. Another more extensive work by her and highly recommended for IPT reading is *Traceability in the U.S. Food Supply: Economic Theory and Industry Studies,* by Elise Golan, Barry Krissoff, Fred Kuchler, Linda Clavin, Kenneth Nelson, and Gregory Price; USDA Economic Research Service, Agricultural Report Number 830, March 2004, http://www.ers.usda.gov/publications/aer830/aer830.pdf, accessed January 25, 2007.

"depth," and "precision" to describe traceability program structures or formatting, other IPT programs and systems perform in generally the same manner using alternative terms. Yet the same object is shared by all programs and systems; namely, to ensure a specific level of tolerance and/or attributes and methodology to traceback or certification processes to ensure specifications.

The reason that IPT is so important is that many of the tolerances and attributes of importance are not evident to the consumer's naked eye. One cannot tell at the grocery store if an item was grown in accordance with organic standards or with unapproved chemicals, if a tomato is derived from transgenic processing, or if a product has mycotoxins in it. For these and many other reasons, IPT programs and systems are essential in protecting our food supply. Traceability or traceback is used by both public and private entities for various reasons, such as to recall mislabeled or contaminated products and for public notifications to alert consumers about food security issues (Golan, Krissoff, and Kuchler 2004).

IPT programs usually key in on specific attributes of interest. Once the attribute or attributes are determined, a program must be established and managed. Responsibility and oversight of the program are key to ensuring compliance.

3.3 SO WHERE DO YOU START WHEN PUTTING TOGETHER AN IPT PROGRAM?

The International Organization for Standards (ISO) defines traceability as "the ability to trace the history, application or location of that which is under consideration." The Canadian Food Traceability Data Standard, developed by the Can-Trace initiative, further defines traceability as being made up of two components: tracking and tracing. Tracking is the capability to follow the path of a specified unit and/or lot of trade items downstream through the supply chain. Traditionally trade items are routinely tracked for availability, inventory management, and logistical purposes. Meanwhile, tracing is the capability to identify the origin of a particular unit located within the supply chain by referring to records held upstream in the supply chain. Units are traced for purposes such as recall and complaints (Miskin 2006).

IPT is information. It represents the data that uniquely identify primary materials, ingredients, processes, additives, and finished products at each step in the supply chain, from seed to the consumer. It also identifies the parties, locations, and shipments involved in the planting, harvest, transportation, transformation, processing, packaging, storage, and distribution of food products. Finally, it records the processes to be validated by auditors to demonstrate compliance with food safety [Hazard Analysis and Critical Control Point (HACCP)], food quality, and IP programs (Miskin 2006).

Regardless of IPT format, there are several challenges all successful systems must overcome. One is the fact that the data must be collected from multiple sources, including animal ear tags, harvest/slaughter records, certificates of authenticity, labels or markings on boxes and pallets, receiving and shipping activities, processing activities, food safety and food quality inspections, packing equipment, and so on. Another issue is that the data may not be stored in a single location. Data storage can be electronic, manual, or both. In the event of a recall, the data may have to be integrated from a combination of manually kept logs, processing records, shipping documents, weight sheets, pick lists, spreadsheets, and/or databases (Miskin 2006).

To be effective, IPT data and information must be shared with suppliers and customers, creating two additional challenges. The first is the lack of standardization between trading partners in how products, parties, and locations are identified. The second is the lack of standardization in how the data are shared. The solution to these two challenges has existed for many years in the downstream portion of the supply chain, where a product is packed and shipped in boxes and pallets. Industry initiatives such as ECR (Efficient Consumer Response) and EFR (Efficient Foodservice Response) provide a solution through the use of bar codes on boxes and pallets and the transmission of electronic messages, all of which use the GS1 data standard (Miskin 2006).

3.4   IPT PROGRAM DESIGN

3.4.1   APPLYING PHILOSOPHY

The concept of traceability takes on a completely different significance when it is extended beyond the farm to embrace the greater agro-industrial sector as a whole. In this case, IPT means the ability to track and retrace all of the stages of process, production, and distribution system and must therefore be viewed as IPT over the entire food chain—from farm field to the consumer's table. It follows that food-chain IPT should be relatively simple when all of the processing is handled by a single organization, but becomes extremely complex for multiple-ingredient products, which call upon several different systems for raw material production, processing, and marketing (Bodria 2003).

It is necessary, in this case, to identify and characterize all of the material flows (raw materials, additives, semi-finished products, packaging materials, etc.) that converge into a given product, as well as all of the organizations involved at each stage, to ensure that the product's history can effectively be retraced to ascertain the causes and responsibilities for any problems or defects. Food-chain IPT is therefore a concept that can be defined as "the identification of the organizations, processes, material flows, and other credence attributes involved in the formation of a product unit that is individually and physically identifiable" (Bodria 2003 and Jorgenson 2004).

3.4.2   IPT—DIFFERENT FROM OTHER TYPES OF PROGRAMS AND SYSTEMS

From the above definition, it follows that IPT is based on two fundamental elements. First is the fact that IPT is, in effect, an allocation of *responsibility*, making it substantially different from other product and process assurance systems such as ISO 9000 for quality and HACCP for safety, which are both designed to control technical aspects. For these two systems (ISO and HACCP), all of the actors involved in the preparation of the product must assume responsibility for the materials used and for the procedures and operating conditions within their competence, so that in case of harmful or defective products the causes can be identified and the appropriate corrective and control actions implemented[*] (Bodria 2003).

The second fundamental element of IPT is the unit of interest size or *lot*, that is to say the unit of product that can be physically and individually identified and that provides the true basis of an effective system for managing emergencies and attributing responsibilities. In fact, the lot makes it possible to identify all of the units that have undergone a given production process so that they can be isolated in the event of quality or food safety problems (Bodria 2003).

As has been mentioned, IPT means the end-to-end *tracking* (raw product forward to end consumer) of ingredients and chain of custody associated with the manufacture and distribution of food products, and end-to-end *tracing* (end consumer back to raw product origins). The implementation of an IPT program will affect a company's strategy, business processes, and technology and will require a disciplined approach. IPT solutions are being elevated to the C-level in food and beverage companies as a critical component of core business strategy. Thus, a company's IPT program and its design play a significant part in overall corporate performance. Programs must take into consideration (1) knowledge of company, (2) program objective and IPT standard(s) used, (3) direct and indirect customer(s) and their trait(s) of interest and credence attribute(s), (4) level of tolerance(s), (5) measure(s) of performance, (6) compliance determinant(s) and third-party involvement, (7) channeling, and (8) alternative approaches (Jorgenson 2004).

3.4.2.1 Knowledge of Company or Understanding One's Company

Most food companies have some tracking capability through their accounting, operational, and recall management systems. However, these programs often prove insufficient and suffer from common shortfalls. These traditionally "back-office" systems are critical to companies for complying with an increasing array of governmental and industry requirements. If properly designed, they can make IPT *at worst* profit-neutral and *at best* profit-fortifying. These systems, therefore, need to be brought out of the back office and into the executive suite so they can take their proper role in a company's competitive strategy (Jorgenson 2004).

Specifically, many back-office systems provide

- Inadequate amounts of data
- Inaccurate data
- Slow response times in the event of crises
- The inability to maintain the identity of individual ingredients throughout processing
- The inability to track ingredients/products that are within close proximity of each other

---

[*] Some argue that ISO and HACCP systems, with their feedback loops, are too focused on critical control points, rather than comprehensive responsibilities. That responsibility is more than a checklist item; it represents spherical conditions that affect ingredient and product. Thus, the scope of responsibility goes beyond quality and safety, which may be part of any product and process assurance program or system.

- The inability to keep track of ingredients and products between incoming ingredients from loading dock and outgoing pallets of finished products
- The inability to track food and ingredients back to their points of origin at the farm, ranch, etc.
- The inability to create a composite picture of the lifecycle of a food product across multiple unrelated enterprises in the supply chain

To overcome shortcomings, innovative IPT programs focus on (Jorgenson 2004)[*]:

- **Technology:** The need for better information in the chain is driving innovation and investment at all points in the chain. **Data collection** is critical; **connectivity** is the key value creation.
- **Traceability:** Regulators have responded to consumer concerns by mandating traceability, "zero tolerance for food safety." Traceability can be used to create value and offset the cost of compliance, which has been estimated to be 0.5–1% of the cost of goods.
- **Transparency:** Value comes from sharing information to improve operations and efficiency among business partners. Successful chains are using transparency to achieve competitive advantage and improve margins.
- **Technology Improves Operations in the Chain:** This helps improve connectivity across activities and ensures managing brand integrity, sourcing for better quality/compliance, certification of products and suppliers, chemical compliance of pesticide usage and residue, and reporting.
- **Observations and Challenges:** Many systems take into consideration that the total supply chain needs to be connected enough, to those that are willing to pay—can businesses respond in 4–24 hours from time of an event; and global definition of "production lot."

3.4.2.2   Establishing Program Objectives and IPT Standard(s) to Be Used

3.4.2.2.1   Establishing Program Objectives

Most IPT programs begin with evaluating and agreeing on not only safety aspects but also the business goals they are intended to support. Potential business goals may include (Jorgenson 2004):

- Gathering the data necessary to support marketing claims (e.g., 100% organic, GMO free, fair-trade products, adherence to humane animal agriculture practices)
- Proving compliance with contractual requirements (e.g., meeting the specifications of raw materials and ingredients that are supplied to food manufacturers)

---

[*] Recent development in electronics and sensors technology has made available data collection systems that can provide the basis for the development of agricultural IPT. Current localization systems based on differential GPS can offer accuracy on the order of 1–2 m, whereas "variable rate" distribution systems and "yield monitoring" systems can easily record what and how much we harvest and distribute.

- Verifying regulatory compliance (e.g., Bioterrorism Preparedness, COOL, USDA National Organic Program standards)

3.4.1.2.2   Determining IPT System Standard(s)

The complex composition of the food chain makes it very difficult to define a single IPT system that can be applied to the broad diversity of food products. It is therefore necessary to define the specific or general standards that provide guidelines for the implementation, management, and surveillance of an IPT program. Such standards should aim to assure the IPT of *each specific ingredient and products* and the *individual actions* (e.g., processes) taken to produce it (as opposed to generic supply chain logistics), and to identify the organizations involved in its formation. Note the emphasis difference of IPT on quality and safety programs (Bodria 2002).

3.4.2.3   Customer(s): Direct and Indirect and Their Traits(s) of Interest—Credence Attributes

To determine customers' wants, a successful IPT program necessitates that some designated "leader" handle the coordination of the supply chain, a role that could presumably, although not necessarily, be filled by the organization that markets the finished product. The leader organization is responsible for tracing the food chain leading to the formation of the product, and for defining operational procedures (audits, lab tests, etc.) to ensure that the causes and responsibilities of any food credence attribute or safety hazard can be identified (Bodria 2002).

3.4.2.3.1 For Example—For Leadership

An IPT standard could be developed along the following lines (Bodria 2002):

- *Identification* and *designation*, as the agents responsible for IPT, of the organizations that handle the processing operations and transfers of primary raw materials or other components significant for the purposes of traceability, and of those that supply secondary materials (process agents, additives, packaging, etc.). This may be done by contracting and forms of testing and auditing. Ownership of responsibility can be defined and agreed upon contractually with built-in forms of checks and balances, incentives, etc.
- *Designation* of a coordinator responsible for defining the operating methods and traceability procedures, and for collecting the relevant documentation and ascertaining compliance. A third party and/or laboratory may fill this area.
- *Documentation* of the material flows within the food chain, recording each passage in qualitative and quantitative terms. Later used for verification by third parties.
- *Management* of lots through every stage of the process, ensuring that they are identifiable and that their traceability is documented at all times.
- A *code of food chain* on each of the documents that accompany the loose or packaged materials entering the production process—an important management process tool for tracking.

- The *marking* of every package that reaches the end consumer or targeted customer with a logo identifying the food chain and with a lot code (e.g., USDA Organic or any other official or recognized third-party certification).
- The *possibility of traversing* the supply chain in both directions: to "trace" (i.e., work back from the finished product to its origins) the nature and history of all the components and "track" (i.e., reconstruct its forward progress) an unsafe raw material to identify the finished product lots that may have been contaminated by it. This track and trace method may be used for whatever trait is desired.

3.4.2.4 Levels of Tolerance(s) for Output Traits or Credence Attributes

In general, there is a direct correlation between increased product purity standards (tolerance levels) and higher IPT costs. Standards for the final product largely determine the complexity of production, handling, processing, testing, and labeling procedures required to maintain the identity of a commodity and therefore the costs associated with the IPT program. The benefits of value-added output traits can only be captured if purity of the product is maintained throughout production and marketing, but the added value must be sufficient to pay for the added IPT costs.

Many believe that the introduction of value-enhanced, identity-preserved crops will further *decommodify* the U.S. commodity handling system. This may result in a shift away from traditional bulk commodity handling practices to a system that tracks and preserves the genetic or process identity of products from seed to end user. In such systems, specialized biotech crops and organic crops may result in greater farm profitability because of higher commodity prices. However, economists disagree on whether these traits or attributes will possess sufficient value to be shared with all participants in the value chain. IP is only successful if it enables all handlers in the value chain to share the increased value achieved by segregation. If a disproportionate burden of IP costs falls on any one group in the handling chain, IP systems may fail economically (Sundstrom et al. 2002).

The burden of maintaining purity and the cost of IP is distributed differently under different conditions. In the case of higher-value commodities, such as corn with higher oil or improved nutritional content, a price premium must offset the increased costs of IP. In other cases, IP is used primarily to ensure the absence of a particular component in a commodity, such as in the marketing of non-GMO foods. In this case, the burden falls primarily on the producer and marketer of the non-GMO product to ensure the purity of the product, along with the substantial additional costs for testing to confirm this.

Although some markets pay a premium for non-GMO certification, in many cases there is little or no price premium for such products because their inherent value is no greater than similar commodities not subjected to IP and testing. Because organic products must also be GMO-free, organic producers face potential additional costs of testing to ensure the absence of GMO traits. The need to test for GMO traits depends entirely upon the regulatory, marketing, and labeling requirements of different countries and product categories, which are largely under development and in flux. In particular, the levels at which threshold tolerances are set for adventitious contamination have a large impact on IP requirements and costs. Thus, it is difficult to determine precise cost-benefit-risk relationships at the present time. No doubt these issues will be settled in the marketplace as the higher potential value for producers and consumers is balanced against the costs of delivering identity-preserved commodities. One thing is clear: the economic success of IP systems depends upon having sufficient market premiums at all points in the value chain (Sundstrom et al. 2002).

3.4.2.4.1 IP Programs (Not Including Traceability)

These programs must not be confused with traceability, which also enables retracing the chain's links all of the way back to the grower. The IP programs draw the guidelines necessary for minimum certification. They guarantee that products remain free of any contamination and, therefore, retain their specific quality. They establish procedures and document evidence that procedures were observed. This involves the handling of information flows.

Within a system, at least one of the major players in the chain must also assume the role of an organizing third party for IP. It is this entity that captures the customer's demand and controls product quality all along the supply chain (see Table 3.1). It is this entity that draws the contracts and controls their good execution. And as the last link before the final user, this entity is responsible for sharing the added value among the participants. The entity controls the chain because of its central position at the joining of two information flows. In the supply chain, the information bearing value is demand; conversely, the information flows produced by the third party come at a cost for the coordinating entity (contracts, control, product separation) but have value for the customer. There is thus a capture of value by means of information (Table 3.1).

TABLE 3.1
Product Differentiating Characteristics

Differentiating Characteristics	Level I Identity Preservation	Level II Specialty Variety	Level III Super Commodity	Level IV Standard Grade
Relative value/premium	High	Medium	Low	None
Buyer control	Variety of production practices, certifications, etc.	Minimum/maximum attributes	Attribute preferences	Grades only
Attribute testing	Typically required by grain buyer	Correlates to cost/value of grain	Efficient consistent	Grade-driven
Producer contracts types	Acreage production bushels	Production bushels normal/open	Normal/open	Normal/open
Producer linkages	High	Moderate	None	None
Minimum segregation	Begins at farm	Begins at farm	Merchandiser/end-user determined	Merchandiser/end-user determined
Production volumes	Low	Moderate	High	Very high

Source: Modified with permission from Value Enhanced Grains Solutions website (http://www.vegrains.org).

3.4.2.5   Measures of Performance—for Consumers, for Firms

When considering IPT performance requirements, a key question should be, "What problem am I trying to solve or what opportunity am I trying to seize?" Defining the opportunity or problem will be essential, especially as to how the answer would be applied to various firm objectives.

In this section we are really talking about two measures of performance, both of which can be written upon at length. The first deals with measuring the performance of the ***output product*** to the desired trait or credence attribute of interest. This is usually for the benefit of the customer and consumer. The second focuses internally on measuring ***IPT requirement procedures*** as they apply to conforming to regulations, contracts, etc. Often firms will look at this internal aspect as a cost versus benefit of IPT procedures to seek ways to leverage opportunities and to minimize inefficiencies. This aspect is often the determining factor as to if a firm will participate with producing products that require IPT systems.

3.4.2.5.1   Output Product

Firms' ancillary programs such as quality and management system programs assist in measuring the continual improvement of a product. Firms have procedures in place that dictate processes, procedures, etc., that must be performed, which in the end help support compliance to whatever measures they are seeking. Many of these other systems also use feedback loops for quicker product improvement and corrections for shortcomings. In addition, firms may have in-house auditors, tests, and analyses (laboratory) to test output product and to confirm procedures, although at other times third parties may inspect for credence attribute performance measures that may entail qualitative on-site visits and quantitative laboratory confirmation. For consumers and customers, it is the output product measure of performance that is most critical. For example, this is where third-party auditors will confirm or deny organic claims as being true or false. Many times customers make their decisions on the basis of laboratory results that provide performance measures as to specific traits, food origins, etc.

3.4.2.5.2   IPT Requirement Procedures

For organizations, many internal aspects must be considered when developing an IPT program. Costs of IPT may be spread through several objects and the cost versus benefit from each objective may be nearly impossible to calculate individually. Here is a list of objectives to consider (Boyle et al., 2004):

- Inventory management
- Regulatory compliance
- Managing raw material to specification
- Documentation to substantiate brand claims
- Recall containment
- Contract compliance
- Brand assurance

Normally, a firm will attempt to attain more than one of these objectives. Fortunately, the IPT capabilities needed to achieve many of these objectives are highly leveragable, providing significant opportunity to build a compelling business case for improving overall traceability capabilities. For example, the information required to track inventory lots to contain a potential food safety incident can also serve as a documentation audit trail that substantiates a brand claim in the market-place (e.g., country of origin).

Studies are being conducted regarding the cost versus benefits of on-farm IPT practices for farmers. In one study, farmers from two organizations are participating in growing ultra-low linolenic non-GMO soybeans (See Chapter 15 survey for results of one farm organization.) In this case, each farmer documents study time/costs of IPT procedures. The goal is to measure how much the additional cost of implementing IPT procedures totals and then compare that cost against the additional premium that is paid for the particular IPT product. Studies such as this hope to illustrate leverage points farmers can take advantage of and weak areas where efficiencies could improve. In the end, it is hoped that by having participants directly involved with this type of study, and if the results are positive, other farmers will participate in IPT programs. From like studies, other cost versus benefits analyses can be performed with elevators, especially for processors and manufacturers.

3.4.2.5.3   Summary of Benefits

The advantages of adopting IPT are to improve supply management, make it easy to trace back for food safety and quality, and to detect any quality problem before the product reaches the market.

Some of the **benefits for organizations** that utilize IPT systems include (Fonsah 2005):

- Minimizing the production and marketing of unsafe and inferior quality goods
- Reducing the costs involved in the distribution system
- Minimizing the cost of recalls
- Reducing the potential for bad publicity
- Reducing liability and increasing revenue of the implementing company

For consumers, the benefits include:

- Verification of standard (e.g., kosher, halal, organic, non-GMO, etc.)
- Enhanced animal welfare
- Improved laborer welfare
- Regional credence attributes

Factors affecting IPT costs include:

- Breadth of traceability and the amount of information to record
- Depth and the number of transactions
- Degree of precision and exactness of the tracking units
- Degree of product transformation and complexity of the system
- Number of new segregation or IP activities

- Number of new accounting systems and procedures
- Technological difficulties of tracking

Factors affecting benefits include:

- Value of coordination along the supply chain
- Size of the market
- The higher the value of the food product
- Likelihood of safety or quality failure
- The penalty for safety/quality failures, in which penalties include loss of market, legal expenses, or government-mandated fines
- The size of the expected premiums

3.4.2.6   Compliance Determinant(s)—Third Parties

To prove IPT compliance, IPT programs must usually follow prescribed proto-cols. This is commonly outlined by contract between parties and/or by regulations. Contracts will typically outline the methodology of verification or certification to ensure compliance. At a minimum a contracting entity can accomplish this through a paper and/or electronic paper trail. In this way the trail from seed to dinner plate will range from being (ideally) continuous and flowing with complete transparency to a (less desirable) more fragmented and laborious compliance trail. As simple and complete as a paper trail may be to conform to specifications, an additional step is becoming more common and is often required—auditing. In much the same way, regulations, regardless of their origins, will declare specific requirements that per-tain to many aspects of food and food safety. Although not all sovereign regulations specifically address IP and/or traceability, many do address IPT such as procedures and systems within other regulations or rules (see Part II for more details on stan-dards that relate to IPT).

3.4.2.6.1   Certification

Certification by third parties is required in nearly all credible IPT programs. It is clear that food-chain IPT must be subjected to surveillance and certifications per-formed by independent bodies that are credible and representative (Bodria 2002). In fact, a false declaration of traceability does not just constitute a deception toward the consumer but is also an act of unfair competition between firms. In the case of voluntary adoption of food-chain IPT, the certification could consist of

- An *international standard* that sets out general implementation guidelines
- Several *certification bodies* accredited by the national standard authorities
- A system for *documenting material flows* suitable for the different product supply chains

Pointedly, auditors, and explicitly third-party auditors, are being used to verify paper/computer trail statements and process claims (farm fields) to enforce contractual obligations such as non-GMO soybeans or to meet public assurance (think USDA Organic Standard). In some cases, in-house auditors provide evaluation toward compliance. Often, an in-house audit is used as a maintenance function of the overall system and is designed for continual quality and IPT improvements of the system. These auditors are usually involved with quality assurance and quality management. In other words, auditors will usually verify that a "process" is either in or out of compliance.

In addition to third-party and internal auditors, laboratory tests are conducted to ensure quality control [QC; think protein or oil content of crop or detection of genetically modified (GM) traits]. This is another way to enforce compliance and an area that is receiving increasing attention. As more new varieties of crops are brought into the public arena, testing must keep up with these newer entrants to test them for safety and approval for trade (see Part III on auditors and laboratories).

3.4.2.7   IPT-*Lite* (Channeling Programs)[*]

IP certification programs can work in two ways to guarantee purity and ensure the value of specific crop traits. A true IPT program is not simply product segregation, but rather a process that results in certification that a product meets specific quality standards. A good example is using pure planting stocks to sample and verify product identity in order to establish confidence in the integrity of the system and the quality of the products. Alternatively, **channeling** is a process-based certification program that focuses on the segregation of large volumes of commodities (Sundstrom et al. 2002).

The emphasis in channeling is on the integrity of the process used to produce the commodity, but the final product may or may not be tested specifically for the quality traits of interest. True IPT programs may cost as much as 5–10 times more to implement and maintain than channeling systems because of more stringent standards and the additional costs of repeated sampling and testing (Sundstrom et al. 2002).

3.4.2.7.1   Channeling Failure

The channeling of agricultural products for specific markets has been used as long as markets have been diversified. Different varieties, grades, and types of products have long been directed to different specific end uses, and there are many successful examples of such market diversification and product segregation, including white and yellow corn and fiber quality grades in cotton. However, the introduction of crops developed using biotechnology and subsequent concerns over their safety have increased the demands on commodity IP systems, and failure to properly preserve the identity of a product can be devastating. For example, StarLink was a hybrid corn variety produced through biotechnology that provided protection from the European corn borer. It was approved by the U.S. Environmental Protection Agency for animal feed but not for human consumption, pending further tests for potential allergenicity. The particular *Bacillus thuringiensis* (Bt) protein produced in StarLink (Cry9C) was not immediately broken down in simulated digestion tests, and because some allergens are also not readily digested, more data were required before it could be approved for human consumption.

---

[*] This is another process-based system and may result in certification.

A strict IP program was to be implemented to ensure that the StarLink grain was only destined for animal feeds, but this program failed in practice. Although only 0.5% of the total U.S. corn acreage was planted with StarLink corn in crop years 1999 and 2000, some of this corn was mixed with corn sold for human uses, and traces of its DNA (but not the Cry9C protein itself) were found in taco shells and related corn products sold in supermarkets in the United States and abroad. Although no danger to human health was anticipated from this low level of exposure, and no adverse effects in humans was ever documented, the USDA and Aventis Crop Science (the developer of StarLink corn) moved quickly to remove contaminated products from the marketplace. Food manufacturers, milling companies, retailers, and seed dealers recalled or withheld from the market all products that were identified as containing StarLink DNA, and StarLink registration has been voluntarily withdrawn. The estimated cost of this IP failure may exceed $1 billion. This incident exposed weaknesses in the grain commodity IP system that must be addressed if biotech or value-added crops are to be grown and marketed with confidence (Sundstrom et al. 2002).

3.4.2.8   Different Approaches toward IPT: Compulsory or Voluntary IPT Standards?

This is a large issue being fought and modeled by various cultures and governments. Placing food-chain IPT procedures within an appropriate regulatory system is one of the main issues of conflict and is a question of primary importance* (Bodria 2002).

3.4.2.8.1   Compulsory

Some organizations, most notably the European Union (EU), appear to favor statutory or mandatory regulations toward IPT. In fact, in its *White Paper on Food Safety*, the EU states that "[T]he competent authorities monitor and enforce this responsibility through the operation of national surveillance and control systems..." In this case, governments rather than private organizations act as the third party and laboratories for tests and verification. It is assumed, but not explicitly said, that generic rules that include nearly step-by-step methodologies would be mandated—the one size fits all. The notion of government and innovation to improve efficiencies would be a limiting factor for the sake of government control and the feel-good aspect of its oversight (Bodria 2002).

Talks about mandatory IPT compliance have been a policy issue for some time. Propositions about enacting a compulsory system that would trace back animal feed to monitor BSE or mycotoxins, improve food safety, monitor food transportation systems, and minimize the risk of tampering have been a global priority of policy-makers. All of these propositions have one thing in common—providing adequate information to consumers (choice) on a variety of food attributes including country-of-origin, animal welfare, GMOs, organic, etc. (Fonsah 2005).

An alternative voluntary route leaves food-chain IPT to the initiative of individual organizations that voluntarily undertake compliance with the rules and

---

* IPT requires high standards. In fact, some countries, especially less developed countries, regard this regulatory measure as a technical barrier to trade.

procedures set out in a standard. Standards, be it government or where there is no government regulation by industry, establish criteria, tests, and audits to monitor IPT compliance. This should be a more nimble structure to meet the changing nature and needs of society and changes due to improved technology (Bodria 2002).

In the compulsory case, IPT is treated as essential for the assurance of product safety and is bound to a legally binding framework of rules in much the same way as HACCP hygiene monitoring. This solution has the advantage of a generalized application of IPT, but it also presents several shortcomings. The HACCP practice has highlighted the difficulty of achieving simultaneous compliance by many differing production systems and firms, as well as sometimes overriding business management decisions (Bodria 2002).

Alternatively, a voluntary system based on a clear definition of IPT that is specified in an international standard implies a free and conscious commitment on the part of the organization's management and therefore leaves less scope for dodges or accusations of excess complexity. In addition, this type of approach makes traceability a selling point to the consumer, making it an element of added value on the marketplace, thereby enhancing the competitiveness of the product. This would be another motivational factor for businesses. Voluntary IPT therefore has the practical effect of making its fair application advantageous to the producers themselves, as well as to the surveillance bodies (Bodria 2002).

With the passage of time, commingling of economic needs, and global ties binding cultures more closely to one another, the noted differences between government and private regulations and compulsory and volunteer IPT systems may become blurred. In the end, cultural and economic dictums should prevail.

3.4.3 Bringing It All Together

The diverse challenges of IPT share a common solution set—the ability to accurately and quickly follow products backward and forward in the supply chain. Breaking down a program into its components can help explain what goes into establishing such a program and how the infrastructure can be used to meet multiple compliance requirements. Regardless of the company in which it operates or the business goals it is designed to support, a meaningful traceability program should incorporate the following components (Boyle 2004):

- *Know your business and establish IPT goals:* An effective IPT program begins with a clear definition of the business goals it is intended to support and the context of these objectives. Hand in hand with goals are the business's weaknesses and strengths. These goals include regulatory compliance, managing brand to sustainable agricultural practices, and social welfare concerns.
- *Design enabling business processes to comply with standards:* The next step is to design business processes that will support the goals and standards by which they will be measured. Many organizations have referred to various QC and quality assurance (QA) systems such as HACCP. However,

none of these programs on their own can fulfill all requirements that IPT programs entail.*

- **Who are the various customers?** This could involve changes in process management to distribution and interactions with suppliers and customers due to each customer's individual trait(s) of interest. It is at this step that the real work begins.

- **Levels of tolerance:** The organization must understand the level of tolerance goals and the management practices and processes that lead to success. The tolerance level of traits or credence attributes may be simple or difficult to measure or observe depending on its characteristics. It is important to know the tolerance range or limits of the contract or regulations that govern the trade, the method of testing the level of tolerance, and how often the testing will occur.

- **Measuring performance:** There are two major areas of performance measures. The first area focuses on output traits or credence attributes of the product (customer focused), and the second includes the measurements used (usually financial cost-benefit accounting methods) to compare total IPT production costs with premium revenues received (company focused). The first deals with the measuring of output traits or credence attributes by recordkeeping, auditing, laboratories, etc. This may be easier than the financial cost-benefit accounting, which is very detailed in analysis of all aspects concerning IPT costs that include equipment, marketing, management, labor, inputs, processing, etc.

- **How to comply with regulations:** Closely tied to measuring performance above, third-party certifiers are essential. Most often these entities must be certified by an authority that can grant licenses or privileges to certify. Many official certifying agencies are certified by national, regional, and international standards organizations such as ISO, HACCP, etc.

3.5 COMPONENTS

This section lists many, but not all, of the components of an IPT system from parent seed breeder to final sale. The components section is divided into four groups. The first group consists of various standards or criteria that an IPT program may follow. The second group comprises the principal components of the food chain. These are the key or primary players involved in grain production; this listing is also similar

---

* Defining relevant units of production. Most inventory systems are designed to track a unit of production as a stock-keeping unit (SKU) or part number to support inventory management and accounting processes. On farms the unit may be the bushel, bin, or wagon. At a grain elevator the unit may be the truckload or train car. This is likely to be inadequate to support a traceability program if the objectives include reducing the financial and brand exposure of a recall, improving precision in operational decision-making, or supporting the integrity of credence attribute or brand claims with more robust documentation. For example, instead of tracking one day's receipts of corn syrup to a plant as a lot discernible from other lots, one might define a lot as a train car of corn syrup, which can be distinguished from every other train car of corn syrup. This small change would significantly improve tracking precision in an environment where 50 train cars of corn syrup are received each day.

in structure to livestock, fruit, and vegetable production. The third group lists the direct support cadre that helps facilitate principal component organizations in meeting compliance. The fourth and last group embraces organizations that indirectly, yet instrumentally, may influence principal component organizations.

3.5.1 Governance Standards and Criteria Specifications Parameters*

Standards are important for food safety and for establishing public confidence in the food chain. Standards come in many forms and each is distinctly unique. However, each outlines specifications that are to be observed. Below are listed the various types of standard formats.

- Public or private
- Formal or informal
- Highly regulated to no or little regulatory oversight
- Sector standards
- Industry standards
- Less defined standards (e.g., animal and labor welfare, sustainable agriculture, etc.)
- Country standards
- Regional standards
- International standards
- Organic standards
- Religious standards

Many of these types of standards and criteria specifications are explained within this publication in other chapters. Organic standards have been the most recently recognized example of IPT.

3.5.2 Principal Components in the Agri-Business Food Chain

To confidently preserve traits of interest (e.g., non-GMO), or ensure credence attributes of interest (e.g., fair labor, animal welfare, food origins), many, if not all, of the below components must be tied together to safeguard the characteristic of importance. The first aspect for any of these components is their in-house IPT program. This includes details of their program, standards to be followed, documentation, inspections, audits, etc. The next major detail is the manner in which the information/data are passed along to the next component.

In this regard, it is in the transparent transfer that identity is preserved and, if need be later, where traceability is done more quickly and efficiently.

- Parent seed companies
- Farmers
- Elevator/cooperatives

---

* See Part II for greater details regarding specific standards.

- Transportation and conveyance equipment—throughout chain; trucks, wagons, etc.
- Storage facilities—temporary and long-term storage
- First-level processor—crusher, extractor, etc.
- Second-level processor—refining, batch processing, etc.
- Warehousing
- Retail—end location before purchase by customer or consumer

For an example, see Appendix B, "Farm IPT Program and its Components." This example provides a glimpse of the general system, procedural component, and system checklist of an on-farm system.

3.5.3 Direct Support Cadre

Organizations that assist or facilitate principal components IPT programs are called direct support cadre. The below listing comprises many of the specialty areas of direct support that embrace the essential infrastructure needs of many of the principal component organizations. Direct support organizations help enable principal players to meet compliance requirements.

- Auditors (see Chapter 7 for greater detail)
- Laboratories (see Chapter 8 for greater detail)
- Software providers (see Chapter 10 for greater detail)
- Training personnel (see Chapter 11 for greater detail)
- Chemical companies—providers of pesticides and fertilizers
- Equipment manufacturers—support of combines, planters, etc.
- Other service providers

Regarding software and software providers—to facilitate traceability, a trail of information must be created at each step of production. Although many farm operations are computerized, there is still much paperwork created. For example, when a farmer receives seed in bags it is tracked to where it is stored, which field it is planted in, when it is harvested, which dryer bins it goes into, etc. All of this documentation must be entered into the computer system on the basis of unique storage locations or tracking numbers. As products become more processed, computer systems based on bar coding technology are playing an increasingly more important role (Mayer 2003).

3.5.4 Ancillary Support

The advisory and policy groups are instrumental in forming and modifying standards and performance criteria. Often these groups include industry participants as well as communities, activists, regulators, etc. The usual focus is to bring forth more acceptable regulations toward achieving the goals of society (consumers) and efficiency in production (producers). Many times these groups attempt to tackle new problems such as the introduction of new processes or new products (think GMO).

Other times the issues may address social concerns of the environment, food safety recalls, labor and animal welfare, etc. Insurance companies play a key role in reducing production costs through the observance of protocols and rules.

- Advisory and policy groups, lobbyists (see Chapter 9 for greater detail)
- Insurance companies (see Chapter 12 for greater detail)
- Others

Chapter 12, "Food Recalls and Insurance," is eclectic in composition, but its patchwork approach of wide ranging topics attempts to provide a clearer picture of negative aspects of not insuring and the resultant aspects of recall.

3.6 ANALYTICAL TECHNIQUES FOR LABORATORY AND FIELD*

Analytical laboratory techniques may ascertain a plant or grain's chemical composition and deoxyribonucleic acid (DNA). This is essential for environmental risk assessment, government regulation, production, and trade, and this is especially important regarding GM crops. At present, DNA- and protein-based assays can analyze chemical composition. DNA analysis has increased because of stringent food labeling and traceability regulations for GM crops. One international event that pushed the issue of GMO safety in the food chain to greater prominence was the detection of unapproved transgenes in corn, the discovery of StarLink corn in human food, and the subsequent recall of hundreds of food products (Auer 2003).

Analytical techniques for tracking chemical composition (protein, oil traits), radioactive isotopes (food/processing origins), genes (DNA), and transgenes (foreign-species DNA) must be chosen based on research questions and a combination of other factors. The accuracy, precision, reproducibility, sensitivity, and specificity of the method used must be understood in relation to the research question. Practical considerations include the cost and time per sample, the chemicals and equipment required, sample handling and processing, adaptability to field conditions, and technical expertise. For product IPT and food labeling activities, methods must be practical for testing points at the farm, during transport, and in food processing. Regardless of the technique, appropriate experimental controls, production processes, and information about parental crop lines and transgenes must be available (Auer 2003).

3.6.1 LABORATORY METHODS

The three most widely used *laboratory methods* are (1) DNA-based molecular techniques to characterize genetic markers, (2) isozyme analysis of protein profiles, and (3) marker genes that produce a selectable phenotype. Information in this section is directly derived and modified from Carol Auer's "Tracking Genes from Seed

---

* This section is of particular importance because it highlights the various types of tests available used in IPT programs. This is different from Chapter 8, which focuses on specific auditing laboratories and their characteristics.

to Supermarket: Techniques and Trends" (2003) and Steve Tanner's "Testing for Genetically Modified Grain." (2001)

3.6.1.1 DNA-Based Molecular Techniques

DNA-based molecular techniques are used to identify genetic markers and describe genetic relationships. They have become a powerful tool for crop breeding, population genetics, and descriptive studies on gene flow. **Polymerase chain reaction (PCR)** is a detection technique that "looks" for specific DNA base sequences (or foreign genes) that have been inserted into the organism's DNA. PCR uses primers to target specific base sequences unique to the foreign DNA and then amplifies these sequences, often a million-fold, through a series of processes. PCR then uses gel electrophoresis to detect the presence of the modified DNA. If the primers contact the target gene, specific bands will be present on the electrophoretic plate; products that do not contain the target gene will not have these bands. Positive and negative controls are analyzed with each set of unknowns for confirmation purposes.

Molecular markers are advantageous because they are abundant in the plant genome, are not affected by environment, can be based on sequences that are selectively neutral, and can provide a high level of resolution between closely related plants. Disadvantages of molecular markers include the requirements for expensive laboratory equipment, costly reagents, and technical expertise.*

- **Advantages of PCR**: PCR is very sensitive, is specific for the target DNA base sequence, can provide semiquantitative results, and may be suitable for processed food.
- **Disadvantages of PCR**: PCR typically takes 2–3 days to analyze; requires relatively expensive expertise, equipment, and laboratory environment; cannot test for an array of genetic modifications; and costs range from $200–$500 per sample.

3.6.1.2 Isozymes

Isozymes are related enzymes that catalyze the same reaction but have different structural, chemical, or immunological characteristics. Isozyme (allozyme) analysis uses the isozyme profile to distinguish between related plant classes, an approach that has been documented for many crop species. Although laboratory equipment and costs are modest, isozyme variation is not always sufficient to discriminate between classes and might not be selectively neutral. Plant samples must be handled carefully to protect enzyme activity, and activity is affected by tissue type, developmental stage, and environmental conditions.

---

* The most useful molecular techniques to describe genetic relationships include amplified fragment length polymorphisms (AFLP), random amplified polymorphic DNA (RAPD), restriction fragment length polymorphism (RFLP), and microsatellite markers. AFLP and RAPD have an advantage in that they do not require prior information about DNA sequences or a large investment in primer/probe development.

3.6.1.3   Marker Genes That Produce a Selectable Phenotype

In experimental research on gene flow, GM crops containing selectable marker genes can simplify the identification of hybrids and the screening of many plants. The most common selectable markers are antibiotic resistance and herbicide resistance, both of which are routinely used in the initial selection of transformed plant cells and plant propagation. Visible markers or reporter genes can be inserted to study gene flow, including green fluorescent protein (GFP). The family of GFP genes provides the advantage of real-time, noninvasive identification of GM plants and pollen in the laboratory or field.*

3.6.2   Rapid Test Kits for the Field

To effectively market biotech and non-biotech crops, the grain and food industries has access to reliable detection methods to measure the value of improved quality attributes and to distinguish biotech from non-biotech crops.

The two most common immunological assays are (1) **enzyme-linked immuno-sorbent assays (ELISA)** and (2) immunochromatographic assays (lateral-flow strip tests).

3.6.2.1   ELISA

ELISA technology has been developed specifically to detect the presence of biotech grain. ELISA can produce qualitative, semiquantitative, and quantitative results in 1–4 hours of laboratory time. ELISAs are commonly used in a variety of assays (mycotoxins, bacteria, pregnancy tests, etc.) and have been used in the USDA Grain Inspection, Packers & Stockyards Administration (GIPSA)'s official inspection system for many years to provide relatively inexpensive, easy to operate, and rapid analyses for mycotoxins. The ELISA approach is fundamentally different from the PCR approach. The PCR technique detects a particular DNA-base sequence; the ELISA technique generally detects a specific amino acid sequence or protein produced as a result of the genetic modification. Using glyphosate-tolerant (Roundup Ready) soybeans as an example, PCR detects the DNA sequences that have been inserted into the soybean DNA, but ELISA detects the specific protein that is expressed as a result of the genetic modification. Clearly, this protein must be uniquely associated with the genetic modification and be sufficiently different from other proteins to avoid a high incidence of false positives.

- *Advantages of ELISA:* ELISA is rapid and can usually be completed in 10 minutes, is generally sensitive, does not require expensive equipment, and can be performed by trained nontechnical personnel.
- *Disadvantages of ELISA:* Some test kits are not available for all biotech grains. As with PCR, tests are usually specific for a particular genetic

---

* For example, tobacco plants expressing GFP under the control of a promoter for anther and pollen expression demonstrated that a hand-held ultraviolet (UV) light can detect transgenic pollen carried by bees. GFP expression could support direct monitoring of pollen movement over different large distances and research on containment strategies. However, government approval would be required before unconfined release of the gene encoding GFP into the environment (Auer 2003).

modification, but it is possible that test kits capable of detecting multiple genetic modifications could be developed. ELISA tests are dependent on the expression of the foreign protein by the plant, which can be influenced by environmental factors. ELISA tests are often not suitable for the testing of processed foods, because the expressed protein may be removed, altered, or destroyed during processing.

3.6.2.2   Lateral-Flow Strip Tests

Lateral-flow strip tests produce qualitative results in 5–10 minutes at any location and for less than $10. However, sufficient protein concentrations must be present for antibody detection and protein levels can be affected by the plant's environment, tissue-specific patterns of transgene expression, protein extraction efficiency, and food processing techniques that degrade proteins.

3.6.2.3   Other Testing Methods

In addition to PCR and protein-based methods, chromatography, mass spectrometry, and near-infrared spectroscopy (NIR) can be used in some situations; for example, with GM crops that have significant changes in chemical composition. However, these methods can fail when alterations in GM crop biochemistry are within the range of natural variation found in conventional crops.

3.6.3   GIPSA'S APPROACH

In November 2000, the USDA established a Biotechnology Reference Laboratory at GIPSA's Technical Center in Kansas City, Missouri. The laboratory helps buyers and sellers manage risks and increase overall market efficiency. The mission of the laboratory is to ensure the reliability of sampling and detection methods for biotechnology-derived grains and to facilitate information exchange. GIPSA provides guidance on sampling of grain consignments, grain IP protocols, an accreditation program for DNA-based testing laboratories, impartial verification of commercially available rapid test kits, and third-party testing for specific biotech events. Much of the information below is from *Proceedings of GEAPS Exchange '01* "Testing for Genetically Modified Grain" (Tanner 2001).

3.6.3.1   Sampling the Lot (Barge, Railcar, Truck, etc.)

Unofficial sampling methods that may have served the commodity system well in the past may not produce satisfactory results in today's marketing systems. Specifically, grain facilities that are attempting to segregate or identify biotech grain are encouraged to review GIPSA sampling procedures. Probability theory can be used to describe risks associated with random samples. Buyers and sellers can use this knowledge to manage marketing risks.* The USDA has extensive procedures for sampling lots. The procedures have been developed for sampling static lots (railcars, barges, trucks) and moving grain streams. These procedures are used for all

---

* Risks associated with nonrandomly selected samples are unknown and therefore cannot be managed.

official sampling and are recommended for obtaining a representative sample for biotechnology-derived grain testing.*

3.6.3.2   Sample Acceptance Plans

Measuring a sample from a lot is a cost-effective means of obtaining information on a lot. Unfortunately, samples will vary in the amount of the constituent of interest. Also, the parameter being measured and the analytical method may introduce variation in measurements. Probability theory can be used to describe the variation. The parameter being measured, the sample size, and the number of samples tested influence the measurement variability. By choosing an appropriate sample size and number of samples tested, buyers and sellers can manage the risks associated with sampling variability.

3.6.3.3   Single Sample: Qualitative Testing

The model of a grain sample is a collection of kernels from a grain lot. One objective of testing may be to estimate the amount of biotechnology-derived kernels in the lot. Qualitative testing will not provide an estimate of the amount of biotechnology-derived kernels in the lot. Qualitative testing produces a positive result if one or more biotechnology-derived kernels are in the sample and produces a negative result if no biotechnology-derived kernels are in the sample. A positive result may mean that one biotechnology-derived kernel was present in the sample or that all kernels in the sample were biotechnology derived.

3.6.3.4   Multiple Samples: Qualitative Testing

A single large sample serves the buyer's interests well. However, some buyers may be willing to accept some low concentrations but unwilling to accept high concentrations. Sellers of lots with low concentrations would like to have high probabilities of these lots being accepted. Decreasing the sample size will increase the chances of a negative result on low concentrations. Unfortunately, decreasing the sample size increases the chance of a negative result with higher concentrations. A single qualitative test may not serve the interests of the buyer and the seller. An alternative is to implement a multiple sample plan.

3.6.3.5   GIPSA's Grain IP Protocols

Protocols to improve confidence that grain shipments meet certain contract specifications are likely to become more widely used. GIPSA has cooperated in the implementation of one such protocol to satisfy Japanese importers of food corn that StarLink

---

* The *Grain Inspection Handbook, Book I, Grain Sampling* (1) contains these instructions and can be obtained by contacting GIPSA or by accessing the GIPSA webpage at http://www.usda.gov/gipsa/strulreg/handbooks/grbook1/gihbk1.htm. *The Mechanical Sampling Systems Handbook* (2) contains information on mechanical sampling systems and can be obtained by contacting GIPSA or by accessing the GIPSA webpage at http://www.usda.gov/gipsa/strulreg/handbooks/msshb/mssh95.pdf. Also, random sample is the desired sample from any lot. However, obtaining a true random sample is often not possible in practice. The procedures developed by GIPSA are designed to provide an approximation of a random sample. GIPSA handbooks refer to these samples as representative samples (see http://www.usda.gov/gipsa/strulreg/handbooks/msshb/mssh95.pdf).

corn is not present in export shipments. Under this protocol, the official inspection system provides official testing of domestic shipments (barge or rail) expected to be exported to Japan. Containers are sampled via official sampling procedures and at least three subsamples of 400 kernels each are tested by rapid test methods. The testing protocol has the goal of rejecting corn with one or more kernels of StarLink in 1,200 kernels for export to Japan. If any subsample tests positive, that barge or railcar is excluded from the identity protocol. All units that test negative are physically sealed and included in the identity protocol. At export port locations official inspection personnel monitor the export elevator's processes for avoiding inadvertent commingling of grain and the inbound and outbound transfer of grain included in the protocol.

3.7   BATCH PROCESSING*

Within IPT's chain of events, one of the most difficult areas to retain IPT is during batch processing; this event may occur often in the development of a food product. Facing many food safety crises, food companies try to limit incurred risk and to reassure consumers. The point is not only to follow the products efficiently, but also to minimize recalls, and the number of batches (lots) constituting a given finished product. For example, a major IPT problem area of concern during processing is characterized as "disassembling and assembling" of bills of material (also known as 3-level bill of materials). Such "dispersion problems" are encountered often in the food industry.

The goal for many processors is to try to control the mixing of production batches to limit the size and consequently the cost and the media impact of batches recalled in case of a problem. Given the 3-level bill of materials (raw materials split into components assembled into recipes), the objective is to minimize the manufacturing batch dispersion to optimize traceability.

Moe (1998) proposes an interesting definition for traceability in the batch production industry: he introduces the notions of chain and internal traceability.

> Traceability is the ability to track a product batch and its history through the whole, or part, of a production chain from harvest through transport, storage, processing, distribution, and sales (hereafter called chain traceability) or internally in one of the steps in the chain for example the production step (hereafter called internal traceability).

Two types of product traceability can be distinguished. Tracing is the ability, in every point of the supply chain, to find origin and characteristics of a product from one or several given criteria. It is used to find the source of a quality problem. Tracking is the ability, in every point of the supply chain, to find the localization of products from one or several given criteria. It is used in case of product recall. The distinction between these two traceabilities is important. Indeed, an effective information system for one of these traceabilities is not necessarily effective for the other.

---

* Reprinted excerpts from Dupuy, C., V. Botta-Genoulaz, and A. Guinet. "Batch Dispersion Model to Optimise Traceability in Food Industry."*Journal of Food Engineering* 70, 2005: 333–339. Copyright 2005 with permission from Elsevier.

However, an effective information system could handle most, if not all, requirements for both traceabilities.

Kim, Fox, and Gruninger (1995) proposed a quality data model in which the concept of the traceable resource unit (TRU) was introduced. A TRU is defined as a homogeneous collection of one resource class (think pork) that is used, consumed, produced, and released from one process to then be moved along in the food chain to its next stage. The TRU is a unique unit; that is to say that no other unit can have the same (or comparable) characteristic from the traceability point of view. More concretely, a TRU corresponds to an identified type of production batch. In the case of discrete processes, the batch identification is generally easy.

3.7.1 Definition of Batch Dispersion

To evaluate the accuracy of traceability in the production process, Dupuy, Botta-Genoulaz, and Guinet (2005) introduced three new measures: downward dispersion, upward dispersion, and batch dispersion. The downward dispersion of a raw material batch is the number of finished product batches that contain parts of this raw material batch. For example, if a reception batch of ham is used in x batches of sausages, then the downward dispersion will be equal to x. The upward dispersion of a finished product batch is the number of different raw material batches used to produce this batch. For example, salami produced with components of two different batches of pork shoulder and three different batches of pork side will have an upward dispersion equal to five. Finally, the batch dispersion of a system is equal to the sum of all raw material downward dispersion and all finished product upward dispersion.

3.7.2 An Industrial Issue: Example—The Sausage Industry

The example comes from a French sausage manufacturing company. The pork industry is interested in improving its traceability. To produce sausage, this company cuts pork meat in components like ham, belly, loin, trimmings, etc. Further in the production process, these meat components are minced and mixed to create minced meat batches. These minced meat batches will be used to produce different types of sausages (Figure 3.1).

Each type of raw material provides components in fixed proportions (a carcass can have only four legs). This is the disassembling (or cutting) bill of material. A component can also come from different raw material types. The finished products (sausages) are composed of several components in given proportions. This is the assembling (or mixing) bill of material. During a working day, the company receives several batches of different types of raw material (ham, side of pork, shoulder, etc.), therefore many batches of component will be created as will many finished product batches (Dupuy, Botta-Genoulaz, and Guinet 2005).

The batch dispersion problem does not concern only the sausage production process. For example, numerous processed foods derived of or having grain ingredients are produced daily. This may concern all of the production processes that associate disassembling and assembling processes and in which traceability optimization is an important factor. This is one, if not the most, difficult portion of any system-wide

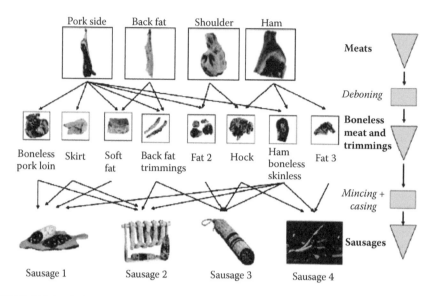

FIGURE 3.1 Industrial case, meat cut, and sausage production. (Reprinted from C. Dupuy, V. Botta-Genoulaz, and A. Guinet, *Journal of Food Engineering* 70, 333–339, 2005.)

program. The preservation of particular traits or attributes as these individual ingredients mix together during several processes along the food chain is a struggle that the food industry is attempting to overcome (Dupuy, Botta-Genoulaz, and Guinet 2005).

REFERENCES

Auer, Carol. "Tracking Genes from Seed to Supermarket: Techniques and Trends." *Trends in Plant Science* 8, no. 12, 2003: 591–597.

Bodria, Luigi. "System Integration and Certification. The Market Demand for Clarity and Transparency—Part 2. Presented at the Club of Bologna, Italy meeting, November 16, 2002. *Agricultural Engineering International: the CIGR Journal of Scientific Research and Development.* Invited Overview Paper. Vol. V. February 2003.

Boyle, Robert D. "Full-Product Traceability: Intelligent Tracking is Key to Supply Chain Integrity." *APICS—The Performance Advantage* 5, 2004: 26–27.

Boyle, Robert, Bill Jorgenson, William Pape, and Jerred Pauwels. "How Do You Measure Traceability Performance?" *Food Traceability Report,* March 2004.

Dupuy, Clement, Valerie Botta-Genoulaz, and Alain Guinet. "Batch Dispersion Model to Optimise Traceability in Food Industry." *Journal of Food Engineering*, 70, no. 3, 2005: 333–339.

Fonsah, Greg. "Tackling Traceability: Having a Traceback Plan in Place Allows Growers to Monitor Produce from the Farm to the Packingshed to the Consumer." *American/ Western Fruit Grower,* 2005: 14.

Golan, Elise, Barry Krissoff, Fred Kuchler. "Food Traceability One Ingredient in a Safe and Efficient Food Supply." *Amber Waves* 2, no. 2, 2004: 14–21.

Jorgenson, Bill 2004. "Technology for the Supply Chain: Benefits, Costs, Perceptions." Presented at the International Food and Agribusiness Management Association (IAMA) Montreux, Switzerland meeting, June 14, 2004.

Kim, Henry M., Mark S. Fox, and Michael Gruninger. "An Ontology of Quality for Enterprise Modeling." Paper presented at the 4th Workshop of Enabling Technologies, Infrastructure for Collaborative Enterprises (WET-ICE '95), Berkeley Springs, WV, 1995.

Mayer, Cheryl. "Tracking & Traceability." *Germination—The Magazine of the Canadian Seed Industry* 7, no. 4, 2003: 10–12.

Miskin, Michael. "Connecting the Supply Chain: Traceability Management Systems Can Do More than Track Problems." *Food in Canada,* 2006: 3–29.

Moe, Tina. "Perspectives on Traceability in Food Manufacture." *Trends in Food Science & Technology*, 1998 9: 211–214.

Sundstrom, Frederick J., Jack Williams, Allen Van Deynze, and Kent J. Bradford. "Identity Preservation of Agricultural Commodities." *Agricultural Biotechnology in California Series.* Publication 8077, 2002: 1–15.

Tanner, Steven N. "Testing for Genetically Modified Grain." *Proceedings of GEAPS Exchange '01.* U.S. Department of Agriculture, GIPSA, 2001.

Section II

Programs and Standards

Part II includes three chapters that highlight the spectrum of programs and standards that incorporate identity preservation and traceability (IPT) fundamentals. The chapters include Chapter 4, covering official seed agencies; Chapter 5, over industrial programs; and Chapter 6, which discusses national, regional, international, organic, and religious standards. Official seed agencies lead this part because of their importance—often it is the origins of seeds upon which the food supply system depends and that have had IPT programs in existence for many years. The industrial programs chapter illustrates how various industries and organizations have implemented IPT programs; many of these programs only incorporate a discrete food segment whereas others include nearly the entire food chain. The last chapter of this section provides an extensive assortment of standards that are incorporated by many food systems of the world.

4 Official Seed Agencies

4.1 INTRODUCTION

Official seed agencies, be they state, national, or organic, stand as the basis of any identity preservation and traceability (IPT) system or program. For nearly a century, these types of agencies have developed seeds for national and international consumption. It is almost a given that these organizations have a pure, highly preserved identity that is prized by their customers. This chapter will offer a sample of official seed agencies, services offered, price listings if available, and any other IPT services of interest such as checklists. In the United States, most states or regions have their own seed and/or crop improvement organizations; the sampling provided is of primarily grain-focused organizations (see Appendix C Official United States and Canadian seed agencies for more information).*

Much if not all of the information provided in this chapter regarding official seed agencies is derived and condensed from their home websites. Information provided is offered as a sample of checklists, application information, etc., of the actual IPT programs that each organization offers. It is recommended to visit their websites for more accurate and up-to-date information regarding any of their programs and government regulations.

Many of these associations came into being to address the very first level of the IPT chain by focusing on progeny seeds and plants. These organizations' form of documentation is much more public than that of private seed companies, which may focus on proprietary tools of IPT.

Included are the national Association of Official Seed Certifying Agencies (AOSCA); state crop improvement associations of Iowa, Minnesota, and Indiana; Canadian Seed Institute (CSI) and CSI's Centre for Systems Integration; and the Canadian Soybean Export Association.

In addition to quality assurance (QA) programs, each crop improvement organization has developed newer programs that address identity preservation (IP). Fees incurred for IP certification are dependent on the end user's traits of interest, degree of traceability, and tolerance levels.

Note that each organization emphasizes particular services that are important to their region and customers. Again, what follows are company/organizational statements from their websites that naturally reflect their views.

---

* For a directory of Association of Official Seed Analysts (AOSA), see http://www.aosaseed.com/membership_directory.htm#Associate%20Members; for AOSCA, see http://www.aosca.org/member%20agencies.html; and for AOSCA international seed authorities, see http://www.aosca.org/international%20seed%20authorities.htm.

4.2   AOSCA

Association of Official Seed Certifying Agencies (AOSCA)
1601 52nd Avenue, Suite 1
Moline, IL 61265
Phone: 309-736-0120
Fax: 309-736-0115
Chet Boruff, Chief Executive Officer, cboruff@aosca.org
Peggy Gromoll, Administrative Assistant, pgromoll@aosca.org
Website: http://www.aosca.org

The AOSCA was established in 1919 (as the International Crop Improvement
Association) and is committed to supporting customers in the production, identification, distribution, and promotion of certified classes of seed and other crop propagation materials. The AOSCA has several international member countries located in
North and South America, Australia, and New Zealand. Their mission is to promote
and assist the advance and development of seed or plant products in local, national,
and international markets by coordinated efforts of official seed certification agencies acting to evaluate, document, and verify that a seed or plant product meets certain accepted standards. They accomplish this through establishing and maintaining
minimum standards for genetic purity, recommending minimum standards for seed
quality for the classes of certified seed, and periodically reviewing agency genetic
standards and procedures to ensure compliance with the U.S. Federal Seed Act.

In cooperation with the Organization of Economic Cooperation and Development
(OECD) and other international organizations, the AOSCA is involved in the development of standards, regulations, procedures, and policies to expedite movement of
seed and encourage international commerce in improved varieties.

4.2.1   Programs and Services

The AOSCA provides a wide range of programs and, with member agencies on three
continents, they offer a broad network of member organizations to coordinate the
delivery of services that enhance and certify the quality of seed and crop propagat-
ing materials. Cooperation among member agencies ensures uniform quality from
field inspection through laboratory testing. Regarding crop certification, AOSCA
agency personnel work with local and national clients on the coordination of programs across regional boundaries. Field inspection services are an integral part of
AOSCA's "systems approach," which includes detailed inspection reports created
and maintained as part of the recordkeeping process. Quality control inspection services may be tailored to best fit individual customer needs for seed certification, QA,
IP, or other programs. The AOSCA's (members only) *Yellow Book* provides more
in-depth information regarding many of its programs such as the IP program for crop
standards and service program publications.

4.2.2   QA

Regarding IPT, the purpose of AOSCA's QA program is to provide a complete service for seed products as varieties, hybrids, brands, or blends that are generally not

marketed as certified seed. System guidelines are very similar to certification guidelines and allow the seed producer to market seed with the assurance to each customer that the seed is of known purity and quality as verified by an unbiased third-party agency. These third-party agents provide coordinated, professional, and unbiased field inspections and laboratory testing for quality control of purity standards related to established descriptors across seeds, seed lots, and years of production in addition to an unbiased record system for use in meeting state, federal, and many foreign seed law requirements.

4.2.3   IP

For the AOSCA, IP refers to the maintenance of a product's specific traits or characteristics through growing, production, and marketing channels. The function of AOSCA's IP certification program is to assist in preserving the genetic and/or physical identity of a product. To use the IP logo, these specific minimum requirements must be met and are designed to assure the customer that the identities of certain traits, physical qualities, or avoidance of specific traits are met. Several AOSCA IP programs have been specifically developed to address transgenic crops and provide a systems approach to ensure that these products meet tolerance levels of genetic material derived from biotechnology.

IP protocol standards include the following(many of these standards are shared in common with other public, private, and nonprofit parent seed organizations and are evident in their published protocols and standards):

1. Eligibility requirements for crop varieties/brands or processes used are such that a detailed description of the morphological, physiological, and other characteristics of the plants and seed that distinguish it from other varieties/brands or processes utilized must be provided to the inspection agency.
2. Applicant's responsibilities:
   (a) Care of equipment (e.g., that all equipment that may come in contact with product is cleaned prior to usage)
   (b) Maintaining identity of product (e.g., each field must be identified by number or other designation, maps must show field identities and locations, inspected crops must be clearly identified at all times, bins identified by bin or lot numbers, bags identified and stored appropriately)
   (c) Record requirements (e.g., field numbers, amount of harvest, assigned bin and lot numbers, records of transfers, and copies of all completed agency documents)
3. Application for field inspection {includes standard applicant information regarding address, fields, variety/brand, type and name of program [99.5% non-genetically modified organism (GMO)], planting date, previous crop, seed source identity, etc.}
4. Establishing source of seed: The inspection agency will be supplied with evidence of the source of seed used to plant each field for inspection.
5. Field inspection of one or more fields will be made each time a crop is harvested and when genetic purity and identity or any other factor affecting product identity can best be determined.

6. Field inspectors will provide a written inspection report for each field inspected. Fields will be passed if conditions are satisfactory, but all or parts of the field will be rejected if program requirements are not met.
7. Product handling requirements include the following: facilities' ability to perform handling without introducing mixtures, identity of the product maintained at all times, records of all program operations are completed and adequate to account for all incoming product and final disposition of product, handlers' program records will be inspected, an authorized inspection agency representative shall take adequate samples, etc.
8. Carryover product: All eligible product not used in the crop year of production must be reported to the agency to remain eligible for future labeling.
9. Labeling: The product meeting specific program requirements may be labeled using the IP logo and clearly state the program name.

Other AOSCA protocols include the Non-GMO Soy Program and Non-GMO Corn Program.

4.2.4 Seed Certification

The purpose of seed certification is to preserve genetic purity and its identity. Seed certification is an official AOSCA agency program enabling seed companies to market genetically pure seed. Certification services are available for field crops, turf grasses, vegetables, fruits, vegetatively propagated species, woody plants, and forbs. Once seed has been certified, it qualifies for the official "blue" certified seed tag and meets state, federal, and international seed law requirements. Requirements for producing certified seed include special land requirements, planting eligible stock, field inspections, proper seed labeling, and meeting standards based on complete lab analysis. Below are several common classes of seed. Some national seed programs have additional classes of seed for specific traits and/or level of purity.

4.2.2.1 Seed Classes
- *Breeder Seed:* Seed directly controlled by the originating or sponsoring plant breeding organization
- *Foundation Seed:* The progeny of Breeder or Foundation Seed handled to maintain *specific* genetic purity and identity.
- *Registered Seed:* The progeny of Breeder or Foundation Seed handled to maintain *satisfactory* genetic purity and identity.
- *Certified Seed:* The progeny of Breeder, Foundation or Registered Seed handled to maintain satisfactory genetic purity and identity.

The program imparts the following:
- Coordinated, professional, and unbiased field inspections and laboratory testing
- An unbiased record system for use in meeting state, federal, and international seed laws

- Seed buyers with assurance that the designated seed has met purity standards related to a known description across seed lots and years of production

4.2.5 ORGANIC CERTIFICATION

Organic is a labeling term that denotes products produced under the authority of the U.S. Organic Foods Production Act. It is based on minimal and restrictive use of off-farm inputs and on management practices that restore, maintain, and enhance ecological functions. Certification includes inspections by trained and qualified inspectors of farm fields and processing facilities, detailed recordkeeping, and periodic testing of soil and water to ensure that growers and handlers are meeting the standards that have been set. Several AOSCA agencies are certified from the U.S. Department of Agriculture (USDA) National Organic Programs and their respective state authorities.

4.3 IOWA CROP IMPROVEMENT ASSOCIATION

Iowa Crop Improvement Association
4611 Mortensen Road, Suite 101
Ames, IA 50014-6228
Phone: 515-294-6921
Fax: 515-294-1897
E-mail: iowacrop@iastate.edu
Website: http://www.agron.iastate.edu/icia

The Iowa Crop Improvement Association (ICIA) is the official seed-certifying agency (a nonprofit organization) for the state of Iowa and its mission is to provide an unbiased source of service and education in production and QA for Iowa agricultural crops. This status, mission, and focus are very similar throughout nearly all U.S. crop improvement associations. The organization was first formed in 1902. In the 1920s, ICIA began providing Iowa with quality, unbiased seed production services and crop performance testing services. The organization was renamed ICIA in 1950 following the merger of several other agricultural organizations.

ICIA's mission (again, similar to other crop associations) is to provide an unbiased source of service and education in production and QA for their state's agricultural crops. ICIA's objectives are as follows:

- To provide mechanisms for conducting domestic and international seed certification and seed QA
- To provide educational and leadership opportunities to influence public policy regarding crop improvement
- To conduct, in cooperation with the Iowa State University College of Agriculture, testing and disseminating information on the adaptation and performance of crop hybrids and varieties
- To coordinate all ICIA activities to be consistent with environmentally sound agricultural practices
- To provide a mechanism for commodity IP

4.3.1 IP Grain Services

The IP program of ICIA promotes assurance that the desired traits in specialty crop production are maintained throughout all production and handling processes. The IP program is designed for application on special-use raw products under agricultural or horticultural production, including those destined for use as food, feed, nutraceuticals, fiber, and unique oils or grain.

ICIA oversees the following points during the production and handling process:

*   Field inspection of production to specified standards
*   Proof of seed stock with plant description or specific trait definition
*   Quantity of harvested product
*   Representative sampling
*   Laboratory evaluation
*   Documented transfers of product
*   IP official labeling of product with labels, imprints, or certificates

4.3.2 Seed Production Services

ICIA currently offers three protocols for offering unbiased, third-party service to seed producers. They are Certified Seed, QA Seed, and Native Species Seed.

Certified Seed is seed produced from approved seed stock, which is used to produce a variety for marketing as Foundation, Registered, or Certified Seed. QA Seed is an alternative for varieties or brands that are not eligible to be or do not need to be marketed as Certified Seed. Native Species Seed is certified based upon the source or geographic origin of the seed's collection source. Services included in these programs include: recordkeeping, field inspection, seed sampling, lab inspection, and labeling.

4.3.3 Iowa Seed Directory

The ICIA offers a yearly *Iowa Seed Directory* that provides information on the production and conditioning of seed in Iowa. The directory has two purposes for its publication: (1) to provide a complete available listing of all fields that have met certification requirements, and (2) to be a useful and convenient resource for prospective buyers of seed who are attempting to locate supplies. The listings in the directory include Approved Conditioners and Certified, QA, and Native Species Seed.

4.4 MINNESOTA CROP IMPROVEMENT ASSOCIATION

Minnesota Crop Improvement Association
1900 Hendon Avenue
St. Paul, MN 55108
Phone: 612-625-7766
Fax: 612-625-3748
Toll free: 1-800-510-6242
E-mail: mncia@tc.umn.edu
Website: http://www.mncia.org

The Minnesota Crop Improvement Association (MCIA) was founded during the 1903 Minnesota State Fair at a meeting in the Territorial Pioneers Log Cabin on the fairgrounds by those interested in the "systematic encouragement for the use of pedigreed seed" and was dedicated to improving the productivity, profitability, and competitive position of its members. MCIA was a founder of the International Crop Improvement Association, which was established at a meeting in St. Paul in 1919 and later became the AOSCA. MCIA is a nonprofit organization that operates on fees charged for services performed. MCIA offers an assortment of certifications of Parent and Foundation Seed, QA programs, education, IP and organic certification services, customized third-party services, and an array of laboratory services. MCIA is Minnesota's official seed-certifying agency and official noxious weed seed, free forage, and mulch certifying agency and is recognized by Minnesota's Department of Agriculture and Agricultural Experiment Station.[*]

4.4.1 IP Grain Certification Program

MCIA provides services to producers, processors, and marketers of IP products to help them develop and implement effective IP systems. IP services offered include the following:

- MCIA acts as an unbiased third-party that checks part or entire IP systems. Checks may include seed sources, planting records, field inspections, harvest records, storage facilities and conditions, product transportation, handling and processing, final product testing, and labeling verification.
- Advantages to utilizing MCIA's IP grain services. For example, for companies with IP systems already in place, MCIA offers process verification services through documented on-site audits and inspections that provide assurance to buyers that IP protocols are being followed. In addition, MCIA offers on-farm field and storage site inspection services to verify that growers are following production practices required by the production system.
- MCIA provides third-party certification under the AOSCA IP standards, which are recognized both domestically and internationally.

Grain producers who intend to merchandise IP crops should consider the following:

- *Know the specifics of the IP product:* Research the market, potential added costs, delivery windows, delivery location, and storage requirements. Verify that the variety used will meet IP requirements. Obtain a contract for the finished product to protect value-added premium.
- *Select eligible fields:* Fields must not have had the same species grown on them the previous year.

---

[*] MCIA's official publication, *The Minnesota Seed* Grower, was established in 1928 and has since been published on a regular schedule. MCIA was also the first U.S. seed certification agency to adopt a computerized system for keeping certification records.

- *Obtain eligible planting stock:* Keep invoices, tags, and bulk certificate samples. Ask that lot numbers be indicated on the invoice at the time of seed purchase.
- *Clean planting equipment thoroughly between IP stock and other plantings:* Remove all seed of other types or kinds and verify seed stock eligibility prior to filling planter. Record the variety and lot number throughout planting; map documentation is recommended. If all seed is not eligible for IP merchandising, document starting and ending points of eligible plantings.
- *Isolate field from non-IP fields:* IP fields should be adequately separated from other fields to guarantee the final product can mechanically be kept separate. If the field is lodged, plants from the IP field must not be in contact with adjacent fields. In cross-pollinated crops, isolation should be sufficient to eliminate potential contamination from foreign pollen.
- *Prepare fields and make certain all quality requirements are met:* These preparations could include any weed, disease, insect, or isolation corrections if required.
- *Attach permanent labels on all bins and storage areas:* Clean all trucks, trailers, bins, and augers before beginning work in IP fields.
- *Clean all harvesting equipment thoroughly:* The first load of grain from an IP field should be dumped into a non-IP load to guarantee equipment is clear of possible contamination.
- *Take samples for any quality testing requirements:* Carefully document all processing details and periodically check bins to verify quality.
- *Implement marketing plan*
- *Arrange inspections:* Third-party inspections must be arranged in advance and are often required to fulfill contract specifications.

4.4.2 MCIA ORGANIC CERTIFICATION SERVICES

MCIA is a USDA National Organic Program (NOP) Accredited Certifying Agent (ACA) for the provision of organic certification services in Minnesota and neighboring states. MCIA currently conducts inspections and certification for organic producers, handlers, processors, and wild crop collectors. The USDA has deemed MCIA compliant with the International Organization for Standardization (ISO) Guide 65 Assessment for U.S. organic certifying agencies. Operations certified organic by MCIA may display the NOP Certified Organic seal on qualified products.

Because organic certification is a process-oriented system covering production, harvesting, handling, processing, packaging, labeling, and transportation, operations certified by MCIA include food handlers, distributors, retailers, agricultural handling facilities, wild crop collectors, and farm and garden producers. Products currently certified by MCIA range from soup to nuts, including coffee roasting, maple syrup, poultry slaughter, seed, fruit and vegetable production, soups, whole grains, wild rice, and others.

4.4.3   Requirements for Approval of Grain Handling Facilities

These requirements are the basis for approving facilities handling grain eligible for AOSCA IP grain certification. General requirements include the following:

1. Copies of U.S. grain standards, standards for the use of the AOSCA IP, and specific IP program standards for the products to be handled must be held on-site.
2. Facility must be inspected annually.
3. Facility maintenance and housekeeping must be adequate to ensure that the quality and identity of the IP products handled are maintained.
4. Storage facilities and grain handling equipment must be adequate to ensure that the quality and identity of the IP products handled are maintained.
5. An IP grain handler's agreement indicating the intention to comply with all IP grain handling requirements must be signed each year by the facility manager.
6. Only approved bins and equipment within the facility may be used to handle IP products.
7. All equipment and storage facilities must be accessible for cleaning.
8. MCIA has the right to inspect facilities and disposition records at any time.
9. Only product handled by an agency-approved IP grain handling facility shall be eligible for IP grain certification.

4.4.4   MCIA IP Grain Certification Fee Schedule
and Organic Fee Schedule

TABLE 4.1
MCIA IP Fees (2006)

Grain Certification Fee Schedule

Membership

General membership fee (June 1–May 31)
.... $50.00

Field Inspection Fees	**Per Acre**
One inspection (min fee per field $40)	........$2.00
Two inspections (min fee per field $55)	.......$2.75
Three inspections (min fee per field $70)	.....$3.50

Late Application Fee (per field) $20.00

Re-Inspections (per field) $40.00

Sampling

For product on which final fees are collected:

IP samples (per trip) $50.00

Service sampling

Actual cost of time ($50 per hour),
  Mileage and expenses (min per trip) $50.00

Final Fees

IP min charge/lot $10.00

Approved Facility Fees

Approved IP grain handling facility
  fee (includes membership) $150.00

Organic Fee Schedule

Deposit $200/application

Inspection $60/hr

Flat rate site fee $120.00

Re-inspection .... $60/hr + expenses

Document services and review $60/hr

Producer Late Fees: June 1 $100/application

  July 1 .... $200/application

  August 1 .... $300/application

(Continued)

TABLE 4.1 *(Continued)*

Inspection.............................$60/hr + expenses	Beans, peas, seed corn, soybeans, wheat
Wild Crop Late Fees: Two Months prior to	(per bushel) $0.06
Harvest $100/Application	*All Other Crops and Processed Products Organic*
1 mo prior to harvest.............$200/application	*Certification Fees: Based on Gross Organic*
At harvest $300/application	*Sales and/or Processing Fees:*
Inspection 60/hr + expenses	First $2,000,000 ....0.5%
Handler Late Fees: Anniversary of Certificate	Amounts over $2,000,000 ....0.1%
$100/Application	Minimum charge....................................$25.00
1 mo after anniversary............$200/application	Sampling:
2 mo after anniversary............$300/application	For product on which final fees are collected:
Inspection 60/hr + expenses	First samples (per trip)..............................$50.00
Certification transfer fee $25.00	Resamples (per trip)
Certification fees will be based on one of the	-First lot...$50.00
following two schedules. Seed and whole grain	-Each additional lot..................................$25.00
organic certification fees:	Service sampling......................$60/hr + expenses
Certification min charge/lot $10.00	(min per trip) $50.00
Sunflowers, grasses & legumes (per lb.).....$0.001	MCIA website for more information; http://www.
Corn grain, barley, oats, rye (per bushel)......$0.03	mncia.org

Source: Data adopted awaiting permission from MCIA.

4.5 INDIANA CROP IMPROVEMENT ASSOCIATION

Indiana Crop Improvement Association
7700 Stockwell Road
Lafayette, IN 47909
Phone: 765-523-2535 or 866-899-2518
Fax: 765-523-2536
E-mail: icia@indianacrop.org
Website: http://www.indianacrop.org

The Indiana Crop Improvement Association (ICIA) was created to deliver unbiased services to customers in the seed, grain, food, and related industries. As a nonprofit, self-supporting agency, ICIA impartially carries out various seed programs including seed certification, IP, QA, and laboratory testing. ICIA's mission is to improve productivity, profitability, and the competitive position of ICIA members by providing services to producers, conditioners, and distributors of plant products, enabling them to provide high-quality plant products to Indiana, the United States, and the world. The association's office and seed laboratory facilities are located in Lafayette, Indiana. Although not on campus, the association has a strong working relationship with Purdue University, and all ICIA full-time staff are associates in the Purdue Agronomy Department.

4.5.1 IP PROGRAMS

ICIA's IP program is an extension of the identification and tracking service provided through its seed certification and QA programs. The identity of a crop is maintained beyond the seed through commercial production in quantities required to meet the end-user's needs. ICIA's IP programs comply with the general IP guidelines for Certified IP products adopted and maintained by the AOSCA.

ICIA tailors IP services to meet specific needs; for example, it can verify that the variety in a field is the variety specified, inspect fields to determine and report on crop conditions, identify and specify the amount of any crop contaminants, estimate yield, etc. Through inspections and auditing, it can also verify the integrity of a product through a particular process, such as tracing a non-GMO raw material through a food plant. However, this does not certify that a crop is organic. IP programs may involve field inspections, seed lab testing, bin inspections, auditing, and issuance of specific labels or certificates. All programs are tailored to meet the specific needs of the customer. ICIA then provides auditing and other services necessary to validate that the customer's quality plan is being followed.

4.5.2 FIELD SERVICES

Indiana Crop Services provides trained personnel located across the state to deliver a wide range of QA field services. Field inspections result in a third-party documentation for seed certification, QA, IP, and other customized services. Additional specific objectives of field inspection are discussed in the following subsections.

4.5.2.1 Seed Certification

The purpose of seed certification, as it is in other states, is to preserve the genetic purity and identity of crop varieties. It is an official system, with standards supported by both federal and state laws, and is designed to help increase the supply and speed the distribution of seed of improved crop cultivars while maintaining the genetic integrity of the product.

Often it takes several years of concentrated effort for a company or an institution to develop a new crop variety. These varieties are released with many different and important genetic traits that influence pest resistance, standability, grain quality, maturity, herbicide tolerance, and yield, to name a few.

Seed certification relies on seed pedigree records, field inspections, laboratory testing, post-season trueness-to-type plot testing, and other agency-approved protocols to help evaluate and perpetuate varietal purity and identity. Seed of varieties must meet the minimum genetic standards in each phase of the program to be labeled and sold with the familiar blue tag as certified seed.

In Indiana, the ICIA has been designated the official seed-certifying agency by the director of the Purdue Agricultural Research Programs at Purdue University.*

4.5.2.2 QA Program

ICIA's QA program provides a uniform, unbiased quality control system and marketing tool for crop seed marketed as brands, varieties, or hybrids. Although it is

---

* In general, in the United States, seed certification is a voluntary program. However, there are crop varieties protected under the U.S. Plant Variety Protection Act under the "Certification Option" provided under Title V of the Federal Seed Act (Federal Seed Branch) that must be sold by variety name only as a class of certified seed.

designed as a complete quality control service for products not using seed certification, the guidelines are similar to those of the certification system. It is readily customized to meet the needs of the customer. ICIA's QA program adds significant value to seed programs as it assists in maintaining product purity and identity in an era of increasingly costly genetic traits.

The ICIA QA program provides the following:

- Coordinated, professional, and unbiased field inspections, laboratory testing, and post-control grow out tests for quality control in seed production, conditioning, and marketing.
- An unbiased record system for use in meeting state and federal seed law requirements, assessing royalties or research fees, establishing a defense for use in avoiding problems, and helping to resolve problems between seed suppliers, growers, and customers.
- A marketing image of sound quality control.
- Assurance to buyers that seed bearing the QA trademark has met purity standards related to a known description across seed lots and years of production.
- The QA program can also be of great assistance in helping describe new products and is frequently used in facilitating wholesale movement of seed.

4.5.2.3   Indiana Crop Lab Services

4.5.2.3.1   Conventional Tests

ICIA offers a full range of professional services that includes an in-house registered seed technologist and services that are also ISO 9001-2000 certified to assure customers that quality is of the utmost importance in delivering service.

4.5.2.3.2   Genetics Lab Services

ICIA offers a full compliment of genetic identification testing services. Polymerase chain reaction (PCR), enzyme-linked immunoabsorbent assay (ELISA), isozyme, and other trait tests are available for use by the seed, grain and related industries. Tests include most *Bacillus thuringiensis* (Bt) events including YieldGard, rootworm, and Herculex. ICIA provides a "stacked" test for Bt and rootworm. ICIA will offer non-GMO testing services to farmers who need to verify commercial products moving into specific markets. Tables 4.2 and 4.3 illustrate services and fees available.

4.5.2.3.3   Laboratory Services

The ICIA Seed Laboratory provides many services for the seed industry. Laboratory services are available only to ICIA members or associate members who are bonafide seed producers. The ICIA laboratory does not provide seed testing for farmers' bin-run grain. Custom molecular marker service programs are also available to assist with plant breeding, quality control, and genetic identification. Note that work may be assessed at an hourly rate of $32.00 per hour.

TABLE 4.2
Indiana (ICIA) Services and Fees (Sample List)

Membership

New member fee (one time)............$500.00

Associate (per yr) $50.00

Approved conditioner (per yr) $50.00

Field Inspection Fees   **Per Acre**

A. Corn: Base application fee............$40.00

(if submitted after June 15)...............$90.00

Per field $10.00 Per acre $7.00

B. Small grains and soybeans; certified:

Base application fee $10.00

(small grains after May 15) $60.00

(soybeans after July 1) $60.00

Per field $10.00 Per acre $2.50 (hybrids) $5.00

Per acre (legumes, grasses)..................$5.00

QA (noncertified inspection):

Base fee ....$2.00 /acre

Second inspection price ....$1.00 /acre

(i.e., Bloom or Roundup or reinspection for purity)

Field applications submitted after inspection is

  underway will be charged an additional $10.00

  fee per field.

C. Phytosanitary inspection fees

With regular inspection-per acre $1.00

Without regular inspection;

 (Per field) $10.00

Minimum (soybeans) $100.00

Minimum (corn, small grains) $50.00

D. Breeder plot inspections

1–5 plots of 5 acres or less $50.00/plot

More than 6 plots $40.00/plot

(Per acre) $1.50

Labeling Fees

For those printed by the ICIA:

A. Small grains and soybeans

1. Certified classes, QA and plain labels

a. Price per tag with analysis data $0.06

B. Corn

1. Certified and QA price per tag $0.07

2. OECD card stock per tag 11

C. Bulk retail sales for all bushels covered by bulk

  retail sales certificates (price per bushel)

D. Price per transfer certificate for

  bulk transfers.................................$5.00

Source: Data adopted with permission from Indiana (ICIA).

TABLE 4.3
Indiana Laboratory Tests and Fees

Warm Germination Test

Alfalfa $10.00

Clover $10.00

Corn $6.50

Grasses $10.00**

Small grains $6.50

Sorghum $6.50

Soybeans $6.00

Sand Germination

Soybeans $15.00

Corn $20.00

Purity Analysis

Alfalfa $10.00**

Clover $10.00**

Corn $5.00**

Small grains $5.00**

Soybeans $5.00**

Separations..................................... hourly rate**

Varietal Analysis

Oats $7.00

Soybeans $7.00**

Bt Seed Testing

Bt Testing (90 Seed Test)..............$70.00

(fewer than 90 seeds--$1.00/seed)

Bt GMO Testing

Cry1ab $70.00

Cry9c $55.00

Corn GMO *"Package"*

(All Bt's, RR, & Liberty)............$165.00

Seed Count per Pound

Corn..$3.00

Soybeans $2.00

Wheat $3.00

Soybean Antibody (ELISA) Test

Lipoxygenase

(Continued)

TABLE 4.3 *(Continued)*

L1 only $5.00/seed		Soybeans (Liberty Link) $20.00	
L1 & L2 only $6.00/seed		Corn (Liberty Link) $25.00	
L1, L2, & L3 ....$7.00/seed		Corn (Roundup Ready) $25.00	
Peroxidase antibody test $30.00/variety		Corn (IMI) $40.00	
Varietal screen $5.00/seed		**PCR Prices**	
Cold Germination Test		Screen (presence/absence of GMO, nonspecific)	
Corn $9.00		Number of tests	Price per test
Soybeans $9.00		1..$250.00	
Saturated Cold Test		2–9 ....$175.00	
Corn/soybeans $18.00		More than 10 ....$125.00	
Accelerated Aging Test		**Electrophroesis Isozyme Purity Test**	
Soybeans $8.00		Dent seed corn $150.00	
Corn $8.00		Popcorn $220.00	
Wheat $8.00		Small grain electorphroesis	
Tetrazolium Test		ID test..$10.00	
Corn $18.00		**GMO ROUNDUP READY SOYBEAN**	
Small grains $18.00		**ELISA TEST**	
Soybeans $12.00		Number of tests	Price per test
Other crops..........................hourly rate**		1–3 ....$160.00	
Waxy Maize Test		4–9 ....120.00	
Corn $20.00		10–19 ....$80.00	
Herbicide Test		20 or more $70.00	
Soybeans (Roundup Ready) $20.00		Breeder seed test $10.00/seed	
Soybeans (STS) $20.00		Commercial lot $100.00/sample	

Source: Data adopted with permission from Indiana (ICIA).

** Work may be assessed at an hourly rate of $32.00 per hour.

4.6   CANADIAN SEED INSTITUTE AND CANADIAN SEED INSTITUTE CENTRE FOR SYSTEMS INTEGRATION

Canadian Seed Institute (CSI) and CSI Centre for Systems Integration
Jim McCullagh, Executive Director
Suite 200-240 Catherine Street
Ottawa, ON, K2P 2G8
Phone: 613-236-6451
Fax: 613-236-7000
Toll-Free: 1-800-516-3300
E-mail: jmccullagh@csi-ics.com
E-mail: csi@storm.ca
Website: www.csi-ics.com
Website: http://www.csi-ics.com/organic/index.asp?lang=en

The Canadian Seed Institute (CSI) is a nonprofit organization founded in 1997 by the Canadian Seed Trade Association (CSTA), the Canadian Seed Growers' Association (CSGA), and the Commercial Seed Analysts Association of Canada (CSAAC). The institute employs independent assessors to evaluate seed establishments using the CSI standard. CSI also provides accreditation and monitoring programs for the Canadian seed industry. Recognized by the Canadian Food Inspection Agency (CFIA), CSI has been authorized to be the single-point contact for all seed organizations, seed laboratories, operators, and graders seeking registration, licensing, or accreditation. Presently, CSI monitors over 1,300 Canadian seed establishments, authorized importers, and accredited seed testing laboratories. CSI's standards are developed to harmonize with other countries' laws and regulations to eliminate many technical barriers faced with international trade.

In this chapter, IP of Canadian parent seeds (other than soybean crops) will be examined. The other two IP programs for grain crops and soybeans will be discussed in the standards chapter within the Canadian Identity Preserved Recognition System (CIPRS) and for soybeans through the Canadian Soybean Identity Preservation Procedure found in Chapter 6.

The *Centre for Systems Integration*, an independent nonprofit division of CSI, was created to simplify certification under multiple programs for Canada's agricultural and forestry segments. The Centre for Systems Integration offers a variety of services, including

- Organic certification that provides access to Canadian, U.S., Japanese, and European markets
- ISO registration, in collaboration with ISO registrars
- Food safety certification, including on-farm and post-farm food safety programs, and *Hazard Analysis and Critical Control Point (HACCP)* certification
- CIPRS certification with the Canadian Grain Commission
- Consultation services in partnership with Aon Management Consulting on how to apply Six Sigma principles

CSI is an independent body that administers the accreditations of both individuals and facilities that handle seed in Canada. The most recent project was working with the Canadian Grain Commission to develop their IP recognition system—CIPRS. To have a facility accredited by the CSI, seed handlers and processors must have a documented quality management system in place that meets all of the elements of CSI's standards. Everything from equipment hygiene to recordkeeping must be covered in the quality management system. An approved auditor must then audit the seed facility. On the basis of the audit report, CSI will determine whether or not to accredit the facility. Only accredited facilities are allowed to process seed of pedigreed status.

The Centre of Systems Integration was created to simplify multiple certification requirements by offering clients the expertise to integrate into one quality management system that can be audited by a single auditor.

The CSI standards combine the process improvement and customer focus of the ISO 9000 series of quality systems standards with the regulatory requirements for documentation and traceability of the Canada Seeds Act. The institute currently has programs in place for seed companies and facilities that wish to store, handle, process, package, test, grade, and import or export seed. In addition to general quality processes, CSI also monitors the technical aspects of the processing industry that ensure the facility is operating according to industry standards and using approved methods and procedures.

In Canada, accredited seed testing laboratories perform seed testing for domestic certification. To qualify as an accredited lab, the facility and the seed analyst must be accredited by the CFIA. Canada is a member of the Association of Official Seed Analysts and the International Seed Testing Association. Canadian participants in both of these organizations play a key leadership role. These international commitments ensure that Canadian laboratory methods and procedures align with continually improving techniques that are used worldwide.

4.6.1   PROGRAMS AND ACCREDITATION

CSI programs and accreditation include the following:

- *Integrated Seed Quality Management System (ISQMS):* ISQMS is directed toward seed businesses that want a quality system that extends from production to retailing.
- *CFIA Phytosanitary Certification Program for Seed:* This program is for exporters shipping seeds in small packages to the United States.
- Approved conditioner for businesses that condition seed.
- Bulk storage facility for businesses that store and/or hold seed in bulk.
- Authorized importer for businesses that import seed.
- Seed testing lab for businesses that conduct seed analysis.
- Organic certification for organizations to meet the U.S. NOP rules.

Table 4.4 describes the programs under which CSI is accredited.

TABLE 4.4
CSI Accreditation Bodies

Accreditation Body	Program
CFIA	Accredited as a "Conformity Verification Body" for assessment activities in support of the seed and plant health programs
Canadian Grain Commission	Accredited as a "Service Provider" to the CIPRS for conducting audits
USDA	ACA under the USDA's NOP
Deutscher Akkreditierungs Rat (DAR)	Accredited to ISO Guide 65 as an inspection body to the European Union Organic Regulation (EEC) 2092/91
National Quality Institute (NQI)/Registrar Accreditation Board (RAB)	CSI auditors accredited as ISO 9000:2001 auditors

Source: Data adopted with permission from CSI.

4.6.2   CSI Programs and Accreditation—ISQMS

The ISQMS program is an industry-driven program that recognizes the extra efforts of seed businesses that have incorporated activities beyond conditioning, storage, and import or export of seed via additional activities required under ISQMS requirements.

4.6.3   CSI Accreditation Lists

Accredited assessor list: http://www.csi-ics.com/pdfs/Assessor%20List_Seed.pdf
Accredited laboratory list: http://www.csi-ics.com/pdfs/facilities/Accredited%20Lab%20List.pdf

4.6.4   CSI Accreditation Fees

Sample listing:
Initial application fee $300.00
Operator and grader evaluation fee..$75.00
Accredited seed testing lab $450.00
CIPRS $500.00
ISQMS $500.00

4.6.5   CSI's USDA NOP Certification

Following an intensive process of review by the USDA, CSI is now allowed to certify organic farms (other than livestock) and handling facilities under the U.S. NOP. This means CSI can assist organic producers and processors to certify their operations. The certification process involves CSI-approved organic inspectors visiting a business to review their organic system plan and check it against the U.S. NOP Rule. The inspector's report is reviewed by CSI, and if the operation is in compliance with the NOP, CSI grants organic certification.

4.6.6   CSI Quality System Assessments and ISO Client Compliance*

CSI requires ISO-registered clients to have the technical components of their quality system assessed by a CSI accredited assessor/technical expert in conjunction with or in addition to an ISO audit. The technical assessment will fulfill the requirements of the agreement between CSI and the client as well as CSI's obligations to CFIA to verify specific technical components of the various programs.

---

* Six Sigma Implementation. In addition to quality systems such as ISO and HACCP, CSI, in cooperation with AON Rath & Strong, assists in the implementation of Six Sigma. Six Sigma stands for Six Standard Deviations (Sigma is the Greek letter used to represent standard deviation in statistics) from mean. The term "Six Sigma" relates to the number of mathematical defects in a process. Six Sigma practitioners focus on systematically eliminating the defects so they can get as close to "zero defects" as possible. Six Sigma methodology provides the techniques and tools to improve the capability and reduce the defects in any process. Generally speaking, companies use Six Sigma to reduce variation in products and processes, but the net effect of any Six Sigma project is what people are really looking for—fewer defects, shorter cycle times, increased capacity and throughput, lower costs, higher revenues, and reduced capital expenditures.

Several CSI clients have taken the initiative to gain ISO registration for business purposes and have requested clarification on the requirement for a CSI assessment in addition to the ISO audit. A committee was assembled by CSI to examine the feasibility of using ISO audits without CSI participation for the purposes of CFIA. *Note: The committee came to the conclusion that the current ISO audits did not investigate and ISO auditors were not trained to evaluate technical issues related to the sampling, grading, handling, testing, importing, or labeling of pedigreed seed.* The committee recommended that a technical assessment be conducted to deal with requirements of the Seeds Act and Regulations or the Canadian Methods and Procedures for Testing Seed, as applicable.

5 Industry Identity Preservation and Traceability Programs

5.1 INTRODUCTION

This chapter provides numerous examples of industry identity preservation and traceability (IPT) programs and includes TraceFish as an industry template. The food industry's numerous and dynamic privately developed IPT programs illustrate the varying scopes and depths utilized by organizations to accomplish their safety and IPT programs. The industry players' vertical and horizontal integration into other aspects of agriculture varies tremendously depending on the companies' mission. Some, such as the seed companies, remain very focused on seed purity and specific traits (genetics). The Grain Elevator and Processing Society (GEAPS) is a nonprofit society that provides guidelines for the grain industry as a whole, whereas others such as National Starch and the American Institute of Baking (AIB) have industry-specific requirements that they abide by. The following is a brief summary of what to expect in this chapter.

- *IP template for grain industry:* TraceFish leads off this chapter because in many ways this organization spearheaded the notion of IPT, which many other industries have followed. TraceFish was also one of the first to incorporate Global Trading Identification Number (GTIN) and batch number systems.
- *Parent seed:* Pioneer's software systems, in-house systems, programs, website, and services.
- *Grains and oilseed supplier—Clarkson Grain's Pure Green system:* Certified by external organizations, offers sales/services for genetically modified organisms (GMOs), but prefers non-GMOs and organic products.
- *Seeds, processed grains, and inventory software supplier—Northland & Pacifica Research:* Externally certified, offers sales/services, and Windows-based software for its GMO, non-GMO, and organic products.
- *MicroSoy Flakes—MicroSoy Corporation:* MicroSoy Corporation is a processor of non-GMO conventional and organic products as well as kosher IP soy products and is certified by the Organic Crop Improvement Association (OCIA), the Ministry of Agriculture, Forestry and Fisheries of Japan Agricultural Standard (JAS), and Star-K.
- *Soya-based food and drinks products—Alpro Soya:* Processor of a unique "whole soya bean process" for making soya milk and other soya products.

- *Grain processor and handler—Cargill's InnovaSure IdP System:* In-house GMO and non-GMO tracking, tracing, and identity preservation (IP) system for throughout the supply chain.
- *Society and extension training—GEAPS and Purdue:* Provides its members with information regarding grain and IPT and is also conduit for Purdue's extension grain education programs.
- *Knowledge-based products and services—John Deere FoodOrigins:* A full service, in-house, tracking, tracing, and IP system for throughout the supply chain.
- *International grain-trading and logistics company—AgMotion's Tracekey system:* In-house web-based software for organic and non-GMO IP and marketing systems.
- *Processor—National Starch's TRUETRACE system:* Full service, in-house, non-GMO corn tracking, tracing, and IP system.
- *Baking institute—AIB:* Offers certification of standards, audits, and technical/analytical services.

What follows are company/organizational statements from their websites and naturally reflect their views.

5.2 TRACEFISH

Petter Olsen
Norwegian Institute of Fisheries and
  Aquaculture Ltd.
N-9291 Tromsø, Norway
Phone: +47-77-62-90-00
Fax: +47-77-62-91-00
E-mail: petter.olsen@fiskforsk.norut.no
Fiskeriforskning
Muninbakken 9-13
Postboks 6122
N-9291 Tromsø, Norway
Phone: +47-77-62-90-00
Fax: +47-77-62-91-00
E-mail: tracefish@fiskeriforskning.no

**Stirling Aquaculture Institute of
  Aquaculture**
University of Stirling
FK9 4LA, UK
Phone: +44-1786-467900
Fax: +44-1786-451462
E-mail: staq@stir.ac.uk
Alistair Lane
European Aquaculture Society
Slijkensesteenweg 4
B-8400 Oostende
Belgium
Phone: +32-59-32-38-59
Fax: +32-59-32-10-05
E-mail: aquaflow@aquaculture.cc

Available at http://www.tracefish.org and http://www.rontec.co.uk/Fish_News_International_Article.htm.

TraceFish (or The Traceability of Fish Products Concerted Action Project) was an undertaking coordinated by Fiskeriforskning (Norwegian Institute of Fisheries and Aquaculture Ltd.).* It began in 2000 with the aim of bringing together companies

---

* TraceFish was funded by the European Commission (EC) under the "Quality of Life and Management of Living Resources" thematic programme project and is an electronic system of chain traceability. It was developed under the patronage of the EC in its Concerted Action project QLK1-2000-00164.

and research institutes to establish common views with respect to which data should follow a fish product through the chain from catch/farming to consumer. Twenty-four companies/institutes were members of the consortium, including major European fish exporters, processors, importers, and research institutes. In collaboration with their Joint Venture partner, Nesco Weighing Ltd., they have developed software writing data to TraceFish XML format and running on a version of the Data Terminal. This enables them to offer full TraceFish implementation.[*]

The premise was that with increasing information demands from buyers and consumers, it was no longer practical to physically transmit all of the relevant data along with the product. A more sensible approach was created to mark each package with a unique identifier and then electronically transmit or extract all of the relevant information such as its source/origin (e.g., the use of the EAN.UCC System for the identification, bar coding, traceability, and e-communications regarding fish and fish products).[†] This was done to ensure that the fish industry did not find itself in the same kind of situation that engulfed the meat industry, which led to loss of sales and customer confidence. Even now the meat industry does not have anything like TraceFish standards in place, although they may want to follow TraceFish's lead to adopt a similar system.

5.2.1   HISTORY—TRACEABILITY IN THE FISHING INDUSTRY

The fishing industry is the last major food source that cannot, in most cases, tell the consumer about the product it is selling. With the food scares involving meat, bovine spongioform encephalopathy (BSE, or mad cow disease), foot and mouth disease (FMD), and the uncertainty about basic commodities such as drinking water and genetically modified (GM) food, concerns about food safety is ever increasing. The media has highlighted these food scares, and processors in the fish industry are no exception. The stories that kept resurfacing about fish farms and the safety of aquaculture-reared fish, including the medications given to the fish during the farming process and fish being caught in waters contaminated by radioactivity and toxic chemicals entering the human food chain, have appeared in several publications.

Although some forms of traceability have been put in place by parts of the industry for some time, there has never previously existed a process by which information has been made accessible throughout the supply and processing chain. Starting January 1, 2005, the European Union (EU) mandated that all fish products sold within the EU be subject to appropriate traceability. The U.S. Food and Drug Administration

---

[*] There are scientific publications underway describing the impact of the TraceFish standard. One of the few papers already published and available discusses the impact of communication standards (TraceFish is taken as an example) on business transaction costs. The reference is Dreyer H.C., R. Wahl, J. Storøy, E. Forås, and P. Olsen. "Traceability Standards and Supply Chain Relationships." Presented at the 16th Annual Conference for Nordic Researchers in Logistics (NOFOMA), Linköbing, Sweden, 2004.

[†] For references to documents or texts concerning TraceFish, the most important one is probably the EAN/UCC Traceability of Fish Guidelines, which can be found at http://www.ean-int.org/Doc/TRA_0403.pdf. EAN/UCC is represented in over 100 countries and over 1 million companies utilize their numbering series. The EAN recommendation for implementation is completely based on TraceFish, and the TraceFish standards and process are referenced numerous times in the EAN guidelines.

(FDA) is also looking to enact similar legislation in the United States in the near future.[*]

Despite the development of TraceFish standards, a complete system for the collection and transmission of traceability data, including software to meet these standards, was not created by the TraceFish consortium. However, a traceability system has already been developed for the Danish fresh fish chain that was in development before the TraceFish project. This research focused on all aspects of the fresh fish chain by using bar codes and serial shipping container codes to identify each resource unit and track each delivery. This research was successful in showing that traceability could be achieved and recognized the fact that system costs for vessels and small firms need to be addressed and more user-friendly interfaces must be developed to promote efficiency.

The TraceFish strategy does not demand perfect traceability; that is, that a particular retail product should be traceable back to a single vessel or farm and batch of origin, or vice versa, from origin to destination. Pragmatically it is recognized that mixing of units is likely to occur at several stages in the distribution chains (e.g., in grading at auction markets prior to sale and in the processing of raw materials into products). Where such mixing occurs, the food business is transforming the trade units. The requirement for traceability is that the business records the IDs of the received trade units that may be input to each created trade unit and vice versa. The particular product is then traceable back to a finite number of vessels or farms and batches of origin, and vice versa.

5.2.2 TraceFish—How It Works

When looking at traceability it is important to distinguish between two different types of traceability: internal traceability and external or chain traceability (as has been mentioned earlier). Internal traceability is within a company or location that is under consideration. In terms of a product, it relates to the origin of materials, the processing history, and the distribution of the product after delivery. On the other hand, chain traceability, or external traceability, is focused on the maintenance of product information from one link in the chain to the next. It describes which data are transmitted and received and how. Chain traceability is between companies and countries and depends on the presence of internal traceability in each link.[†] When

---

[*] The EC does not explicitly demand fish traceability according to TraceFish. It demands "one-up, one-down" traceability from January 1, 2005, and TraceFish is currently the only standard for this type of traceability. This means that organizations can meet the EU requirements with ad hoc solutions or proprietary systems, but if they want to implement and gain the benefits from standardized exchange of traceability information they will have to use TraceFish. The main tangible benefit from using a standard way of electronically communicating (in this case TraceFish) is that participants can send and receive messages to anyone else who supports the standard; they do not require them to be on the same system or use the same software as you. The alternative would be to base traceability on paper-based forms or "unstructured" electronic messages, which would mean significant need for repunching and, in practice, loss of information and increased response time if something happens.

[†] This increased focus is already being seen, with major retailers shifting focus to meet those demands. One of those is the Carrefour group. This French chain identifies "risky" fishing areas before deciding on whether to buy certain fish products or not. This shifted focus can also be seen with more and more

the process of establishing TraceFish was completed three standards existed. For full-chain traceability, the standards describe:

- What data should be recorded how and where in the *captured* fish chain
- What data should be recorded how and where in the *farmed* fish chain
- How these data should be coded, transmitted, or made available in electronic form
- What (existing) electronic standard should be chosen to aid in the dissemination of these data

TraceFish produced three standards that were developed for industry use. They are not the only way of achieving full-chain traceability, but they are the only ones accepted by CEN and EAN.* The standards establish where, what, and how data should be recorded in the farmed and wild-caught fish chain for full-chain traceability. They also identify how modern electronics and software can be used to transmit data through the chain and the standards to be used to successfully obtain the data if and when required. These standards are formatted on a pull-system, rather than a push-system basis. This means that only the minimum amount of necessary data is pushed along the chain. Most data are held at the individual point of action, whether that be a boat, auction, transport company, or processor. The only data pushed forward are the information required for labeling purposes or for commercial use by users further down the chain.

All commercially sensitive information is held at the point of action and is only accessible by those parties who have authority to do so (e.g., food standards agencies). The standards are based on a GTIN and a batch number. The GTIN is unique; the first part is issued by the EAN (ID of supplier) and the supplier allocates the second part (ID of product). The batch can be as big or as small as the organization sees fit, or as much as they are prepared to risk having to destroy should the product be recalled.

Throughout the project, Nesco contributed to the Technical Consortium and Technical Work Group by providing its "Traceway" Integrated Traceability System. Traceway is not just a piece of hardware or a software package but a blend of both, creating an integrated traceability system compliant with the EU standard but also designed for individual application and the customers' specific requirements. Traceway is a collection of building blocks, put together and configured for an individual process, be it onboard ships or docks, at an auction, during transportation,

---

(Continued)
producers investing in ecocertification (producing seafood that is good for the ocean ecosystem) or organic certification (producing organic seafood). It can also be seen in some supermarkets.

* Three TraceFish standards are publicly available. The two fish industry standards are sold and distributed through CEN, the European Committee for Standardization, http://www.cenorm.org. The titles are CWA 14659, *Traceability of Fishery Products—Specification of the Information to be Recorded in Farmed Fish Distribution Chain* and CWA 14660, *Traceability of Fishery Products—Specification of the Information to be Recorded in Captured Fish Distribution Chain*. The third standard is the technical (XML) TraceFish standard, which is distributed and maintained directly by the members in the TraceFish technical group and has been freely distributed. The latest "official" version of the technical standard is called *Traceability of Fish Products—Specification of Information Encoding*.

at processing, at the fish farm, or during packaging for the retailer or end user. The whole idea behind the Traceway System is to keep the process as simple as possible so as to enable the information to be accessed as easily as possible and when required, and for the component parts to be compatible throughout the whole chain of supply.

Although virtually every distribution chain is different, they all appear to be made up of several characteristic components or **building blocks**. The types of business identified in this document for captured fish distribution chains are:

- Fishing vessels
- Vessel landing businesses and auction markets
- Processors, transporters, and storers
- Traders and wholesalers
- Retailers and caterers

5.2.3 TraceFish Certification—Future Goals

TraceFish is not a label. The two fish industry standards mentioned have status as voluntary industry agreements in the form of guidelines and principles. There has been talk of making the two TraceFish CWA standards certifiable, but this work has not started. One of the reasons for this is that the proliferation of the standards is biggest "upstream," in connection with catch/farming and primary processing. Labels are more relevant downstream from secondary processing to consumer. The technical standard is certifiable; this is inherent in XML. The TraceFish XML schemas specify what it takes for messages to be "well formed" and "valid," and this requirement is absolute. Thus, it is possible and likely that the solution providers that support TraceFish (Maritech, Akvasmart, TraceTracker, FarmControl, Hugtak, C-Trace, Nesco, etc.) will market their applications as "TraceFish compatible." This means that the software can send and receive messages in XML format as specified in the TraceFish technical standard. At least two of the solution providers above have indicated that they will also use TraceFish XML to internally exchange traceability information between their own applications.*

5.2.4 Key Notion with TraceFish and Other Traceability Systems

When it comes to identifying the trade units, producers may affix the identifier to the trade unit any way they want, including human readable on the label, human readable in accompanying documentation, in a bar code, or in (or linked to) a radio frequency tag. Neither the TraceFish CWA standards nor the TraceFish technical

---

* As indicated above, several organizations have worked very closely with those who develop software suites or applications (enterprise resource planning, or ERP-type, in particular) for use in the fish industry such as Maritech, Stein-Erik Joellanger (stein.joellanger@maritech.no), Akvasmart (technical), Elin Loevtangen (elovtangen@akvasmart.no), Akvasmart (managerial), Rune Loenne (rloenne@akvasmart.com), TraceTracker (technical), Steinar Kjaernsroed (steinar.kjaernsrod@tracetracker.com), TraceTracker (managerial), Ole-Henning Fredriksen (ole-henning.fredriksen@tracetracker.com), FarmControl/Hugtak, Stefán T. Höskuldsson (stefan@hugtak.is), C-Trace, Alan Steele (alan.steele@ctrace.co.uk), and Gordon Norman at Nesco (g.norman@nesco-weighing.co.uk).

standard has any requirements with respect to the nature of the data carrier; however, what is important is the structure and makeup of the unique identifier. Both TraceFish CWA standards explicitly state that identification of trade units must be based on the "GTIN+" concept.* TraceFish acknowledges that *this is the most important and also the strictest TraceFish requirement*. It is not uncommon for a company to produce dozens or hundreds of trade units every day, each marked identically. This violates the most fundamental TraceFish principle: that each single trade unit must receive a unique number to identify it. Even if it is from the same production batch and has all of its properties in common with another trade unit, it must have its own unique number. The reason for this is referential integrity, in particular so that if initially identical trade units take different routes or have different history (temperature, delivery, r-packaging, destination, application, etc.) there is a mechanism to identify exactly what happened to each trade unit and where each trade unit went.

5.2.5   REGARDING FOOD SAFETY

TraceFish is a standard for documentation, not for food safety in itself. TraceFish standardizes what should be recorded and transmitted, and to some degree how measurements should be taken. It does not standardize thresholds or safety limits; this is the responsibility of national or international food safety legislation. TraceFish does set a standard for what it is required, recommended, and possible to record. TraceFish believes the benefits of using their system includes reduced information loss; better payment for better quality; enabling of remote auctions; tailoring and marketing of products with particular properties; less frequent, quicker, and smaller recalls; documentation of liability; reduced cost of information logistics; better production control; and the enabling of value-adding data such as more accurate estimates of remaining shelf life.

5.2.6   CHALLENGES OF AQUACULTURE

This is an industry that trades globally in a vast range of finfish and shellfish species and their byproducts and is hugely diverse in comparison to other protein sources. There are hundreds of different species of fish captured around the world, often with specialized fisheries, fish handling, and food safety requirements. Fish are pursued and captured in the wild by independent fishermen. This encompasses enormous variability in comparison to the controlled farming, often monoculture, of other protein sources. A similarly wide range of live, chilled, frozen, processed, and added-value fishery products are then produced and traded within the various distribution chains, again often with specialized food handling and food safety requirements. There is a huge and complex international trade in the raw materials and primary and secondary processed products.

According to TraceFish, to ensure perfect traceability at all stages of the marketing process, fisheries and aquaculture products have to be accompanied by a document

---

* GTIN plus is a numbering system to uniquely identify each particular trade unit [e.g., the production batch and serial number (AI 10) or the date and time of production (AI 11)].

indicating the information described above as well as the Latin name of the product. The EAN.UCC System enables cost-efficient, timely, and accurate transfer of commercial information, production method, and catch area by means of standard data structures, bar codes, and electronic messages.

5.3 PIONEER HI-BRED INTERNATIONAL, INC.—MARKETPOINT

Pioneer Hi-Bred International, Inc.
Resource Connection
P.O. Box 1000
Johnston, IA 50131-0184
Phone: 515-270-3200
Fax: 515-270-3581

MarketPoint
115 Summit Drive
Exton, PA 19341
Phone: 610-594-1880
Fax: 610-594-1881
Toll Free: 877-365-1903

Available at http://www.pioneer.com and http://www.pioneer.com/marketpoint/traceability/default.htm.

Parent seed companies typically develop and sell new seed varieties to farmers and to other parent seed companies. These parent seed companies' IPT systems are well developed and overseen by official seed agencies. It is the parent seed industry, and often joint cooperation with other organizations such as universities, that provides the starting point for seed (specifically grains for this book) IPT.

Historically, parent seed companies have come into being from the outgrowth of universities' extension programs, which intended to develop improved seed varieties in cooperation with smaller, family-size seed companies, some which were started at the turn of the twentieth century.* Seed company spokespeople often cite that they have been in the IPT business since their beginnings. Yellow corn was always grown, harvested, and marketed differently from white corn. Over time, other aspects and grain traits became increasingly more important, such as the development of *Bacillus thuringiensis* (Bt) corn and Roundup Ready soybeans. Since the late 1990s, food-chain IPT has taken on greater importance in how agriculture has managed and viewed itself. In addition, much more has and is taking place after the parent seed company stage to ensure that grains retain their particular identity and that they can also be traced back through the system. To aid in this, several parent seed companies have expanded the scope of their IPT programs to include the farmer and beyond, as we will see with the following systems.

5.3.1 PIONEER HISTORY

In May 1926, Henry A. Wallace and eight associates created the Hi-Bred Corn Company, one of the first companies to develop, produce, and sell hybrid corn. These hybrids delivered some of the best agronomics and performance available for their time. Each new generation of hybrids was selected and bred to raise the

---

* For a more complete overview of the parent seed industry's history and development see Gregory S. Bennet, *American Hybrid Corn History: A Century of Yields,* M.A. Thesis, Iowa State University, 2001.

performance bar and deliver even greater value to farmers. In fact, since 1926, the United States average corn yield has increased five-fold. Advancements in farm equipment, production practices, fertility programs, and genetics have all contributed to this bounty.

In response to market demands for high-quality grains and specialized traits in seed and grain, there became an increased need to reduce commingling or "contamination" of grains. To overcome this challenge, Pioneer began their own traceability and IP grain systems to help ensure "through acknowledgement, processes, and documentation, which distinct steps were taken to help prevent commingling of Pioneer-brand grain and oilseeds." In addition to confronting commingling issues, their Traceability Center provides value-chain customers with the processes and tools they need in the areas of risk management and food safety. The Traceability Center highlights coordinated quality crop systems that help protect seed purity and grain identity to meet grower and end-use customer requirements.

Pioneer's *MarketPoint* resource is a web-based tool that links grower customers or end-use customers (livestock producers, grain and oilseed processors, and export customers) to custom information and services from Pioneer. It offers products and systems focused on grain quality education, IPT, product stewardship, agronomic reporting, and more.

Pioneer's *Market Opportunity Center* is housed within the Pioneer MarketPoint website. The Market Opportunity Center provides grain production management tools that allow searches for real-time market opportunities, coordinate production agreements, and track supply information throughout the growing season.

The Pioneer *GrowingPoint* website is designed to deliver comprehensive, value-added information over the web to farm operators. It includes in-depth information from a producer's perspective on agronomy, technology, and profitable business practices. All of the information and electronic tools on the site are designed to help farm operators make profitable growing decisions, keep them electronically connected to their sales professional, and allow Pioneer to provide additional value to its most loyal customers.

Pioneer also offers their *Crop Production Systems*, which incorporate IPT programs.[*] Crop Production Systems are custom solution packages based on Pioneer brand seed grown under specific direction, to support growing, harvesting, and delivering high-quality grain and oilseeds. The offering to livestock, grain, and oilseed customers includes custom IP solutions based on Pioneer brand seed. These IP solutions range in complexity on the basis of customer needs, from providing a level of segregation to providing full traceability with verification. Quality Crop Systems are custom solution packages that link Pioneer growers to downstream opportunities. The Components of Quality Crop Systems include the following:

- Trace products from seed to delivered grain
- Document traceability
- Provide grain production management tools
- Reduce or eliminate paperwork
- Improve grain or trait quality

---

[*] Pioneer also offers the Pioneer Grain Stewardship Education Program to growers.

- Reduce variability in grains received
- Manage adequate supply volumes
- Provide grower training certification
- Provide production, inventory, and delivery management

For example, their program helps by providing "tips" on numerous aspects of farming such as their Insect Resistance Management (IRM) program and the farmer's legal obligation, as outlined in Pioneer's Technology Agreement (TA), to maintain a "refuge" (see sample below).

- *Minimum refuge area:* Each farm is required to maintain a minimum refuge of non-Bt corn acres.
  - Minimum of 20% of corn acres in the northern Corn Belt (non-cotton growing) region
  - Minimum of 50% of corn acres in the southern cotton-growing region
- *Refuge distance:* Pioneer recommends the refuge be placed within one-quarter mile of the YG/LL field, if at all possible. The U.S. Environmental Protection Agency (EPA) requires the refuge no further than one-half mile from the YG/LL field. (YG = YieldGard/Cry1Ab corn borer resistance and LL = Liberty Link/Glufosinate herbicide tolerance).
- *Insecticide use:* The refuge may be treated with insecticides if needed, but sprayable Bt insecticides must not be used.
- *Buffers/isolation:* Because of the pollination of a corn plant, it is not possible to completely eliminate crosspollination. Pollen from YG/LL plants may be transmitted to non-YG/LL corn fields around the YG/LL fields. To minimize crosspollination concerns, Iowa State University (ISU) professors Roger Ginder and Robert Wisner suggest a separation distance of 660 ft, similar to that used by the seed industry as a separation for seed production.
- *Notification of neighbors:* The Quality Grains Initiative at ISU suggests that growers discuss their planting intentions with their neighbors and try to work together to maximize each other's grain marketing options.
- *Auditing:* EPA requires Pioneer to conduct an annual survey of growers to understand concerns around the IRM plan.
- *Planter clean out:* Pioneer encourages cleaning the planter before and after planting the YG/LL products (see the ISU planter clean-out tips at http://www.extension.iastate.edu/Publications/PM1847.pdf).
- *Harvesting clean out:* Pioneer recommends following a clean-out procedure on the combine and other transportation equipment after harvest of the YG/LL hybrids.
- *Storage:* Pioneer recommends following a clean-out procedure on storage bins and related equipment used for the YG/LL hybrids.

Abbreviations used with corn hybrids include *Bt*, transgenic corn borer protection; LL, Liberty Link/Glufosinate herbicide tolerance; IR, IMI, IMT, PT, Imidazolinone resistant (Pursuit, Resolve, Contour); YG, YieldGard/Cry1Ab corn borer resistance; and RR, Roundup Ready/Roundup herbicide tolerance.

5.3.2   PIONEER'S IDENTITY-PRESERVED CHECKLIST PRE-HARVEST AGREEMENT

An example of an IPT system is the 2006 Low Linolenic Soybean Program in Indiana, Michigan, and Ohio. Production contracts with Bunge required that low linolenic soybeans must be identity preserved and planting, harvesting, and transportation equipment must be cleaned prior to use.

- Premium of $0.35 per bushel for harvest delivery.
- Premium of $0.40 per bushel for on-farm storage.
- Contract is buyer's call.
- Crushing will be done at Bunge facilities in Bellevue, Ohio, and Marion, Ohio.
- Planning to have delivery periods to the Bunge facilities at Bellevue and Marion, Ohio.

Pioneer encourages that customers see their Pioneer sales professional for the latest program, premium, and variety information. Pioneer brand low linolenic soybean varieties qualify for all applicable Pioneer brand product purchasing discounts.

It should be noted that although it is unclear what the penalties for lying or noncomplying will be in regards to IPT programs, Chapter 12 points toward considerations for truthfulness and compliance. See Tables 5.1a and 5.1b for an example of Pioneer's preharvest agreement checklist.

5.4   CLARKSON GRAIN COMPANY, INC.

Clarkson Grain Company, Inc.
P.O. Box 80
320 East South Street
Cerro Gordo, IL 61818
Phone: 217-763-2861
Toll free: 800-252-1638
E-mail: info@clarksongrain.com
Available at http://www.clarksongrain.com.

TABLE 5.1a
Pioneer's Pre-Harvest Agreement Checklist

1. Verification of seed:
   seed invoice/bag tag/bag sticker
2. Separate seed storage
3. Planter cleaned prior to planting IP crop
4. Field identification/location
5. Field sign placed
6. Isolation—what's planted near IP crop
7. Verify acres planted
8. Crop protection usage:
   pesticide/herbicide/insecticide
9. Periodic production estimates:
   pre-harvest/post-harvest/in storage bin
10. Harvest equipment cleaned: combine
    (cleaned/flushed)/wagons/trucks
11. Collect and submit grain samples:
    from combine/from dryer/from bin
12. Grain drying:
    dryer cleaned/flushed prior to IP crop
    crying temperature less than 140°
13. Storage bin:
    bin cleaned prior to IP crop,
    bin tag, or sign
14. Delivery equipment:
    equipment cleaned prior to IP crop
15. Truck identification license, pre-assigned
    scale ticket, truck sign)
16. Grower signature

TABLE 5.1b

Identity-Preserved Checklist Post-Harvest Confirmation (Sample)

Grower name, contract number, previous year crop, acres planted, corn variety

The grower hereby certifies and warrants that the following procedures and practices were followed:

At time of planting—Certified seed of contract variety was used.

I purchased the required amount of certified seed, of the contracted variety, to plant the contracted acreage and have proof of the variety and the amount purchased.

I thoroughly cleaned the planter prior to planting and all corn seed of other varieties and other crop types were removed to ensure purity of contracted variety was maintained.

During harvest and storage—I confirm my combine was thoroughly cleaned to remove seeds of other corn varieties and other crop types prior to combining an IP variety field.

All delivery equipment used to deliver IP variety corn was inspected by grower for cleanliness and cleaned thoroughly prior to filling.

If stored on farm, storage bin was thoroughly cleaned, of sound quality, and clearly identified as storing an IP-variety corn prior to filling with IP corn variety.

I am aware that a delivery sample of my IP-contracted corn may be retained for inspection and genetic identification, if required.

At all times—To the best of my ability, I ensure that the above identity-preserved variety is not contaminated with corn of other varieties at any time during the production, harvest, or storage periods.

Source: Adopted awaiting permission from Pioneer MarketPoint Resource.

Traditionally commodity markets buy grain by grade standards, which focus on physical features that have little to do with value. Value depends on factors such as protein, sugar, starch structure, taste, and texture—features that depend primarily on choice of genetics.

Clarkson Grain (CG) also realizes that value-added products move through supply chains with IP protocols (e.g., organic, non-GMO, and Kosher) that require verifications as requested by clients/contracts. CG tracks materials from seed to farm and field and on to their clients.

In 1991, CG began supplying organic grains and oilseeds to Japanese buyers, with loads ranging from tons to hundreds of thousands of tons. Today they supply organic raw and intermediate materials to customers around the world. They operate 25,000 tons of dedicated commercial organic storage backed by several times that in farm storage.

For organic certification, CG certifies its facilities, products, and activities with Quality Assurance International (QAI), and OCIA to secure access into American, Asian, and European markets. It respects and recommends several other certifiers and regularly buys from farmers using most National Organic Program (NOP) certifiers.

CG also supplies grains, oilseeds, and related ingredients to people making foods and feeds. They select, produce, and handle materials to optimize clients' process yield, quality factors including taste and nutrition, and security and access to markets from conventional to organic. They contract with approved, qualified farmers in 20 U.S. states and 3 countries to produce selected hybrids and varieties. They produce ingredients in plants they own or control and offer several products on exclusive arrangements.

CG coins their IP programs as "IdP." CG helps clients identify features they desire such as functional, biochemical, or physical properties; seed source; production culture (organic, chemically restricted); absence of genetically engineered traits; and traceability. CG then applies segregation and verification protocols by internal and third-party inspectors to deliver complying materials. CG owns, operates, and contracts storage, handling, and shipping facilities to IdP conventional as well as organic grains and oilseeds. They are a licensed grain dealer and warehouse, not a broker. For very disciplined IdP programs, CG uses multilayered inspections, laboratory tests, and audits to verify integrity and likelihood of GMO material. Control points include seed delivery, field visits, harvest samples, farm bin samples, delivery trucks, and warehouse samples from commercial storage. Third-party verifiers report directly to the buyer.

CG emphasis is primarily on organic and non-GMOs, not on GMOs or transgenic crops. They note that much of the world remains sensitive about genetically engineered crops. CG respects clients' concerns. Although CG cannot guarantee 100% non-GMO materials, it offers disciplined programs that "absolutely" minimize GMO presence within its Pure Green program; up to 99.9% for non-GMO corn and non-GMO soy.

To assure compliance, they use professional third-party verification approved by their clients. CG overlaps security steps so failure of one or two does not jeopardize supply integrity. Inspection starts with seed before planting and continues through production, harvest, storage, conditioning, and shipment.

5.4.1 CG Supply Contracts

Contracts to deliver selected raw and intermediate products are designed to meet customer standards. CG states that their products exceed commonly accepted U.S. Department of Agriculture (USDA) grades.

- CG ships on customer's schedule or call or CG's own monitoring of customer inventory.
- CG helps select varieties, hybrids, or qualities that optimize market success. CG provides segregation needed to maximize market access and help protect from food problems.
- CG offers fixed price or fixed margin contracts for organic materials.
- CG offers contemporary pricing choices on conventional raw materials, including exchanging futures.
- CG prefers annual or quarterly supply contracts over spot contracts. This offers better control over quantity, quality, and security.
- CG understands that delivered materials must work for the customer.
- CG works with any responsible third-party verifiers and accepts any reasonable laboratory tests as long as they are conducted before shipment.
- CG also offers conditioning, packaging, shipping, and fumigation choices.

- *Farmer contracts:* CG contracts with farmers to produce, store, condition, and deliver selected varieties of organic, transitional, and conventional grains and oilseeds listed in their product catalogue. CG seeks preferred, qualified farmers, where "qualified" means that they have appropriate infrastructure, soils, and location.
- *Delivery choices:* CG contracts crops freight on board (FOB) farm or delivered but take title only upon delivery to a transfer location controlled by CG. Each year, farmers first contracting get first choice in selecting delivery time. Within date ranges, delivery is on their call, not on producer's convenience.
- *Delivery time adjustments:* When buyers change production schedules, CG has to adjust delivery schedules. This is not as convenient as the graded commodity market. For that reason, CG pays premiums and storage and tries their best to accommodate producer needs.
- *Organic certification:* CG recommends using NOP-authorized certifiers capable of meeting both JAS and the International Federation of Organic Agriculture Movements (IFOAM) requirements. Combines: CG prefers rotary.
- *Storage:* CG prefers bins with full air floors served by independent handling equipment and computerized fan controllers.

5.5   NORTHLAND SEED & GRAIN, NORTHLAND ORGANIC, AND PACIFICA RESEARCH

Northland Seed & Grain Corporation/
Northland Organic
495 Portland Avenue
St. Paul, MN 55102
Phone: 651-221-0855
Fax: 651-221-0856
E-mail: soybean@northlandorganic.com

Pacifica Research
202 "E" Street, #C
Brawley, CA 92227
Toll free: 800-536-5130
Fax: 760-344-8952
E-mail: pacifica@pacificaresearch.com

Available at http://www.northlandorganic.com/index.html and http://www.pacific research.com.

5.5.1   NORTHLAND SEED & GRAIN CORPORATION

Northland Seed & Grain Corporation (based in St. Paul, Minnesota) consists of Northland Seed & Grain (non-GMO) and Northland Organic Foods (certified organic) and specializes in the development, production, and international distribution of both conventional non-GMO and certified organic specialty variety seeds, grains, food ingredients, and animal feed. This is accomplished by offering premium quality IP non-GMO seeds, soybeans, and grains, as well as processed products such as flours, meals, feeds, and oils to its customers. Northland works in collaboration with third-party inspection/certification agencies and public and private laboratories to carefully monitor every step of production, ensuring its customers the lowest possible levels of

GMOs. Northland's method of creating and preserving the identity of non-GM foods begins with seed production and growing and extends all of the way through harvesting, processing, packaging, and transportation. Northland's strict tracking protocol and identification system makes it possible to trace products from the seed-breeding phase all of the way to the customer's door.

Northland Seed & Grain is an established producer and global supplier of IP, non-GMO seeds, raw materials, and ingredients to the food and feed industries. Northland's IP non-GMO products are sold under its IP PURE brand name and include specialty variety soybean seeds for sowing, food and feed grade whole grains and soybeans, soy meal, soymilk powder, oil (soy, sunflower, safflower, canola), lecithin (fluid, granules, powder), and flour (soy, wheat, oat). They are certified by QAI, OCIA, and JAS. QAI is an independent third-party certifier of organic food systems and has been the foundation of domestic and international organic food trade. OCIA is one of the world's oldest, largest, and trusted organizations in the organic certification industry.

In November 2003, GeneScan USA (also known as Eurofins, see chapter 8) and Northland Seed & Grain Corporation announced that Northland had chosen GeneScan IP Certification service and GeneScan's worldwide network to third-party certify Northland's Non-GMO IP PURE Program according to the GeneScan General Standard. Northland Seed & Grain was one of the first U.S.-based companies to begin the GeneScan IP certification process. GeneScan Analytics GmbH currently has IP programs in place in South America, China, and Europe.

In conjunction with certified crop inspection agencies and private laboratories, Northland Seed & Grain utilizes a strictly controlled growing, processing, packaging, and transportation program to insure IP seed variety purity and to provide premium, non-GMO food products.

Since the introduction of GMOs, Northland Organic Foods and its sister company, Northland Seed & Grain, have been pioneers in the development of unique and reliable programs. Northland's specialty seed-breeding program specializes in the development of traditional cross-bred, certified organic, identity-preserved, and non-GMO seeds and grains that are ideal for food manufacturing purposes. Northlands' strict non-GMO certification program ensures the integrity and non-GMO purity of all of its seeds, grains, and food products. By carefully monitoring all levels of production—from the seed selection and growing to the processing, packaging, and transportation—Northland guarantees its customers the highest quality products.

5.5.2 NORTHLAND ORGANIC FOODS

Northland Organic Foods is a leading producer, supplier, and international distributor and broker of organic premium quality, identity-preserved, non-GMO certified organic soybeans, wheat, corn, rice, and other cereal grains as well as certified organic commodities such as seeds, oils, meals, flours, and feeds.

Northland Organic Foods certified organic products include the following:

*   Whole soybeans
*   Edamame (U.S. grown)

- Whole grains (wheat, corn, barley, millet)
- Soy meal
- Soy oil
- Canola oil
- Oleic safflower oil
- High oleic sunflower oil
- Soy beverage powder
- Soy flour
- Wheat flours
- Oat flour

5.5.2.1 Example: Soybeans (Non-GMO)

Northland's innovative seed program offers a wide variety of specialty soybeans that are ideal for producing tofu, soy sauce, soymilk, natto, sprouts, soy oil, meal, flour, and animal feeds. Inspections are performed by independent certification agencies and samples are taken for analysis by private labs to further verify compliance to Northland's rigid standards. Northland's step-by-step programs include the following:

1. Pre-planting phase
   - Northland develops and markets only certified, identity-preserved, non-GMO seeds suitable for food use.
   - Northland contracts with carefully selected experienced growers.
   - Field and seed-lot histories are tracked to further guarantee seed variety purity.
2. Growing phase
   - Inspections are conducted by crop inspection agencies recognized by the U.S. government to ensure varietal purity, plant characteristics, and clear isolation.
   - Private laboratories analyze seed and plant tissue samples to confirm that they are non-GMO.
   - Additional inspections and sampling are conducted by a certified crop inspection agency just prior to harvest.
3. Post-harvest phase
   - Proper storage and transportation guidelines are followed to ensure product segregation.
   - All commodities are processed and packaged in accordance with Northland's strict non-GMO quality-control program.
   - All Northland products are processed at certified cleaning plants, mills, and presses.
   - Laboratory testing includes genetic (deoxyribonucleic acid, or DNA) testing to ensure non-GMO purity.
   - Additional samples of cleaned products are kept for library samples and future laboratory analysis.

5.5.3 Pacifica Research

Pacifica Research is a software publishing company that provides Windows-based software that specializes in:

- Seed inventory control
- Flower inventory control
- Agricultural accounting
- General accounting
- Hay brokerage
- Entomology

Pacifica Research's Seed Inventory Control software was developed "by seedsmen for seedsmen" to handle the unique challenges of the seed industry. Pacifica Seed Inventory Control is a multiuser, real-time business management system that has been used for more than 15 years. It is capable of providing reliable and up-to-the-minute information for production and marketing decisions. Its interactive modules function effortlessly as a fully integrated system. Pacifica addresses the many facets of producing, purchasing, selling, seed pricing, and inventory by variety, lot, and sublot.[*] Pacifica's Seed Inventory Control software:

- Allows customers to print package labels, tags, and bar codes for instant inventory adjustments by location and package
- Provides tools to conduct accurate performance evaluations of sales, staff, and customers
- Can print detailed forecasting and projection reports
- Handles purchases, sales, and adjustments in any unit of measure including pounds, ounces, kilograms, grams, per seed, per thousand, per 10M, per 100M, per acre, per hectare, per bushel, or selected personal measurement
- Allows lots to be split into sublots representing multiple locations, package sizes, treatments, selling prices, or costs without losing original lot identity
- Permits lots to be flagged and reported via specified attributes such as stock seed, consignment, stop sale/rejected/returned, production, coated, blended, etc.
- Keeps track of lot attributes such as germ, purity, grower, vendor, treatment, seed count, coating type, etc.
- Contains costs at the lot level and may be accrued against acquisition, freight, production, conditioning, processing, and overhead
- Includes production management and grower's accounting

---

[*] Every detail—from buying/selling with automatic unit conversion, to package labels, tags, and bar codes—is stored online in a single, powerful database for instantaneous retrieval.

5.6 MICROSOY CORPORATION

MicroSoy Corporation
300 East Microsoy Drive
Jefferson, IA 50129
Phone: 515-386-2100
Fax: 515-386-3287
E-mail: info@microsoyflakes.com

Japan Office: Kearny Place Honmachi 7F
6-13, 1-chome
Awaza Nishi-Ku, Osaka 550-0011 Japan
Phone: 06-6110-7005
Fax: 06-6110-7006
E-mail: microsoy@aurora.ocn.ne.jp

Available at http://www.microsoyflakes.com/index.htm.

MicroSoy Corporation is located in the heart of soybean country, Jefferson, Iowa. Since 1991, MicroSoy has been producing MicroSoy Flakes through a patented technology. MicroSoy Flakes are produced using a mechanical process of dehulling, cracking, and flaking without the use of solvents or additives. This technology preserves all of the natural goodness of soy.

At MicroSoy, they believe that quality and safety are important criteria in selecting a food or beverage, which not only tastes good but is also good for our health. They process only non-GMO soybeans. Each of their products carries the non-GMO seal certified by Cert-ID (see Chapter 8). Their organic lines of products are certified by OCIA and JAS. MicroSoy products are kosher-certified by the Star-K organization.

MicroSoy products are certified non-GMO (conventional and organic), identity-preserved, and dehulled soybean flakes. The product line includes Instant Soy-Oatmeal Hot Cereal, MicroSoy Crumbles, MicroSoy Cookies, Super Spuds (Instant Mashed Soy-Potato), and Whole Grain Soy Cereal Bars.

5.6.1 EXAMPLE OF LOWERING POTATO CARBOHYDRATES

MicroSoy has developed an innovative product/ingredient to reduce the carbohydrates and increase the protein content of potatoes while preserving the potato flavor. This is accomplished through a special type of soy ingredient—MicroSoy Flakes.*

MicroSoy Flakes are made from farm-delivered, cleaned, identity-preserved, certified, non-GMO soybeans. Once the soybeans pass through their quality checks, they are mechanically processed (dried, cracked, and rolled into thin flakes) without the use of solvents or additives. The flakes can be used as is, untoasted, or can be toasted depending on the customer preference and application needs.

The untoasted MicroSoy Flakes receive very little heat during the process, preserving the wholesome quality of the soybeans. The untoasted MicroSoy Flakes have a natural yellow appearance. Untoasted MicroSoy Flakes are ideal for soymilk, tofu, hummus, mashed potatoes, and processed meat applications.

MicroSoy's toasting procedure removes the "beany" flavor from the MicroSoy Flakes, resulting in a smooth texture and sweet-nutty flavored product. The toasted products are ideal for cereal, yogurt, ice cream toppings, piecrust, and other bakery products. See Table 5.2 for additional product information.

---

* As featured in *Prepared Foods Magazine* May 2004, p 86.

TABLE 5.2
MicroSoy's Flakes Product Chart (2008)

Product Name	Product Code	Description	Application
Toasted MicroSoy	Thickness available:[a] 0.2 mm (TSX02 or TSO02) 0.6 mm (TSX02 or TSO02) 1.2 mm (TSX12 or TSO12) Grits (TSXGR or TSOGR)	Toasted Full fat No "beany" flavor Smooth texture and sweet-nutty flavor	Soy crumbles, pancake mix, hot and cold cereal, mashed soy potatoes, pie crusts, salad sprinkles, power bar, soy cream cheese, and many other food applications.
Untoasted MicroSoy Flakes for ingredients	Thickness available: [a] 0.2 mm (IGX02 or IGO02) 0.6 mm (IGX06 or IGO06) 1.2 mm (IGX12 or IGO12)	Untoasted Full fat Rehydrates fast Contains all the natural components of soybeans	Soup, hummus, egg replacement, dall, keema (ethnic food applications), and many other food applications.
MicroSoy Flakes for soymilk	SMX02 or SMO02	Untoasted, rehydrates fast for more efficient soymilk making, high isoflavone level	Soymilk products
MicroSoy Flakes for tofu	TMX02 or TMO02	Untoasted Rehydrates fast for more efficient tofu making	Tofu products
Whole soybeans	WBO or WBX	Cleaned whole soybeans	Soymilk and tofu
Soybean chips	BCO or BCX	Cleaned whole soybeans	Soymilk and tofu

[a] Thickness specified based on average thickness.
Source: With permission from MicroSoy.

5.7 ALPRO SOYA

Belgium
Alpro NV
Vlamingstraat 28
8560 Wevelgem
Phone: +32-56-43-22-11

Netherlands
Alpro Soya Nederland BV
Hoge Mosten 22
4822 NH Breda
Phone: +31-76-596-70-70

Alpro NV—divisional headquarters
Kennedy Park 8
8500 Kortrijk
Phone: +056-43-22-11

Alpro Belgium
Prins Albertlaan 12
8870 Izegem
Phone: +32-51-33-22-11

France

Sojinal

Route de Merxheim 8

68500 Issenheim

Phone: +33-3-89-745553

Germany

Alpro GmbH

Münsterstrasse 306

40470 Düsseldorf

Phone: +49-211-550-49-811

United Kingdom Alpro UK Ltd.

Latimer Business Park,

Altendiez Way

NN15 5YT Burton Latimer

Phone: +44-1536-720600

Available at http://www.alpro.com, http://www.alprosoja.com, and http://www.alprosoya.co.uk/homepage/_en-UK/index.html.

Alpro is Europe's pioneer in the development of mainstream soya-based food and drink products for the general market. Since 1980, Alpro has been championing a healthier, more sustainable way of producing tasty products that utilize the soybean's unique nutritional value. Alpro, per their advertisements, has been dreaming of a healthier world, a place where people can live without disturbing the Earth's balance by simply doing business in a healthy and fair way. With this focus and comprehensive approach to production and marketing, they are selling wholesome food products in three European countries.

The company employs over 650 people in 5 countries, and they are continuing to grow, especially as the market recognizes the unique value of the brand and what they stand for. Alpro believes that there is room for tasty, wholesome products that respect the consumer's right to healthy food and a sustainable approach to developing and selling that food; that it is not just what they sell that is important, but it is also a question of how they produce it.

Alpro promotes a natural, transparent, sustainable approach to farming and food business. They originally started in 1934; however, it was not until the 1980s that their pilot plant perfected a unique and natural process for making soya milk. At the same time Europe was experiencing a resurgence of vegetarian food, increased demand for cholesterol-free food, and a solution to cow's milk protein allergies.

In 1989, Alpro built one of Europe's largest and most modern production units for soya food on the basis of the ultra-high temperature (UHT) process in Wevelgem, Belgium. In 1996, Alpro took over Sojinal and thereby acquired an extra soya milk production unit in Issenheim, France. In 2000, Alpro built a new soya milk factory in Kettering, United Kingdom.

Alpro cites several points for their success:

- They base themselves on a unique "whole soybean process" that, unlike other processes, uses no chemicals during extraction. Furthermore, Alpro uses no GM soybeans. To ensure this, as well as maintaining the highest quality levels "they trace the production from the farm to the shop, traceability that guarantees totally waterproof controls." The result is a pure, natural, and chemical-free product that fits perfectly with their target market.
- Nearly 35 people work on quality control daily, maintaining standards that earned them International Organization for Standardization

(ISO) 9001 and Hazard Analysis and Critical Control Points (HACCP) certification.

To guarantee ISO and HACCP standards, Alpro carries out stringent quality checks during each of the production phases. In their in-house laboratory, their products are subjected to several bacteriological analyses. The result is a product with an extremely high bacteriological purity.

Alpro uses a full traceability system (ISO certified) of all raw materials that is based on more than 15 years of continuous organic and identity-preserved certification experience. They incorporate a complete HACCP for all process steps that offers direct contact with their farmers. Their system offers the capability for a full recall of product and product can be traced to the following:

- *The sourcing of beans:* Documentation of IP, varieties, area grown, and GMO status.
- *Transportation:* Documentation of IP, defined cleaning procedures, third-party sampling before and after transport.
- *Testing of non-GMO status:* Documentation of IP, third-party polymerase chain reaction (PCR) testing, and purchase condition less than 0.1% GMO versus 1.0% legal.
- *Storage at grain terminal:* Documentation of IP, dedicated silos, protocols for cleaning and transfer, and audits of storage facilities.
- *Transport of dehulling facility:* Documentation of IP, dedicated trucks with log book, dedicated silos, protocols for cleaning, and audits.

5.8 CARGILL—INNOVASURE IDP

Cargill, Inc.
Corporate Headquarters
P.O. Box 9300
Minneapolis, MN 55440-9300
Phone: 952-742-7575

Available at http://www.cargilldci.com/innovasure/index.shtm.

Cargill's unique InnovaSure IP services help insure IPT characteristics that farmers and their customer's desire. InnovaSure services of IP, what they term IdP, allow Cargill to provide its customers an established system for tracking, tracing, and IP throughout the supply chain. With InnovaSure, from seed selection to farm, the IdP services utilize leading-edge technologies and stringent IdP protocols to provide the ingredients and traits desired. Below is an example of Cargill's InnovaSure IdP services, which include corn seed selection, storage and handling, processing, and distribution.

5.8.1 Innova Sure Corn Seed Selection

Quality and traceability start with parent seed development and careful seed selection.

- Evaluation of all commercially available varieties each year to develop a list of approved seeds that will deliver the best performance.
- Only non-genetically enhanced varieties are currently included on the approved hybrid list used in the Indiana mill location. Genetically enhanced varieties with specific starch properties are included on the approved hybrid list for the Illinois mill location.
- When selecting hybrids, they match the starch properties of the corn with the functional applications of customers using Cargill laboratories to conduct the evaluations.
- All hybrids come from seed suppliers that have demonstrated that they meet their stringent IdP protocols.
- Cargill performs PCR testing on seed lots utilized in the Indiana mill to maximize the integrity of the hybrids.

5.8.2   INNOVASURE STORAGE AND HANDLING

Detailed handling techniques are critical to ensuring the reliability of InnovaSure identity-preserved products throughout the supply chain. The InnovaSure system includes detailed measures for maintaining the integrity of the grain during storage, handling, and transportation.

- Growers use separate storage bins for all identity-preserved grain. These bins are carefully cleaned between crops to minimize the possibility of carryover from a previous crop.
- Cargill operates its own elevators dedicated exclusively to the handling and storage of identity-preserved grains. At their elevators and mills they test the deliveries of corn, including tests for genetic enhancement at the Indiana mill location.
- Grain is again tested when it reaches their mills. They test deliveries for foreign material and food grade traits

5.8.3   INNOVASURE PROCESSING

Several control protocols are used in InnovaSure's mills to ensure high IdP integrity.

- To confirm quality, they take frequent samples and conduct rigorous testing of whole corn, yellow goods, and other corn products while in process.
- They test for several quality criteria such as granulation size, fat content, and foreign material.
- Their mills only process InnovaSure IdP corn.

5.8.4   INNOVASURE DISTRIBUTION

Distribution of InnovaSure products follows documented identity-preserved protocols to ensure accountability.

- Trucks and rail cars are cleaned and inspected before InnovaSure products are loaded.

- InnovaSure personnel grade the contents and test for genetic enhancement if required by the customer.
- Pending test results, cars are sealed and products are shipped.
- InnovaSure includes a certificate of analysis, plus a statement confirming the product's identity.

5.9 GEAPS AND PURDUE DISTANCE LEARNING PROGRAM

Grain Elevator and Processing Society (GEAPS)
301 4th Avenue South, Suite 365
P.O. Box 15026
Minneapolis, MN 55415-0026
Phone: 612-339-4625
Fax: 612-339-4644
E-mail: info@geaps.com

Available at http://www.geaps.com.

GEAPS was founded in 1927 and is comprised of approximately 2,800 individual members, including 36 local chapters across North America.[*] It is the only individual membership organization in the grain operations industry—an international professional society dedicated to providing its members with forums to generate leadership, innovation, and excellence in grain-related industry operations. As a professional society, it is one of the primary information resources for the world of grain-handling operations. In this way GEAPS also promotes the use of IPT systems and programs to help in the development of value-added products, improve quality, and expand their customer base. Plans are underway to expand industry operations of ISO 22000 measures and procedures into various aspects of IPT.

In the early 1990s, GEAPS undertook a comprehensive strategic plan review. The organization's updated objectives are to:

- Provide international and local forums for the collection, analysis, and exchange of information affecting the grain-related industries
- Advance the educational and professional qualifications of its members
- Promote and encourage safe, efficient, and environmentally responsible operations
- Promote and encourage the preservation and improvement of product quality during handling, storage, and processing
- Promote and encourage the development and application of operations technology
- Represent member interests in the development, interpretation, and implementation of government regulations and industry consensus standards
- Communicate with the trade media and general public concerning issues of interest to GEAPS members and the grain-related industries
- Coordinate its activities with other allied industry organizations in pursuit of GEAPS' mission

---

[*] GEAPS began as the Society of Grain Elevator Superintendents (SOGES).

GEAPS membership benefits include:

- *In-Grain* member newsletter
- Virtual reference library
- GEAPS exchange
- Alerts and news
- DirectaSource buyers guide and member directory

GEAPS *In-Grain* online publication informs its members of:

- Government affairs
- Grades and weights issues
- Membership activities
- News about members
- Available resources
- Industry news
- Safety, health, and environment issues
- Operations features
- GEAPS committees at work
- Learning opportunities

5.9.1 GEAPS AND PURDUE DISTANCE LEARNING PROGRAM

GEAPS and Purdue University have developed online grain operation educational programs. These jointly produced, internet-based "distance-learning" programs utilize educational material provided by GEAPS and other sources, are organized into curriculums, and offer students in formal course format training for the grain-handling operations industry. They plan to develop other classes in cooperation with ISU and its Iowa Grain Quality Lab, which will include training on ISO 22000 and its associated IPT quality-control measures.

The 5-week online courses were developed under the guidance of Dr. Dirk Maier, a professor of agricultural engineering at Purdue and a long-time GEAPS member. GEAPS created a task force of members who oversaw course development and offer input and advice. Purdue posts materials on the Internet and manages student enrollment and progress.

GEAPS provides much of the educational material and is expected to target all seven of GEAPS' "core competencies" of grain operations. The organization's top priorities for educational programming are:

- Handling systems and operations technology management
- Agribusiness environment and management practice
- Property and risk-casualty management
- Grain-handling equipment management
- Human resources management
- Facility operations management
- Grain-quality management

Sessions cover grain facility components such as storage options; site selection; budgeting; receiving, weighing, sampling, conveying, grain distribution, and cleaning systems; and other major planning and design components and considerations. For example, the class, titled GEAPS 510 "Grain Facilities Planning and Design I," is offered to GEAPS members for $350 and nonmembers for $400. The fee includes class materials and tuition. CDs containing the course lectures and other documents are e-mailed or mailed to students. Online registration is also available. These programs are a cooperative effort of GEAPS, Purdue University's Cooperative Extension Service, and the Departments of Agricultural and Biological Engineering, Entomology, and Botany and Plant Pathology.

5.10   JOHN DEERE—FOODORIGINS

John Deere—FoodOrigins
Amy Bantham, Director
46 Waltham Street, 6th Floor
Boston, MA 02118
Phone: 617-239-1120 ext. 2548
E-mail: amybantham@foodorigins.com

John Deere—FoodOrigins (Europe)
Jean-Louis Duval, Business Development
    Director (Europe)
15 Rue de Dagny
77240 Cesson, France
Phone: 33-164-10-84-85
E-mail: jean-louisduval@foodorigins.com

FoodOrigins
5401 Trillium Boulevard, Suite 225
Hoffman Estates, IL 60192
Toll free: 877-774-8660
E-mail: info@FoodOrigins.com

John Deere Agri Services
Elm Building
5500 Trillium Boulevard,
    Suite EC101B
Hoffman Estates, IL 60192
Phone: 847-645-8900
Fax: 847-645-9490
E-mail: JDASInformation@JohnDeere.com

Available at http://www.deere.com/en_US/deerecom/agriservices/inc.html and http://www.deere.com/en_US/deerecom/usa_canada.html.

FoodOrigins is a division of John Deere Shared Services and provides customized business solutions and technology for the global food supply chain to increase profitability, promote food safety, and achieve efficiencies. FoodOrigins builds on John Deere's leadership in production agriculture and its heritage of innovative engineering by connecting producers to processors, manufactures, marketers, retailers, and government.

FoodOrigins envisions an agri-food industry in which products, livestock, crops, and fibers are individually tracked from source to usage. Value traceability allows product attributes and performance data to be recorded and reported to increase producer profitability, product consistency, management efficiency, and overall food safety. They accomplish this through enhanced business performance by providing services and technology infrastructure needed to integrate business activities across food supply chain.

John Deere recently merged several technology-based operating units: AGRIS Corporation, GeoVeritas, John Deere FoodOrigins, John Deere Global Ag Services, and Agreen Tech into a business entity named John Deere Agri Services. John Deere Agri Services develops and provides knowledge-based products and services to meet the needs of a wide array of customer groups in the agri-food and fiber supply chain. Thus FoodOrigins is part of a large consortium of technology companies that John Deere has joined together.

FoodOrigins starts with the seed manufacturer and goes through all of the intermediate steps and ends up as a consumer end product. FoodOrigins focuses on sharing information with its member participants in and along the food chain and follows the product through its various transformations. In the simplest sense, FoodOrigins provides a set of services that allow companies, on the basis of the economic need, to trace food as an individual unit of production (e.g., a bin of wheat or 300 gallons of tomato paste) across transformations from beginning to end. At present, FoodOrigins is working in the grains and oilseed, meat and livestock, and fruit and vegetable areas.

There are three benefits for companies in sharing this information through the chain. They all relate to the ability to tie what has happened to the product and to trace it. As John Deere promotes:

- Good supply chain management suggests that companies should know a great deal about the product that they are buying or the product that their provider/supplier is buying so that they can practice better supply chain and operations management.
- Anytime a food marketer makes a claim about a food, their credibility is derived from how they can support that claim. In Europe, the consumer is much more interested in the food they are eating (e.g., farm management practices, chemical used, etc.) and where it comes from. In the United States they believe that because they have good governmental practices, their food is good, and that is generally true. However, U.S. consumers may become more like European consumers and demand more accountability or transparency.
- There are many companies that are becoming more sustainable in their agricultural practices. At its most basic level, sustainable agriculture is making sure that food production is fair to the participants who grow food, and that those producers are doing growing food in a way that is renewable and reusable so that they do not deplete natural resources. For example, Starbucks has been very vocal about the fact that they practice good sustainable agriculture.

FoodOrigins provides customized solutions that use information technology to connect information and advance partnership in the food supply chain. It provides solutions for grains and oilseed, livestock, fruits and vegetables, and regulatory compliance. As John Deere refers to it, they provide "grain corridor management" for farmers, elevators, and processors in the wheat, corn, and feed sectors. For livestock it puts forward animal identification and tracking systems from ranch to retail; for fruit and vegetable challenges their system helps establish information, inspection, and reporting for fresh and processed fruits and vegetables; and for regulatory compliance resolution they offer recordkeeping, reporting, and certification to meet U.S., EU, and other global requirements.

FoodOrigins understands that for every product there is a physical production process (e.g., wheat to flour to bread) and a related flow (e.g. wheat variety to flour attributes to bread quality and yield). FoodOrigins' Solutions connect information in the supply chain to simplify tracking and recordkeeping, making it easier for agri-businesses to improve performance and meet changing governmental guidelines. Overall it provides data collection tools for data collection ranging from traditional manual methods to customized software applications, data connectivity regarding information about individual units of production (animals, bushels, lots) across companies in the supply chain, and tools and services in the form of procedures for data analysis and reporting to authorized supply chain partners and regulatory agencies.

5.10.1 GRAINS AND OILSEEDS PRODUCT LINE SOLUTIONS

FoodOrigins' solutions help to track, trace, and report on grain and grain products from the farm to the table. Their solutions have been shown to generate value for participants through improved operational efficiency, greater consistency and logistics savings, and improved supplier collaboration. This new value can offset the cost of complying with new regulations.

5.10.2 FOODORIGINS FOR FARMERS

IPT has resulted in an increase in farmer income due to marketing to targeted buyers and growing crops according to mill and/or bakery specifications. Over time, the farmer can leverage traceability programs to build dedicated, long-term supplier partnerships with large food manufacturers and retailers in the supply chain.

5.10.3 FOODORIGINS FOR MILLING

FoodOrigins offers a management application designed to help mills improve customer relations. Automated data entry supports one-step regulatory traceability and provides online tracking for analysis and new value creation. The Bin Management Traceability component allows for web-enabled management of raw materials and processing ingredients as they are stored, combined, and transferred. It also tracks inventory, completes the calculation of blends, and documents related activities online.

5.10.4 FOODORIGINS FOR BAKING

Another management application designed to help bakeries capture and manage information related to incoming flour as it is processed into final goods and shipped to retailers is FoodOrigins solutions for baking. Automated data entry supports one-step regulatory traceability and provides online tracking for analysis and new value creation.

5.10.5 FOODORIGINS MARKET RESULTS

Between January and April 2001, FoodOrigins conducted an in-market beta test in partnership with one of the largest flourmills in the United States, a leading bread bakery and wheat farm from a farmers' cooperative. This exercise verified FoodOrigins' ability to capture and store data, documented the specific sources of value, and quantified economic benefits (both savings and new revenue) for all supply chain partners.

5.10.5.1 Flourmill

The beta test for this flourmill showed combined projected benefits of $0.34 per bushel, or a projected aggregate savings of $4.2–6.8 million over 1–3 years that stemmed from

- 2% yield improvements from identified optimal kernel structures and increased grain consistency (wheat procurement)
- Efficiency gains from increased grain consistency and improved specifications compliance (manufacturing)
- Savings from easier regulatory compliance and streamlined order processing (administration)
- Purchasing efficiencies from identified optimal grain mixes closer to the farm source (logistics)

5.10.5.2 Bakery

The bakery documented yield ranges of 315–333 loaves of bread per individual dough lot and waste ranges of 5–12 loaves per lot. Using FoodOrigins' traceability system, the bakery identified the ingredients and flour recipe that produced the optimal yield of 333 loaves and minimum waste of 5 loaves, resulting in a 5.7% increase in productivity.

5.10.6 Overall FoodOrigins Benefits

FoodOrigins helps agri-business and governments answer marketplace questions about food processing and safety within a changing global environment.

New Value	**Regulatory Compliance**
Brand assurance	Animal identification
Access to export markets	Grain and feed tracking
Production efficiencies	Pesticide residuals
Ingredient consistency	Environmental practices
Quality improvements	Product and ingredient origin
	Food safety and security

5.11 AGMOTION INTERNATIONAL TRADING COMPANY

AgMotion, Inc. – Corporate Office	**AgMotion, Inc. – Specialty Grains Division**
1000 Piper Jaffray Plaza	444 Cedar Street, Suite 1000
444 Cedar Street	St. Paul, MN 55101
St. Paul, MN 55101	Phone: 651-225-7500
Phone: 651-293-1640	

Available at http://www.agmotion.com.

Founded in August 2000, AgMotion is a holding company with AgMotion Technologies and two other firms under its corporate umbrella. The other two consist of U.S. Commodities, an international grain-trading company that is AgMotion's biggest component, and Northstar Commodity, a grain futures broker specializing in risk-management consulting (similar to the type of consulting a brokerage firm offers investors). Northstar also offers MarketMaster pricing software as part of its Managed Grain program, the aim of which is to help growers improve their margins and grain elevators' smooth product flow. AgMotion employs 55 individuals, with two offices in the Twin Cities and four other offices in North America. Recently, AgMotion passed the milestone of $100 million in annual sales and with its software has faced little direct competition thus far.

In 2001, the St. Paul-based grain-trading firm claimed a 982% increase in revenues and attributed a portion of this increase to their newly designed web-based software developed in-house that reduces logistics costs. AgMotion Specialty Grains combines years of international trading and logistics experience with advanced technology to support environmentally responsible organic and non-GMO agriculture worldwide. With their trading partners, they have strengthened the organic and non-GMO food chain by enhancing relationships between producers and customers. This is accomplished through quality assurance, business integrity, and effective communication.

5.11.1 AGMOTION SPECIALTY GRAINS SERVICES

AgMotion Specialty Grains buys and sells organic grains and feedstuffs, non-GMO, and identity-preserved products around the world. They offer value to their customers by combining years of global agricultural commodity brokerage and marketing expertise with advanced technology, logistics, and all of the other tools to market and distribute identity-preserved grains into world markets. Through its Tracekey™ feature, AgMotion offers customers the opportunity to trace grains from their origin to their end users. AgMotion services offer improved and more cost-effective traceability, visibility, and quality assurance over traditional marketing methods.

5.11.2 GRAIN ORIGINATION

AgMotion's U.S. Commodities, working with their Northstar Commodity, provides growers more options for managing price and basis risk in one flexible program. Growers who enroll bushels with Northstar Commodity's Managed Grain program and designate U.S. Commodities as the buyer of the grain get the expertise of Northstar managing their board risk without margin calls. AgMotion Specialty Grains is engaged in the production and distribution of the following products:

- Edible beans
- Non-GMO and organic grains
- Pulses/legumes
- Edible and sprouting seeds
- Non-GMO and organic oilseeds
- Sugar
- Non-GMO and organic animal feeds
- Other specialty grain products

Additionally, AgMotion Specialty Grains offers a comprehensive selection of JAS-certified organic and conventional non-GMO soybeans including Vinton 81, HP204, and a wide range of other newly developed and emerging high-protein varieties. These products are used extensively in Japan and the United States for the production of tofu, soy sauce, soymilk, and other soy-based foods.

For example, AgMotion software allows grain shipment information to be entered only once, instead of two or more times, into a system that facilitates logistics and payment of all related expenses. According to AgMotion, the software can be integrated into a business' current operating systems and software. The package also includes the TraceKey traceability feature, essential to any sale in the growing organic food industry, because it lets food processors know the source of the products they buy. The feature also helps stores and buyers trace products back to growers, a potentially important security feature.

5.12   NATIONAL STARCH—TRUETRACE

National Starch and Chemical Company
10 Finderne Avenue
P.O. Box 6500
Bridgewater, NJ 08807-0500
Phone: 908-685-5000; 800-743-6343 (toll-free)
Fax: 908-685-5355

Available at http://www.nationalstarch.com and http://www.foodinnovation.com.

National Starch and Chemical Company, a member of the ICI Group, is a worldwide manufacturer of adhesives, specialty polymers, electronic materials, and specialty starches. They have 9,500 employees across a global network of 154 manufacturing and customer service centers located in 37 countries on 6 continents and sales of $3.29 billion.

National Starch, a subsidiary of National Starch and Chemical Company, has expanded its crop IP program and implemented a broader, documented identity-tracing program to verify the non-GMO status of the company's food ingredients. The program, named TRUETRACE, provides customers with traceability for National Starch's food ingredients at all stages of their development, from seed to crop to production and distribution. The program covers all of the company's food ingredients made from corn grown in the United States. Protecting corn varieties from adventitious contamination and providing traceability has become ever more challenging because farmers in the Corn Belt of the United States have been greatly increasing their acreage of GM corn crops over the last few years. Currently, between one-third and one-half of the corn acreage in the Corn Belt states are being used to grow GM corn, and that is projected to increase considerably in the next few years. TRUETRACE adheres to the guidelines of the British Retail Consortium/Food and Drinks Federation (BRC/FDF) Technical Standard for the Supply of Identity Preserved Non-GM Food Ingredients and Products[*] (Mayer 2003).

---

[*] National Starch promotes that this standard represents the best practices available for ensuring the proper segregation and documentation of non-GM corn and provides for non-GM IPT that meets or exceeds regulations in major markets worldwide.

5.12.1   HOW **TRUETRACE** WORKS

Growers in National's TRUETRACE program grow non-GM corn exclusively or take special precautions to isolate GM corn from non-GM corn to avoid cross-contamination. These growers provide National with extensive documentation of their seed varieties, field locations, and equipment cleaning, which are all subject to periodic audits. Corn delivered to National Starch manufacturing facilities can thus be traced to the original farm on which it was grown and the seed varieties used in production. According to National Starch,

> National Starch is able to provide the TRUETRACE program because of its direct, long-standing relationships with corn growers in its primary contracting areas, and because it has a team of experts in plant science, agronomy, supply chain logistics, and regulatory affairs. This infrastructure and the know-how make it possible for us to offer this quality assurance program to our customers.

5.12.2   NATIONAL STARCH RECEIVES NON-GMO SEAL OF APPROVAL

In 2005, according to the Philippe Nuttal reports, inspection company SGS (see Chapter 7 on auditors) certified both of the National Starch Food Innovation's corn-starch factories, confirming they turn out non-GMO products that meet the desired quality standards. National Starch wanted to be open with their customers and prove that an independent organization had come in and verified their processing.

The traceability goes all of the way through from the farmers' field to the finished product. They noted that occasionally they receive batches that are contaminated and speculate that this generally comes about during the transportation of the corn when, for example, a truck has not been cleaned properly and there are still traces left from the last batch of GM corn.

National Starch notes that their traceability program comes at a price, adding 5–15% to the cost of the production of corn. A significant percentage of National Starch's customers are companies that specialize in organic or health foods, although some are more mainstream firms.

5.13 AIB INTERNATIONAL

AIB International – World Headquarters
1213 Bakers Way
P.O. Box 3999
Manhattan, KS 66505-3999
Phone: 785-537-4750
Fax: 785-537-1493

AIB International – UK Headquarters
P.O. Box 11
Leatherhead, Surrey
KT22 7YZ, UK
Phone: +44-1372-360-553
Fax: +44-1372-361-869

Available at http://www.aibonline.org.

AIB International promotes themselves as the "Gold Standard Certification Program" for food processing companies that meets food quality and safety standards with high-value technical and educational programs. The program is designed to enhance product quality and reduce food safety risks through audits, inspections, and methodological services. Their technical experts conduct assessments, employee training, and formal audits that verify compliance to the certification program requirements.

5.13.1 History

In 1919, AIB International became a corporation by the North American wholesale and retail baking industries as a technology transfer center for bakers and food processors.* Its original and current mission is to "put science to work for the baker" in all of the programs, products, and services provided by AIB to baking and general food production industries worldwide. Although AIB's history has been traditionally linked with North American wholesale and retail baking, they currently serve many segments of the food industry worldwide. AIB currently has more than 900 members in many countries, ranging from international food ingredient and foodservice companies to small single-unit traditional and artisan retail bakeries.

AIB is headquartered in Manhattan, Kansas, home of Kansas State University and one of the major centers for wheat and related grain product research and development. The institute works closely with local grain science and trade organizations and maintains links and working relationships with many other food production and equipment, food safety, trade development, food legislation groups, and university food science research programs in the United States and abroad.

AIB's Food Safety Audit Program, which began shortly after World War II, has always been in great demand by food industry producers, distributors, and warehouses. Its Food Safety and Hygiene audit services are recognized worldwide as the "standard" against which other food safety programs are to be judged.

More than 7,000 facilities in 70 countries currently subscribe to AIB programs. AIB International does not sell any chemicals, pesticides, or equipment and has no conflict of interest with any facility being inspected. All reports and services are confidential and reports are not released to or discussed with an external party unless a release form has been signed.

5.13.2 AIB Program

The three elements of AIB's comprehensive quality protocol program include Good Manufacturing Practice (GMP) audit qualification, HACCP validation and verification, and quality systems evaluation. Their program reduces the need for customer audits, other third-party audits, and laboratory evaluation of quality assurance

---

* AIB was the first long-term industry commitment to instruction in the basics of bakery science supplied by the Bachman School of Baking, sponsored by the Fleischmann Company and conducted at the Fleischmann Laboratories in New York City from 1911 to 1942. This school provided much needed knowledge in the area of fermentation and predicted the interest in baking education on the part of members of the so-called "Allied Trades" that would later become important to the continued success of the AIB.

systems by non-expert auditors. It also offers a marketing advantage because products meet strict quality criteria and customer specifications.

5.13.3   FOOD SECTOR PROGRAMS

The basis for an AIB International food safety/hygiene audit is their standards. Their in-depth analysis includes an optional rating system that provides management with an index of how well a facility is complying with food safety regulations as well as with the established internal standards set by the individual company. Companies may also write their own standard and have an audit of their factories against this standard. To assist in establishing effective food safety guidelines, AIB International publishes standards that detail the various components for developing a comprehensive food safety and hygiene program. These food sector standards include

- Agricultural crops
- Dairy plants
- Food safety
- Fresh-cut produce
- Food contact packaging manufacturing facilities
- Food distribution centers
- Fresh produce and fruit packinghouses
- Non-food contact packaging manufacturing facilities

5.13.4   AGRICULTURAL CROP STANDARDS

These standards contain the criteria for agricultural field managers to evaluate the food safety risks and to determine levels of compliance with Good Agricultural Practices (GAP) in their management programs. Details are given for areas such as

- Field evaluations
- Cleaning practices
- Employee practices
- Pest control programs
- Management of agrichemicals
- Documentation of crop safety programs
- Maintenance of buildings, fields, and water supplies

The *AIB Consolidated Standards for Agricultural Crops* were published as a tool to help field managers to evaluate the food safety risks within their operations and to determine levels of compliance with the criteria in the standards. These criteria are derived from Good Agricultural Practices, The United States Federal Food Drug and Cosmetic Act (1938); Good Manufacturing Practices, OSHA; CFR Title 21, Part 110 (1986); U.S. Military Sanitary Standards; and the U.S. Federal Insecticide, Fungicide, and Rodenticide Act.

5.13.5 Audit Services

AIB has established food safety programs to meet the needs of a variety of companies, large and small. AIB auditors are involved in all steps of the food supply chain. To meet the needs of increased customer demand, AIB has expanded its food safety audit program to include

- Food safety audits
- HACCP accreditation
- Quality systems evaluation
- Production quality
- Occupational safety
- Integrated Quality System Certification Program
- Certification schemes
- BRC Global Standard, ISO 9000, Food Audits Feed Materials Assurance Scheme (FEMAS)

5.13.6 Food Safety Audits

Trained food safety auditors conduct the food safety audit. Food processors that participate in the in-plant audit program receive a complete examination and technical assistance in all areas that affect product integrity, regulatory exposure, and pesticide use.

The following food safety GMP audits follow their published standards or can be customized to meet specific needs: GMP audits, agricultural audits, allergen audits, food security audits, and retail audits.

5.13.7 Agricultural Audits

This program consists of on-site third-party verification of the supplier's food safety program. In addition, the following areas are of primary importance:

- Review of documentation pertaining to adequacy of the produce safety program
  - Adjacent land use
  - Ranch/farm/land history
  - Fertilizer use
  - HACCP program
  - Water quality
- Pest control and management of agrochemicals
- Operational methods and personnel practices as applied to the GAP
- Maintenance for produce safety program
- Cleaning practices

5.13.8 Research and Technical Services

The AIB research and technical services offer the following:
- Analytical services

- Calibration services
- Production quality audits
- Technical assistance
- Food labeling
- Pilot plant capabilities
- Product quality evaluation
- Customized technical consulting
- Quality control and predictive technologies
- Research and development
- Ingredient and equipment product testing

5.13.9 ANALYTICAL SERVICES

AIB provides one-stop convenience for food manufacturers needing technical laboratory testing of ingredients, formulas, and finished products. Testing services include

- *Grain and flour analysis:* Damaged starch, qualitative enrichment, physical tests, granulation, viscosity, solvent retention capacity, single kernel characterization, etc.
- *Allergen testing:* Peanut, egg, milk, almond and gluten
- *Bake tests/product evaluation:* Breads, tortillas, cakes, and cookies
- *GMO testing:* PCR and ELISA
- *Microbiology:* Standard plate count, yeast and mold, salmonella, staphylococcus
- *Mycotoxins:* Aflatoxin, ochratoxin, etc.
- *Nutrition labeling:* Actual NLEA required analyses or by database
- *Physical dough testing:* Alveograph, amylograph, extensigraph, etc.
- *Proximate analysis:* Moisture, protein, ash, fat, and resistant starch
- *Toxins and residues:* Pesticide residues, chlorinated hydrocarbon, organophosphate
- *Vitamins and minerals:* Vitamins A, C, B1, B2, folic acid, niacin; calcium, etc.

A working agreement with ISO 9001:2000 Certified CII Laboratory Services (see chapter 8) allows AIB to offer a complete range of laboratory services at low cost. Other services include analytical services, audits, calibration services, consulting, food labeling, pilot plant capabilities, predictive technology, product quality evaluation, product testing, research and development, and technical assistance. AIB also offers a variety of other services that are incorporated within an IP program that includes Centurion near-infrared (NIR) instrument calibrations (Centurion is an independent calibration, calibration maintenance, and monitoring service for NIRs), food labeling and nutritional information, and ingredient statement assistance and package compliance review services. Table 5.3 is a sample of AIB's price list.

TABLE 5.3
AIB Price List (as of 2006)[a]

Category	Test	Includes	Price ($)
Allergens and GMOs	Allergen (test kit)—quantitative	Peanut, nuts, egg, milk, soy flour, gluten	80
	GMO	35S/GA21	195
	GMO	35S/NOS	195
	GMO	Other GMO tests	Call
Fats and oils	Cholesterol		85
	Color—Lovibond method (lipids)		Call
	Fat—GC (AOAC 996.06)	Total (saturated, monosaturated, polyunsaturated, trans fats)	140
	Fatty acid profile		165
	Free fatty acids in fats		20
	Free fatty acids in foods		35
	Glycerol		55
	Hexanal		Call
	Insoluble impurities		30
	Iodine value		60
	Moisture and volatiles (of lipids)		50
	Neutral oil and loss (of lipids)		Call
	Omega 3, 6 fatty acids		140
	OSI (AOM)		100
	Acrylamide	LC-MS/MS	220
	Aflatoxin—ELISA		25
	Aflatoxin—HPLC	B1, B2, G1, G2	70
	Chlorinated hydrocarbon		85
	Dichlorvos		120
	Fumonisin—ELISA		30
Toxins and residues	Fumonisin—HPLC		110
	Ochratoxin—ELISA		30
	Ochratoxin—HPLC		70
	Ochratoxin—TLC		70
	Pesticide multiresidue analysis screen	PAM/LUKE methodology: approximately 200 chemicals from organohalogens, organophosphates, organonitrogens, N-methyl carbamates	Call

[a] OSI, oil stability index; AOM, active oxygen method; ELISA, enzyme-linked immunoabsorbent assay ; HPLC, high-performance lipid chromatography; TLC, thin layer chromatography; LC-MS, lipid chromatography-mass spectrometry.
Source: Data used with permission from AIB.

REFERENCES

Mayer, Cheryl. "Tracking & Traceability." *Germination—The Magazine of the Canadian Seed Industry* 7, no. 4, 2003:10–12.

6 Standards

6.1 INTRODUCTION TO U.S. STANDARDS

According to Golan et al. (2004), in the United States, private sector food firms have developed a substantial capacity to trace. Traceability systems are a tool to help firms manage the flow of inputs and products to improve efficiency, product differentiation, food safety, and product quality. Firms balance the private costs and benefits of traceability to determine the efficient level of traceability. In cases of market shortcomings in which the private sector supply of traceability is not socially optimal, the private sector has developed several mechanisms to correct the problem, including contracting, third-party safety/quality audits, and industry-maintained standards. The best targeted government policies for strengthening firms' incentives to invest in traceability are aimed at ensuring that unsafe or falsely advertised foods are quickly removed from the system while allowing firms the flexibility to determine the manner. Possible policy tools include timed recall standards, increased penalties for distribution of unsafe foods, and increased food-borne illness surveillance. In this way, government rules and policies establish various goals and penalties for firms and private industry to achieve and avoid. However, government lets firms and industry determine the methods and implementation to achieve the various goals set out by government. In this way, government lets free market economics decide the level of demand and most efficient methods and technologies to use.

To this end, standards used for U.S. grain, oilseed, and organic production will be highlighted in this chapter within the U.S. Food and Drug Administration (FDA) Bioterrorism Act (Registration & Record Maintenance); U.S. Department of Agriculture (USDA; general programs) Grain Inspection, Packers & Stockyards Administration (GIPSA)'s Process Verified Program (PVP) and Verification Point Services; and the USDA's **National Organic Program** (NOP) rules and standards used for agricultural products produced, stored, processed, exported, imported, etc. within the United States.

Each section will have a short history of the organization, purpose, scope, and important rules and regulations as they apply toward identity preservation and traceability (IPT). What follows are organizational/agency statements from their websites and naturally reflects their views.

6.2 FDA STANDARDS*

Changes in food safety have been swift, and for many in the food chain it has been disruptive. The food safety events that have caused these changes are well documented. Many feel that the United States has been slow in providing guidance to industry and in resolving consumer uncertainty about the food that they eat. The FDA, for

---

* Available at http://www.fda.gov and http://www.fda.gov/oc/bioterrorism/bioact.html.

the time being, is the primary driver behind many of the fundamental food safety regulations in the United States. The Bioterrorism Act of 2002 is the basis for implementing increased accountability of nearly all aspects of food production, processing, transportation, etc. Below is a compressed overview of the Bioterrorism Act and how it affects industry.

The FDA is responsible for protecting the public health by assuring the safety, efficacy, and security of human and veterinary drugs, biological products, medical devices, our nation's food supply, cosmetics, and products that emit radiation. The agency has long been a leader in research to improve the detection of adulterated food products through the efforts of its cadre of top-notch scientists and public health experts and its partnerships with outside academic institutions, private companies, food consortia, and other government agencies.

6.2.1 OVERVIEW OF THE BIOTERRORISM ACT

The events of September 11, 2001 reinforced the need to enhance the security of the United States. Congress responded by passing the Public Health Security and Bioterrorism Preparedness and Response Act of 2002 (the Bioterrorism Act), which President George W. Bush signed into law on June 12, 2002. The Bioterrorism Act is divided into five titles:

- Title I: National Preparedness for Bioterrorism and Other Public Health Emergencies
- Title II: Enhancing Controls on Dangerous Biological Agents and Toxins
- Title III: Protecting Safety and Security of Food and Drug Supply
- Title IV: Drinking Water Security and Safety
- Title V: Additional Provisions

The FDA is responsible for carrying out certain provisions of the Bioterrorism Act, particularly Title III, Subtitle A (Protection of Food Supply) and Subtitle B (Protection of Drug Supply). For this book, Title III, Subtitle A (Protection of Food Supply) will be expanded upon.

6.2.2 PLANS FOR IMPLEMENTING THE ACT

Title III (Safety of Food and Drug Supply) is divided into the following:

- Subtitle A (Food Supply Protection)
- Section 301 (Security Strategy)
- Section 302 (Food Adulteration)
- Section 303 (Detention)
- Section 305 (Registration)
- Section 306 (Records Maintenance)
- Section 307 (Prior Notice)

The key to IPT within the Bioterrorism Act is primarily found in its Registration and Record Maintenance sections.

6.2.3 Registration and Record Maintenance

The Bioterrorism Act requires that all facilities, regardless of size, *domestic* and *foreign*, that manufacture, process, pack, or hold food, including animal feed, dietary supplements, infant formula, beverages (including alcoholic beverages and bottled water), and food additives comply with the regulations that require them to have (1) *registration* with the FDA and (2) *establish and maintain records* to identify *the immediate previous source* and *immediate subsequent recipient* of food. (Note: Nothing is mentioned regarding internal records that would match incoming inputs such as ingredients to outgoing products going to subsequent recipients.)

6.2.4 Registration of Food Facilities*

Information provided to FDA under this *final rule* helps the agency promptly identify and locate food processors and other establishments in the event of deliberate or accidental contamination of the food supply. Except for specific exemptions, the registration requirements apply to all facilities that manufacture, process, pack, or hold food, including animal feed, dietary supplements, infant formula, beverages (including alcoholic beverages and bottled water), and food additives.†

6.2.4.1 Who Must Register

Owners, operators, or agents in charge of *domestic* or *foreign* facilities that manufacture/process, pack, or hold food (subject to FDA's jurisdiction) for human or animal consumption in the United States must register with the FDA.

6.2.4.2 Foods Subject to FDA's Jurisdiction

Foods subject to the FDA's jurisdiction include "(1) articles used for food or drink for man or animals, (2) chewing gum, and (3) articles used for components of any such article." Food contact substances and pesticides are not "food" for purposes of the interim final rule. Thus, a facility that manufactures/processes, packs, or holds a food contact substance or a pesticide is not required to register with FDA.‡

---

* Key to understanding the rule: The *final* rule on Registration of Food Facilities (70 FR 57505, October 3, 2005) confirms the *interim final* rule entitled "Registration of Food Facilities Under the Public Health Security and Bioterrorism Preparedness and Response Act of 2002" (68 FR 58894, October 10, 2003) as corrected by a technical amendment (69 FR 29428, May 24, 2004) and responds to comments submitted in response to the request for comments in the interim final rule.

† In arriving at the interim final rule, the FDA worked closely with the Bureau of Customs and Border Protection (CBP) to ensure the new regulations promote a coordinated strategy for border protection. FDA and CBP continue to collaborate intensely on making the new safeguard of prior notice as efficient and effective as possible.

‡ Definition from Section 201 (f) of the Federal Food, Drug, and Cosmetic Act (FD&C Act) applies. Food contact substances, as defined in § 409(h)(6) of the FD&C Act, and pesticides regulated by EPA, as defined in 7 U.S.C. § 136(u).

6.2.4.3 Examples of Regulated Food within the Scope of the Rule

- Raw commodities for use as food or components of food
- Food and food additives for man or animals
- Dietary supplements and dietary ingredients
- Bakery goods, snack food, candy, and chewing gum
- Dairy products and shell eggs
- Infant formula
- Fruits and vegetables
- Fish and seafood
- Beverages (including alcoholic and bottled water)
- Canned and frozen foods
- Live food animals
- Animal feeds and pet food

6.2.4.4 Facilities Exempted from the Rule

Facilities exempt from the rule include:

- Farms.[*]
- Foreign persons, *except* for foreign persons who transport food in the United States.
- Restaurants are excluded entirely.[†]
- Persons performing covered activities with food *to the extent* that the food is within the *exclusive* jurisdiction of the USDA; that is, facilities handling only meat, poultry, or egg products. Foods that FDA does not regulate include:
  - Foods to the extent they are under the *exclusive* jurisdiction of the USDA under the following:
    - Federal Meat Inspection Act (21 U.S.C. 601 *et seq.*)
    - Poultry Products Inspection Act (21 U.S.C. 451 *et seq.*), or
    - Egg Products Inspection Act (21 U.S.C. 1031 *et seq.*)

---

[*] *Farms*, i.e., facilities in one general physical location devoted to the growing and harvesting of crops, the raising of animals (including seafood), or both. Washing, trimming of outer leaves, and cooling of produce are considered part of harvesting. The term "farm" also includes facilities that pack or hold food, provided that all food used in such activities is grown, raised, or consumed on that farm or another farm under the same ownership, and facilities that manufacture/process food, provided that all food used in such activities is consumed on that farm or another farm under the same ownership. A farm-operated roadside stand that sells food directly to consumers as its primary function would be exempt from registration as a retail food establishment.

[†] *Restaurants*, i.e., facilities that prepare and sell food directly to consumers for immediate consumption, including pet shelters, kennels, and veterinary facilities that provide food directly to animals. Facilities that provide food to interstate conveyances, such as commercial aircraft, or central kitchens that do not prepare and serve food directly to consumers are not restaurants for purposes of the rule. A combination restaurant/retail facility is excluded entirely if sales of food it prepares and sells to consumers for immediate consumption are more than 90% of its total food sales.

- Persons who manufacture, process, pack, transport, distribute, receive, hold, or import food for personal consumption.*
- Persons who receive or hold food on behalf of specific individual consumers and who are not also parties to the transaction and who are not in the business of distributing food (e.g., concierge in an apartment building).
- Persons who manufacture, process, pack, transport, distribute, receive, hold, or import food packaging (the outer packaging of food that bears the label and does not contact the food), *except* for those persons who also engage in a covered activity with respect to food.
- Private residences of individuals, although food may be manufactured/processed, packed, or held there.
- Non-bottled water drinking water collection and distribution establishments and structures, such as municipal water systems.
- Transport vehicles that hold food only in the usual course of their business as carriers.
- Nonprofit food establishments, which are charitable entities that meet the terms of §501(c)(3) of the Internal Revenue Code and that prepare or serve food directly to the consumer or otherwise provide food or meals for consumption by humans or animals in the U.S. central food banks, soup kitchens, and nonprofit food delivery services are examples of nonprofit food establishments.
- Fishing vessels that harvest and transport fish. Such vessels may engage in practices such as heading, eviscerating, or freezing fish solely to prepare the fish for holding onboard the vessel and remain exempt.

6.2.4.5 Electronic Registration

In 2003, the FDA announced further steps to use modern technology to provide new protections for the U.S. food supply. First, FDA announced its new electronic registration system for food facilities, foreign and domestic. This registration system, available online at http://www.cfsan.fda.gov/~furls/ovffreg.html and designed to bolster the safety and security of the U.S. food supply, helps with quick identification and notification of food processors and other facilities involved in any deliberate or accidental contamination of food. Second, FDA issued a report to the U.S. Congress on its progress toward developing more rapid, easier, and less costly tests to detect food contamination.

FDA's registration system, one of the key provisions of the Bioterrorism Act, requires domestic and foreign food facilities to register with the agency. As a result, FDA will have for the first time an official roster of foreign and domestic food facilities, allowing timely notification and response in the event of a food safety threat. This new system will permit 400,000 facilities to register worldwide in 60 days and

---

* *Retail food establishments*, i.e., groceries, delis, and roadside stands that sell food directly to consumers as their primary function; meaning that annual sales directly to consumers are of greater dollar value than annual sales to other buyers. An establishment that manufactures/processes, packs, or holds food and the primary function of which is to sell food directly to consumers, including food that the establishment manufactures/processes, from that establishment is a retail food establishment and is not required to register.

will give the FDA new capabilities to work with everyone involved in the U.S. food supply to keep it safe and secure.*

6.2.5 RECORD MAINTENANCE

In December 2004, the FDA published a *final rule* requiring food firms to *establish and maintain records* that would allow the FDA to conduct an effective and efficient traceback investigation to protect the U.S. human food and animal feed supply in the event that the agency has a reasonable belief that an article of food is adulterated and poses a threat of serious adverse health consequences or death to humans or animals. The economic impact of the final rule includes

- Approximately 707,672 total facilities covered
- 597,172 domestic facilities that manufacture, process, pack, transport, distribute, receive, hold, or import food in the United States
- 110,500 foreign facilities that transport food in the United States

6.2.5.1 Requirements for Who Must Establish and Maintain Records

Domestic persons in the United States that manufacture, process, pack, transport, distribute, receive, hold, or import food; *foreign* persons that transport food; and persons who place food directly in contact with its finished container must establish and maintain records. For these regulations, the term "persons" include individuals, partnerships, corporations, and associations. These records identify the immediate previous source of all food received as well as the immediate subsequent recipient of all food released.

Records must be retained at the establishment where the activities covered in the records occurred or at a reasonable accessible location (see Table 6.1). To minimize the burden on food companies affected by the final rule, companies may keep the required information in any format, paper or electronic. All businesses covered by this rule must comply within 12 months from the date the rule is published in the *Federal Register*, except small and very small businesses. For the record retention period, see Table 6.1.

6.2.5.2 Information That Must Be Included in Records

- *Lot code specified:* Lot code information is required by the FDA to be maintained and linked to specific batches of production. For bulk receipts (flour, oil, etc.) scale ticket numbers are a unique identifier and must also be linked to production.
- *Manufacturing/processing:* Manufacturers and processors must link their ingredient lot numbers to production batch lot numbers.

---

* When FDA has a reasonable belief that an article of food is adulterated and presents a threat of serious adverse health consequences or death to humans or animals, any records or other information to which FDA has access must be available for inspection and copying as soon as possible, not to exceed 24 hours from time of receipt of the official request. The records access authority applies to records required to be established and maintained by the final rule or any other records a covered entity may keep to comply with federal, state, or local law or as a matter of business practice.

TABLE 6.1
FDA Record Retention Periods

Food Having Significant Risk of Spoilage, Loss of Value, or Loss of Palatability within . . .	Nontransporter Records	Transporter Records
60 days	6 months	6 months
>60 days but within 6 months	1 year	1 year
>6 months	2 years	1 year
All animal feed, including pet food	1 year	1 year

Source: FDA's The Bioterrorism Act of 2002 website. Available at http://www.fda.gov/oc/bioterrorism/bioact.html.

- *Packaging:* All food contact packaging must be linked to specific batches of product manufactured by lot identifier.[*]
- *Unique identifiers:* Bulk "food" (animal or human) has identity as defined via "other identification" documentation (scale tickets, etc.) and thus must be isolated and traced through the elevator to meet the FDA specificity requirements.

6.2.5.3   Records Excluded from Records Access

Recipes, financial data, pricing data, personnel data, research data, and sales data are excluded from these requirements. A recipe is defined as the formula, including ingredients, quantities, and instructions necessary to manufacture a food product. Therefore, records relating only to the ingredients of a food product and not the other two components of a recipe are *not* excluded.

6.2.5.4   Excluded from the Requirement to Establish and Maintain Records but Not the Record Availability Requirements for Existing Records

Entities subject only to the record access and prohibited act provisions include:

- Fishing vessels not engaged in processing
- Retail food establishments that employ ten or fewer full-time equivalent employees
- Nonprofit food establishments
- Persons who manufacture, process, pack, transport, distribute, receive, hold, or import food is subject to the record availability requirements with respect to its packaging (the outer packaging of food that bears the label and does not contact the food)

---

[*] PathTracer is an example of software that can link ingredients—bulk or bagged/dry or liquid, trace elements, and packaging into specific batches. See Chapter 10 for more information.

- Persons who manufacture, process, pack, transport, distribute, receive, hold, or import food contact substances other than the finished container that directly contacts the food
- Persons who manufacture, process, pack, transport, distribute, receive, hold, or import the finished container that directly contacts the food, except for those persons who place food directly in contact with its finished container

6.2.5.5 Additional Partial Exclusions

- Persons who distribute food directly to consumers (the term "consumers" does not include businesses) are excluded from the requirement to establish and maintain records to identify the immediate subsequent recipients (they *are* subject to the requirements to identify the immediate previous sources).
- Persons who operate retail food establishments that distribute food to persons who are not consumers must establish and maintain records to identify the immediate subsequent recipients only to the extent that the information is reasonably available.

6.3   USDA GENERAL*

In 1862, when President Abraham Lincoln founded the USDA, he called it the "People's Department." In Lincoln's day, 58% of the population were farmers who needed good seeds and information to grow their crops. Generally speaking, the USDA is responsible for the safety of meat, poultry, and egg products.

Many of the USDA's mission areas overlap into regions that promote IPT. GIPSA is a primary illustration of not only promoting grain quality and sales, but also IPT principles. Other USDA agencies also play a part in the larger scheme of IPT programs or systems, even if they do not a directly indicate an IPT purpose or goal. They accomplish this by providing information and structure, which helps their customers, farmers, elevators, processors, etc. to better integrate IPT programs.

6.3.1   USDA Agencies

6.3.1.1   Agricultural Marketing Service

Agricultural Marketing Service (AMS) facilitates the strategic marketing of agricultural products in domestic and international markets while ensuring fair trading practices and promoting a competitive and efficient marketplace. AMS constantly works to develop new marketing services to increase customer satisfaction and includes six commodity programs: cotton, dairy, fruit and vegetable, livestock and seed, poultry, and tobacco.

6.3.1.2   Agricultural Research Service

Agricultural Research Service (ARS) is USDA's principal in-house agricultural research and information agency.

---

* Available at http://www.usda.gov/wps/portal/usdahome.

6.3.1.3 Animal and Plant Health Inspection Service

Animal and Plant Health Inspection Service (APHIS) provides leadership in ensuring the health and care of animals and plants. The agency improves agricultural productivity and competitiveness and contributes to the national economy and the public health.

6.3.1.4 GIPSA

GIPSA facilitates the marketing of livestock, poultry, meat, cereals, oilseeds, and related agricultural products. It also promotes fair and competitive trading practices for the overall benefit of consumers and U.S. agriculture. GIPSA ensures open and competitive markets for livestock, poultry, and meat by investigating and monitoring industry trade practices (see next section for more information).

6.3.2 Food Safety

Food Safety ensures that the nation's commercial supply of meat, poultry, and egg products is safe, wholesome, and properly labeled, and packaged. This mission area also plays a key role in the President's Council on Food Safety and has been instrumental in coordinating a national food safety strategic plan among various partner agencies including the Department of Health and Human Services (DHHS) and the U.S. Environmental Protection Agency (EPA).

Another AMS organization is the Science and Technology Program. It provides centralized scientific support to AMS programs, including laboratory analyses, laboratory quality assurance, coordination of scientific research conducted by other agencies for AMS, and statistical and mathematical consulting services. In addition, the Science and Technology Division's Plant Variety Protection Office issues certificates of protection for new varieties of sexually reproduced plants. The program also conducts a structure to collect and analyze data about pesticide residue.

6.4 USDA GIPSA

**Grain Inspection, Packers, and
  Stockyards Administration**
Stop 3601
1400 Independence Avenue SW
Washington, DC 20250-3601
Phone: 202-720-0219
Fax: 202-205-9237

**Process Verified
  Program Manager**
1400 Independence Avenue SW
Room 2409-S, Stop 3630
Washington, DC 20250
Phone: 202-720-0228
Fax: 202-720-1015

Available at http://www.gipsa.usda.gov/GIPSA/webapp?area=home&subject=landing&topic=landing.

The USDA PVP and Verification Point Services are examples of pluralism between government and industry to support food security and customer demands that promote IPT principles.

6.4.1 HISTORY

GIPSA was established in 1994 as part of the reorganization of the USDA. The formation of the agency resulted from the joining of two previously independent agencies, the Federal Grain Inspection Service (FGIS) and the Packers and Stockyards Administration (P&S). Today, GIPSA is part of USDA's Marketing and Regulatory Programs, which work to ensure a productive and competitive global marketplace for U.S. agricultural products.

The U.S. Congress established FGIS in 1976 to manage the national grain inspection system, which was initially established in 1916, and to institute a national grain-weighing program. The goal of creating a single federal grain inspection entity was to ensure development and maintenance of uniform U.S. standards, to develop inspection and weighing procedures for grain in domestic and export trade, and to facilitate grain marketing.*

Today's Packers and Stockyards Program is the progeny of the P&S, which was established in 1921 under the Packers and Stockyards Act. The organization was instituted to regulate livestock marketing activities at public stockyards and the operations of meat packers and live poultry dealers.†

6.4.2 MISSION

GIPSA's mission is to facilitate the marketing of livestock, poultry, meat, cereals, oilseeds, and related products and promote fair and competitive trading practices for the overall benefit of consumers and U.S. agriculture. In doing so, they serve a diverse group of customers, including

- Grain, livestock, and poultry producers
- Stockyards, livestock market agencies, and dealers
- Meat packers, brokers, wholesalers, and distributors
- Poultry growers and live poultry dealers
- Foreign grain buyers
- Grain and commodity handlers, processors, millers, and exporters
- Other federal and state agencies
- Authorized state and private inspection and weighing agencies
- Academic and research institutions
- The general public

6.4.3 HOW IPT WORKS FOR GRAIN, RICE, AND LEGUMES

The U.S. grain, rice, and other commodities flow from farm to elevator to destinations around the world. GIPSA's FGIS helps move the nation's harvest into the

---

* The agency's FGIS has headquarters units in Washington, DC, and Kansas City, Missouri, and field offices located in export and domestic markets in the United States and eastern Canada. GIPSA also oversees the official inspection and weighing system, a network of Federal, State, and private entities that provide inspection and weighing services to customers nationwide.

† GIPSA's Packers and Stockyards Program has a headquarters office in Washington, DC; regional field offices in Atlanta, Georgia; Denver, Colorado; and Des Moines, Iowa; and a cadre of resident agents located throughout the country.

marketplace by providing farmers, handlers, processors, exporters, and international buyers with sampling, inspection, process verification, weighing and stowage examination services that accurately and consistently describe the quality and quantity of the commodities being bought and sold.

In response to changing consumer demands, the market is adopting a variety of new marketing mechanisms, such as identity preservation (IP), to augment traditional marketing approaches. GIPSA's goal is to add value in this evolving marketplace by augmenting, not supplanting, existing marketing practices.

To this end, GIPSA published an Advance Notice of Proposed Rulemaking in the *Federal Register* (Vol. 67, No.151, August 6, 2002, p. 50853) seeking public comment on USDA's roles in facilitating the marketing of grains, oilseeds, fruits, vegetables, and nuts. Respondents recommended (1) continuing existing programs to standardize testing methodology and component testing, and (2) building on the success of its *process verification programs* for fruits, vegetables, and livestock by developing similar programs for grains, oilseeds, and related agricultural commodities.

As just mentioned above, the verification procedures verify the process by which a product or service is produced, handled, and processed rather than verifying the contents of the final product. The scope of a process may range from seed purchase to a final product on grocery shelves or a segment in between. However, more extensive processes create a greater need for other technical experts to assist GIPSA. Therefore, GIPSA will seek opportunities to partner with other organizations already performing such services.

6.4.4   GENERAL CERTIFICATION

Official inspections result in the issuance of official certificates. Certificates report the grade of the grain inspected on the basis of characteristics such as test weight, moisture, cleanliness, and damage. Certificates are issued for the various grains for which standards exist under the U.S. Grain Standards Act, as amended, and for rice, pulses, and miscellaneous processed commodities covered by Part 68 of the regulations under the Agricultural Marketing Act of 1946, as amended.

Certificates are the final product in the chain of official inspection services. They document the official procedures followed (e.g., date, location of the inspection or weighing process) and provide specific service results factor by- factor or by service requested.

6.4.4.1   Types of Official Certificates

- *Export Grain Inspection Certificate:* Mandatory export inspection
- *Export Grain Weight Certificate:* Mandatory export weighing
- *Grain Inspection Certificate (Official Sample Lot Inspection):* Domestic lots
- *Certificate:* Warehouseman's sample lot inspection
- *Certificate:* Submitted sample inspection
- *Stowage Examination Certificate:* Certifies results of an official stowage examination
- *Inspection Certificate:* Official commercial sample lot inspection
- *Certificate:* Official commercial submitted sample inspection

6.5 USDA PVP

The USDA PVP (a more intense IPT program) provides suppliers of agricultural products or services the opportunity to assure customers of their ability to provide consistent quality products or services. This is accomplished by having their documented production, manufacturing, or service delivery processes verified through independent, third-party audits. The program supports the market's increased use of IP and similar activities that add value and provides a way to capture the full value of one's products. PVP is available to any grain or oilseed farmer, handler, or processor, large or small, whether the value-adding activity is IP, testing, product branding, or any other marketing goal.

Under the PVP, GIPSA provides independent, third-party certification of the written quality practices and production processes used to provide consistent quality products. *Important note*: The quality practices and production processes are not GIPSA standards or rules. Individuals and organizations, such as farmers, handlers, and processors develop and implement quality management systems based on internationally recognized standards and value-adding processes to satisfy their customers' expectations. Prior to granting certification (1) GIPSA performs a desk audit to evaluate conformance to specified quality management requirements. The agency then (2) confirms the implementation of the written quality management system and manufacturing processes through an onsite audit. (3) Additional periodic, announced, and unannounced audits, including document reviews, major system audits, and surveillance audits, are performed to verify continuing conformance. Through this program, GIPSA verifies the *processes* used to ensure quality, not the quality of the final product.

PVP suppliers are able to make marketing claims (such as production and manufacturing practices or service provision) and market themselves as "USDA Process Verified." They also receive a USDA Certificate of Conformance for use in product marketing, and their approval will be posted, with permission, on GIPSA's website to further substantiate their certification.

At the present time, the PVP uses the International Organization of Standards (ISO) 9000 series standards for documented quality management systems as a format for evaluating program documentation to ensure consistent auditing practices and promote international recognition of audit results.

6.5.1 GIPSA Auditors

GIPSA quality auditors are each fully trained ISO 9000 Lead Auditors with more than 6 years of experience as an auditor. All of the auditors have an agricultural or food processing background that includes grain production, handling, or processing, as well as related commodity experience.[*]

6.5.2 Cost

PVP is user-fee funded. GIPSA charges an hourly fee for all review and audit services, and for travel costs at the government-approved reimbursement rate. The exact

---

[*] GIPSA also maintains a team of trained technical experts who can accompany auditors when specific expertise is required to complete the audit.

cost of service varies based on the scope of the process being audited, the number of participating parties, and other factors. Detailed cost estimates are provided prior to providing service.

6.5.3 PROCEDURES

To operate an approved USDA PVP, suppliers must submit documented quality management systems to the FGIS PVP and successfully pass an audit according to

- FGIS Directive 9180.79 1-21-06
- PVP audit checklist
- Process Verification Points and use of the Process Verified Shield
- Requesting service

6.5.4 VERIFICATION SERVICE (DIRECTIVE **9180.79** AS OF JANUARY **31, 2006**)

This directive establishes official procedures for obtaining and performing verification services for all products assigned to GIPSA and services associated with marketing of these products.* PVP provides independent third-party verification that processing or marketing claims are clearly defined and verified. The services are provided under the authority of the Agricultural Marketing Act of 1946 (AMA), as amended, and the Code of Federal Regulations (CFR) 7, Part 868, and this directive.

The directive provides a framework for determining whether a *processing or marketing claim*, referred to in this document as a *verification point*, can be accepted under PVP. It also provides guidelines for use of the verification points, the USDA Shield (logo), and the term, "USDA Process Verified."

Important note: Verification points are processing, handling, service, or marketing claims made by an organization that USDA has certified under the PVP. The claims are used for advertising or promotional purposes and demonstrate that verification points

- Add value to the product or service or uses practices *beyond* normal business activity
- Are substantive, verifiable, and repeatable
- Are within the scope of the PVP
- Are not requirements of a regulation or law, PVP requirements, or a standard under which the organization generally operates

From the FGIS PVP Form 002 (March 1, 2006), version 5, the Audit Report and Checklist includes 135 Musts, 9 Documented Items, and 24 Recorded Items.

---

* The programs offered do not seek to compete with or duplicate programs already existing in the private sector. Rather, they are intended to complement those programs by offering an independent, internationally respected source of verification activities. At the same time, the programs will have sufficient safeguards to ensure the integrity of their results.

6.5.4.1   Examples of Verification Points

Allowable verification points can include:

- Source verification, IP, and traceability to specific points within a system
- Adherence to a recognized standard that is not otherwise required by industry or regulation
- A unique production or handling practice
- A service with a unique characteristic for that type of operation or outside of normal business practice
- A quantifiable characteristic such as size, weight, age, or grade
- Documentation, monitoring or auditing that is unique to the company and outside of normal business practice
- A characteristic, practice, or requirement that is specifically requested by a customer or consumer

Non-allowable verification points may include:

- Adherence to Good Manufacturing Practices (GMPs) when it is a requirement
- Conformance to PVP requirements
- Objectives of the Quality Management System
- Compliance to industry rules and regulations

6.5.4.2   Auditing the Verification Points

- Verification points must be clearly stated in the quality manual. The claims will be reviewed during the adequacy audit to establish that they meet the above requirements. Applicants must provide appropriate information to establish the validity of the claims.
- Each verification point will be audited during the on-site audit to verify that the claims are accurate and repeatable.

6.5.4.3   Adequacy Audit (Document Review)

All audits will be conducted in conformance to ISO 19011 guidelines for quality and/ or environmental management systems auditing.

- The assigned auditor will conduct a complete adequacy audit of the applicant's quality system documentation to ensure that each element of the specific quality system is in compliance.
- If the documentation is adequate, the auditor will arrange to conduct an on-site audit.

6.5.4.3.1   Audit Reports

- Upon completion of the on-site audit, the auditor will prepare a detailed report of the audit observations, findings, and recommendations. The report will include, at a minimum, the name, address, and the organizational structure of the business.

- Auditors will itemize any significant findings of nonconformance in the finding section of the audit report and assign a tracking number to each nonconformance. Auditors will classify each itemized non-conformance as a *continuous improvement point*, *a minor nonconformance*, or a *major nonconformance* according to the following definitions:
- *Continuous improvement point (CIP):* An observation made by an auditor that is not a nonconformance, but an area where the operation might improve.
- *Minor nonconformance:* A nonconformance that, although it needs to be corrected in a timely manner, does not compromise the integrity of the product/quality system.
- *Major nonconformance:* A nonconformance that compromises the integrity of the quality system to the extent that approval should be denied, revoked, or delayed until corrective action can be completed. Any absence or complete breakdown in a required element will be considered a major nonconformance. An accumulation of minor nonconformances also may result in the assignment of a major nonconformance for an audit.

6.5.4.3.2   Approval

- In most verification programs, approval decisions will be made by the verification programs manager after a review committee has reviewed the applicable audit reports and made a recommendation to grant or deny approval.
- Organizations meeting all verification program requirements will be issued a certificate of conformance valid for 1 year from the date of the on-site audit. Information regarding the organization's status will be posted on the GIPSA website.

6.5.4.3.2.1   An Example of PVP Assistance   Companies are encouraged to create a quality manual that describes its processes and procedures. Some important points to remember:

- A quality manual must describe the company's processes and procedures as they relate to the PVP. It can reference existing procedures, instructions, etc. within the quality manual.
- The quality manual should say "what" to do and "how" to do it. It is not sufficient to simply state to do something.
- Use what is already in place as long as it meets the PVP requirements. It can reference existing procedures, work instructions, forms, etc. within the quality manual.
- The PVP requires ten documented procedures, at a minimum.

6.6   USDA NOP

Mark Bradley, Associate Deputy Administrator USDA-AMS-TMP-NOP
Room 4008 South Building, Ag Stop 0268
1400 Independence Avenue, SW
Washington, DC 20250-0020
Phone: 202-720-3252
Fax: 202-205-7808

Available at http://www.ams.usda.gov and http://www.ams.usda.gov/nop/indexIE.htm.

The U.S. Congress passed the Organic Foods Production Act (OFPA) of 1990. The OFPA required the USDA to create "national standards governing the marketing of certain agricultural products as organically produced products," to assure consumers that "organically produced products meet a consistent standard" and to facilitate "interstate commerce in fresh and processed food that is organically produced." The OFPA and the NOP regulations require that agricultural products labeled as organic originate from farms or handling operations certified by a state or private entity that has been accredited by USDA. On December 21, 2000, the AMS, an agency within the USDA, published a final rule that implemented OFPA. The combination of OFPA and the final rule created the NOP. The NOP is a marketing program housed within the USDA AMS. Neither the OFPA nor the NOP regulations address food safety or nutrition.

The NOP developed national organic standards and established an organic certification program on the basis of recommendations of the 15-member National Organic Standards Board (NOSB). The NOSB is appointed by the Secretary of Agriculture and is comprised of representatives from the following categories: farmer/grower, handler/processor, retailer, consumer/public interest, environmentalist, scientist, and certifying agent. In addition to considering NOSB recommendations, USDA reviewed state, private, and foreign organic certification programs to help formulate these regulations.*

Organic crops are raised without using most conventional pesticides, petroleum-based fertilizers, or sewage sludge-based fertilizers. Animals raised on an organic operation must be fed organic feed and given access to the outdoors. They are given no antibiotics or growth hormones.

NOP labeling standards are based on the percentage of organic ingredients in a product.†

- Products labeled "100% organic" must contain only organically produced ingredients.

---

* NOP certificates dates: According to 205.404 of the NOP, certification bodies cannot indicate validity dates on the NOP certificates. However, operations shall be certified under the act until the operation's next anniversary date of certification. The same paragraph requires the certifying agent to indicate the "effective date of certification" (first date of certification) instead of the last date of certification. Because of this, NOP-certified operators often face difficulties to prove the validity of their certificate. In case of such a situation, operators can advise inquirers to search for presently certified companies or products at the certifying agent's website. If it does not help, operators can contact their certifier, who will issue a declaration.

† The use of the seal is voluntary, so it is possible for organic products to not have an USDA organic seal.

- Products labeled "organic" must consist of at least 95% organically produced ingredients.
- Products meeting the requirements for "100% organic" and "organic" may display the USDA Organic seal.

A common question is, "Does *natural* mean *organic?*" No. *Natural* and *organic* are not interchangeable. Other truthful claims, such as free-range, hormone-free, and natural, can still appear on food labels. Only food labeled "organic" has been certified as meeting USDA organic standards.

6.6.1 SUMMARY OF NOP FINAL RULE

The final rule establishes the NOP under the direction of the AMS. This national program has four important parts.

1. The NOP will *facilitate domestic and international marketing* of fresh and processed food that is organically produced and assure consumers that such products meet consistent, uniform standards.
2. This program establishes national standards for the production and handling of organically produced products, including a *national list of substances approved for and prohibited from use* in organic production and handling.
3. The final rule establishes a *national-level accreditation program* to be administered by AMS for state officials and private persons who want to be accredited as certifying agents. Under the program, certifying agents will certify production and handling operations in compliance with the requirements of this regulation and initiate compliance actions to enforce program requirements.
4. The final rule *includes requirements for labeling products as organic and containing organic ingredients*. This final rule also provides for importation of organic agricultural products from foreign programs determined to have equivalent organic program requirements.

6.6.2 HOW NOP WORKS

Organic food is produced by farmers who emphasize the use of renewable resources and the conservation of soil and water to enhance environmental quality for future generations. Organic meat, poultry, eggs, and dairy products come from animals that are given no antibiotics or growth hormones.* Organic food is produced without using most conventional pesticides, fertilizers made with synthetic ingredients or sewage sludge, bioengineering, or ionizing radiation. Before a product can be labeled organic, a government-approved certifier inspects the farm where the food is grown to make sure the farmer is following all of the rules necessary to meet NOP standards. Companies that handle or process organic food before it gets to

---

* The act allows use of animal vaccines in organic livestock production if the product is not on the national list because the vaccine would not be a synthetic material.

local supermarkets or restaurants must also be certified. *Note:* The USDA makes no claims that organically produced food is safer or more nutritious than conventionally produced food. Organic food differs from conventionally produced food in the way it is grown, handled, and processed.

6.6.3 NOP Regulations*

The NOP regulations prohibit the use of genetic engineering, ionizing radiation, and sewage sludge in organic production and handling. As a general rule, all natural (nonsynthetic) substances are allowed in organic production and all synthetic substances are prohibited. The National List of Allowed Synthetic and Prohibited Non-Synthetic Substances, a section in the regulations, contains the specific exceptions to the rule.

- Major Statutes Organic Food Production Act of 1990, 7 U.S.C. §§ 6501-6522 Regulations
- National Organic Program Regulations, 7 C.F.R. Part 205

6.6.3.1 Requirements on Who Needs to Be Certified

Operations or portions of operations that produce or handle agricultural products that are intended to be sold, labeled, or represented as "100% organic," "organic," or "made with organic ingredients" or food group(s) must be certified by NOP.

6.6.3.2 Certification Standards

Certification standards establish the requirements that organic production and handling operations must meet to become accredited by USDA-accredited certifying agents. The information that an applicant must submit to the certifying agent includes the applicant's organic system plan. This plan describes (among other things) practices and substances used in production, recordkeeping procedures, and practices to prevent commingling of organic and non-organic products. The certification standards also address on-site inspections.

6.6.3.3 Accreditation Standards

Accreditation standards establish the requirements an applicant must meet to become a USDA-accredited certifying agent. The standards are designed to ensure that all organic certifying agents act consistently and impartially. Successful applicants will employ experienced personnel, demonstrate their expertise in certifying organic producers and handlers, prevent conflicts of interest, and maintain strict confidentiality.

6.6.3.4 Exempt and Excluded Operations

This regulation establishes several categories of exempt or excluded operations. An exempt or excluded operation does not need to be certified. However, operations that

---

* For NOP policies, procedures, and reference documents, see http://www.ams.usda.gov/nop/NoticesPolicies/MasterList.html (accessed December 16, 2005).

qualify as exempt or excluded operations can voluntarily choose to be certified. A production or handling operation that is exempt or excluded from obtaining certification still must meet other regulatory requirements contained in this rule as explained below.

6.6.3.4.1   Exempt Operations

1. A production or handling operation that has $5,000 or less in gross annual income from organic sales is exempt from certification. This exemption is primarily designed for those producers who market their product directly to consumers. It will also permit such producers to market their products directly to retail food establishments for resale to consumers. The exemption is not restricted to U.S. producers. However, as a practical matter, NOP does not envision any significant use of the exemption by foreign producers because (1) the products from such operations cannot be used as ingredients identified as organic in processed products produced by another handling operation, and (2) it is unlikely that such operations will be selling their products directly to consumers in the United States.

2. A retail food establishment or portion of a retail food establishment that handles organically produced agricultural products but does not process them is exempt from all of the requirements in these regulations.

3. A handling operation or portion of a handling operation that handles only agricultural products containing less than 70% organic ingredients by total weight of the finished product (excluding water and salt) is exempt from the requirements in these regulations, except the recordkeeping provisions of Section 205.101(c), the provisions for prevention of contact of organic products with prohibited substances in Section 205.272, and the labeling regulations in Sections 205.305 and 205.310. The recordkeeping provisions maintain an audit trail for organic products. The prevention of contact with prohibited substances and the labeling requirements protect the integrity of organically produced products.

4. A handling operation or portion of a handling operation that uses the word "organic" only on the information panel is exempt from the requirements in these regulations, except the recordkeeping provisions of Section 205.101(c), the provisions for prevention of contact of organic products with prohibited substances as provided in Section 205.272, and the labeling regulations in Sections 205.305 and 205.310. The recordkeeping provisions maintain an audit trail for organic products. The prevention of contact with prohibited substances and labeling requirements protect the integrity of organically produced products.

As noted above, exempt handling operations producing multi-ingredient products must maintain records as required by Section 205.101(c). This would include records sufficient to (1) prove that ingredients identified as organic were organically produced and handled, and (2) verify quantities produced from such ingredients. Such records must be maintained for no less than 3 years, and the operation must allow representatives of the secretary and the applicable state programs governing the state official access to the records during normal business hours for inspection and copying to determine compliance with the applicable regulations.

6.6.3.4.2   Excluded Operations

1. A handling operation or portion of a handling operation that sells organic agricultural products labeled as "100% organic," "organic," or "made with..." that are packaged or otherwise enclosed in a container prior to being received or acquired by the operation, remain in the same package or container, and are not otherwise processed while in the control of the handling operation, is excluded from the requirements in these regulations except for the provisions for prevention of commingling and contact of organic products with prohibited substances in Section 205.272. The requirements for the prevention of commingling and contact with prohibited substances protect the integrity of organically produced products.

2. A retail food establishment or portion of a retail food establishment that processes on the premises of the retail food establishment raw and ready-to-eat food from certified agricultural products labeled as "100% organic," "organic," or "made with..." is excluded from the requirements in these regulations, except for the provisions for prevention of contact of organic products with prohibited substances as provided in Section 205.272 and the labeling regulations in Section 205.310. The prevention of commingling and contact with prohibited substances and labeling requirements protect the integrity of organically produced products.

Excluded retail food establishments include restaurants, delicatessens, bakeries, grocery stores, or any retail outlet with an in-store restaurant, delicatessen, bakery, salad bar, or other eat-in or carryout service of processed or prepared raw and ready-to-eat food.

There is clearly a great deal of public concern regarding the handling of organic products by retail food establishments. NOP has not required certification of retail food establishments at this time because of a lack of consensus as to whether retail food establishments should be certified, a lack of consensus on retailer certification standards, and a concern about the capacity of existing certifying agents to certify the sheer volume of such businesses. Retail food establishments, not exempt under the act, could at some future date be subject to regulation under the NOP. Any such regulation would be preceded by rulemaking with an opportunity for public comment.

No retailer, regardless of this exclusion and the exceptions found in the definitions for "handler" or "handling operation," may sell, label, or provide market information on a product unless such product has been produced and handled in accordance with the act and these regulations. Any retailer who knowingly sells or labels a product as organic, except in accordance with the act and these regulations, will be subject to a civil penalty of not more than $10,000 per violation under this program.

6.6.4   How Farmers and Handlers Become Certified

An applicant will submit specific information to an accredited certifying agent. Information will include type of operation; history of substances applied to land for the previous 3 years; organic products being grown, raised, or processed; and the applicant's organic plan, which includes practices and substances used in production.

The organic plan also must describe the monitoring practices to be performed to verify that the plan is effectively implemented, the recordkeeping system, and the practices to prevent commingling of organic and non-organic products and to prevent contact of products with prohibited substances.

Applicants for certification should keep accurate post-certification records for 5 years concerning the production, harvesting, and handling of agricultural products that are to be sold as organic. These records should document that the operation is in compliance with the regulations and verify the information provided to the certifying agent. Access to these records must be provided to authorized representatives of the USDA, including the certifying agent.

6.6.5 Inspection and Certification Process

Certifying agents will review applications for certification eligibility. A qualified inspector will conduct an on-site inspection of the applicant's operation. Inspections will be scheduled when the inspector can observe the practices used to produce or handle organic products and talk to someone knowledgeable about the operation.

The certifying agent will review the information submitted by the applicant and the inspector's report. If this information shows that the applicant is complying with the relevant standards and requirements, the certifying agent will grant certification and issue a certificate. Certification will remain in effect until terminated, either voluntarily or through the enforcement process. Annual inspections will be conducted of each certified operation, and updates of information will be provided.

6.6.6 Compliance Review and Enforcement Measures

The rule will permit USDA or the certifying agent to conduct unannounced inspections at any time to adequately enforce the regulations. The OFPA also requires that residue tests be performed to help in enforcement of the regulations. Certifying agents and the USDA conduct residue tests of organically produced products when there is reason to believe that they have been contaminated with prohibited substances. If any detectable residues are present, an investigation will be conducted to determine their source.

6.6.7 An Organic System Plan Contains Six Components

The six components of an organic system plan are:

1. The organic system plan must describe the practices and procedures used, including the frequency with which they will be used, in the certified operation.
2. It must list and characterize each substance used as a production or handling input, including the documentation of commercial availability, as applicable.

3. It must identify the monitoring techniques that will be used to verify that the organic plan is being implemented in a manner that complies with all applicable requirements.

4. It must explain the recordkeeping system used to preserve the identity of organic products from the point of certification through delivery to the customer who assumes legal title to the goods.

5. The organic system plan must describe the management practices and physical barriers established to prevent commingling of organic and non-organic products on a split operation and to prevent contact of organic production and handling operations and products with prohibited substances.

6. The organic system plan must contain the additional information deemed necessary by the certifying agent to evaluate site-specific conditions relevant to compliance with these or applicable state program regulations. Producers or handlers may submit a plan developed to comply with other federal, State, or local regulatory programs if it fulfills the requirements of an organic system plan.

In the first element, practices are tangible production and handling techniques, such as the method for applying manure, the mechanical and biological methods used to prepare and combine ingredients and package finished products, and the measures taken to exclude pests from a facility.

By requiring information on the frequency with which production and handling practices and procedures will be performed, the final rule requires an organic system plan to include an implementation schedule, including information on the timing and sequence of all relevant production and handling activities. For example, the plan will include information about planned crop rotation sequences, the timing of any applications of organic materials, and the timing and location of soil tests. Livestock management practices might describe development of a rotational grazing plan or addition of mineral supplements to the feed supply. A handling operation might identify steps involved in locating and contracting with farmers who could produce organic ingredients that were in short supply.

The second element that must be included in an organic system plan is information on the application of substances to land, facilities, or agricultural products. This requirement encompasses both natural and synthetic materials allowed for use in production and handling operations. For natural materials that may be used in organic operations under specific restrictions, the organic plan must detail how the application of the materials will comply with those restrictions. For example, farmers who apply manure to their fields must document in their organic system plans how they will prevent that application from contributing to water contamination. A producer and handler who base the selection of seed and planting stock material under Section 205.204 or an agricultural ingredient under Section 205.301 on the commercial availability of that substance must provide documentation in the organic system plan.

The third element is a description of the methods used to evaluate its effectiveness. Measured through regular tallies of bushels or pounds, this would include provisions for analyzing soil organic matter levels at periodic intervals.

The fourth element is a description of the recordkeeping system used to verify and document an audit trail. A livestock operation must trace each animal from its entrance into through removal from the organic operation.

The fifth element included in an organic system plan pertains to split production or handling operations. This provision requires an operation that produces both organic and non-organic products to describe the management practices and physical barriers established to prevent commingling.

The final element regards the accreditation process, which provides an assurance that certifying agents are competent to determine the specific documentation they require to review and evaluate an operation's organic system plan.

6.6.8   CERTIFICATION—DOMESTIC AND FOREIGN

The USDA accredits state, private, and foreign organizations or persons to become "certifying agents." Certifying agents certify that production and handling practices meet the national standards. See http://www.ams.usda.gov/nop/CertifyingAgents/Accredited.html for a comprehensive list of the USDA Accredited Certifying Agents (ACAs) organized alphabetically by state for domestic ACAs and by country for foreign ACAs.

6.6.8.1   Imported Organic Products

Imported agricultural products may be sold in the United States if they are certified by USDA-ACAs. USDA has ACAs in several foreign countries. In lieu of USDA accreditation, a foreign certifying agent may receive recognition when USDA has determined, upon the request of a foreign government, which the foreign certifying agent's government is able to assess and accredit certifying agents as meeting the requirements of the USDA NOP. The USDA is working with New Zealand, the United Kingdom, Spain, Canada, Israel, and Denmark on this type of agreement.

6.6.8.2   Organic Philosophical Challenge

Many organic producers believe that organic production should be done in a manner consistent with biodiversity (sustainable) that must preserve or protect biodiversity and that "industrial organic farms" are not sustainable. Preservation of biodiversity is a requirement in many existing organic certification standards, including the Codex guidelines. Thus, industrial organic farms should not be considered or certified as organic.

Note: It is particularly important to remember that organic standards are process-based. Certifying agents attest to the ability of organic operations to follow a set of production standards and practices that meet the requirements of the act and the regulations. This regulation prohibits the use of excluded methods in organic operations. The tested presence of a detectable residue of a product of excluded methods alone does not necessarily constitute a violation of this regulation. As long as an organic operation has not used excluded methods and takes reasonable steps to avoid contact with the products of excluded methods as detailed in their approved organic system plan, the unintentional presence of the products of excluded methods should not affect the status of an organic product or operation.

NOP regulation § 205.105 of allowed and prohibited substances, methods, and ingredients in organic production and handling includes:

1. Synthetic substances and ingredients, except as provided in § 205.601 or § 205.603
2. No synthetic substances prohibited in § 205.602 or § 205.604
3. Non-agricultural substances used in or on processed products, except as provided in § 205.605
4. Non-organic agricultural substances used in or on processed products, except as otherwise provided in § 205.606
5. Excluded methods, except for vaccines, provided that the vaccines are approved in accordance with § 205.600(a)
6. Ionizing radiation, as described in FDA regulation, 21 CFR 179.26
7. Sewage sludge

6.7 INTRODUCTION TO CANADIAN STANDARDS

This following sections review various aspects of Canadian IPT programs as they apply to grains and oilseeds. As with the U.S. standards sections, each section will have a short history of the organization, its purpose and scope, and important rules and regulations as they apply toward IPT.

National systems, such as Canada's, are very detailed and extensive in their approach toward IPT. In many ways Canada is much further ahead in providing rules and regulations to guide its firms and industry without burdening them with explicit "how to do it" rules. The top three participants in Canadian IPT include (1) the Canadian Grain Commission's Canadian Identity Preserved Recognition System (CIPRS) program quality management system and audit producers, (2) the Canadian Soybean Export Association and its soybean IP standard and procedure, and (3) Can-Trace and its technology guidelines and standards.

The Canadian Soybean Export Association and its Soybean IP Standard and procedure are especially important and helpful in its explanation of describing not only the standard, but also recommendations of good practices and appropriate documentation. This is on par with Europe's EurepGap standards and procedures. What follows are company/organizational/agency statements from their websites and naturally reflects their views.

6.8 CANADIAN GRAIN COMMISSION

General inquiries
600-303 Main Street
Winnipeg, Manitoba R3C 3G8
Phone: 204-983-2770
Toll-free: 800-853-6705
Fax: 204-983-2751
E-mail: contact@grainscanada.gc.ca

Jim McCullagh, Executive Director
Canadian Seed Institute
200-240 Catherine Street
Ottawa ON K2P 2G8
Phone: 613-236-6451
Fax: 613-236-7000
E-mail: csi@storm.ca

Laura Anderson, Program Manager
Canadian Grain Commission
303-303 Main Street
Winnipeg, Manitoba R3C 3G8
Phone: 204-983-2881
Fax: 204-983-5382
E-mail: landerson@grainscanada.gc.ca

**Jo-Anne Sutherland, Certification
    and Accreditation Advisor**
Canadian Grain Commission
303-303 Main Street
Winnipeg, Manitoba R3C 3G8
Phone: 204-984-6979
Fax: 204-983-5382
E-mail: jsutherland@grainscanada.gc.ca

**Len Seguin, Chief Grain Inspector
    for Canada**
Canadian Grain Commission
900-303 Main Street
Winnipeg MB R3C 3G8
E-mail: lseguin@grainscanada.gc.ca

Available at http://www.grainscanada.gc.ca/main-e.htm.

The *Canadian Seed Institute (CSI)* is a not-for-profit organization established by Canadian seed associations to ensure delivery of consistent, cost-effective monitoring and quality-assurance programs for the Canadian seed industry. CSI provides national accreditation services to the industry, establishing the foundation of the Canadian quality-assurance system for seed certification.

Since passing the *Canada Grain Act in 1912*, Canada has had a quality-assurance system administered by a regulatory agency, the Canadian Grain Commission (CGC). Through quality and safety testing procedures, CGC insures the quality of grains and issues the globally recognized Certificate Final for supplying domestic and world markets with safe, high-quality grain, oilseeds, and pulses.

The *Canadian Grain Commission (CGC)* is a federal government agency and operates under the authority of the Canada Grain Act. The head office is in Winnipeg and has approximately 700 employees. Its annual budget comes partly from fees from services and partly from Parliament. CGC reports to Parliament through the Minister of Agriculture and Agri-Food Canada.

CGC offers several services to the grain industry as grain moves its way from the producer's field to markets. For the CGC, identity-preserved agricultural production involves maintaining the unique traits or quality characteristics of a crop from seed through transportation, handling, and processing. These traits can involve anything from high sugar content for snack soybeans to high-colored durum for the pasta market or unique oils for industrial uses. It is really about agricultural companies working with end processors to identify market needs and then ensuring the processes are in place to meet those needs.

In Canada, grain is most often wheat, and wheat is often turned into bread, whole wheat bread, crusty bread, French bread, Italian bread, bannock, pita bread, and tortillas. Canadian grain products also include pasta, noodles, mustard, licorice, sprouts from mustard, flax, beans, and chick peas; oils from canola, flax, sunflowers, corn, and wheat germ; soups from barley, wheat, lentils, and peas; and porridge, muffins, cakes, biscuits, cookies, crackers, couscous, hummus, kasha, and beer.

Canadian grain is graded by its visual characteristics, similar to the United States. Grades are carefully established to describe the processing qualities of the grain. The Certificate Final issued for each export shipment of grain is internationally recognized and accepted as Canada's assurance that what its customers buy is what they are expecting to buy.

6.8.1   FEDERAL GOVERNMENT SPONSORSHIP

The development of CIPRS is supported by Agriculture and Agri-Food Canada under the Canadian Adaptation and Rural Development Fund and the Agri-Food Trade Program.

6.8.2   CGC/CSI PARTNERSHIP

CIPRS is a joint project of CSI and CGC. This partnership brings together the expertise of the CSI in standard development and conformity assessment, and the international reputation of the CGC as a credible and trusted organization with a mandate for grain quality certification.

There is a growing market demand for the development of quality-assurance systems to programs. CIPRS is a new tool for the industry to provide assurance of specific quality attributes to domestic and international buyers. CIPRS is a voluntary program.

6.8.3   CIPRS PROGRAM

Canada has maintained an enviable reputation for supplying domestic and world markets with safe, high-quality grains, oilseeds, and pulses. In a marketplace with ever-increasing demands for unique product specifications and traceability, there are many new opportunities for agricultural products. A key factor in capitalizing on these opportunities is industry's ability to deliver products with better quality-assurance systems. Although industry is taking the lead in implementing these systems, CGC has developed a new voluntary pilot program to oversee and officially recognize those programs to maximize their acceptance in global markets. CIPRS is a new tool the industry can use to provide third-party assurance of the processes they are using to deliver the specific quality attributes their domestic and international buyers are demanding (see Figure 6.1).

6.8.3.1   System Development Format

The development of the system encompasses various tasks.

- CGC quality management system (QMS) standard for IP programs
- Accreditation program development, which includes:
  - QMS standard for service bodies
  - Training and assessment of auditors
  - Auditing the auditors' protocols
  - Audit protocols for IP programs

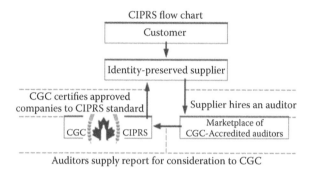

FIGURE 6.1 CIPRS flow chart. (Reprinted with permission from CIPRS.)

Key participants include components from farm fields to world markets.

6.8.3.2 Distribution Points for Canadian Grain

Canadian grain distribution points include the following:

- *Country elevator* is the primary collection point where farmers deliver their crops. There are many country elevators throughout the crop-producing areas of Canada.
- *Terminal elevator* is a port grain handling facility designed to load lakers for shipment through the St. Lawrence Seaway or freighters for shipment to overseas export destinations.
- *Transfer elevator* is a port grain handling facility designed to unload lakers, railcars, or trucks and transfer the grain to export freighters.
- *Processing plant* are facilities where IP products are cleaned, sorted, bagged, and loaded into containers.
- *Lakers* are vessels small enough to transport grain through the St. Lawrence Seaway from Thunder Bay to transfer elevators along the St. Lawrence River.
- *Freighters* are large ocean-going vessels with a total capacity of up to 60,000 metric tons, designed to ship large volumes of bulk grain in holds.
- *Container vessels* are large ocean-going vessels designed to accommodate containers.

6.8.3.3 How CIPRS Works

CIPRS certifies companies selling products through identity-preserved programs that have effective quality management systems for the production, handling, and transportation of specialty grains, oilseeds, or pulses. These systems provide full documentation and traceability from seed to export vessel or domestic end user.

CIPRS is based on QMS that document and itemize processes to control production from farmer to labeling and shipping. It is an integrated approach to ensuring a company has the system in place to produce and certify a specific product for the customer. CIPRS ensures that a company's QMS meets the standard created by the CGC. The standard is designed to be compatible with QMS such as ISO.

CIPRS requires that companies selling products through IP programs have effective QMS for the production, handling, and transportation of specialized grain products. These systems maintain and provide full documentation and traceability from seed to export vessel or domestic end user.

6.8.3.4 Program Components

- The CGC's CIPRS QMS Standard for IP programs sets out what the IP program must do, focusing on the need to identify and meet customer requirements.
- *Conformity assessment:* Third-party audits are conducted on IP programs by CGC-accredited auditors to ensure that the standard is being met.
- Certificate of Recognition is the buyer's assurance that the IP process is operating as it should and that it meets the CGC standard.

The CGC standard for IP programs is a national Canadian standard that can be applied to all crop types distributed through any Canadian supply chain. It provides the measuring stick against which IP programs can be assessed. If the IP program measures up, it will be recognized by the CGC with an official certificate. This CGC Certificate of Recognition brands Canadian IP programs that can deliver on what they promise.

6.8.3.5 Crop-Specific Standards

Some commodity organizations have developed crop-specific IP standards with additional controls along the supply chain to satisfy the needs of their markets. One example is the Canadian Soybean Exporters' Association's IP standard (see next section).

Just as the CIPRS provides added assurance that individual IP programs can deliver on what they promise, verification against a crop-specific IP standard provides assurance that the additional controls are in place. This dual recognition provides further branding of the Canadian product in international markets. The service delivery model will also apply to these crop-specific standards, keeping auditing costs to a minimum.

6.8.3.6 Certificate Assures Quality

The CGC's Certificate Final is issued after samples taken as an ocean-going vessel is loaded have been officially inspected. The Certificate Final provides buyers with an added level of assurance that the shipment will meet their quality expectations. See Table 6.2 for an example of CGC's IP program.

TABLE 6.2

CGC CIPRS Program Quality Management System and Audit Procedures

Program Quality Management System and Audit Procedures

Stage of Production/ Distribution	Control Points	Quality System Requirements	Audit Procedures
All stages	IP quality manual	• Up-to-date version • Defined personnel responsibilities and authorities • Personnel training plans • Defined product quality requirements as specified by customer • Defined variety purity of GM testing methods and sensitivity • Location of testing in supply chain identified • Crop production and handling plans • IP product handling plan • Transportation plan • Nonconforming product plan	Review of manual, ensuring that the testing, production, handling, and transportation plans are consistent with the quality requirements of the standard
Crop production and handling	Personnel	• Farmer contracts	Review of contracts
	Seed	• Use of seed specified in the production plan—either seed stock traceable to grower or certified seed	Review of: • Seed purchase invoices • Certified seed tags
	Planting	• Isolation distance from adjacent fields and previous land use consistent • Planters and seed drills are cleaned before planting a new crop • Traceability from seed to field	Review of farmer records; for example: • Field maps • Field history records • Planting equipment clean-out records • Planting records
	Production	• Weed, insect, and disease control consistent with crop production plan • Field inspections during growing season	Review of: • Input records • Field inspection reports

(Continued)

TABLE 6.2   (Continued)

Program Quality Management System and Audit Procedures

Stage of Production/ Distribution	Control Points	Quality System Requirements	Audit Procedures
Crop production and handling	Harvesting and on-farm storage	• Combines and trailers cleaned before harvesting • Storage bins cleaned before harvest • Equipment used to load and unload storage bins cleaned before using • Any contaminated crop will be disposed of as indicated in the crop production plan • Traceability from field to storage bin • Traceability from storage bin to mode of transport	Review of: • Equipment and bin clean-out records • Bin maps • Disposal of nonconforming product records • Storage records • Shipping records
Transportation	Farm to receiving elevator or processor	• Defined processes for cleaning and inspection of mode of transport • Mode of transport cleaned before use	Review of: • Bills of lading • Documented cleaning procedures Cleaning and inspection records
Grain handling	Personnel	• Defined processes • Assigned responsibilities and authorities • Competent staff	Review of: • Quality manual and documented procedures • Training records On-site audit of: • Processes • Staff understanding of processes and responsibilities
	Receiving	• Defined processes for cleaning and flushing facility before receipt of IP product • Sample taken and stored and information on source recorded • Testing for conformance to required quality attributes • Defined process for nonconforming product	Review of: • Documented cleaning and flushing procedures • Documented procedures for handling of nonconforming product On-site audit of: • Sample storage and records • Bills of lading • Scale tickets • Receiving processes

(Continued)

TABLE 6.2 *(Continued)*

Program Quality Management System and Audit Procedures

Stage of Production/ Distribution	Control Points	Quality System Requirements	Audit Procedures
	Handling	• All movements of IP product through facility are recorded • Traceability from mode of transport to storage bin	Review of: • Product movement records • Storage records
	Storage	• Storage bins cleaned before used for IP product • Stored IP product periodically checked to ensure continued conformity to quality requirements • Storage bins meet all physical requirements necessary to maintain quality of IP product • Packaging is clean and consistent with IP product handling plan • Defined process for nonconforming product	Review of: • Bin cleaning records • Quality check records • Documented procedures for handling of nonconforming product On-site inspection of: • Storage bins • Packaging
	Shipping	• Defined processes for cleaning and inspection of mode of transport • Mode of transport cleaned before use • Traceability of storage bin to mode of transport	Review of: • Documented cleaning procedures • Cleaning and inspection records • Shipping records

Source: Data used with permission from CGC.

6.8.3.7 Links to International Systems

Work is underway to link CIPRS to other international programs. The standard is compatible with ISO and other QMS. Negotiations are taking place with the goal of having the program cross-recognized with other national standards overseas.

6.8.3.8 Publications for Certification

• *CGC IP-STAN 1.0.0:* CGC's QMS standard for identity-preserved programs

- *CGC Guide 1.0.0:* Guidance document for the CGC QMS standard for identity-preserved programs
- *CGC IP-QSP 1.1.0:* Certification of an identity-preserved program under the CIPRS

6.8.3.9 Publications for Accreditation

- *CGC ASP-STAN 2.0.0:* General Requirements for Accredited Service Providers, May 3, 2004—revision 1.0, April 18, 2005
- *CGC IP-QSP 2.1.0:* Accreditation and Monitoring of Approved Service Providers, May 3, 2004—revision 2.0, February 20, 2006
- Application for accreditation form, Adobe PDF
- Application for accreditation form, Microsoft Word document
- Fee schedule
- *CGC IP-STAN 1.0.0:* QMS Standard for Identity-Preserved Programs—revision 3, February 20, 2006

To become recognized under the CIPRS, the following steps are to be observed:

- Develop an identity-preserved QMS in line with the CIPRS standard.
- Be audited by an independent CGC-accredited auditor who will submit an audit report to the CGC.
- Await the CGC review of the audit report and decision on certification.
- If the review is successful, the company's program will be certified.

The following is an example of CGC IP-STAN 1.0.0—QMS STD (index summarized and condensed):

CGC QMS Standard for Identity-Preserved Programs

Index of chapters and subchapters

1.0 General Requirements	2.3 Quality Records
2.0 Documentation Requirements	2.3.1 Quality System Records
2.2 Control of Documents	2.3.2 Process Control Records

Sample: Records shall include the following, where applicable, and any other records deemed essential to process control by the company and/or its suppliers:

- Field maps
- Grower contracts
- Field history records
- Planting records
- Both internal and external field inspection reports
- Harvest records
- Equipment clean-out records
- Stock seed tags

- Sampler declarations
- Testing records
- Storage records, bin records
- Any nonconformance reports
- Pertinent supplier records
- Past assessment reports
- Shipping records
- Bills of lading

Continuation of chapters and subchapters

2.3.3 Storing Records
2.3.4 Customer Records
2.3.5 Record Disposal
3.0 Management Responsibility
3.1 Management Commitment
3.2 Customer Focus
3.3 Quality Policy
3.4 Planning
3.4.1 Quality Objectives
3.4.2 QMS Planning
3.5 Responsibility, Authority, and
     Communication
3.6 Management Review
4.0 Resource Management
4.1 Provision of Resources
4.2 Human Resources
4.2.1 Employee Training
4.2.2 Training Records
4.3 Infrastructure and Work Environment

5.0 Product Realization
5.1 Planning of Product Realization
5.2 Customer Related Processes
5.2.1 Determination of Requirements
     Related to the Product
5.2.2 Review of Requirements Related
     to the Product
5.2.3 Customer Communication
5.3 Purchasing
5.4 Production and Service Provision
5.4.1 Control of Production and
     Service Provision
5.4.2 Planting
5.4.3 Cross-contamination
5.4.4 Harvesting
5.4.5 Transportation
5.4.6 Discharge and Storage at
     Collection Points
5.4.7 Identification and Traceability

The company shall establish and maintain procedures to ensure that all IP grain handled by the company is controlled and identified. The identification and traceability system shall be such that a product can be traced through the entire production and distribution system. The identification and traceability system shall be such that segregation is maintained between different product types.

5.4.8 Storage and Packaging
5.5 Control of Monitoring and Measuring
     Devices
6.0 Measurement, Analysis, and Improvement
6.1 General
6.2 Monitoring and Measuring
6.2.1 Customer Satisfaction
6.2.2 Internal Audit
6.2.3 Monitoring and Measurement of Product

6.2.4 Monitoring and Measurement
     of Processes
6.3 Control of Nonconformances
6.4 Analysis of Data
6.5. Improvement
6.5.1 Continual Improvement
6.5.2 Corrective Action
6.5.3 Preventive Action
7.0 Monitoring

6.8.4 Accredited Service Providers as of October 2005

Canadian Seed Institute
200-240 Catherine Street
Ottawa, Ontario K2P 2G8
Phone: 613-236-6451
E-mail: jmccullagh@csi-ics.com

Intertek Agri Services
960 C Alloy Drive
Thunder Bay, Ontario P7B 6A4
Phone: 807-345-5392
Fax: 807-345-4032
E-mail: Chris.Bazaluk@Intertek.com

NSF-ISR
360 Main Street, Suite 2300
Winnipeg, Manitoba R3C 3Z3
Phone: 204-944-3625
E-mail: partridge@nsf-isr.org

6.8.4.1 Fee Schedule

Initial and Recertification fee[a]	$500.00
Accreditation application (one-time)	$500.00
Auditor training and assessment (per person)[a]	$500.00
Initial and re-accreditation (every 3 years)[a,b]	$1,000.00
Annual accreditation fee[a]	$3,000.00
On-site assessments (per day, as required)[a]	$500.00

[a] Plus travel costs
[b] For applicants who are not currently accredited as compliant with ISO Guide 62

Note: Additional fees may be charged if the scope of the accreditation/certification is expanded (i.e., additional sites).

6.9 CANADIAN SOYBEAN EXPORT ASSOCIATION

Canadian Soybean Export Association
180 Riverview Drive, P.O. Box 1199
Chatham, Ontario, N7M 5L8

Ontario Soybean Growers
Michelle McMullen, Industry Opportunities Coordinator
Research Park Centre, Suite 205
Guelph, Ontario, N1G 4T2
Phone: 519-767-2472
Fax: 519-767-2466
E-mail: mmcmullen@soybean.on.ca

Available for CSEA members at http://www.canadiansoybeans.com/members.html and http://www.canadiansoybeans.com.

The Canadian Soybean Export Association (CSEA) is a voluntary association of members of the Canadian soybean industry working as a team to promote the export of Canadian soybeans and soya products into world markets. The CSEA's IP Soybean Procedures are very extensive and detailed and may be used as a reference when developing an IPT system that originates from the farm.*

6.9.1 IP SOYBEAN PROCEDURE

Canada's food-grade soybean customers have praised the Canadian soybean industry's Canadian Soybean IP Procedure. The standard is a minimum guideline that outlines IP procedures for all stages of soybean production, including planting, growing, harvesting, processing, and shipping. The IP Soybean Procedure is available at http://www.canadiansoybeans.com/soybeanstandard.pdf.

6.9.2 CSEA SOYBEAN PRODUCT LIST

- SQWH
- Crush
- Organic
- Sprouts
- IP tofu
- IP miso
- IP natto
- IP soymilk
- Non-GMO food soybean
- Non-GMO crush soybean
- Soybean meal/oil
- Organic soybean
- Meal/oil
- Non-GMO soybean meal/oil
- Non-GMO full-fat soyflour
- Soyflour and soynuts
- Roasted soybeans
- Isolates proteins and concentrates

SQWH, special quality white hilum; Organic, organic/transitional; IP, identity-preserved; Sprouts, soybeans suitable for the sprout market; and Non-GMO, non-GMO. See website at http://www.canadiansoybeans.com/products.html for more information.

6.9.3 CSEA-APPROVED IP PROCEDURES

Excerpts from the CSEA-Approved IP Procedures (second revision February 21, 2003):

---

* CSEA was formally begun in 1995, and its membership consists of personnel from the Canadian soybean industry in Ontario, Quebec, Manitoba, and British Columbia. Members include personnel from industry sectors such as soybean exporters, traders, research, and provincial and federal government officials.

Table of Contents

Recommendations on Good Practice (center column) are not part of the official CSEA IP Standard. They are additional suggestions for the IP program but are not enforced. See Table 6.3 for CSEA IP standards.

6.10   CAN-TRACE

GS1 Canada
885 Don Mills Road, Suite 222
Toronto, Ontario M3C 1V9
Phone: 416-510-8039; 800-567-7084 (toll-free)
Fax: 416-510-1916
E-mail: info@can-trace.org

Norm Cheesman
Director, Can-Trace
Phone: 416-510-8039 ext. 2309
E-mail: norm.cheesman@gs1ca.org

Available at http://www.can-trace.org.

In 2003, the Canadian food industry joined together and developed a program to identify industry requirements for a national all-product, whole-chain food traceability (tracking and tracing) standard. The goal of this initiative was to develop and verify an information (data) standard necessary to establish traceability on the basis of international standards. Its implementation would be voluntary. The initiative was given the name of Can-Trace, which today has participation from over 25 national trade associations and government organizations.

Can-Trace is a collaborative and open initiative committed to the development of traceability standards for all food products grown, manufactured, and sold in Canada. GS1 Canada is the initiative's secretariat. GS1 Canada and Can-Trace-sponsoring associations are continuously approaching various organizations to join and support this expanding initiative. GS1 Canada [formerly the Electronic Commerce Council of Canada (ECCC)] is the Secretariat to this initiative.*

Can-Trace is an industry-led national initiative for establishing food traceability in Canada. The Agricultural Policy Framework (APF) of the federal provincial and territorial governments agreed to the objective, which would allow for 80% of the domestic product available at the retail level to be traceable through the agri-food continuum.

---

* Can-Trace receives a portion of its funding support from Agriculture and Agri-Food Canada through the Canadian Food Safety and Quality Program (CFSQP). In addition, federal and provincial governments are observers on the Can-Trace Steering Committee. The initiative was created to help meet the objectives of the Agriculture Policy Framework. Federal and provincial government representatives participate in the Can-Trace Steering Committee and in the different working groups.

TABLE 6.3
CSEA IP Standards

Minimum Level	Recommendations on Good Practice[a]	Documentation
1.0 Seed Standards		
1.1 Certified seed accredited to Association of Official Seed Certification Agencies (AOSCA) standards or equivalent. Equivalent seed must be produced under a controlled system similar to the Canadian Seed Growers' Association (CSGA) pedigreed seed increase system. "Bin run" seed not to be used. Bin run—grain retained from a previous crop that is used as seed for planting.	1.1.1 Grower should retain certified seed tag for each bag of seed. Seed lot traceability is recommended.	1.1.2 Grower must be able to produce certified seed tag for each lot of seed purchased to produce the quantity of IP soybeans being contracted or delivered. Grower must retain his/her invoice or receipt of purchase for all quantities of IP seed purchased. The contracting party must have sufficient documentation to prove that the seed purity and identity has been maintained.
2.0 Planting		
2.1 Planter must be thoroughly cleaned and inspected prior to planting IP soybean variety. This must be done regardless if grower uses his/her own equipment or uses a custom planter.	2.1.1 Grower should endeavor to plant IP soybean crop before planter is used on other soybean crops. IP seed bags should be stored separately from other soybean seed and other crop seed prior to planting. Grower should refer to cleaning procedures as detailed by equipment manufacturer if available.	2.1.2 Growers must detail cleaning procedure used and sign this document to authenticate that they have implemented the procedures described. (Note: No mention of training.)
2.2 Approved isolation distance for the IP crop must be used. The CSGA isolation standard for certified soybean seed is 3 m between another soybean and another pulse crop (Bean, Fababean, Lentil, Lupin, Pea, or Peanut). There is no isolation distance necessary between soybeans and crops of Barley, Buckwheat, Canaryseed, Flax, Oat, Rye, Triticale, and wheat, providing the crops do not overlap.	2.2.1 Grower should endeavor to leave a minimum of 1 m of isolation between an IP soybean field and fields of crops that do not require the 3-m isolation.	2.2.2 Proper isolation distance must be documented at time of field inspection.

(Continued)

TABLE 6.3 *(Continued)*

Minimum Level	Recommendations on Good Practice[a]	Documentation
2.3 Grower must have records of previous crop(s) grown on IP soybean field.	2.3.1 Grower should keep detailed field maps and history of crops grown.	2.3.2 Grower must be able to provide a written history of previous crop.
3.0 Field Season		
3.1 A second- or third-party field inspector must inspect the IP field during the growing season to confirm that isolation distances have been met and there is proper control of volunteer crops and weeds. The field inspector must also verify that the crop looks uniform as detailed in the variety description. (Inspection confirms previous paperwork and records.)	3.1.1 If the IP crop is not being grown under contract (in which case the contracting party should conduct the field inspection) the grower should arrange for a qualified individual, at arms length from the operation of the farm, to conduct the field inspection.	3.1.2 The field inspection report must document that isolation distances have been met, there is proper control of weeds and volunteer crops, and that the soybean variety appears to be characteristically uniform for the appropriate growth stage. The inspector and the grower must sign and date this report.
4.0 Harvest		
4.1 Combine must be thoroughly cleaned and inspected prior to harvesting IP soybean variety. This is to be done regardless if grower uses his own equipment or uses a custom combine.	4.1.1 Grower should attempt to harvest IP soybean crop before combine is used on other soybean crops. Grower should refer to equipment manufacturer cleaning procedures.	4.1.2 Grower must detail cleaning procedure used and sign this document to authenticate that they have implemented the procedures described.
4.2 Equipment used to transfer soybeans must be thoroughly cleaned and inspected prior to transferring IP soybean crop. This is to be done regardless if grower uses his/her own equipment or uses custom harvesting.	4.2.1 Grower should endeavor to harvest IP soybean crop before transfer equipment is used on other soybean crops.	4.2.2 Grower must detail cleaning procedure used and sign this document to authenticate that they have implemented the procedures described.
4.3 Conveyance vehicles/ equipment used to transport IP soybeans at harvest must be thoroughly cleaned and inspected prior to transporting IP soybean crop. This is to be done regardless if the grower uses his/her own equipment or custom trucking.	4.3.1 Grower should try to arrange for conveyance vehicles/ equipment that has only been used recently to transport clean substances such as grain or food items. It is critical that all grain and meal residue is cleaned from the inside of the truck. Ideally the truck or hopper should be covered.	4.3.2 Grower must inspect truck and sign a document to authenticate that the truck/ hopper was cleaned prior to loading.

(Continued)

TABLE 6.3 *(Continued)*

Minimum Level	Recommendations on Good Practice [a]	Documentation
5.0 On-Farm Storage		
5.1 Grower must maintain record of what was stored in their bin prior to filling with IP soybean crop.	5.1.1 Grower should keep full records with crop type and dates when bins were loaded unloaded and cleaned.	5.1.2 Grower must keep written records of what crop was in their storage bin prior to filling with IP soybeans.
5.2 Storage bin must be thoroughly cleaned and inspected prior to loading.	5.2.1	5.2.2 Grower must sign a document indicating that their bin was thoroughly cleaned and inspected prior to filling.
5.3 Storage bins used to store IP crops must be visually identified so that all persons working in farm operation are aware that each bin should only be used for a particular IP crop.	5.3.1 Grower should put a sign on or otherwise visually identify any storage bin that will be used for IP soybean crop. All persons working in farm operation should be made aware that the storage bin is only to be used for the IP crop.	5.3.2 Grower must sign a document indicating that any storage bin used for an IP soybean crop was visually identified.
5.4 Equipment used to unload storage bin must be thoroughly cleaned and inspected prior to usage.	5.4.1	5.4.2 Grower must sign a document indicating that equipment used to unload storage bin was cleaned and inspected prior to usage.
6.0 Transportation		
6.1 Conveyance vehicles/ equipment must be thoroughly cleaned and inspected prior to loading. This must be done regardless if grower uses his/ her own equipment or uses custom trucking.	6.1.1 If possible, grower should try to arrange for hopper/ trucking equipment that has only been used recently to transport clean substances such as grain or food items. It is critical that all grain and meal residue is cleaned from the inside of the truck. Ideally the truck or hopper should be covered.	6.1.2 Grower must inspect truck and sign a document to authenticate that the truck/ hopper was cleaned prior to loading.
6.2 Trucker must present documentation verifying the IP soybean variety and name of the grower.	6.2.1 Trucker should be carrying a completed bill of lading. The producer, trucker, and receiver should sign the bill of lading. The trucker should also carry any additional documentation required by the receiving elevator.	6.2.2 Grower must fill out documentation for the trucker that identifies the IP soybean variety being delivered and the grower name.

(Continued)

TABLE 6.3   *(Continued)*

Minimum Level	Recommendations on Good Practice[a]	Documentation
7.0 Elevator Receiving		
7.1 Elevator must have an IP manual that details their full IP procedures for receiving, storage, processing, and loading.	7.1.1 All procedures should be described in detail. All relevant staff should be trained in IP procedures and have access to the manual for reference.	7.1.2 Manual must be available for inspection by auditing authority.
7.2 Incoming loads must be identified and verified as an IP crop or a non-IP crop. The crop must be identified as IP, SQWH, or crush. SQWH and crush soybeans are not qualified for IP certification. The crop is not unloaded as IP unless its identity is verified.	7.2.1 Receiving procedures should be detailed in IP manual.	7.2.2 Scale tickets for incoming loads must indicate variety name and unloading/ storage details for all crops.
7.3 Any non-IP loads that are received into the elevator must be tracked and accounted for.	7.3.1 Elevator should have detailed documentation showing which bins were used to store non-IP loads. Elevator should be able to show documentation demonstrating the end use for the non-IP soybeans.	7.3.2 Elevator must have detailed documentation for storage and tracking of non-IP loads that were received into the elevator.
7.4 Elevator must take a sample from each load of IP soybeans received.	7.4.1 If requested by grower, at time of delivery, the elevator should supply half of this sample for the grower to keep.	7.4.2 Elevator must retain documentation detailing variety name, moisture, weight, and grade details for each load.
7.5 Elevator pit/conveyor/legs must be thoroughly cleaned and inspected prior to receiving IP crops. Alternatively they could also be dedicated to a specific IP crop.	7.5.1 Cleaning procedures should be detailed in IP manual.	7.5.2 Elevator must have documentation to authenticate that pit/conveyor/legs have been cleaned and inspected prior to receiving a specific IP crop. Records must include the date and the name of the employee who conducted the inspection.

(Continued)

TABLE 6.3 *(Continued)*

Minimum Level	Recommendations on Good Practice[a]	Documentation
8.0 Elevator Storage		
8.1 Elevator must keep detailed storage history. Records must indicate what crop or variety was stored in their bin/silo prior to it being used to store an IP soybean crop.	8.1.1 Elevator should keep full records with crop type, variety name, and dates when bins were loaded, unloaded, and cleaned. All tonnage loaded and unloaded should be recorded.	8.1.2 Elevator must have detailed storage history records. Records must indicate what crop or variety was stored in their silo/bins prior to it being used to store an IP soybean crop.
8.2 Storage bins/silos must be thoroughly cleaned and inspected prior to loading with IP grain.	8.2.1 Cleaning procedures should be detailed in IP manual.	8.2.2 Elevator must have records documenting that silo was thoroughly cleaned and inspected prior to loading with IP grain. Records must include the date and the name of the employee who conducted the inspection.
8.3 Equipment used to load/unload bins and silos must be cleaned and inspected prior to being used for IP crop.	8.3.1 Cleaning procedures should be detailed in IP manual.	8.3.2 Elevator must have records documenting that all equipment used to load/unload bins and silos with IP soybean crop was cleaned and inspected prior to use. Records must include the date and the name of the employee who conducted the inspection
8.4 Elevator must identify all bins/silos that are used to store IP soybean variety. Bins used to store SQWH and crush soybeans must also be identified. All elevator staff should be aware of and have access to bin/silo designation.	8.4.1 Current elevator schematic should be available at pits and all other pertinent spots in elevator.	8.4.2 Elevator must have detailed bin and silo maps/schematics indicating which crop and variety is to be stored in each bin.
9.0 Processing		
9.1 Conveyors/augers/legs must be cleaned when transporting different IP varieties and different crops.	9.1.1 All transferring equipment should be shut down and cleaned prior to switching IP varieties, non-IP soybean varieties, or other crops. Cleaning procedures should be detailed in IP manual.	9.1.2 Elevator must have records showing that all transferring equipment was cleaned and inspected prior to processing IP crop. Records must include the date and the name of employee who conducted the inspection.

(Continued)

TABLE 6.3 *(Continued)*

Minimum Level	Recommendations on Good Practice[a]	Documentation
9.2 All processing equipment must be thoroughly cleaned and inspected prior to processing IP crop.	9.2.1 All processing equipment should be shut down and cleaned prior to switching IP varieties or to other crops. Cleaning procedures should be detailed in IP manual.	9.2.2 Elevator must have written records showing that all processing equipment was thoroughly cleaned and inspected prior to processing IP soybean crop. Records must include the date and name of the employee who conducted the inspection.
9.3 Elevator must have documentation detailing the flow of IP grain through the processing system.	9.3.1 Elevator should record tonnage when grain is transferred to different bins and the tonnage that is transferred to processing equipment.	9.3.1 Elevator must have written records detailing origin bin(s) used for unloading raw grain for processing and destination bins used for storing the processed grain. Any bin movements prior to processing must be recorded.
10.0 Loading		
10.1 All containers/vessels/trucks must be inspected and cleaned as required prior to loading.	10.1.1 Inspection/cleaning procedures should be detailed in IP manual. The IP manual should detail procedures for rejection of container/vessels/trucks if they are not suitable for food use.	10.1.2 Elevator or exporter must have written records showing that containers/vessels/trucks have been inspected and cleaned as required prior to loading with IP grain. Records must have date and the name of the employee who conducted the inspection.
10.2 Elevator must have documentation detailing the flow of IP grain handled through the elevator.	10.2.1 Elevator should record tonnage when grain is transferred to different bins and the tonnage that is unloaded from the elevator.	10.2.2 Elevator must have written records detailing bins/silos used for storing IP grain that has not been processed but has been stored and unloaded from the elevator.
10.3 Elevator must document grain that exits the elevator system.	10.3.1 Elevator should record loading details for all soybeans—IP and non-IP—moving through the elevator system.	10.3.2 Elevator must document and retain full records for all containers, trucks, and railcars loaded from the facility. Records must include container, truck, or railcar identification number, identification of the grain (IP varieties), and the quantity loaded. The bin that the grain has been loaded from must be recorded.

(Continued)

TABLE 6.3 *(Continued)*

Minimum Level	Recommendations on Good Practice[a]	Documentation
11.0 Audit Standards		
11.1 The grower must retain grower documentation unless requested by the elevator. Documentation must be retained for a minimum period subject to the requirements of the HACCP standard. Rule of thumb for length of time to keep HACCP records is 3 years.	11.1.1	11.1.2
11.2 Elevator/exporter must have retained records to support an annual audit.	11.1.1	11.1.2 Elevator/exporter must declare on their sales contracts if they are selling soybeans under the CSEA IP standard.
11.3 All documentation must be retained for a minimum period subject to the requirements of the HACCP standard. Rule of thumb for length of time to keep HACCP records is 3 years.	11.2.1	11.2.2
12.0 Nonconforming Product		
12.1 The elevator/exporter shall ensure procedures exist to investigate the cause of potential and actual nonconformity. Nonconforming product— includes any product that qualified as IP but because of adventitious or intentional mixing no longer meets IP requirements.	12.1.2 IP manual should detail how employees will inform the correct individual in the chain of command about nonconforming product.	12.1.3 The elevator/exporter must have a written protocol detailing how they will address a situation in which they have nonconforming product.
12.2 If the exporter has nonconforming product they must show in their documentation that they have a procedure to address the situation. This must include either documentation for disposal, customer acceptance, or alternate non-IP sales arrangements.	12.1.2 The elevator/exporter should develop a corrective action procedure.	12.1.3 The exporter must have documentation showing that the nonconforming product has either been disposed of, that the customer has been informed and accepted the nonconformance, or that alternate non-IP sales arrangements were made.

[a] Bold indicates emphasis added.

Source: Reprinted with permission from CSEA.

- Voluntary
- Includes all stakeholders in the food supply chain (primary producers, processors, distributors, retailers, intermediaries, government, and consumers)
- Includes all commodity groups

The objective of Can-Trace is to define and develop minimum information require-ments for national whole-chain all-product traceability standard on the basis of the globally recognized EAN.UCC System* (see Chapter 10 on software providers GS1 EAN.UCC for more information). Specifically, this voluntary standard establishes the minimum data elements required to be collected, kept, and shared between trad-ing partners. *The key point with Can-Trace* is that it must be internationally com-patible, whole-chain in scope, capable of accommodating multiple commodities, and flexible enough to enable integration and leveraging of other systems. The EAN. UCC system together with ISO formed the foundation for Can-Trace standards.

Integration may be done within HACCP (Hazard Analysis Critical Control Points) and "HACCP-based systems." Many participants believe that traceability works best as part of an on-farm food safety program. It was noted that consumers are more concerned about nutrition and food safety than traceability, that traceability is a way for companies to support labeling claims, and that traceability requires buy-in at all levels of the supply chain.

Third-party verification may be done through integration into existing programs or systems similar to HACCP. *Can-Trace does not require verification*, but some buyers may require verification. The system is industry driven, and industry will ensure that accurate records are kept. *Can-Trace has noted that there are no plans at this time to take traceability to the consumer and that traceability will end at the back door of retail or food service.*

Can-Trace's minimum requirements leverage existing data capture and management systems when implementing a traceability program. Some Canadian primary-producer food manufacturers, processors, distributors, and retailers already have significant invest-ment in product identification schemas and information technology systems. Identified systems leverage these investments to control cost and speed implementation.

The decision to focus on beef, pork, produce, and seafood as a first priority was the result of input received from industry and governments during public consulta-tions held across Canada in late 2003. The basic traceability data elements common to these four commodities (beef, pork, produce, and seafood), referred to as "manda-tory" data elements, will likely apply to all foods. What may be added in the future as a result of food industry experience with implementing this standard are "optional" elements that are specific to a particular food. The current focus of working groups

---

* The EAN.UCC (European Article Number. Universal Code Council) System, which is used world-wide, standardizes bar codes, EDI transactions sets, XML schemas, and other supply chain solutions for more efficient business. GS1 is the custodian of these standards. The issue of how this standard will be implemented in a business setting or in a particular food sector falls outside of the current mandate of Can-Trace. See the Can-Trace website for updates on a companion document being developed that will provide guidance to users as to how the various data elements should be used in documents and physical markings.

is on single-ingredient products. The long-term objective is to develop minimum data requirements for all commodities and multi-ingredient products produced and sold in Canada.

Another important note: Traceability requirements of primary producers and of their raw material providers have not been included within the scope of this standard at this time. For example, a primary producer may receive inputs such as fertilizer, herbicides, feed, and biologicals that contribute to the growing/raising of a commodity. These traceability requirements are in the process of being covered by on-farm food safety and quality programs. No assumption should be made that the exclusion of raw material providers from this release reflects a lack of recognition of their importance within the supply chain.

6.10.1 CAN-TRACE DRIVERS

Traceability has come to the forefront of public discussion in the agri-food sector in recent months for several reasons:

- International market pressures from trading partners
- Regulatory programs in Canada in the beef sector at both the federal level and in the province of Quebec
- The APF in Canada, an initiative of the federal, provincial, and territorial governments to establish food traceability targets
- Legislation and regulations in the United States and Europe concerning both animal health, security, and food safety

As a result of these and other factors, more companies and organizations have begun to develop traceability systems for their particular sector or supply chain requirements. However, without the benefit of a national or international standard for food traceability, such efforts are proprietary and do not necessarily cover the depth and breadth of the entire supply chain.

The food industry has realized that significant benefits could be derived from a single national traceability data standard, such as minimizing the cost of a food recall for all components of a supply chain, support for food quality programs, and supply chain improvement. Until the Can-Trace initiative got underway, no such standard existed. In an increasingly competitive economy, the industry was not willing to continue supporting multiple systems or standards for traceability.

The Can-Trace Standards Working Group developed this standard. Their mandate was to develop the minimum information (or data) requirements that need to be "collected, kept, and shared" at each "hand off" point in the supply chain to establish traceability.

6.10.2 CAN-TRACE TECHNOLOGY GUIDELINES

The Can-Trace technology and standards take into consideration that for companies selling to mass merchants and grocers, the future is now. E-commerce has brought revolutionary changes to the way business is conducted. Every traditional business process has been impacted. The business areas affected include but are not limited to data synchronization, data communication, and product identifier (RFID, UPC, GTIN, etc.) changes. Large and small businesses are being affected. Implementation of these technologies, some would argue, is a key to long-term competitiveness. A major goal of Can-Trace is to identify how physical markings and documents (paper-based or electronic) can be used to capture and communicates this through the various data elements in the Can-Trace Canadian Food Traceability Data Standard (also referred to as the CFTDS), version 2.0.

6.10.2.1   Key Assumptions and Methodology

The Can-Trace standard requires participants in the food supply chain (primary producers, processors, wholesalers/distributors, and retail stores/food service operators), where appropriate, to keep on record, share, and collect from other trading partners certain minimum data elements to enable whole-chain traceability on the basis of a one-up/one-down model. Data need to be synchronized between partners through the supportive technologies, physical markings and document exchange that will be reviewed in this report.

6.10.2.2   Recommendations Regarding Supporting Documents

Figure 6.2 illustrates the supporting documentation (physical or electronic) that is used to store and communicate the specific data elements between participants in the supply chain. For example, an advanced shipping notice (ASN) document would carry almost all of the recommended Can-Trace mandatory data elements. The purpose of the chart is to identify which documents carry which particular information. For example the Shipping/Transportation Document and Receiving Confirmation/ Exceptions must carry receiver identifier, lot number, product description, product identifier, quantity, shipment identifier, unit of measure, and sender identifier to ensure an accurate exchange of traceability information between trading partners and, thus, a completely traceable food supply chain.* In addition, the item set-up transaction must carry receiver identifier, product description, product identifier and sender identifier information.

Although the CFTDS presents what minimum information is necessary to move between trading partners to ensure traceability within the food supply chain, the report for Technology Guidelines (see website) is an attempt to present how that information should be exchanged between partners as it applies to physical markings and supporting documents. The report is intended to provide guidance to those interested in establishing traceability systems based on the Can-Trace standard, or those who currently have such a system in place (Figure 6.2).

6.10.3   CFTDS VERSION 2.0

This standard defines the minimum data that are needed to support a one-up/one-down traceability model. Under a one-up/one-down system, each participant within

---

* For the purpose of establishing technical consistency, lot number, product description, product identifier, and an identifier of the sender were deemed the minimum mandatory physical markings for trade units.

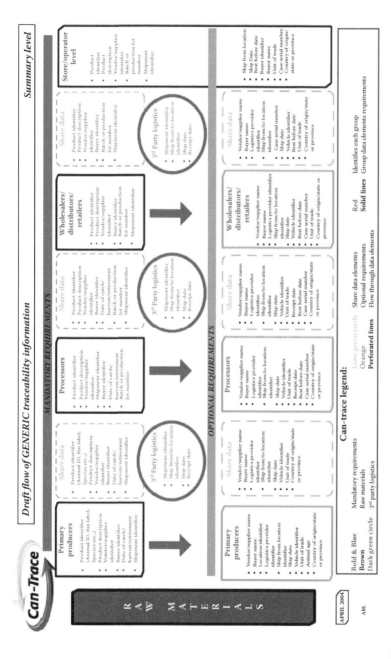

FIGURE 6.2 Draft flow of GENERIC traceability information. Available at http://www.can-trace.org/REPORTS/tabid/81/Default.aspx. (Reprinted with permission from Can-Trace.)

the food supply chain is responsible for maintaining records about the products they receive, their use (i.e., the link between inputs and outputs), and where they were shipped to or sold. The CFTDS addresses information flowing from the primary producer end of the supply chain up to delivery to the back door of the retail or food-service operation. The store shelf or end consumer is therefore beyond the scope of this standard.

6.10.3.1 Principles

CFTDS was developed based on the following principles:

- The standard is voluntary.
- The standard is "whole-chain" in its applicability.
- The standard references data requirements, not technology or systems specifications.
- The data standard is based on global standards (GS1 and ISO).
- The standard is not meant to replace existing systems but to complement them.

6.10.3.2 Important Considerations

CFTDS is not a technology standard. This is a standard that sets out the minimum information or data elements needed to effectively track and trace food products for a variety of food safety, quality, and supply chain improvement applications. The Can-Trace Standard applies to both domestic and imported products.

To be most effective, a traceability program for an organization should be integrated into existing business systems, logistical processes, quality programs, and food safety programs such as HACCP. This standard provides the basis upon which to build the traceability component.

Effective tracking and tracing require the linking of information and product flow. This linkage is necessary so that a product may be tracked from point of origin to the back door of the retail store or foodservice operator. Conversely, this linkage also ensures that product can be traced back through the supply chain.

In a one-up/one-down model, no single supply chain partner holds all of the information. Each partner keeps information regarding production inputs and needs to keep and share information regarding production outputs.

6.10.3.3 Important Definitions

- *Traceability:* Can-Trace uses the ISO definition of traceability (which appears in ISO 9000/2000): "Traceability is the ability to trace the history, application, or location of that which is under consideration." For additional clarity, Can-Trace further defines traceability as being composed of two components: tracking and tracing.
- *Tracking:* Tracking is the ability to follow the path of a specified unit and/or lot of trade items downstream through the supply chain as it moves between trading partners. Trade items are tracked routinely for availability, inventory management, and logistical purposes. In the context of this standard, the focus is on tracking items from the point of origin to the point of use.

- *Tracing:* Tracing is the ability to identify the origin of a particular unit located within the supply chain by reference to records held upstream in the supply chain. Units are traced for purposes such as recall and complaints.
- *Lot number:* A lot number is a number or code assigned to uniquely represent a batch or group of inputs, products, animals, and/or outputs.* The company or individual creating the goods generally assigns the number.
- *Supply chain:* A set of approaches utilized to efficiently integrate suppliers and clients (comprised of stores, retailers, wholesalers, warehouses, and manufacturers) so food products are produced and distributed in the right quantities, to the right locations, and at the right time to minimize system-wide costs while satisfying service-level requirements.
- *Supply chain roles:* By the time a product has moved from the grower to the retail store level, that product may have gone through several transformations. Each transformation will have involved several different role players. Every role player has a responsibility to collect, keep, and share information to enable one-up/one-down traceability.
- *Primary producer:* The primary producer may be the farmer, fisherman, or grower.
- *Processor:* The processor typically receives input from a primary producer and transforms that product. Examples of processors include a slaughterhouse (abattoir) or a packer that consolidates produce from several growers. A food supply chain may comprise more than one processor.
- *Carrier/third-party transporter:* The carrier or third-party transporter is responsible for the handling or delivery of product.
- *Wholesaler/distributor:* The wholesaler or distributor provides raw or finished product such as fresh fish or meat to the retailer. The retailer then distributes to each individual store.
- *Retail/store/foodservice operator:* The store and foodservice operator have the final relationship with the consumer. The foodservice operator may be an individual restaurant, an extended-care facility, healthcare provider, or hospitality service such as a hotel chain.

Each of the above roles in the supply chain needs to keep or share the mandatory elements and, depending on requirements of their sector, may need to keep and share some of the optional elements.

- *Mandatory and optional data elements:* The data that must be exchanged between trading partners to accomplish traceability are critical. It should be noted that although the CFTDS is a voluntary standard, compliance with the standard mandates the use of twelve data elements, hence the term "mandatory data elements," which are the minimum required to establish traceability. Optional data elements include data that may be used in addition to the

---

* A lot is defined as a set of units of a product that has been produced, processed, and/or packaged under similar circumstances. Note: The lot is determined by parameters established beforehand by the organization.

mandatory data. These data elements can support other business objectives such as food quality or marketing programs, but they are not essential to establishing traceability.

- *Production inputs:* These are the products/trade units that are received by a trading partner in the food supply chain. Because the scope of Can-Trace does not include agricultural inputs (e.g., fertilizers, feeds, etc.) production inputs at the level of primary production are limited to the animals, plants, or their products that are produced at that level. It is critical for traceability that the link between input and output be recorded and kept.
- *Production outputs:* The products/trade units that have been produced and/or shipped from a trading partner in the food supply chain and may include animals (including fish) plants, and their products as well as foods produced from these products/trade units. Again, it is critical for traceability that the link between input and output be recorded and kept.

6.10.3.4 Basic Elements of Traceability

- *Product, party, and location identification:* To track and trace a product through the whole supply chain, every raw material harvested from farm or sea and every food product moving from one level to another in the chain must be uniquely identified. Each role in the life of the product must also be uniquely identified. There are many ways to assign identifiers.
- *Linking of information:* To ensure the continuity of the flow of traceability information, each trading partner must pass on information about the identified lot or product group to the next partner in the production chain.[*]
- *Recording of information:* Effective traceability requires each role to record and archive data at each step of the supply chain.
- *Sharing of information:* To ensure the continuity of the flow of traceability information, each stakeholder must pass on information about the identified product, party, or location to the relevant member in the supply chain.
- *Data types:* There are two types of data required for traceability: master and transactional data. *Master data* seldom changes and applies to product, party, and location data. Examples include product description, receiver identifier, location, etc. *Transactional data* are unique to each individual transaction such as lot number and shipment date.

---

[*] The information necessary for traceability is classified as collect data, keep data, and share data (data storage and data exchange can be a direct function of trading partners or may be managed indirectly through a third party). It is imperative that the links between the received and the processed products and between the processed and the shipped products (resulting from a product transformation) are recorded. Within a company, the control of all of these links and accurate recordkeeping make it possible to connect what (information and products) has been received (production inputs) and what (information and products) has been produced and/or shipped (production outputs).

- *Generic mandatory data requirements:* There is no need to duplicate existing records for traceability. For example, a shipment identifier serves as a reference to other data elements such as ship-from location identifier, ship-to location identifier, receipt date, and ship date, etc.
- *Generic optional data elements:* Depending on the commodity, these are the generic optional data elements. *Note:* This list provides some examples of optional data elements and is not an exhaustive list; certain sectors and/or programs may have additional requirements that are not listed here.

Optional data elements include:

- Animal age (beef)
- Best-before date
- Receiver name
- Contact information
- Sender name
- Supplier license number (seafood—this is mandatory at the primary-producer level)
- Logistics provider identifier
- Shipping container serial number
- Vehicle identifier
- Date of pack/harvest/catch/retirement
- Country, province, or state or origin

6.11   INTRODUCTION TO EUROPEAN STANDARDS

This following highlights two prominent European standards, European Commission (EC) Standards and EurepGap, including organizational/agency statements from their websites, and thus naturally reflects their views.

The key EU regulations include Regulation (EC) No. 178/2002, which established the European Food Safety Authority (EFSA), from which Regulation (EC) 1829/2003 (concerning genetically modified food and feed) and Regulation (EC) 1830/2003 (concerning the traceability and labeling of genetically modified organisms (GMOs) and the traceability of food and feed products produced from GMOs) and its amending Directive 2001/18/EC are derived. At the end of this section are other food issue concerns that are addressed by labeling.

6.11.1   INTERPLAY BETWEEN REGULATION (EC) No. 1829/2003 & (EC) No. 1830/2003

Regulation (EC) No. 1829/2003 on genetically modified food and feed was developed alongside (EC) No. 1830/2003 on traceability and labeling of GMOs and the traceability of food and feed products produced from GMOs, both on September 22, 2003. The two regulations are intended to operate in tandem and rely on each other for certain requirements. Notably, the regulation provides traceability requirements for all food and feed products that fall under the scope of Regulation (EC) No. 1829/2003. These traceability requirements are of fundamental importance when labeling of the final product relies on information transmission in the absence of detectable genetically modified (GM) material in products.

Similarly, the labeling requirements for food and feed products produced from GMOs, subject to the traceability requirements under Article 5 of the regulation, are provided for by Chapter II, Section II and Chapter III, Section II of Regulation (EC) No. 1829/2003. In addition, Regulation (EC) No. 1829/2003 lays down threshold values for food and feed products below which adventitious traces of such products are exempted from its labeling requirements. The same thresholds have been utilized by the regulation to provide the same exemption from its own labeling and traceability requirements, ensuring a coherent and consistent EC approach.

6.11.2   EurepGAP

EurepGAP offers a variety of services, although it is somewhat more restrictive in its scope regarding the food supply chain. Most prominent are their Integrated Farm Assurance (IFA) Program and Farm Assurance Schemes (schemes include food safety; environmental protection; occupational health, safety, and welfare; and animal welfare where applicable). These are primarily on-farm systems that farmers follow to meet prescribed demands of their suppliers. They also offer systems that utilize a benchmarking framework to achieve certification on the basis of ISO Guide 65. EurepGAP also accredits bodies to conduct accreditation for its benchmarking procedures.

6.12   EUROPEAN UNION STANDARDS

Office of the EC
Commission of the European Community
200 Rue de la Loi
B-1049 Brussels
Belgium
Phone: 0032-2-235-1111

Available at http://europa.eu.

The European Union (EU; founded in November 1993 by the Maastricht Treaty[*]) is a union of over 25 independent states (also known as an intergovernmental and supranational union) based on the European Communities to enhance political, economic, and social cooperation. Prior to this, the organization was formerly known as the EC or the European Economic Community (EEC).[†]

---

[*] This was formally known as the Treaty on EU.
[†] The treaty led to the creation of the Euro and introduced the three-pillar structure (the Economic and Social Policy pillar; the Common Foreign and Security Policy, or CFSP pillar; and the Justice and Home Affairs pillar).

For centuries, Europe was the scene of frequent and bloody wars. Several European leaders became convinced that the only way to secure a lasting peace between their countries was to unite them economically and politically. So in 1950, several countries began integrating the coal and steel industries of Western Europe. As a result, in 1951, the European Coal and Steel Community (ECSC) was set up with six members: Belgium, West Germany, Luxembourg, France, Italy, and the Netherlands. The power to make decisions about the coal and steel industry in these countries was placed in the hands of an independent, supranational body.

The ECSC was such a success that, within a few years, these same six countries decided to go further and integrate other sectors of their economies. In 1957, they signed the Treaties of Rome, creating the European Atomic Energy Community (EURATOM) and the EEC. The member states set about removing trade barriers between them and forming a "common market."

The EU currently has a common single market consisting of a customs union, a single currency managed by the European Central Bank (so far adopted by 12 of the 25 member states), the Common Agricultural Policy, a common trade policy, and a Common Fisheries Policy.

The institutions of the EU include the following:

- Council of Ministers
- European Parliament
- EC
- European Court of Justice

6.12.1 THE EC

The EC is the institution responsible for ensuring that the measures in the treaties are carried out. The EC has a relatively small administrative staff, based mainly in Brussels, which is divided into Directorates-General (DGs). Each DG covers a particular subject area. The duties of the EC include administering EU funds and investigating complaints of breaches of EU laws by member states.

The EU, as a major global trader of food and feed, has entered into international trade agreements and contributed to the development of international standards that underpin food law. It also supports the principles of free trade in safe food and feed following fair and ethical trading practices. This is of enormous importance to citizens in Europe and around the world whether they are politicians, traders, or consumers.

6.12.1.1 Integration Means Common Policies

Economic and political integration between the member states of the EU means that these countries have to make joint decisions on many matters. So they have developed common policies in a very wide range of fields: from agriculture to culture, from consumer affairs to competition, and from the environment and energy to transport and trade. *Note:* The aim of the agricultural policy is no longer to produce as much food as cheaply as possible, driven by postwar scarcity, but to support farming methods that produce healthy, high-quality food and protect the environment.

The need for environmental protection is now taken into account across the whole range of EU policies.

6.12.2   EU Notion of Food Safety

Every European citizen is entitled to a varied diet of safe and wholesome food. Citizens are entitled to clear and accurate information on the composition, manufacturing processes, and use of foodstuffs. With a view to guaranteeing a high level of public health, the EU and its member states have placed food safety high up on the European political agenda. The EU's involvement is nevertheless focused more directly on the key areas of the Common Agricultural Policy (CAP), internal markets, the protection of consumers, public health, and measures to protect the environment.

The EU is second only to the United States as a global exporter of agricultural products. With more than 370 million consumers, the European market is one of the largest in the world and will grow even more with the enlargement toward the countries of Central and Eastern Europe. In the wake of the food-related crises experienced during the 1990s, the EC has become aware of the need to establish and enforce stricter safety standards across the entire food chain. The EU's White Paper on food safety, published in January 2000, introduced a more preventive policy to deal with potential food-related risks and to improve, at the European level, the capacity for reacting rapidly to any emerging risk.

6.12.2.1   Background: CAP and Consumer Protection Policy

Originally devised to reduce the shortages of the post-war period, the CAP went into effect in 1962 with the primary objective of ensuring food self-sufficiency for Europe's citizens. In the 1970s, this objective was attained and even exceeded for most agricultural products. The emphasis on high productivity in the agricultural sector and the food industry has shifted toward greater concern for satisfying the needs and requirements of consumers with regards to the safety and quality of products.

6.12.2.2   1990: Food Crises Mark a Turning Point

The food crises of the 1990s, such as bovine spongioform encephalopathy (BSE, or mad cow disease), marked a turning point in the policy of consumer protection and food safety. The crises highlighted the limitations of EU legislation and caused a strong reaction on the part of the public authorities. The adoption of sectoral directives had resulted in differing approaches and levels of application in the member states, with legal gaps remaining unfilled in some areas.

6.12.2.3   A New Departure: The White Paper on Food Safety

The public debate triggered by the White Paper led to the groundbreaking move toward the complete overhaul of legislation in this area. The EC announced the development of a legal framework covering the entire food chain, "from the farm to the fork," through a comprehensive and integrated approach, with a provision being made for the creation of the EFSA. With a view to creating true uniformity throughout the EU, the White Paper emphasized the need for greater harmonization of national control systems extended to the external borders of the EU with an eye to its forthcoming enlargement. It also advocated regular dialogue with consumers and professionals to restore confidence on both sides. Lastly, the White Paper stressed the need to provide citizens with clear and accurate information on the quality, potential risks, and composition of foods.

Adopted at the end of January 2002, Regulation (EC) No. 178/2002 is the linchpin of the new legislation governing food safety, forming the basis of the new approach. It formally establishes the EFSA along with a standing committee on food chain and animal health to replace the eight existing committees.[*] Moreover, with a view to restoring confidence, the EU's consumer protection policy will place stronger emphasis on the harmonization of national laws. Lastly, the process of recasting food safety legislation will give rise to benchmark legal texts focusing on all of the areas of activity connected with food safety.

The general objectives of food safety policy are:

- To ensure a high level of protection of human and animal health by means of increased controls throughout the food chain.
- To place quality at the forefront of concerns. The concept of quality as an intrinsic element of food safety comprises two aspects: (1) non-negotiable quality in terms of the safety of the food we eat and minimum requirements for protecting the environment and animal and plant species; and (2) relative or subjective quality making a foodstuff truly unique as a result of taste, appearance, smell, production methods, and ease of use.
- To restore the confidence of consumers. To this end, the safety of foodstuffs is enhanced through stricter monitoring and control procedures, with the further requirement that consumers be given clear and accurate information on all aspects of food safety. The EC conformity marking and specific elements such as the eco-label or protected geographical indications and designations of origin are among the initiatives placing quality, consumer protection, and the defense of traditional production methods at the center of concerns.

6.12.3 EC/178/2002—Procedures for Food Safety

6.12.3.1 Summary of Why This Was Done

The White Paper on food safety emphasized the need for a policy underpinned by a sound scientific basis and up-to-date legislation. The general overhaul of EU legislation is designed to restore consumer confidence in the wake of recent food-related crises, with all of the interested parties having a part to play: the general public, nongovernmental organizations, professional associations, trading partners, and

---

[*] The rapid alert system for human food and animal feed is reinforced. The EC has special powers allowing it to take emergency action when the member states alone are unable to contain a serious risk to human or animal health or the environment.

international trade organizations. Ultimately, to the goal was to define at the EU level a common basis for measures governing human food and animal feed.

With a view to adopting a comprehensive, integrated "farm to table" approach, legislation covers all aspects of the food production chain: primary production, processing, transport, and distribution through to the sale or supply of food and feed. At all stages of this chain, the legal responsibility for ensuring the safety of foodstuffs rests with the operator. A similar system applies to feed business operators.

6.12.3.2 EFSA

EFSA enhances the current scientific and technical support system. Its main task is to provide assistance and independent scientific advice and to create a network geared to close cooperation with similar bodies in the member states. It assesses risks relating to the food chain and will inform the general public accordingly. EFSA ("the Authority") provides scientific advice and scientific and technical support in all areas impacting on food safety. It constitutes an independent source of information on all matters in this field and ensures that the general public is kept informed.[*]

To operate effectively, the EFSA has been entrusted with six key tasks.

1. To provide independent scientific advice on food safety and other related matters such as animal health and welfare, plant health, GMOs, and nutrition
2. To give opinions on technical food issues to shape policies and legislation relating to the food chain
3. To collect and analyze information on any potential risk and data on dietary exposure to control and monitor safety throughout the food chain
4. To identify and give warning of emerging risks as early as possible
5. To assist the EC in emergencies by providing scientific advice within ad hoc crisis management units
6. To establish a permanent dialogue with the general public and inform it of potential or emerging risks

6.12.4 GENERAL FOOD LAW—TRACEABILITY REGULATION EC/178/2002

The identification of the origin of feed and food ingredients and food sources is of prime importance for the protection of consumers, particularly when products are found to be faulty. Traceability facilitates the withdrawal of foods and enables consumers to be provided with targeted and accurate information concerning implicated products.

Regulation EC/178/2002 defines traceability as the ability to trace and follow food, feed, and ingredients through all stages of production, processing, and distribution. The regulation contains general provisions for traceability (applicable from January 1, 2005) that cover all food and feed and all food and feed business operators

---

[*] The Authority is endowed with legal personality. The Court of Justice of the European Communities has jurisdiction in any dispute relating to contractual liability. The General Principles of Food Law (Articles 5 to 10) entered into force on February 21, 2002 and must be followed when measures are taken. Existing food law principles and procedures must be adapted by January 1, 2007 in order to comply with the general framework established by Regulation EC/178/2002.

without prejudice to existing legislation on specific sectors such as beef, fish, GMOs, etc. Importers are similarly affected because they will be required to identify from whom the product was exported in the country of origin. Unless specific provisions for further traceability exist, the requirement for traceability is limited to ensuring that businesses are at least able to identify the *immediate supplier* of the product in question and the *immediate subsequent recipient*, with the exemption of retailers to final consumers (one step back, one step forward).

Member states must develop effective monitoring systems and are required to establish measures and penalties for contraventions of the regulation. Member states are also expected to pay attention to international food safety standards in their national policies and to support international processes to develop further rules on food and feed safety. The main source of international food safety standards is the Codex Alimentarius. The preamble of the regulation recognizes that it may take time for states to adapt their food laws, and Article 4.3 gave them until January 1, 2007 to do so.[*]

6.12.4.1   Guiding Influence

Under Regulation 178/2002, food may not be placed on the market that is: "(a) injurious to health; (b) unfit for human consumption" (Article 14.2). And feed may not be placed on the market that may: "have an adverse effect on human or animal health; make the food derived from food producing animals unsafe for human consumption" (Article 15.2).

6.12.4.2   Risk Analysis

The regulation establishes the principles of risk analysis in relation to food and establishes the structures and mechanisms for the scientific and technical evaluations that are undertaken by the EFSA.[†]

---

[*] The EU or its constituent states are also members of other international organizations, the task of which is to promote animal health or food safety through international trade. The most important are the Codex Alimentarius, a Rome-based body under the auspices of the U.N. Food and Agriculture Organization (FAO) and the Office *International* des Épizooties (OIE) based in Paris.

[†] Depending on the nature of the measure, food law (and in particular measures relating to food safety) must be underpinned by strong science. The EU has been at the forefront of the development of the risk analysis principles and their subsequent international acceptance. Regulation EC 178/2002 establishes in EU law that the three interrelated components of risk analysis (risk assessment, risk management, and risk communication) provide the basis for food law as appropriate to the measure under consideration. Clearly not all food law has a scientific basis (e.g., food law relating to consumer information or the prevention of misleading practices does not need a scientific foundation). Scientific assessment of risk must be undertaken in an independent, objective, and transparent manner on the basis of the best available science. Risk management is the process of weighing policy alternatives in the light of results of a risk assessment and, if required, selecting the appropriate actions necessary to prevent, reduce, or eliminate the risk to ensure the high level of health protection determined as appropriate. In the risk management phase, the decision-makers need to consider a range of information in addition to the scientific risk assessment. For example, these include the feasibility of controlling a risk, the most effective risk reduction actions depending on the part of the food supply chain where the problem occurs, the practical arrangements needed, the socioeconomic effects, and the environmental impact. Regulation EC/178/2002 establishes the principle that risk management actions are not just based on a scientific assessment of risk but also take into consideration a wide range of other factors legitimate to the matter under consideration.

6.12.4.3   Transparency

Food safety and the protection of consumer interests are of increasing concern to the general public, nongovernmental organizations, professional associations, international trading partners, and trade organizations. Therefore, the regulation establishes a framework for the greater involvement of stakeholders at all stages in the development of food law and establishes the mechanisms necessary to increase consumer confidence in food law.

This consumer confidence is an essential outcome of a successful food policy and is therefore a primary goal of EU action related to food. Transparency of legislation and effective public consultation are essential elements of building this greater confidence. Better communication about food safety and the evaluation and explanation of potential risks, including full transparency of scientific opinions, are of key importance.

6.12.4.4   General Food Law—Precautionary Principle

Regulation EC/178/2002 (Article 7) formally establishes the Precautionary Principle as an option open to risk managers when decisions have to be made to protect health, but scientific information concerning the risk is inconclusive or incomplete in some way.[*]

The Precautionary Principle is relevant in those circumstances in which risk managers have identified that there are reasonable grounds for concern that an unacceptable level of risk to health exists, but the supporting information and data may not be sufficiently complete to enable a comprehensive risk assessment to be made. When faced with these specific circumstances, decision-makers or risk managers may take measures or other actions to protect health on the basis of the Precautionary Principle while seeking more complete scientific and other data. Such measures have to comply with the normal principles of nondiscrimination and proportionality and should be considered as provisional until such time that more comprehensive information concerning the risk can be gathered and analyzed.

Regulation (EC) No. 178/2002 lays down five general principles:

1. *The food chain as a whole must be taken into consideration.* It is vital that a high level of food safety be ensured at all stages of the food chain, from primary production through to the consumer, in the interest of overall effectiveness.
2. *Risk analysis is a fundamental component of food safety policy.* Three separate procedures are necessary: risk assessment based on scientific evidence, risk management through the intervention of public authorities, and the provision of information to the general public on any risks. *If the available scientific data are not sufficient to evaluate the risk fully, the*

---

[*] General Food Law Procedures—Regulation EC/178/2002 has different procedures in matters of food safety. In particular, it provides for the creation of the Rapid Alert System for Food and Feed (RASFF) and the adoption of emergency procedures, crisis management, a regulatory committee, and modus operandi (see http://europa.eu.int/eur- lex/pri/en/oj/dat/2002/l_031/l_03120020201en00010024.pdf for more information).

application of the Precautionary Principle is desirable for the purpose of ensuring a high level of protection.

3. *Responsibility now lies with all operators in the food sector.* All operators in the sector are responsible for the safety of the products that they import, produce, process, and place on the market or distribute. If a risk arises, the operator concerned must take the necessary restrictive measures without delay and inform the authorities accordingly.

4. *Products must be traceable at all stages of the food chain.* Using appropriate systems for collecting information, operators must be able to identify any person or business supplying them with a foodstuff or to whom they supply their products.

5. *Citizens are entitled to clear and accurate information from the public authorities.* They should be consulted openly and transparently throughout the decision-making process. This approach ties in with the principles of EU consumer policy recognizing people's right to information, education, and representation.

6.12.4.5 General Obligations in the Food Trade

Food and feed imported with a view to being placed on the market or exported to a third country must comply with the relevant requirements of EU food law.

6.12.4.6 General Requirements of Food Law

Food must not be placed on the market if it is unsafe (i.e., if it is harmful to health and/or unfit for consumption). Feed must not be placed on the market or given to any food-producing animal if it is unsafe. At all stages of the food production chain, business operators must ensure that food and feed satisfies the requirements of food law and that those requirements are being adhered to.

6.12.4.7 Essential

The traceability of food, feed, food-producing animals, and all substances incorporated into foodstuffs must be established at all stages of production, processing, and distribution. To this end, business operators are required to apply appropriate systems and procedures.

6.12.4.8 Important Legislation for GM Food and Feed, and Traceability and Labeling of GMOs (and Their Products)

Regulation (EC) 1829/2003 is of the European Parliament and of the Council of September 22, 2003 on GM food and feed. Regulation (EC) 1830/2003 of the European Parliament and of the Council of September 22, 2003 concerns the traceability and labeling of GMOs, the traceability of food and feed products produced from GMOs, and amends Directive 2001/18/EC. Regulation (EC) 65/2004 of January 14, 2004 establishes a system for the development and assignment of unique identifiers for GMOs. Regulation (EC) 641/2004 of April 6, 2004 details rules for the implementation of Regulation (EC) 1829/2003 of the European Parliament and of the Council with regards to the application for the authorization of new GM food and feed, the notification of existing products, and the adventitious or technically unavoidable presence of GM material that has benefited from a favorable risk evaluation. Directive 2001/18/EC of the European Parliament and of the Council of March 12, 2001 is on the deliberate release into the environment of GMOs and repeals Council Directive 90/220/EEC.

6.12.5 Summary of Provisions for Regulation (EC) No. 1829/2003

This regulation aims to harmonize food safety rules and procedures across the EU to

- Promote free trade in the internal market
- Protect human, animal, and plant health
- Protect the environment
- Protect consumers' interests

Regulation (EC) No. 1829/2003 aims to harmonize national rules on GM food and feed. It established a common EU marketing authorization procedure and outlines labeling requirements; labeling will assist consumers in making informed choices. The authorization procedure includes safety assessments for the protection of human and animal health and the environment.

6.12.5.1 Principles of Regulation (EC) No. 1829/2003

The regulation stipulates that the products to which it applies must not

- Have adverse effects on human health, animal health, or the environment
- Mislead the consumer or user
- Differ from the food/feed they are intended to replace to such an extent that their normal consumption would be nutritionally disadvantageous for human beings (and for animals in the case of GM feed)
- In the case of GM food and feed, harm or mislead the consumer by impairing the distinctive features of the animal products

The regulation puts in place a centralized, uniform, and transparent EU procedure for all applications for placing on the market, whether they concern the GMO itself or the food and feed products derived from them.

This means that business operators may file a single application for the GMO and all of its uses; a single risk assessment is performed and a single authorization is granted for a GMO and all of its uses (cultivation, importation, processing into food/feed or industrial products). If one of these uses concerns food, all of the uses (cultivation, processing into industrial products, etc.) may be treated under Regulation (EC) No. 1829/2003.*

---

* Although risk assessment of food and feed is to be primarily based on scientific evidence, societal, economic, ethical, and cultural factors may also be taken into account. The regulation also incorporates the precautionary principle, allowing states to take action to protect public health when scientific uncertainty remains about risk. Article 7 Precautionary Principle: In specific circumstances in which, following an assessment of available information, the possibility of harmful effects on health is identified but scientific uncertainty persists, provisional risk management measures necessary to ensure the high level of health protection chosen in the EC may be adopted pending further scientific information for a more comprehensive risk assessment.

The principle of traceability is extremely important and is to be applied at all stages of the food chain. This includes food and feed business operators keeping records of who supplied the product and whom it is subsequently sold to, and the requirement of accurate food labeling throughout the food chain. Labeling and packaging must not mislead consumers. The rules are to apply equally to food being exported from and imported into the EU. Under Regulation 178/2002, emergency measures can be taken to stop unsafe products from reaching the market, or to remove unsafe products from the market.*

Regulation (EC) No. 1829/2003 requires that all foods containing, consisting of, or produced from GMOs must be labeled as GM. These labeling provisions are closely connected to the requirements of Regulation (EC) No. 1830/2003 *concerning the traceability and labeling of GMOs and the traceability of food and feed products produced from GMOs* and amending *Directive 2001/18/EC.*

Very similar authorization procedures and labeling requirements apply for feed containing, consisting of, or produced from, GMOs. For both food and feed, a threshold is set at 0.9% for an allowable presence of "adventitious or technically unavoidable" traces of approved GMOs.

6.12.5.2   Feed Additives

GM feed additives must comply with the provisions of Regulation (EC) No. 1831/2003 *on additives for use in animal nutrition* in addition to the authorization procedures of Regulation 1829/2003.† *Note:* Before the entry into force of the regulation on GM food and feed, there was no EC legislation governing feed derived from GMOs. Feed containing GMOs or consisting of such organisms was subject to Directive 90/220/EEC. Hence, several GMOs have been authorized as products containing GMOs or consisting of such organisms for use in feed, in accordance with Directive 90/220/EEC; these are chiefly maize varieties, rape varieties, and one soya variety.‡

6.12.6   SUMMARY OF PROVISIONS FOR REGULATION (EC) No. 1830/2003

Regulation (EC) No. 1830/2003 involves the traceability and labeling of GMOs and the traceability of food and feed products produced from GMOs. The labeling and traceability requirements of the regulation extend to products that are placed on the

---

* These emergency measures were applied to imports of GM corn gluten feed and brewers' grains that were contaminated by a GM maize variety that is not approved within the EU. See Decision 2005/317/EC on emergency measures regarding the non-authorized genetically modified organism Bt10 in maize products and "Illegal GM Maize Fear Sparks EU Ban on U.S. Animal Feeds" from *The Guardian,* Online Edition, April 16, 2005.

† Products authorized shall be entered into a public register of GM food and feed (http://europa.eu.int/comm/food/dyna/gm_register/index_en.cfm). Authorizations will be granted for a period of 10 years, subject where appropriate to a post-market monitoring plan. Authorizations are renewable for 10-year periods.

‡ These GM feed products that could be legally placed on the market in the EU according to the rules in place before Regulation 1829/2003, and other feed products that did not require special approval at the time they were placed on the market, were gathered in the EC register of GM food and feed. Until April 18, 2004, GM food was regulated as novel food, and foods derived from 18 GM events have been approved so far (essentially maize and soy derivatives, oilseed rape oil, and cottonseed oil). There was no specific legislation covering GM feed, but nine GM events.

market and that contain or consist of GMOs. The regulation also includes provisions for the traceability of food and feed products produced from GMOs.[*]

The purpose of having procedures allowing traceability of GMOs is to facilitate monitoring, risk management, and possible withdrawal of products for the protection of human and animal health and the environment. The purpose of labeling these products is to allow "operators" (defined below) and consumers to have adequate information to make informed choices. The two issues of traceability and labeling are linked because the systems should be mutually supportive; for example, traceability should assist in verification of the accuracy of labeling.

The objectives for traceability under the regulation are to facilitate the following:

- Control and verification of labeling claims
- Targeted monitoring of potential effects on the environment, when appropriate
- Identification and withdrawal of products that contain or consist of GMOs should an unforeseen risk to human health or the environment be established

To ensure traceability and labeling, the provisions of the regulation require operators to transmit and retain specified information for the above GM product types at each stage of their placing on the market. Notably:

- Operators are required to have systems and procedures in place to identify to whom and from whom products are made available.
- *For GMOs intended for deliberate release into the environment (e.g. seeds)*, operators are required to transmit specified information on the identity (unique identifier) of the individual GMOs a product contains.
- *For GMOs intended for food, feed, or for processing*, operators may either transmit the specified information detailed above or transmit a declaration that the product shall only be used as food or feed or for processing, together with the identity of the GMOs that *have been used* to constitute the mixture.
- *For food and feed produced from GMO(s)*, operators are required to inform the next operator in the chain that the product is produced from GMOs.
- Operators are required to retain the information for a period of 5 years and make it available to competent authorities on demand.
- Thresholds have been established below which adventitious or technically unavoidable traces of certain GMOs and GM material in food, feed, and processing products do not require labeling or tracing.

---

[*] Adopted on September 22, 2003, following the publication of Commission Regulation (EC) No. 65/2004 establishing a system for the development and assignment of unique identifiers for GMOs, and fully applicable on April 16, 2004. This differentiated treatment is in line with the Cartagena Protocol on Biosafety (an international agreement on transboundary movements of GMOs).

Transmission and retention of the above information is intended to reduce the need for sampling and testing of products, which is not an obligatory requirement for the operators under the regulation. Nevertheless, to facilitate a coordinated approach for inspection and control by the member states, EC has developed technical guidance on sampling and testing methods.* A new regulation was adopted in 2004 that established the unique identifier system—Regulation (EC) No. 65/2004 establishes a system for the development and assignment of unique identifiers for GMOs.

Directive 2001/18/EC is amended by this regulation with Article 4(6) being removed and a paragraph being inserted in Article 21 establishing a threshold of 0.9% for "adventitious or technically unavoidable" traces of GMOs. Products containing traces below this threshold do not have to meet the requirements of this regulation. In addition, specific labeling for food containing, consisting of, or produced from GMOs is provided for in Regulation (EC) No. 1829/2003. In some cases, food produced from GMOs (e.g., some refined oils) does not differ from a physicochemical point of view from products of non-GM origin. The labeling of such products relies on a dedicated system of traceability established by Regulation (EC) No. 1830/2003.

The labeling requirements do not apply to food containing material that contains, consists of, or is produced from GMOs in a proportion no higher than 0.9% of the food ingredients considered individually, or food consisting of a single ingredient, provided that this presence is adventitious or technically unavoidable.

6.12.6.1 Labeling and Traceability

The traceability rules make it mandatory of the operators concerned (i.e., all persons who place a product on the market or receive a product placed on the market within the EC) to be able to identify their supplier and the companies to which the products have been supplied.

The traceability requirement varies depending on whether the product consists of or contains GMOs (Article 4 of Regulation (EC) No. 1830/2003) or has been produced from GMOs (Article 5 of Regulation (EC) No. 1830/2003). Hence, two hypotheses must be distinguished:

1. ***In the case of a product consisting of or containing GMOs:*** Operators must ensure that the following two particulars are transmitted in writing to the operator receiving the product:
   a. An indication that the product or some of its ingredients contains or consists of GMOs

---

* In view of the requirements of the regulation, it is important to note that at the time of finalizing this report, a decision on documentation requirements for GMOs intended for food, feed, or for processing to be used in international trade was adopted under the Cartagena Protocol on Biosafety (Third Meeting of the Parties, March 13–17, 2006, Curitiba, Brazil). According to this decision, parties to the protocol must take measures to ensure that documentation accompanying international shipments of GMOs in commercial production includes the identity of the GMOs contained in the shipment, when their precise identity is known. In cases in which the identity of GMOs in a shipment is not precisely known, documentation should make clear that the shipment "may contain" GMOs, together with the identity of the GMOs that may be contained in the shipment.

    b. The unique identifier(s) assigned to those GMOs in the case of products containing or consisting of GMOs

In the case of products consisting of or containing mixtures of GMOs to be used only and directly as food or feed or for processing, the information relating to the unique identifiers may be replaced by a declaration of use by the operator, accompanied by a list of the unique identifiers for all those GMOs that have been used to constitute the mixture. Operators must ensure that the information received is transmitted in writing to the operator receiving the product.

  2. ***In the case of products produced from GMOs:*** Operators must ensure that the following particulars are transmitted in writing to the operator receiving the product:
    a. An indication of each of the food ingredients that are produced from GMOs
    b. An indication of each of the feed materials or additives which are produced from GMOs
    c. In the case of products for which no list of ingredients exists, an indication that the product is produced from GMOs

In addition to traceability requirements, products consisting of or containing GMOs and food products produced from GMOs that are authorized under the procedure set out in Directive 2001/18/EC (Part C) or under Regulation (EC) No. 1829/2003 are subject to the labeling requirements laid down in Regulation (EC) No. 1829/2003 and Regulation (EC) No. 1830/2003. Labeling informs the consumer and user of the product, hence allowing them to make an informed choice.

Generally speaking, for all prepackaged products consisting of or containing GMOs, Regulation (EC) No. 1830/2003 requires that operators indicate on a label, "This product contains GMOs" or "This product contains GM [(name of organism(s)]." In the case of non prepackaged products offered to the final consumer or to mass caterers (restaurants, hospitals, canteens, and similar caterers) these words must appear on or in connection with the display of the product.*

GM foods that are delivered as such to the final consumer or mass caterers (restaurants, hospitals, canteens, and similar caterers) must be labeled in accordance with Article 12 of Regulation (EC) No. 1829/2003, regardless of whether DNA or proteins derived from genetic modification are contained in the final product or not. The labeling requirement also includes highly refined products such as oil obtained from GM maize.

The same rules apply to animal feed, including any compound feed that contains transgenic soya. Corn gluten feed produced from transgenic maize must also be labeled, in compliance with Article 25 of Regulation (EC) No. 1829/2003, so

---

* Of the 18 GMOs authorized in accordance with Directive 90/220/EEC, 8 are authorized for the purpose of use in feeding stuffs.

as to provide livestock farmers with accurate information on the composition and properties of feed. Therefore, GM food and feed are subject to the specific labeling requirements imposed by the GMO legislation.[*]

6.12.6.2 Exemption from the Traceability and Labeling Requirements

Conventional products (i.e., products created without recourse to genetic modification) may be accidentally contaminated by GMOs during harvesting, storage, transport, or processing. This does not only apply to GMOs. In the production of food, feed, and seed, it is practically impossible to achieve products that are 100% pure. Taking this into account, the legislation has laid down limits above which conventional food and feed must be labeled as products consisting of GMOs, containing GMOs, or produced from GMOs. However, these conventional products "contaminated" by authorized GMOs are not subject to traceability and labeling requirements if they contain traces of these (authorized) GMOs below a limit of 0.9%, provided the presence of this material is adventitious or technically unavoidable. This is the case when operators demonstrate to the competent authorities that they have taken adequate measures to avoid the presence of this material.

Regarding the question of whether meat or milk of an animal fed with GM feed should they be labeled as "GM": In line with the general EU rules on labeling, Regulation (EC) No. 1829/2003 does not require labeling of products such as meat, milk, or eggs obtained from animals fed with GM feed or treated with GM medicinal products. These products are also not subject to traceability requirements.

Regarding the new regulation and the allowed presence of traces of GM materials that have received a favorable scientific assessment, but are not yet formally approved: The adventitious or technically unavoidable presence of GM material in products placed on the market in the EU can occur during cultivation, handling, storage, and transport. This situation already exists and affects products originating in the EU and third world countries.[†] This is not a problem unique to GMOs. In the production of food, feed, and seed, it is practically impossible to achieve products that are 100% pure. Regulation (EC) No. 1829/2003 acknowledges this fact and defines the specific conditions under which a technically unavoidable presence of GMOs not yet formally authorized could be permitted.

The scientific committees advising the EC have already assessed several GMOs. These committees have indicated that the GMOs do not pose a danger to the environment and health, but their final approval is still pending. *The rules allow the presence of these GMOs in a food or feed up to a maximum of 0.5%, above which it is prohibited to put the product on the market.*

---

[*] See in particular Directive 2000/13/EC on the approximation of the laws of the member states relating to the labeling, presentation, and advertising of foodstuffs. See also Directive 96/25/EC on the circulation of feed materials, amending Directives 70/524/EEC, 74/63/EEC, 82/471/EEC, and 93/74/EEC and repealing Directive 77/101/EEC.

[†] The EC has published a list of GM material that has not been authorized but that has had a favorable scientific assessment. This list is found at http://www.europa.eu/index_en_htm. To access the list, type "non-authorized CM food" in the search tool.

Article 47 of Regulation (EC) No. 1829/2003 on GM food and feed provides that

1. The presence in food or feed of material that contains, consists of, or is produced from GMOs in a proportion no higher than 0.5% shall not be considered to be in breach of Article 4(2) or Article 16(2), provided that
   a. This presence is adventitious or technically unavoidable.
   b. The GM material has benefited from a favorable opinion from the EC scientific committee(s) or the Authority before the date of application of this regulation.
   c. The application for its authorization has not been rejected in accordance with the relevant EC legislation.
   d. Detection methods are publicly available.

2. In order to establish that the presence of this material is adventitious or technically unavoidable, operators must be in a position to demonstrate to the competent authorities that they have taken appropriate steps to avoid the presence of such materials.

In accordance with Article 4(8) and 5(4) of Regulation (EC) No. 1830/2003, the traceability and labeling requirements laid down in Article 4(1) to 4(6) and the traceability requirements laid down in Articles 5(1) and 5(2), respectively, shall not apply to traces of the GMOs listed below under paragraph (a) that are present in the products concerned in a proportion no higher than 0.5% provided that these traces are adventitious or technically unavoidable.*

6.12.7   Other Food Issues—Food Origin, Animal Welfare, Contaminated Food, and Environment

6.12.7.1   Origin Labeling

Common labeling requirements (name, composition, durability, etc.) applicable to all foodstuffs are laid down in horizontal legislation (Directive 2000/13/EC and related texts). In that framework, origin is normally not considered as necessary information to enable consumers to make an informed choice, because that origin is not an important element to characterize or to identify the product (such as for example biscuits, breakfast cereals or soft drinks). In addition, the consumer can have some information on the origin by the compulsory identification (name and address) of the manufacturer or packager, or of a seller established within the EC. However, origin

---

* Thresholds for GM impurities in conventional seeds: Legislation on seeds has always recognized that a 100% purity is not possible, which is why thresholds have been set that take into account the fact that plants are grown in an open field; that cross-pollination is a natural phenomenon; and that one cannot control wind and insects, which contribute to this. For example, certified soya beans may have up to 1% impurities of another soy variety. Impurities can arrive through cross-pollination, dissemination of volunteers and at harvest, transport, and storage.

or provenance shall be indicated in a case in which consumers could be misled on the true origin of the product.

Because of a decision in the past that there exists a specific need to inform consumers, specific labeling provisions are included in vertical legislation applicable to products ranging from fruits and vegetables to meat, eggs, fish, wine, honey, and chocolate. These rules often result from specific composition or quality standards, but may also request mandatory indication of origin or provenance and that information being deemed necessary for consumer choice regarding such foodstuffs, generally basic products, the characteristics/quality of which are influenced by their origin. In these cases, detailed rules for indicating that origin are laid down within the legislation concerned. Research has shown that

1. Consumers would be interested in the origin of fresh meat, in addition to beef, because they feel that meat from their country is "safer."
2. Origin is also associated with quality in the case of certain products such as delicatessen, cheese, and wine (but this need is already taken on board through the existing legislation).
3. Consumers have difficulties in identifying food produced in compliance with certain animal welfare standards because the information on labels is inappropriate, unclear, or missing.

There is at present much debate about consumer attitudes to origin, both for food (milk, poultry, meat) and non-food (textile, shoes) products. There is also renewed producer interest in using local (EU, national, or regional) origin as a selling point.

6.12.7.2 Animal Welfare Labeling

When questioned, consumers have expressed a preference for simple, symbolic labeling (such as color coding and logos) rather than textual information. The Community Action Plan on the Protection and Welfare of Animals, adopted in January 2006, foresees as one of the five main areas of action the introduction of standardized animal welfare indicators to classify the hierarchy of welfare standards applied (from minimum to higher standards). On this basis, options for labeling will be explored in a systematic manner.

Labeling related to animal welfare conditions makes particular sense if there are different standards allowed by EC legislation; for example, for eggs when the different types of production could compete on the market in relation to the quality of welfare achieved. A similar approach could be taken in the legislation for other products of animal origin.

The Amsterdam Treaty's "Protocol on Protection and Welfare of Animals" lays down new rules concerning action by the EU. It officially recognizes that animals are sentient beings and requires the European institutions to take into account animal welfare requirements when formulating and implementing European legislation. EU legislation on animal protection aims to spare animals any unnecessary suffering in three main areas: farming, transport, and slaughter. In collaboration with the competent authorities of the member states, the Food and Veterinary Office (FVO) carries out spot checks to ensure that EU legislation is being complied with.

6.12.7.3 Contamination of the Food Chain

The contamination of foodstuffs represents a real risk for food safety. It may come from several sources, such as environmental pollution, the production chain, or products used in packaging. The EU has therefore introduced a wide range of legislative measures designed to protect foodstuffs. General arrangements have been made to deal with the presence of contaminants in a human food by setting maximum levels. The EU initially turned its attention to prohibiting and limiting the use of certain chemical products and to the classification, labeling, and packaging of dangerous substances and preparations, including fertilizers and pesticides covered by separate arrangements. The protection afforded by existing legislation was subsequently enhanced by measures involving risk assessment, tests on chemical substances, and exports and imports.

6.12.7.4 Environmental Factors

Food safety policy forms part of a more general, horizontal strategy for sustainable development. There is an inextricable link with certain environmental factors that have a greater or lesser impact on the quality of products intended for human and animal consumption. The main environmental factors have to do with waste management, atmospheric pollution, water quality (safety, drinking water, nitrate content), and the protection of nature and biodiversity.

6.12.7.5 Consumer Information, Education, and Health Monitoring

Information is a basic principle of consumer policy and has become even more vital in the wake of the recent food crises. Details that help the consumer to make an informed decision are found on the packaging and labeling, such as geographical indications and designations of origin, labels, indications of price, and the composition of products.

6.13 EurepGAP

GLOBALGAP (formally EurepGAP) Secretariat
c/o FoodPLUS GmbH
HRB 35211 Koeln
Kristian Moeller Spichernstr, Managing Director
55-50672 Koeln, Germany
Phone: +49-0-2 21-5 79 93-66; Fax: +49-0-2 21-5 79 93-66-89
E-mail: stollenwerk@globalgap.org

Available at http://www.globalgap.org.

In 1997, to restore confidence regarding the safety of food products, large grocery chains in Holland and England and some international suppliers began to establish a new institution that insisted on strict environmental and production criteria that farmers would have to meet if they wanted to sell their products in member supermarkets. The name of this system is EurepGAP, which is an acronym for "Euro Retailer Produce Working Group adopting standards of Good Agricultural Practice." It has subsequently evolved into an equal partnership of agricultural producers and their retail customers.

As of September, 2003, EurepGAP had over 200 member companies from around the world. There are over 12,000 certified growers in more than 20 countries with a combined production capacity covering 975,000 acres. The largest numbers of certified growers are in the Netherlands and the United Kingdom, followed by Spain, South Africa, Israel, and Belgium. In Holland, 100% of supermarkets are participating in EurepGAP and 85% of all fruits and vegetables sold in Dutch retail stores are covered by the EurepGAP protocols.

In responding to the demands of consumers, retailers, and their global suppliers EurepGAP has created and implemented a series of sector specific *farm certification standards*, which are divided into Module Stages (see illustration below). The aim is to ensure integrity, transparency, and harmonization of global agricultural standards. This includes the requirements for safe food that is produced respecting worker health, safety, and welfare, as well as environmental and animal welfare issues.

EurepGAP certification is contingent upon completion and verification of a checklist that consists of 254 questions, 41 of which are considered "Major Musts" and 122 of which are considered "Minor Musts." Another 91 are "shoulds," which are recommended but not required practices. The EurepGAP protocols reach backward down the food chain and direct farmers how to manage their farms. These protocols are so broad based that they cover environmental issues, animal welfare issues, employment issues, sustainability, and any other social or economic factors of concern to supermarkets.[*]

One of the most important recent developments in agriculture is the growing demand for traceability. Starting January 1, 2005, every country of the 16 or 17 countries in the EU will be required to provide full traceability of every food product sold in the EU. Whoever touches the product, be it farmer, shipper, processor, grocery store, etc., will have to provide traceability one step back and one step forward.

Traceability means that a buyer or consumer can track food products and how they were handled all of the way back to the farm and even before, to the seed supplier and the chemical supplier. In some European supermarkets, a customer can take a package of meat to the barcode reader and see a picture of the farm where the steak or chicken came from. That barcode contains data that gives consumers access to information about the animal's parents, where it was born, what medical attention it received, and where it was slaughtered. The store manager is therefore able to immediately isolate any food safety or quality problem.

The demand for traceability is fueled by two trends: consumer concerns about food and environmental safety and the need to identify and segregate higher value specialty crops. With experts predicting that perhaps 30% of total production over

---

[*] The EurepGAP protocols are slowly being implemented. As of January 1, 2004 it applies to fresh fruits, vegetables, and flowers; however, they are not yet uniformly followed throughout Europe. In the Netherlands 100% of their supermarkets are abiding by this program.

the next 6 years will be in value-added crop varieties rather than undifferentiated commodities, the need for traceability systems can only grow (Glassheim et al. 2005).

EurepGAP, like others in the EU, has been driven by the desire to reassure consumers of food safety. Following food safety scares such as BSE (mad cow disease), pesticide concerns, and the rapid introduction of GM foods, consumers throughout the world are asking how food is produced and need reassuring that their food is safe and sustainable. Food safety is a global issue and transcends international boundaries. Many EurepGAP members are global players in the retail industry and obtain food products from around the world. For these reasons a need has arisen for a commonly recognized and applied reference standard of Good Agricultural Practice that has at its center a consumer focus.

Technically speaking, EurepGAP is a set of *descriptive documents suitable to be accredited to internationally recognized certification criteria such as ISO Guide 65*. Representatives from around the globe and all stages of the food chain have been involved in the development of these documents. In addition, the views from stakeholders outside of the industry including consumer and environmental organizations and governments have helped shape the protocols. This wide consultation has produced a robust and challenging but nonetheless achievable protocol that farmers around the world may use to demonstrate compliance with Good Agricultural Practices. The standards are openly available and free to obtain from the EurepGAP website.

EurepGAP members include retailers, producers/farmers, and associate members from the input and service side of agriculture. Governance is by sector-specific EurepGAP steering committees that are chaired by an independent chairperson. The technical and standards committees working in each product sector approve the standard and the certification system. These committees have 50% retailer and 50% producer representation, creating an effective and efficient partnership in the supply chain. The work of the committees is supported by FoodPLUS, a nonprofit company based in Cologne, Germany.

6.13.1   GOALS OF EUREPGAP

By adhering to good agricultural practices, the risks within agricultural production are reduced. EurepGAP provides the tools to objectively verify best practice in a systematic and consistent way throughout the world. This is achieved through their protocol and compliance criteria. EurepGAP's scope is concerned with practices on the farm: *once the product leaves the farm it comes under the control of other codes of conduct and certification schemes relevant to food packing and processing*. In this way, the whole chain is assured right through to the final consumer.

Another key goal of EurepGAP is to provide a forum for continuous improvement. The technical and standards committees, consisting of producer and retail members, have a formal agenda to review emerging issues and carry out risk assessments. This is a rigorous process that follows the principles of HACCP and involves experts in their field leading to revised versions of the protocol.

6.13.2 Integrated Farm Assurance Program

The EurepGAP Technical and Standards Committee for Integrated Farm Assurance (IFA) has evaluated and approved the new version, the General Regulations, Control Points and Compliance Criteria, and the Checklist for IFA (see Figure 6.3).

6.13.3 EurepGAP IFA

The importance of the IFA program is summarized below:

- It provides controlled and more efficient production of agricultural raw materials.
- It is the farmers' response to globalization.
- It reassures and improves confidence in agricultural products.

The objectives of EurepGAP IFA are:

- To facilitate mutual recognition through transparent benchmarking
- To boost worldwide participation in farm assurance
- To encourage continuous improvement
- To provide performance and integrity measurement for assurance schemes (e.g., certification, accreditation)

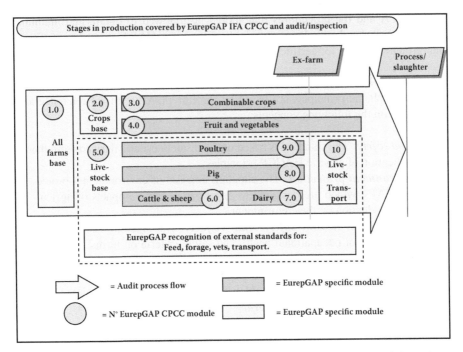

FIGURE 6.3 EurepGAP module interaction. (Reprinted with permission from GLOBALGAP; please see GLOBALGAP for the most updated version.)

Milestones for IFA include:

- Reducing duplication of audits at the farm level
- To see IFA become the preferred global reference standard for farm assurance schemes at the pre-farm gate (agricultural production)
- To see IFA become a common buyer standard for all sources of supply irrespective of the country of origin

6.13.4 GENERAL REGULATIONS OF IFA VERSION 2.0, MARCH 2005

6.13.4.1 Terms of Reference—"The Global Partnership for Safe and Sustainable Agriculture"

To respond to consumer concerns on food safety, animal welfare, environmental protection, and worker health, safety and welfare, the IFA

- Encourages adoption of commercially viable Farm Assurance Schemes, which promote the minimization of agrochemical and medicinal inputs within Europe and worldwide
- Develops a Good Agricultural Practice (GAP) framework for benchmarking existing assurance schemes and standards, including traceability

6.13.4.2 Scope

The EurepGAP document explains the structure of certification to EurepGAP's standard for IFA and the procedures that should be followed to obtain and maintain certification. It details the duties and rights of the EurepGAP Secretariat, certifiers, and farmers applying for certification.

6.13.4.3 Objectives

EurepGAP scheme principles are based on the EurepGAP terms of reference and specifically on the following concepts:

- *Food safety:* The standard is based on food safety criteria derived from the application of generic HACCP principles.
- *Environment protection:* The standard consists of environmental protection GAPs, which are designed to minimize negative impacts of agricultural production on the environment.
- *Occupational health, safety, and welfare:* The standard establishes a global level of occupational health and safety criteria on farms, as well as awareness and responsibility regarding socially related issues; however, it is not a substitute for in-depth audits on corporate social responsibility.
- *Animal welfare (where applicable):* The standard establishes a global level of animal welfare criteria on farms.

EurepGAP provides the standards and framework for an independent, recognized third-party certification of farm production processes based on EN45011/ISO Guide 65. (Certification of the production process, cropping, growing, or production of certified products ensures that only those that reach a certain level of compliance with established GAPs set out in the EurepGAP descriptive documents are certified.)

The scheme covers the whole agricultural production process of the certified product, from when the animal enters the production process or the plant is in the ground (origin, and seed control points) to nonprocessed end product. *Note:* No manufacturing, slaughtering, or processing is covered. The objective of EurepGAP certification is to form part of the verification of GAPs in the whole (farm) production chain.

6.13.4.4 Rules

These general regulations establish the rules applicable to certifying bodies (CBs) approved by the EurepGAP Secretariat to the scope of EurepGAP IFA for granting, maintaining, and removing EurepGAP IFA certification. A certificate holder can be any of the following:

- An individual farmer applying for EurepGAP certification
- A farmer group applying for EurepGAP certification
- An individual farmer that is working under a scheme that has successfully benchmarked to EurepGAP
- Farmer group that is working under a scheme that has successfully benchmarked to EurepGAP

6.13.4.5 Compliance Levels for EurepGAP Certification

Compliance with EurepGAP IFA consists of three types of control points that the applicant is required to undertake to obtain EurepGAP recognition: "Major Musts," "Minor Musts," and "Recommendations." These control points must be fulfilled as outlined in the following subsections (see also Chapters 11 and 12 of General IFA Regulations, Version 2.0—March 2005 under "Sanctions and Noncompliances").

6.13.4.6 Options and Verification for EurepGAP Certification*

Farmers can achieve EurepGAP certification under any one of the four options described below.

- *Option 1—individual certification:* Individual farmer applies for EurepGAP certificate, for one or more modules.
- *Option 2—group certification:* Farmer group applies for EurepGAP group certificate, *for one or more modules.*

---

* References:
(i)   EurepGAP Equivalent Certification System Owner Agreement.
(ii) ISO19011:2002 "Guidelines for Quality and/or Environmental Management Systems Auditing."
(iii) ISO IEC Guide 7-1194 Guidelines for Drafting Standards Suitable for Use of Conformity Assessments.
(iv) ISO/IEC Guide 65–1996. General Requirements for Bodies Operating Product certification Systems.
(v)   ISO/IEC Guide 2.
(vi) ISO 8402.

- **Option 3—benchmarking:** Individual farmer applies for EurepGAP benchmarked scheme certificate, *for one or more modules.*
- **Option 4—benchmarking:** Farmer group applies for EurepGAP benchmarked scheme certificate, *for one or more modules.*

6.13.5   Benchmarking System Procedure of IFA, All Scopes, Version 1.2, June 2005

6.13.5.1   Background and Justification

The recognition of other farm assurance schemes via benchmarking is one of EurepGAP's core objectives. To improve perceived and actual integrity and transparency of the system, the EurepGAP Technical and Standards Committee (TSC) "Fruit & Vegetables" has approved this benchmarking procedure for EurepGAP. The EurepGAP Steering Committee (SC) decided to appoint external, recognized, and competent organizations to undertake the technical review and witness audits ("physical benchmarking").

Accreditation bodies currently achieve their accreditation through the EurepGAP process. This is a key criterion for an applicant's technical expertise, qualifications, and independence in accreditation systems (ISO Guide 65) within the agricultural field. The accreditation was designed to identify an organization that delivers the desired public and industry credibility, with the global resources, technical and organizational competence, and efficiency to handle the EurepGAP benchmarking procedure in an industry-affordable manner. EurepGAP has received applications from the Joint Accreditation System of Australia and New Zealand (JAS-ANZ) and from Deutsches Akkreditierungssystem Prüfwesen GmbH (DAP/Germany).

Certification of a product (a term used to include a process or service) is a means of providing assurance that it complies with specified standards and other descriptive documents. Certification is applicable to all companies and organizations interested in applying for EurepGAP recognition via the benchmarking process and to all available and future EurepGAP scopes [Fruit & Vegetables, Flower & Ornamentals, IFA, Integrated Aquaculture Assurance (IAA), (Green) Coffee, etc.].

6.13.5.2   Equivalent Certification System

A certification system that has achieved accreditation under ISO/IEC Guide 65 (EN45011) with an accreditation body that is a member of the International Accreditation Forum (IAF) and is signed up to the multilateral agreement (MLA) concerning ISO IEC Guide 65. The certification system must be operated only by CBs that have achieved the above accreditation directly for the equivalent standard, in which the certification system has successfully completed the equivalence procedures set out in this standard. See Appendix D for a listing of EurepGap Accredited Bodies, which also includes membership and CB fees, the DAP/Germany accreditation system benchmarking fee schedule, and a JAS-ANZ benchmarking fee schedule.

6.14   INTRODUCTION TO INTERNATIONAL STANDARDS

The following sections call attention to international standards that have become cornerstones of national and international trade, safety, and quality systems. Many of the standards in this chapter are directly used within many nations' food, safety, and quality programs. As will be illustrated, these systems extend well beyond IPT. However, for any organization that already uses any one or a combination of these systems, the ability to include or add IPT within their operations and supply chain would be nearly transparent and easy to accomplish.

Before the larger systems are explored, Section 6.15 will provide a short narrative of other systems that are applicable to the food industry, which includes ISO 9000, the Total Quality Management (TQM) approach, and Deming's Management Program and Quality Control.

The major international guides toward food traceability, safety, and quality programs include Codex Alimentarius (Codex) and the Food and Agricultural Organization of the United Nations (FAO)/World Health Organization (WHO) food standards; ISO 22000; and HACCP standards, HACCP web, and HACCP training providers sections. What follows are company/organizational/agency statements from their respective websites and naturally reflects their views.

6.14.1 CODEX

Codex provides an overall forum for international participation that has resulted in programs and agreements in such areas as guidelines on nutrition labeling, food labeling (country of origin), and standards on food labeling (traceability).

6.14.2 ISO 22000

ISO 22000 is a more recent IPT tool within the ISO format and structure. This standard brings together fragmented national, international, and industry HACCP and food safety standards into one food safety management system.

6.15 OTHER INTERNATIONALLY RECOGNIZED SYSTEMS

In addition to systems that will be expanded upon later in this chapter (the list below is not complete), there are many systems used throughout the world that are designed to meet local and international requirements and are found within various official and private agreements and programs. Below are just a few systems that are well known; however, time and space prohibit more information about each of them to be presented here. The most recognized systems are the ISO 9000 Series, TQM, and Deming's Management Program and Quality Control.

ISO 9000* is the fundamental standard of ISO and was the basis for ISO 22000 development. *The ISO 9000 series has turned out to be one of the best international quality management systems developed. ISO 9001 can be used for internal application by organizations, certification, or contractual purposes.*

The ISO 9000 series was first released in 1987, a first revision was published in 1994, and in 2000 the modification to ISO 9001:2000 was released. Since then, only three main standards persist: ISO 9000:2000, which includes a descriptive approach

---

* Excerpts and modified with permission from http://www.ourfood.com/HACCP_ISO_9000.html by Karl Heintz Wilm.

to quality management as well a revised vocabulary; ISO 9001:2000, which includes the quality management system requirements; and ISO 9004:2000, which includes guidelines for performance improvement moving toward TQM. It is not intended for certification or contractual use.[*] They rely on the following eight principles:

1. Customer-focused organization
2. Involvement of people
3. System approach to management
4. Factual approach to decision-making
5. Mutually beneficial supplier relationship
6. Continual improvement
7. Process approach
8. Leadership

Included in this system are standards for documentation of the system, control of documents, and control of records to show management commitment, customer focus, and quality policy. The 2000 revision is an attempt to harmonize ISO 9000 (quality) with ISO 14001 (environment) and BS 8800 (health) so that an organization can handle quality, environment, health, and safety within one system.

ISO 9000:2000 also offers supply chains, quality assurance, and beginnings of traceability in agriculture. Figure 6.4 illustrates a normal top-level flow diagram for a grain farming operation. This chart demonstrates the flow and linkages between activities, controls, and records that support a quality management system such as ISO 9000:2000.[†]

6.15.1 How Certification Is Obtained

The interested company makes a contract with a certifying agent. It takes approximately 1–2 years to obtain the certificate depending on the complexity of the company.[‡] The single revised ISO 9001:2000, which contains a single quality management requirement standard, is applicable to all organizations, products, and services.[§]

---

[*] The standard ISO 9000 dates back to 1989 and was accepted in Europe under the Number EN 29000 as European norm. The different standardization organizations have integrated the ISO series under different denominations. The European standardization organization CEN (C, Conformité; E, Européen) has created the denomination "DIN EN ISO 9000 ff" published in English, German, and French. This standard contains the norms and the procedure to obtain the ISO 9000 certificate.

[†] Reg Clause: Iowa State University (ISU)–CIRAS; Washington, DC; January 27–28, 2003. Reg developed this excellent flow chart that pictorially describes, from planting to market, an ISO 9001 view of quality control. Management review is referenced. Each functional area links to next levels of documentation. The ISU Crop Management Database program can support the recordkeeping. See http://www.farmfoundation.org/projects/documents/RegClause.pdf (accessed January 9, 2007).

[‡] There are many organizations that are accredited to give out certificates, such as DGS (Deutsche Gesellschaft zur Zertifizierung von Qualitäts management systemen) and TÜV (Technischer Überwachungs Verein).

[§] The ISO-9001:2000 quality system aims to enhance customer satisfaction. This includes the processes for continual improvement of the quality system and the assurance of conformity to the customer and applicable regulatory requirements. In global business, the certification according ISO 9000 turned out to be an imperative duty. The HACCP concept should be integrated in the quality system, thereby fulfilling hygiene regulations.

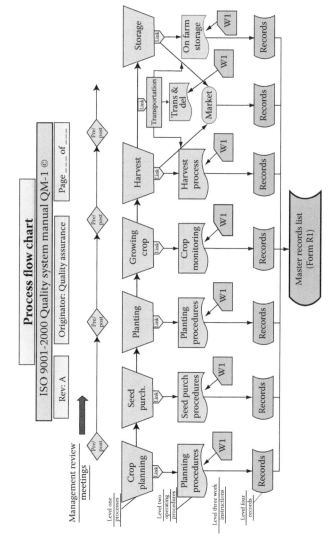

FIGURE 6.4 ISO 9001 process flow chart. (Reprinted with permission from CIRAS.)

207

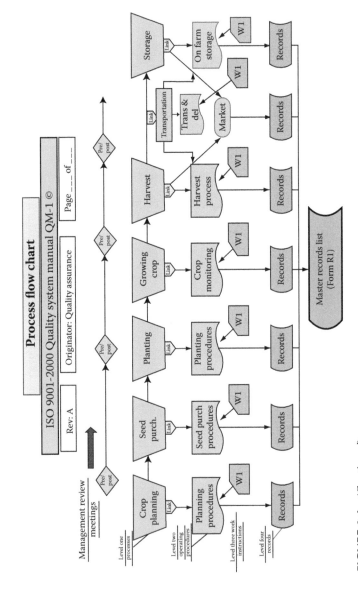

FIGURE 6.4 (*Continued*)

6.15.1.1 TQM

TQM can be installed after ISO 9000. TQM attempts to unite all of the different phases of the activities of a company, from the financial and managerial processes to production and technical details. With growth, international business organizations have to integrate modifications in the basic business structure concerning the rapidly changing international market. ISO 9000 is the basic activity that supports TQM.

In the past, quality control and quality improvement were considered the responsibility of one department or a discrete part of an enterprise. In TQM, every part of the enterprise is tied together.

6.15.1.2 Deming's Management Program and Quality Control

W. Edwards Deming influenced worldwide quality control. He stressed the need to "drive out fear," to stop relying on inspection for insuring quality, and to focus on building cooperation and not competition within an organization. The philosophy of Deming has been successful in the United States. The German website, http://www.deming.de, tries to bring these ideas to the German-speaking area. The British Deming Association also propagates the philosophy, which is based on Deming's 14 points. See the British Deming Association website for additional information (http://www.deming.org.uk and http://www.deming.org/theman/teachings02.html).

6.16 CODEX AND FAO/WHO FOOD STANDARDS

Viale delle Terme di Caracalla
00100 Rome, Italy
Phone: +39-06-5705-1
Fax: +39-06-5705-4593
E-mail: codex@fao.org

U.S. Codex Office
USDA Food Safety and Inspection Service
Room 4861 South Building
1400 Independence Avenue SW
Washington, DC 20250-3700
Phone: 202-205-7760
Fax: 202-720-3157
E-mail: uscodex@fsis.usda.gov

Available at http://www.codexalimentarius.net/web/index_en.jsp, http://www.fsis.usda.gov/regulations_&_policies/Codex_Alimentarius/, and http://www.codexalimentarius.net.

The Codex Alimentarius Commission (CAC) implements the Joint FAO/WHO Food Standards Program, the purpose of which is to protect the health of consumers and to ensure fair practices in the food trade. The *Codex Alimentarius* (Latin, meaning "Food Law or Code") is a collection of over 230 internationally adopted food standards and also includes codes of practice, limits for pesticide residues, and evaluations of additives and veterinary drugs. The main aims of Codex are to protect the health of consumers and to facilitate international food trade through harmonization of science-based standards.[*] The CAC has expressed the view that codes of practice

---

[*] See http://www.codexalimentarius.net for more information. Scientific disciplines, together with consumers' organizations, production and processing industries, food control administrators, and traders.

might provide useful checklists of requirements for national food control or enforcement authorities. The publication of Codex is intended to guide and promote the elaboration and establishment of definitions and requirements for foods, to assist in their harmonization, and, in doing so, to facilitate international trade. Codex brings together a conglomeration of principles and standards to meet its goals of protecting the public. IPT have become increasingly more important as food safety issues have increased with time.

Although IPT is not usually directly mentioned by name, its concepts and follow-up modifications to standards and rules are usually of prime concern to address public opinion of food safety. Throughout the information provided in the following, Codex bolsters and strengthens the importance of its IPT rules and regulations. Although the information provided may seem fragmented, the collection is to highlight how IPT concepts are becoming more prevalent in Codex standards.

6.16.1 Introduction to Standards and the Standards Process

Codex hopes to create standards that immediately protect consumers, ensure fair practices in the sale of food, and facilitate trade. This involves a process that includes specialists from numerous food-related fields.

6.16.2 History of Codex

6.16.2.1 Ancient Times

Evidence from the earliest historical writings indicates that governing authorities were concerned with codifying rules to protect consumers from dishonest practices in the sale of food. Assyrian tablets described the method to be used in determining the correct weights and measures for food grains, and Egyptian scrolls prescribed the labeling to be applied to certain foods. In ancient Athens, beer and wines were inspected for purity and soundness, and the Romans had a well-organized state food control system to protect consumers from fraud or bad produce. In Europe during the Middle Ages, individual countries passed laws concerning the quality and safety of eggs, sausages, cheese, beer, wine, and bread. Some of these ancient statutes still exist today.*

6.16.2.2 Trade Concerns

The different sets of standards arising from the spontaneous and independent development of food laws and standards by different countries inevitably gave rise to trade barriers that were of increasing concern to food traders in the early twentieth century. Trade associations that were formed as a reaction to such barriers pressured governments to harmonize their various food standards so as to facilitate trade in

---

* Codex precursors: *Codex Alimentarius Austriacus*, 1897–1911 and *Codex Alimentarius Europeaus*, 1954–1958. In the Austro-Hungarian Empire between 1897 and 1911, a collection of standards and product descriptions for a wide variety of foods was developed as the *Codex Alimentarius Austriacus*. Although lacking legal force, it was used as a reference by the courts to determine standards of identity for specific foods. The present day Codex Alimentarius draws its name from the Austrian code.

safe foods of a defined quality. The International Dairy Federation (IDF), founded in 1903, was one such association. Its work on standards for milk and milk products later provided a catalyst in the establishment of the CAC and in the setting of its procedures for elaborating standards.

6.16.2.3 Consumer Concerns

In the 1940s, rapid progress was made in food science and technology. With the advent of more sensitive analytical tools, knowledge about the nature of food, its quality, and associated health hazards also grew quickly. There was intense interest in food microbiology, food chemistry, and associated disciplines, and new discoveries were considered newsworthy. Articles about food at all levels flourished, and consumers were bombarded with messages in popular magazines, in the tabloid press, and on the radio.

At the same time, as more and more information about food and related matters became available, there was greater apprehension on the part of consumers. Whereas previously consumers' concerns had extended only as far as the "visibles" (i.e., underweight contents, size variations, misleading labeling, and poor quality), they now embraced a fear of the "invisibles," (i.e., health hazards that could not be seen, smelled, or tasted, such as microorganisms, pesticide residues, environmental contaminants, and food additives). With the emergence of well-organized and informed consumers' groups, internationally and nationally, there became a growing pressure on governments worldwide to protect communities from poor quality and hazardous foods.

When FAO and WHO were founded in the late 1940s, there was heightened international concern about the direction being taken in the field of food regulation. Post-World War II hardships fragmented food and agricultural food systems that had previously been in place locally and across the world. Countries were therefore acting independently and there was little, if any, consultation among them with a view to harmonization.

6.16.2.4 Scientific Base

According to Codex, the second half of the nineteenth century saw the first general food laws adopted and basic food control systems put in place to monitor compliance. During the same period, food chemistry came to be recognized as a reputable discipline, and the determination of the purity of a food was primarily based on the chemical parameters of simple food composition. When harmful industrial chemicals were used to disguise the true color or nature of food, the concept of "adulteration" was extended to include the use of hazardous chemicals in food. Science began providing tools with which to disclose dishonest practices in the sale of food and to distinguish between safe and unsafe edible products.

6.16.2.5 Desire for Leadership

Food regulators, traders, consumers, and experts were looking increasingly to FAO and WHO for leadership in unraveling the maze of food regulations that were impeding trade and providing mostly inadequate protection for consumers. In 1953, the governing body of WHO, the World Health Assembly, stated that the widening use of chemicals in food presented a new public health problem and it was proposed that the two organizations should conduct relevant studies. One such study identified the use of food additives as a critical factor.*

6.16.3 Single International Reference Point

6.16.3.1 Near Present Day

Although FAO and WHO furthered their involvement in food-related matters, a variety of committees set up by international non-governmental organizations (NGOs) also began working in earnest on standards for food commodities. In time, the work of those NGO committees was either assumed by or continued jointly with the appropriate Codex Alimentarius Commodity Committees and, in some cases, the non-governmental committees themselves became Codex committees.

Today's CAC was established in 1962 by the FAO of the United Nations and the WHO†, which is a United Nations body that sets international food standards and related texts such as codes of practice under the Joint FAO/WHO Food Standards Program. With more than 170 member countries, plus the EC, the CAC is a worldwide forum on food safety, consumer protection, and fair practices in the food trade. Codex is a continuously updated guide for governments and other interested parties on the regulatory framework needed for food control systems, food safety, and consumer protection. The international standards contained in the Codex Alimentarius are recognized as benchmarks by the World Trade Organization (WTO).

Over 400 standards, guidelines, and codes of practice have been accepted and adopted to date on the following:

- Food labeling and crop hygiene
- Commodities
- Food safety assessment for food derived from biotechnology
- Methods of analysis and sampling, food inspection, and certification procedures

Codex has become the global reference point for consumers, food producers and processors, national food control agencies, and the international food trade. The

---

* As a result, FAO and WHO convened the first Joint FAO/WHO Conference on Food Additives in 1955. That conference led to the creation of the Joint FAO/WHO Expert Committee on Food Additives (JECFA), which after more than 50 years still meets regularly. JECFA's work continues to be of fundamental importance to the CAC's deliberations on standards and guidelines for food additives, contaminants, and residues of veterinary drugs in foods. It has served as a model for many other FAO and WHO expert bodies, and for similar scientific advisory bodies at the national level or where countries have joined together in regional economic groupings.

† The FAO was established in 1945, and the Who in 1948. The two organizations began to undertake joint work on food issues in 1950, when a joint meeting of experts was held to discuss nutrition. Work on food safety issues was encouraged by several factors including concern about the health effects of food additives, increased public awareness of food safety, and increased international trade requiring harmonization of standards. The need for such a system has increased since then for many reasons, including recent food safety scares and the issue of GM foods. The main purposes of this program are protecting the health of the consumers, ensuring fair trade practices in the food trade, and promoting coordination of all food standards work undertaken by international governmental and NGOs.

code has had an enormous impact on the thinking of food producers and processors as well as on the awareness of the end users—the consumers. Its influence extends to every continent, and its contribution to the protection of public health and fair practices in the food trade is immeasurable.

The Codex system presents a unique opportunity for all countries to join the international community in formulating and harmonizing food standards and ensuring their global implementation. It also allows them a role in the development of codes governing hygienic processing practices and recommendations relating to compliance with those standards.

The significance of the food code for consumer health protection was underscored in 1985 by the U.N. Resolution 39/248, from which guidelines were adopted for use in the elaboration and reinforcement of consumer protection policies.*

Codex has relevance to the international food trade. With respect to the ever-increasing global market in particular, the advantages of having universally uniform food standards for the protection of consumers are self-evident.†

6.16.4 How it Works—Standards, Codes of Practice, Guidelines, and Other Recommendations

Codex standards usually relate to product characteristics and may deal with all government-regulated characteristics appropriate to the commodity or to only one characteristic. Maximum residue limits (MRLs) for residues of pesticides or veterinary drugs in foods are examples of standards dealing with only one characteristic.

There are *Codex general standards* for food additives, contaminants, and toxins in foods that contain general and commodity-specific provisions. *The Codex General Standard for the Labeling of Prepackaged Foods* covers all foods in this category. Because standards relate to product characteristics, they can be applied wherever the products are traded.

Codex methods of analysis and sampling, including those for contaminants and residues of pesticides and veterinary drugs in foods, are also considered Codex standards. *Codex codes of practice,* including codes of hygienic practice, define the production, processing, manufacturing, transport, and storage practices for individual foods or groups of foods that are considered essential to ensure the safety and suitability of food for consumption. For food hygiene, the basic text is the *Codex General Principles of Food Hygiene*, which introduces the use of the HACCP food safety management system. A code of practice on the control of the use of veterinary drugs provides general guidance in this area.

---

* The guidelines advise that "When formulating national policies and plans with regard to food, governments should take into account the need of all consumers for food security and should support and, as far as possible, adopt standards from the ... Codex Alimentarius or, in their absence, other generally accepted international food standards."

† It is not surprising, therefore, that the Agreement on the Application of Sanitary and Phytosanitary Measures (SPS Agreement) and the Agreement on Technical Barriers to Trade (TBT Agreement) encourage the international harmonization of food standards. Products of the Uruguay round of multinational trade negotiations, these agreements cite international standards, guidelines, and recommendations as the preferred measures for facilitating international trade in food. As such, Codex standards have become the benchmarks against which national food measures and regulations are evaluated within the legal parameters of the WTO agreements.

Codex guidelines fall into two categories:

- Principles that set out policy in certain key areas
- Guidelines for the interpretation of these principles or for the interpretation of the provisions of the Codex general standards

There are free-standing *Codex principles* covering

- Addition of essential nutrients to foods
- Food import and export inspection and certification
- Establishment and application of microbiological criteria for foods
- Conduct of microbiological risk assessment
- Risk analysis of foods derived from modern biotechnology

Interpretative Codex guidelines include those for food labeling, especially the regulation of claims made on the label. This group includes guidelines for nutrition and health claims; conditions for production, marketing, and labeling of organic foods; and foods claimed to be halal (see Chapter 6, Section 6.26). There are several guidelines that interpret the provisions of the Codex Principles for Food Import and Export Inspection and Certification, and guidelines on the conduct of safety assessments of foods from DNA-modified plants and microorganisms.*

Codex has helped to create greater global and national awareness by encouraging broader community involvement. They have done this through the establishment of scientifically sound standards. The benefits and goals of Codex are to increase consumer protection. Codex has been supported in its work by the now universally accepted maxim that people have the right to expect their food to be safe, of good quality, and suitable for consumption. Outbreaks of food-borne illness can damage trade and tourism and can lead to loss of earnings, unemployment, and litigation. Poor quality food can destroy the commercial credibility of suppliers, nationally and internationally, whereas food spoilage is wasteful, costly, and can adversely affect trade and consumer confidence.

To facilitate one of its objectives, Codex has developed guidelines on nutrition labeling to ensure that nutrition labeling is effective. Since the late 1990s and early 2000, a new area of concern has been focused on animal feed and foods derived from biotechnology. Consumer concerns in the wake of the BSE (mad cow disease) crisis of the early 1990s led Codex to take up the question of the safety of feed for food-producing animals. The CAC went even further than responding to the immediate crisis, and the resulting Code of Practice on Good Animal Feeding takes into account all relevant aspects of animal health and the environment to minimize risks to consumers' health. It applies to the production and use of all materials destined

---

* Particular guidance is given on avoiding the use of certain genes/combinations of genes. For example antibiotic resistance genes that will be expressed in the end product and genes from known allergenic sources are to be avoided (unless their safety has been proven). It is suggested that attention should also be given to issues such as the effects of nutritional modifications on human health, possible immunological effects, and whether the gene can be transmitted to human gut bacteria. There is recognition that new genomic knowledge should make the effects of genetic modifications easier to predict, and also that safety assessments may have to be reviewed in light of future scientific knowledge.

for animal feed and feed ingredients at all levels, whether produced industrially or on a farm. It also includes grazing or free-range feeding, forage crop production, and aquaculture.

To address many of the food safety concerns, Codex has proposed principles for traceability/product tracing as a tool within a food inspection and certification system.

6.16.5 CODEX LABELING RULES

Food labeling is the primary means of communication between the producer and seller of food on one hand, and the purchaser and consumer on the other. The Codex Alimentarius standards and guidelines on food labeling published in various volumes of the Codex Alimentarius are now collected and republished in this compact format to allow their wide use and understanding by governments, regulatory authorities, food industries and retailers, and consumers. In the Codex Standard on Food Labeling (Article 4.6), traceability in the form of a lot or batch numbering system was introduced more than a decade ago. The objective of the lot or batch numbering system is understood as meeting the need for better information on the identity of products and can therefore be a useful source of information; for example, when food is the subject of dispute concerning labeling claims or constitutes a health hazard to consumers. In other words, traceability is not necessarily confined to questions of product safety.

Article 4.5 of the Codex Standard for Food Labeling provides that the country of origin of the food shall be declared if its omission would mislead or deceive the consumer. Country-of-origin labeling is not safety-related and the only possible way to control such labeling is through adequate traceability based on paper documentation.

The Codex scorecard (i.e., the number of rules/regulations that they govern as of July 1, 2005):

- Commodity standards—202
- Commodity-related guidelines and codes of practice—38
- General standards and guidelines on food labeling—7
- General codes and guidelines on food hygiene—5
- Guidelines on food safety risk assessment—5
- Standards, codes, and guidelines on contaminants in foods—14
- Standards and guidelines on sampling, analysis, inspection, and certification procedures—22
- Maximum limits for pesticide residues—2,579, covering 213 pesticides
- Food additive provisions—683, covering 222 food additives
- Maximum limits for veterinary drugs in foods—377, covering 44 veterinary drugs

6.16.6 COMMODITY STANDARDS

By far the largest number of specific standards in the Codex Alimentarius is the group called "commodity standards." The major commodities included in the Codex are

- Cereals, pulses (legumes), and derived products including vegetable proteins
- Sugars, cocoa products, chocolate, and other miscellaneous products
- Processed and quick-frozen fruits and vegetables
- Meat and meat products
- Soups and broths
- Fats, oils, and related products
- Fruit juices, fish, and fishery products
- Fresh fruits and vegetables
- Milk and milk products

6.16.7 ADDITIONAL GROUPS THAT PARTICIPATE IN IPT PROGRAMS

The Joint FAO/WHO Meetings on Pesticide Residues (JMPR) and the Joint FAO/WHO Expert Committee on Food Additives (JECFA), have for many years produced internationally noted data that are widely used by governments, industry, and research centers. Their input into the work of the CAC is of fundamental importance, and the publications resulting from their activities are acclaimed international references. The safety assessments and evaluations performed by JECFA, like those performed by JMPR, are based on the best scientific information available, comprising inputs from many authoritative sources. JECFA was established in 1955 to consider chemical, toxicological, and other aspects of contaminants and residues of veterinary drugs in foods for human consumption.

The Joint FAO/WHO Expert Meetings on Microbiological Risk Assessment (JEMRA) began its work in 2000. JEMRA aims to optimize the use of microbiological risk assessment as the scientific basis for risk management decisions that address microbiological hazards in foods. Its assessments and other advice contribute to the development of Codex standards, codes of hygienic practice, and other guidelines in the area of food hygiene and provide the scientific basis for this work.

The International Atomic Energy Agency (IAEA) provides advice and support on levels of radionuclide contamination in foods and on food irradiation. The World Organization for Animal Health (OIE) provides advice on animal health, animal diseases affecting humans, and the linkages between animal health and food safety.

6.17   ISO AND ISO 22000

ISO, Central Secretariat
1, Rue de Varembé, Case postale 56
CH-1211 Genève 20, Switzerland
Phone: + 41-22-749-01-11
Fax: + 41-22-733-34-30
E-mail: central@iso.org

Available at http://www.iso.org/iso/en/ISOOnline.frontpage (accessed July 17, 2006) and http://www.iso.org/iso/en/aboutiso/introduction/fifty/friendship.html.

ISO 22000 is one of the more recent developments in IPT implementation concepts. As we will see in the following, it grew from increased food security needs and is based on well-established ISO 9000 fundamentals. This new standard, ISO 22000, is making headway to becoming one of the premier IPT standards for others to adopt. This section will discuss the general history of ISO and ISO 22000.

6.17.1 ISO Standardization System

ISO is a global network that identifies what international standards are required by businesses, governments, and society; develops them in partnership with the sectors that will put them to use; adopts them by transparent procedures based on national input; and delivers them to be implemented worldwide. ISO standards condense an international consensus from the broadest possible base of stakeholder groups. ISO standards include features such as quality, ecology, safety, economy, reliability, compatibility, interoperability, efficiency, and effectiveness. They facilitate trade, spread knowledge, and share technological advances and good management practices.[*]

6.17.2 History of ISO[†]

ISO was born from the union of two organizations. One was the ISA (International Federation of the National Standardizing Associations), established in New York in 1926 and administered from Switzerland. The other was the U.N. Standards Coordinating Committee (UNSCC), which was established in 1944 and administered in London.

Despite its transatlantic birthplace, the ISA's activities were mainly limited to continental Europe and it was therefore predominantly a "metric" organization. The standardizing bodies of the main "inch" countries—Great Britain and the United States—never participated in its work, although Britain joined just before the World War II. Attempts were made to keep the ISA going when war broke out in 1939, but as international communication broke down, the ISA president shut down the organization. The secretariat was closed, and stewardship of the ISA was entrusted to Switzerland.

---

[*] ISO, a NGO, is a federation of the national standard bodies of 149* countries, one per country, from all regions of the world, including developed, developing, and transitional economies. Each ISO member is the principal standard organization in its country. ISO has a current portfolio of 15,036* standards that provide practical solutions and achieve benefits for almost every sector of business, industry, and technology. They make up a complete offering for all three dimensions of sustainable development: economic, environmental, and social. ISO's work program ranges from standards for traditional activities (e.g., agriculture and construction); through mechanical engineering, manufacturing, and distribution; to transport, medical devices, the latest in information and communication technology developments; and to standards for services. *As of March 1, 2005.

[†] Excerpts and modified with permission from *Friendship among Equals*, by Jack Latimer and his interview with Willy Kuert, Swiss delegate to the London Conference in 1946. The conference of national standardizing organizations, which established ISO, took place in London October 14–26, 1946. This article can be found on ISO's website at http://www.iso.org/iso/en/aboutiso/introduction/fifty/pdf/foundingen.pdf.

Although the war had brought the activities of one international standardization organization to an end, it brought a new one into being. The UNSCC was established by the United States, Great Britain, and Canada in 1944 to bring the benefits of standardization to bear on the war effort and the work of reconstruction. Britain's ex-colonies were individual members of the organization; continental countries such as France and Belgium joined as they were liberated. Membership was not open to Axis countries or neutral countries. The UNSCC was administered from the London offices of the International Electrotechnical Commission (IEC). The IEC was founded in 1906.

In October 1945, UNSCC delegates agreed that the UNSCC should approach the ISA with a view to achieving and forming an organization that they provisionally called the "International Standards Coordinating Association." On October 14, 1946, at the Institute of Civil Engineers in London, the conference was held that included 25 countries that were represented by 65 delegates. The UNSCC agreed to cease functioning as soon as ISO was operational; the ISA concluded that it had already ceased to exist in 1942. Representatives wanted to have an organization open to every country that would like to collaborate, with equal duties and equal rights.[*]

6.17.3 WHY AND HOW ISO 22000:200X FOOD SAFETY MANAGEMENT STANDARD WAS DEVELOPED[†]

A traceability system is a useful tool to assist an organization operating within a feed and food chain to achieve defined objectives in a management system. The choice of a traceability system is influenced by regulations, product, the product's characteristics, and customer expectations. The complexity of the chain traceability system may vary depending on the features of the product and the objectives to be achieved. *Note: A traceability system on its own is insufficient to achieve food safety.* The implementation by an organization of a traceability system depends on technical limits inherent to the organization and products (i.e., nature of the raw materials, size of the lots, collection and transport procedures, processing and packaging methods) and cost-benefits of applying such a system.

ISO 22000 is international and defines the requirements of a food safety management system covering all organizations in the food chain from farmers to catering, including packaging. In recent times, there has been a worldwide proliferation of

---

[*] From Jack Latimer's interview with Willy Kuert. When ISO first began, there was a lengthy discussion about languages. Naturally enough for that time, English and French were proposed first. Then the Soviet delegates wanted to have Russian treated in exactly the same way as English and French. The group came back and said that the Soviet Union was prepared to translate all of the documents and to send translations to every member of the new organization. However, the Soviet Union wished to have no distinction between Russian and English and French. Then there was a very interesting discussion about finance. A committee had been set up to prepare a formula for deciding membership fees. But eventually a formula was found that depended on the population of each country and its commercial and economic strength.

[†] From the Draft International Standard ISO/DIS 22005. Traceability in the feed and food chain— general principles and basic requirements for system design and implementation. ISO 22005 was prepared by ISO Technical Committee ISO/TC 34: food products.

third-party HACCP and food safety standards developed by national standards organizations and industry groups including the United Kingdom's own British Retail Consortium. The idea of harmonizing the relevant national standards on the international level was initiated by the Danish Standards Association (DS). ISO 22000 aims to harmonize all of these standards.

The standard has the following objectives:

- Comply with the Codex HACCP principles
- Harmonize the voluntary international standards
- Provide an auditable standard that can be used either for internal audits, self-certification, or third-party certification
- The structure is aligned with ISO 9001:2000 and ISO 14001:1996
- Internationally provide communication of HACCP concepts

The ISO 22000 provides definitions on related terms and describes a food management system, including:

- General system requirements
- Definition of management responsibility and commitment
- Documentation requirements
- Definition of responsibility and authority
- Calling for a food safety team, communication, and contingency preparedness and response
- Gives a review on management, resource management, provision of resources, human resources, realization of safe products, product and process data, hazard analysis, design of the **critical control point (CCP)** plan, operation of the food safety management system, control of monitoring and measuring devices, measurement, analysis, and updating of the FSM system
- System verification, validation, and updating
- Correspondence between ISO 22000:200x and ISO 9001:2000

For a greater understanding of traceability in feed and food chain, see Appendix E regarding the general principles and basic requirements for system design and implementation.* A sample of ISO's costs is shown in Table 6.4.

6.18 HACCP STANDARDS, HACCP WEB.COM, AND HACCP TRAINING PROVIDERS

The main focus of HACCP (pronounced "hassip") and its relationship to IPT can be best seen through HACCP Principles #6—recordkeeping and #7—verification

---

* From the Draft International Standard ISO/DIS 22005. Traceability in the feed and food chain—general principles and basic requirements for system design and implementation. ISO 22005 was prepared by ISO Technical Committee ISO/TC 34: food products.

TABLE 6.4

ISO 22005 Publication Cost

ISO/DIS 22005

Ed. 1	Current Stage 40.60	TC 34	
		Product	Price in CHF
Traceability in the feed and food chain—		ISO/DIS 22005 PDF version (en)	64,00
General principles and basic requirements		ISO/DIS 22005 PDF version (fr)	64,00
for system design and implementation		ISO/DIS 22005 paper version (en)	64,00
		ISO/DIS 22005 paper version (fr)	64,00

ISO/DIS 22005 publication

Traceability in the feed and food chain—General principles and basic requirements for system design and implementation

Publication target date:			December 15, 2006
Product:		Size	Price
	ISO/DIS 22005 PDF version (en)	158 kb	CHF 64,00

Source: Information used with permission of ISO–ANSI.

(see below). HACCP is important because many standards and quality systems that deal with IPT either directly include HACCP standards or HACCP concepts in their design. HACCP is highly regarded on a global level, and understanding its concepts and interactions with CCPs helps ensure that a company will have an active and effective IPT program. The HACCP principles are considered by many as a naturally adaptive tool for implementing a successful IPT program and address many concerns that face food safety.[*]

6.18.1 HACCP History—Standards[†]

Traditionally, industry and regulators have depended on spot-checks of manufacturing conditions and random sampling of final products to ensure safe food. However, inspection and testing are like a photo snapshot. They provide information about the product that is important only for the specific time the product was inspected and

---

[*] Much of the HACCP information is derived from many resources: http://www.ces.ncsu.edu/depts/foodsci/ext/pubs/haccpprinciples.html, http://www.fsis.usda.gov/oa/background/keyhaccp.htm, http://www.bulltek.com/english_site/iso9000_introduction_english/haccp_english/body_haccp_english.html, http://www.ourfood.com/HACCP_ISO_9000.html, http://www.cfsan.fda.gov/~lrd/bghaccp.html, and http://www.haccpweb.com/index.html (accessed July 14, 2006).

[†] HACCP is endorsed by the U.N. Codex Alimentarius, U.S. FDA and USDA, EU, Canada, Australia, New Zealand, Japan, and many other countries and trade organizations. In addition, the European hygiene rule defined in the paper 94/356/EG demands for an HACCP concept that can be integrated in a QMS.

tested. What happened before or after is unknown. However, this approach tends to be reactive rather than preventive and can be less efficient than the HACCP system. From a public health and safety point of view, traditional methods offer little protection or assurance.

The drive behind modern HACCP programs first began as a natural extension of GMPs that food companies had been using as a part of their normal operations. A system was needed that enabled the production of safe, nutritional products for use by the National Aeronautics and Space Administration (NASA) starting in the late 1950s to feed future astronauts who would be separated from medical care for extended periods of time. Without medical intervention, an astronaut sickened by food-borne illness would prove a very large liability and could possibly result in the failure of entire missions. Food products could not be recalled or replaced while in space.[*]

6.18.1.1 How HACCP Was Created

Beginning in 1959, the Pillsbury Company embarked on work with NASA to further develop a process stemming from ideas used in engineering systems development known as Failure Mode and Effect Analysis (FMEA). Through the thorough analysis of production processes and identification of microbial hazards that were known to occur in the production establishment, Pillsbury and NASA identified the critical points in the process at which these hazards were likely introduced into product and therefore should be controlled.[†]

The establishment of critical limits of specific mechanical or test parameters for control at those points, the validation of these prescribed steps by scientifically verifiable results, and the development of recordkeeping by which the processing establishment and the regulatory authority could monitor how well process control was working all culminated in what today is known as HACCP. In this way, an expensive or time-consuming testing procedure is not required to guarantee the safety of each piece of food leaving an assembly line, but rather the entire process has been seamlessly integrated as a series of validated steps.

In 1971, the HACCP approach was presented at the first American National Conference for Food Protection. In 1973, the U.S. FDA applied HACCP to low-acid canned foods regulations, although if you read those regulations carefully you will note that they never actually mention HACCP. From 1988 to the present day, HACCP principles have been promoted and incorporated into food safety legislation in many countries around the world.

---

[*] http://en.wikipedia.org/wiki/HACCP (accessed January 5, 2007).

[†] One of the primary forces behind the expanded use of HACCP principles has been the proliferation of new food pathogens. For example, between 1973 and 1988, bacteria not previously recognized as important causes of food-borne illness (e.g., *Escherichia coli* O157:H7 and *Salmonella enteritidis*) became more widespread. There also is increasing public health concern about chemical contamination of food (e.g., the effects of lead in food on the nervous system). Another important factor is that the size of the food industry and the diversity of products and processes have grown tremendously, in the amount of domestic food manufactured and the number and kinds of foods imported. At the same time, FDA and state and local agencies have the same limited level of resources to ensure food safety. The need for HACCP in the United States, particularly in the seafood and juice industries, is further fueled by the growing trend in international trade for worldwide equivalence of food products and the CAC's adoption of HACCP as the international standard for food safety.

Beginning in 1996, the USDA established a detailed Pathogen Reduction/Hazard Analysis of Critical Control Point (PR/HACCP) program under the Food Safety and Inspection Service (FSIS) to regulate the production of raw meat products by large-scale facilities. There is currently no HACCP requirement in the United States for food processors such as supermarket deli or butcher departments that purchase from certified producers.

HACCP is a systematic methodology for analyzing food processing and identifying undesirable/hazardous inclusion of chemical, physical, or biological agents into foods. It is an expectation, if not a requirement, that organizations operating within the food supply chain identify, analyze, and act to prevent, eliminate, or reduce to acceptable levels inclusions of hazards. HACCP helps organizations to significantly reduce harmful contamination. Many of its principles already are in place in such places as the FDA-regulated low-acid canned food industry.

HACCP was introduced as a system to control safety as the product is manufactured rather than trying to detect problems by testing the finished product. This new system is based on assessing the inherent hazards or risks in a particular product or process and designing a system to control them. Specific points where the hazards can be controlled in the process are identified.

The HACCP system has been successfully applied in the food industry. The system fits in well with modern quality and management techniques. It is especially compatible with ISO systems such as the ISO 9000 quality-assurance system, ISO 22000 Food Safety Management System (FSMS), and just-in-time delivery of ingredients. In this environment, manufacturers are assured of receiving quality products matching their specifications. There is little need for special receiving tests and usually time does not allow for extensive quality tests.*

More specifically, HACCP is a process-control system designed to identify and prevent microbial and other hazards in food production. It includes steps designed to prevent problems before they occur and to correct deviations as soon as they are detected. Such preventive control systems with documentation and verification are widely recognized by scientific authorities and international organizations as the most effective approach available for producing safe food.

Key note: HACCP is a tool that can be useful in IPT, in addition to its primary purpose of the prevention of food safety hazards. Although extremely important, HACCP's food safety mission is only one part of a multicomponent food safety system. HACCP doctrine is very clear that HACCP is merely a tool and is not designed to be a stand-alone program. To be effective, other tools must include adherence to GMPs, use of sanitation standard operating procedures, and personal hygiene programs.

---

* The ISO 22000 Series provides for a full management system, fusing requirements of ISO 9001 with HACCP principles/plans (as they relate and can be referenced through ISO 15161). Organizations may opt to implement best global practices for planning, identification of hazards, acting, and improving food safety processing through the HACCP Management System (HACCP MS) or ISO 22000:2005, FSMS, or alternatively ISO 9001:2000, which applies ISO 15161 guidelines. For laboratory services, the applicable international standard is ISO/IEC 17025.

6.18.1.2 European Regulation and Small Businesses

The EU introduced new food hygiene regulations on January 1, 2006 that require all food businesses within the EU, except primary producers, to operate food safety management procedures on the basis of HACCP principles. Significant flexibility has been included to allow small businesses to comply. HACCP systems are not readily applicable to food businesses like retail caterers, and the flexibility allows alternatives to HACCP that achieve the same outcome of safe food being produced. The U.K. Food Standards Agency has produced an adapted simplified version of HACCP for small caterers and retailers called Safer Food Better Business (SFBB) that uses this flexibility and is an example of how quality systems and HACCP principles can be creatively adapted for small businesses and different situations.* See Appendix F for a listing of recognized HACCP training providers.

6.18.2 THE SEVEN PRINCIPLES OF HACCP

The seven principles of HACCP are

1. Hazard analysis
2. CCP identification
3. Establishment of critical limits
4. Monitoring procedures
5. Corrective actions
6. Recordkeeping
7. Verification procedures

HACCP's main premise—HACCP is unique for its introduction of CCPs. Many other standards and rules have mimicked or adopted this HACCP notion. According to HACCP, a CCP is a point in the production line where a risk of hygiene may be put under control or eliminated. With appropriate measures at that point the risk can be avoided, eliminated, or reduced to an acceptable level. Examples of CCPs are

- Income of raw materials
- Storage and cooling of food
- pH of food
- Cleaning and disinfection
- Recipes, handling, and processing of food

---

* HACCP is endorsed by such scientific and food safety authorities as the National Academy of Sciences and the NACMCF, and by such international organizations as the CAC and the International Commission on Microbiological Specifications for Foods. HACCP offers several advantages. Most importantly, HACCP (1) focuses on identifying and preventing hazards from contaminating food; (2) is based on sound science; (3) permits more efficient and effective government oversight, primarily because the recordkeeping allows investigators to see how well a firm is complying with food safety laws over a period rather than how well it is doing on any given day; (4) places responsibility for ensuring food safety on the food manufacturer or distributor; and (5) helps food companies compete more effectively in the world market, reducing barriers to international trade.

- Defrost, heating, warm-hold phase, and cooling
- Distribution of food in restaurant, fast food
- Correct separation between clean and unclean sectors
- Hygiene of the surroundings and hygiene of the foodstuff

6.18.2.1   Principle #1—Hazard Analysis

Hazards* are conditions that may pose an unacceptable health risk to the consumer. A flow diagram of the complete process is important in conducting the hazard analysis, and measures to control those hazards are identified. The significant hazards associated with each specific step of the manufacturing process should be listed.[†]

6.18.2.2   Principle #2—Identify CCPs

A CCP is a point, step, or procedure in a food's process production—from its raw state through processing and shipping to consumption by the consumer—at which control can be applied and, as a result, a food safety hazard can be prevented, eliminated, or reduced to acceptable levels. Examples would be cooking, cooling, packaging, metal detection, acidification, or drying steps in a food process.

6.18.2.3   Principle #3—Establish Critical Limits

All CCPs must have preventive measures that are measurable and quantified. A critical limit is the maximum and/or minimum value [or operational boundaries (limits)] to which a physical, biological, or chemical hazard must be controlled to prevent, eliminate, or reduce that hazard to an acceptable level. The criteria for the critical limits are determined ahead of time in consultation with competent authorities. If the critical limit criteria are not met, the process is "out of control," thus the food safety hazard(s) are not being prevented, eliminated, or reduced to acceptable levels.[‡]

6.18.2.4   Principle #4—Monitor the CCPs

Monitoring is a planned sequence of measurements or observations to ensure the product or process is in control (critical limits are being met). Many governing bodies require that each monitoring procedure and its frequency be listed in the HACCP plan. It allows processors to assess trends before a loss of control occurs. Adjustments can be made while continuing the process. The monitoring interval must be adequate to ensure reliable control of the process.[§]

---

* Hazards may be biological, such as a microbe; chemical, such as a toxin; or physical, such as ground glass or metal fragments.
[†] Preventive measures (temperature, pH, moisture level, etc.) to control the hazards are also listed.
[‡] For a cooked food, for example, this might include setting the minimum cooking temperature and time required to ensure the elimination of any harmful microbes.
[§] Such procedures might include determining how and by whom cooking time and temperature should be monitored.

6.18.2.5 Principle #5—Establish Corrective Action

HACCP is intended to prevent product or process deviations. However, should loss of control occur, there must be definite steps in place for disposal of the product and for correction of the process. These must be pre-planned and written. These are actions to be taken when monitoring indicates a deviation from an established critical limit. The final rule requires a plant's HACCP plan to identify the corrective actions to be taken if a critical limit is not met.*

6.18.2.6 Principle #6—Recordkeeping

The HACCP system requires that all plants maintain certain documents, including hazard analyses, written HACCP plan, and records documenting the monitoring of CCPs, critical limits, verification activities, and the handling of processing deviations. This must include all records generated during the monitoring of each CCP and notations of corrective actions taken.†

6.18.2.7 Principle #7—Verification

This would include records of hazards and their control methods, the monitoring of safety requirements, and action taken to correct potential problems. Validation ensures that the plans do what they were designed to do; that is, they are successful in ensuring the production of safe product. Plants are required to validate their own HACCP plans. Verification has several steps and may include such activities as review of HACCP plans, CCP records, critical limits, and microbial sampling and analysis.‡

6.18.3 HACCPWEB.COM

HACCPweb.com is a website and a separate entity from HACCP. Throughout the food industry, some companies use HACCPweb.com to help with their food safety strategies. Whether it is primary producers (juice, baking, meat, or seafood), catering, or retail, website subscribers take an active role in ensuring the safety of the food they handle.

HACCP is legally required by food businesses throughout the United States and Europe; unfortunately, it is often expensive and difficult to implement. HACCPweb. com was developed by One World Learning Ltd., a company set up in 2001, with the objective of making it easier for companies to adhere to food safety regulations.

HACCPweb.com is another avenue to achieving HACCP-compliant status. For many organizations, compliance with HACCP is challenging. However, the notion is that by using HACCPweb.com, HACCP compliance will be as easy as switching on

---

* If, for instance, a cooking step must result in a product center temperature between 165°F and 175°F, and the temperature is 163°F, the corrective action could require a second pass through the cooking step with an increase in the temperature of the cooker. Corrective actions are intended to ensure that no product injurious to health or otherwise adulterated as a result of the deviation enters commerce.

† Usually, the simplest recordkeeping system possible to ensure effectiveness is the most desirable; for example, testing time and temperature-recording devices to verify that a cooking unit is working properly.

‡ FSIS is requires that the HACCP plan include verification tasks to be performed by plant personnel. FDA and USDA are proposing umbrella regulations that will require HACCP plans of industry.

a computer. By using the online software at HACCPweb.com, companies can design their own HACCP plan by customizing the website's template procedures and training staff with computer-based training solutions.

HACCPweb.com customers have access to online HACCP software that takes clients through the seven principles of HACCP. Once signed up as a customer, the website's templates may be downloaded and may be modified to suit individual activities. HACCPweb.com system designs are in accordance with USDA, FDA, Canadian Food Inspection Agency (CFIA), U.K. Food Standards Agency (FSA), and Codex Alimentarius guidelines.

The National Advisory Committee on Microbiological Criteria for Food (NACMCF) and Codex Alimentarius have defined what constitutes a HACCP system and how it should be implemented. The HACCPweb.com course is designed according to Codex and NACMCF guidelines. The course enables students to fully participate in the development of HACCP. HACCPweb.com software is built into the course, enabling the user to build their HACCP plan as they study the seven principles. The HACCPweb.com course helps attendees to become in-house HACCP experts.

6.18.3.1   Prices for HACCPweb.com

HACCPweb.com services are available for 1 month (31 days) at the following rates. The standard package includes (1) the HACCPweb.com online application, (2) the online training course, and (3) the template prerequisite procedures. The cost of this standard package is normally $325 (U.S., or Sterling £189).* This is an e-learning course with voice-over and interactive exercises to reinforce learning. The HACCPweb.com software is built into the course to allow the user develop their HACCP plan as they study the course. The course provides the knowledge to enable participants to fully participate in the HACCP development process. Template procedures cover a range of topics including sanitation. The procedures are in rich text format, enabling them to be opened and altered by word processors such as Microsoft Word.

6.19   INTRODUCTION TO INTERNATIONAL ORGANIC STANDARDS

The following sections will discuss international organic standards, beginning with alternative systems, in which biodynamic agriculture as yet another alternative to traditional and organic farming will be discussed. The section on biodynamics provides an overview of organic farming by Dr. Delate and examines how orgainc farming has been at the forefront of IPT. The last two sections discuss the International Federation of Organic Agriculture Movements (IFOAM) and the International Organic Accreditation Services (IOAS), which became the first, and by all accounts the most well known, international organic organization. Finally, gaining membership and increasingly meeting the needs of smaller farmers of the Americas is the Organic Crop Improvement Association (OCIA). What follows are company/

---

* The HACCP course takes 8–10 hours to complete. The course covers the role of HACCP; prerequisite programs; microbiological, chemical, and physical hazards; the seven principles of HACCP; and HACCP implementation.

organizational statements from the websites of these organizations and naturally reflects their views.

6.20 ALTERNATIVE STANDARDS

In addition to systems that will be expanded upon later in this chapter, it is important to note that this list is not complete, there are many organic systems used throughout the world to meet local and international requirements, and these are outlined within various agreements.

6.20.1 Biodynamic Agriculture

Of particular notoriety is biodynamic agriculture. Going one step beyond organic production methods are biodynamic practices. Biodynamics is based on the 1924 work of Austrian scientist and philosopher Rudolph Steiner. He was interested in restoring the health of the soil, which European farmers were describing as "becoming depleted following the introduction of chemical fertilizers at the turn of the century." In addition to degraded soil conditions, farmers noticed "a deterioration in the health and quality of crops and livestock." Steiner looked at plants as being only one part of a connected system in which natural energy forces from the Sun, the Moon, the soil, and the air influence crop yield and quality. Biodynamics is a quest for balance between crops and their immediate or far-off environments.

According to biodynamic advocates, there are significant differences between integrated agriculture, organic agriculture, and biodynamic agriculture.

- *Integrated treatment* is a method of chemically fighting pests based on intervention thresholds set by models. In the case of bud eaters, if over 15% of the plants have at least one bud affected, an insecticide is justified. Synthetic chemicals are used to fight diseases or pests. It does not look into the causes of disease.
- *Organic treatment* is a protection method based on the sole use of natural products. Similar to integrated treatment, it does not look into the causes of disease.
- *Biodynamics* uses natural products, not just to combat disease, but also to respect the balance between the crop and its environment and to channel existing energies toward the crops. Accordingly, the plants' natural defenses are strengthened and the imbalance causing disease disappears.

The purpose of biodynamics is to give the soil new vitality, making it the living support of the crops. The crops will then send their roots deep to find a favorable environment of water, minerals, and trace elements. Fertilizing is based on a compost of dung and straw. The compost is allowed to decompose for 1 year before being buried in winter to allow the earth to absorb it. The soil is revitalized by this organic matter. The soil's fauna eat up the organic matter and grow and aerate the soil in the process.

Although some of the "dynamic" nonphysical forces of biodynamic agriculture (e.g., timing planting to correspond to lunar cycles) seem outside of the realm of most Western producers, some of the "biological" practices (i.e., green manures, cover cropping, composting, companion planting, crop rotation, and community-supported agriculture) are accepted and used by mainstream farmers. See the Biodynamic Agricultural Association's website for more information at http://www.biody-namic.org.uk; the Biodynamic Farming and Gardening Association at http://www. biodynamics.com/biodynamics.html; or the National Sustainable Agriculture Information Service on Biodynamic Farming & Composting Preparation at http://attra.ncat.org/attra-pub/biodynamic.html.

6.21   ORGANIC FARMING (OVERVIEW)

6.21.1   How Organic Farming Promotes IPT Principles

Dr. Kathleen Delate[*] of Iowa State University wrote about many of the essential requirements bound to organic farming in her paper *Fundamentals of Organic Agriculture*. This work highlights the commonality of organic production and what it entails. Although not all organic systems around the world (accredited and nonac-credited) are identical, they generally hold to strong principles of ecology: non-use of synthetic chemicals, environmentally friendly processes, animal welfare, nutrient cycling, efficient energy use, laborer welfare, and how farming interacts with society. Below are excerpts and expansions from Dr. Delate's work.

6.21.2   What Is Organic Agriculture?

According to the NOSB of the USDA, organic agriculture is

> [A]n ecological production management system that promotes and enhances biodi-versity, biological cycles, and soil biological activity. It is based on minimal use of off-farm inputs and on management practices that restore, maintain, or enhance eco-logical harmony. The primary goal of organic agriculture is to optimize the health and productivity of interdependent communities of soil life, plants, animals, and people.

Generally, products labeled as "organic" meet strict legal requirements, including certification by a third party. The organic requirements that permit a product to be deemed organic through its certification process are what promotes IPT principles.

Although the term "organic" may be defined by law (depending upon country and/or region's rules and regulations), the terms "natural" and "eco-friendly" may not be as well defined. Labels that contain those terms may imply some organic meth-ods or processes were used in the production of the foodstuff, but do not guarantee

---

[*] Delate, Kathleen. *Fundamentals of Organic Agriculture*. Ames, IA: Iowa State University Extension, Publication PM 1880, May 2003. Available at http://www.extension.iastate.edu/Publications/PM1880. pdf (accessed January 23, 2007). Kathleen Delate is an associate professor in the Department of Horticulture, 147 Horticulture Hall, Iowa State University, Ames, IA 50011-1100; Phone: 515-294-7069, E-mail: kdelate@iastate.edu.

complete adherence to recognized or certified organic practices as defined by a law. Some products marketed as "natural" may have been produced with synthetic or manufactured products (those not considered to be organic), such as "natural beef." Although eco-labels are promoted and advertised by producers interested in lowering synthetic inputs and farming with ecological principles in mind (biodiversity, soil quality, biological pest control), eco-labels are not regulated as strictly as USDA organic labels. This is where the intricacies of differing IPT programs are most noted. Specific IPT programs that focus on customers that demand, for example, fair market prices or plants grown in specific conditions may be met with tailored IPT programs that do not meet organic IPT specifications.

6.21.3   Organic History—a Short U.S. Perspective

Organic agriculture is the oldest form of agriculture. Farming without the use of petroleum-based chemicals (fertilizers, herbicides, and pesticides) was the only option in agriculture until after World War II. The war brought with it technologies that were useful and seemed advantageous at the time to agricultural production.[*]

For example, ammonium nitrate used for munitions during World War II evolved into ammonium nitrate fertilizer, and organophosphate nerve gas production led to the development of powerful insecticides. These technical advances since World War II resulted in significant economic benefits as well as environmental and social detriments.

Organic agriculture seeks to use those advances that consistently yield benefits (e.g., new varieties of crops, precision agriculture technologies, and more efficient machinery) but discard those methods that have led to negative impacts on society and the environment (e.g.,pesticide pollution and insect pest resistance). Organic farming is considered a systems approach in which interactions between components (crops, animals, insects, soil) are as important as the whole farm itself.

Instead of using synthetic fertilizers, organic farmers use crop rotations, cover crops, and compost to maintain or enhance soil fertility. Also, instead of using synthetic pesticides, organic farmers use biological, cultural, and physical methods to limit pest expansion and increase populations of beneficial insects. GMOs such as herbicide-resistant seeds and plants, as well as GMO-derived product ingredients such as GM-lecithin, are disallowed in organic agriculture because they constitute synthetic inputs and pose unknown risks.

Although U.S. organics is discussed below, many of its main tenants are shared by organic growers around the world.

6.21.3.1   U.S. Statistics

The USDA reported on organic production statistics in the United States (Coreene, 2007) and found that for the first time all 50 states in the United States had some certified organic farmland. U.S. producers dedicated over 4.0 million acres of farmland,

---

[*] This is not to suggest that everyone globally grew plants and animals in the same manner.

2.3 million acres of cropland, and 1.7 million acres of rangeland and pasture to organic production systems in 2005.* Over 40 states also had some certified organic rangeland and pasture in 2005, although only 4 states—Alaska, Texas, California, and Montana—had more than 100,000 acres. USDA lifted restrictions on organic meat labeling in the late 1990s, and the organic poultry and beef sectors are now expanding rapidly. The data set at http://www.ers.usda.gov/Data/Organic provides information on organic operations and acreage for crops and livestock (over 40 commodities), with some tables dating back to 1992. Data for 2000–2005 include the number of certified operations by state.

6.21.3.2 Philosophy

The motivations for organic production include concerns about the economy, the environment, and food safety. Although all organic farmers avoid synthetic chemicals in their operations, they differ in how they achieve the ideal system. Organic farmers span the spectrum. Some completely avoid external inputs by creating on-farm sources of compost for fertilization and encourage the activity of beneficial insects through conservation of food and nesting sites. Others import their fertility and pest management inputs. The philosophy of "input substitution" is discredited by many longtime advocates of organic agriculture. A truly sustainable method of organic farming would seek to eliminate, as much as possible, reliance on external inputs.

6.21.3.3 Organic Certification—Legalities and Logistics

When the U.S. Congress passed the OFPA in 1990, it was heralded by many as the first U.S. law to regulate a system of farming.† OFPA requires that anyone selling products as "organic" must follow a set of prescribed practices that includes avoiding synthetic chemicals in crop and livestock production and in the manufacturing of processed products. Organic certification agencies were established in the United States to provide the required third-party certification. Some states, including Iowa, followed suit and established their own organic laws. In 1990, Iowa passed Chapter 190, adopting the definition of organic as prescribed in OFPA and establishing penalties for producers falsely identifying their products as organic. Iowa allows private certification agencies to operate in addition to its own certification program. This system is in contrast to that of California, which relies on a single private certifier, California Certified Organic Farmers (CCOF), and that of Washington, which requires all farmers to be certified through the state. On average, inspection fees average $250 per year per farm to support the independent inspection structure. Additional fees are based on sales or individual acreage, depending on the agency.

---

* California remains the leading state in certified organic cropland, with over 220,000 acres, mostly for fruit and vegetable production. Other top states for certified organic cropland include North Dakota, Montana, Minnesota, Wisconsin, Texas, and Idaho. The U.S. organic industry continues to grow at a rate of 20% annually. Industry estimates placed it at $10 billion in 2001. Worldwide consumption of organic products has experienced tremendous growth, often surpassing the U.S. figures of 20% annual gain.

† This law can be accessed at http://www.ams.usda.gov/nop.

6.21.3.4   The Short Definition of Understanding U.S. Organic Labeling*

The U.S. NOP standards permit four different types of organic labeling. Foods labeled "100% Organic" must contain 100% organically produced ingredients. Products labeled "organic" must contain at least 95% organic ingredients. Packages that state, "Made with organic ingredients" must contain at least 70% organic ingredients. Packages that claim their products have some organic ingredients may contain more than 30% of conventionally produced agricultural ingredients and/ or other substances. Added water and salt are not counted as organic ingredients. The use of the USDA Organic Seal can only be used on the 100% and 95% organic products.

6.21.3.5   Required Certification Practices for Crops

To sell a product as organic, the crop must have been raised on land that had no synthetic chemical (including fertilizers, herbicides, insecticides, or fungicides) inputs applied for three years prior to its harvest. In addition, no GMO crops (e.g., Roundup Ready soybeans and Bt corn) are allowed in organic production.[†] Only naturally occurring materials are allowed in production and processing operations and all treatments must be noted in farm records.

6.21.3.6   Premium Prices Realized by Organic Farmers through IPT Practices

According to the Organic Alliance (http://www.organicalliance.org), organic premiums range from 20% to 400% above conventional prices, depending on the season and availability of the product. As an example, premium prices for organic carrots have ranged from 27% in the summer growing season to 200% in the winter months. Most consumers relate their willingness to pay premium prices for food raised without synthetic chemicals to their concerns about food safety and the environment.[‡]

According to *The Organic & Non-GMO Report* (January 2007, p. 9), organic watchdog publication *Organic Monitor* says that selling organic products in different markets is increasingly challenging for organic producers, as global demand continues to soar but differing national standards impede international trade. The three major trading blocks—North America, Europe, and Asia—are becoming more segregated as they increase in size. As markets and demand expand, manufacturers of organic foods and beverages find themselves unable to sell their products as organic in countries with different organic standards, resulting in production difficulties and excess bureaucracy.

---

* See USDA NOP standards in Section 6.6 for expanded information.

† Split operations, meaning conventional and organic fields, are located on the same farm and are allowed by Iowa law, but they require special care. For example, a border of 25 feet is recommended between organic and conventional fields in mixed operations.

‡ In addition to premiums, although many European countries financially support their farmers' organic production practices, the United States has made small gains in this area. In Iowa, the Natural Resources Conservation Services (NRCS) offers organic farmers $50/acre during their transition to organic farming through the Environmental Quality Indicators Program (EQIP) and through the new organic cost-share programs within the 2002 Farm Bill. Check with local NRCS or FSA offices regarding deadlines and required documents. Other conservation practices used on organic farms (e.g., riparian buffer strips, filter strips, and crop rotations) also may qualify for cost sharing.

The three major regulations that govern organic standards are the U.S. government's NOP, the EU standards, and Japan's JAS. "These standards are non-equivalent and quite separate. The main problem is that they don't recognize each other. There's a global shortage of organic products and U.S. producers should be able to sell their products in Europe and European products should be available in the United States," said the director of *Organic Monitor,* Amarjit Sahota.

6.22   IFOAM AND IOAS

IFOAM	IOAS
Charles-de-Gaulle-Str. 5	40 1st Avenue West, Suite 104
53113 Bonn, Germany	Dickinson, ND 58601
Phone: +49-0-228-926-50-10	Phone: 701-483-5504
Fax: +49-0-228-926-50-99	Fax: 701-483-5508
E-mail: headoffice@ifoam.org	E-mail: info@ioas.org

Available at http://www.ifoam.org/index.html and http://www.ioas.org.

Organic standards have long been used to create an agreement within organic agriculture about what an "organic" claim on a product means, and to some extent, to inform consumers. During the 1940s, regional groups of organic farmers and their supporters began developing organic standards. Currently there are hundreds of private organic standards worldwide. In addition, organic standards have been codified in the technical regulations of more than 60 governments.

Third-party organic certification was first instituted in the 1970s by the same regional organic farming groups that first developed organic standards. In the early years, the farmers inspected one another on a voluntary basis, according to quite a general set of standards. Today, third-party certification is a much more complex and formal process. Although certification started as a voluntary activity, the market began to demand it for sales transactions and now it is required by the regulations of many governments for any kind of an organic claim on a product label. In 1972, the founding members of IFOAM aimed to establish a communication network among organic agricultural communities that were emerging in multiple countries on several continents. Since its inception, IFOAM has also provided an international system to define and document the integrity of organic production and processing and to support the trade of organic products. This international system is now known as IFOAM's Organic Guarantee System (OGS).[*]

---

[*] The 30-year-plus history of IFOAM has proven that the proponents of organic agriculture embody an impressive agent of social and ecological revolution. It all started in 1972 when the president of the French farmers' organization, Nature et Progrès, conceived of a worldwide appeal to come together to ensure a future for organic agriculture. From there, people working in alternative agriculture banded together from, initially, as far apart as India and England. The German-speaking countries and France were also sites of the youngest IFOAM activities. Canada also produced key early participation, and by the 1980s IFOAM had leaders in the United States, attracted involvement from African agents of organic agriculture, and launched a unique and fruitful relationship with the FAO).

IFOAM is a grassroots and democratic organization that currently unites 750 member organizations in 108 countries. IFOAM has established official committees and groups with very specific purposes, from the development of standards to the facilitation of organic agriculture in developing countries. IFOAM is important to understand because many of the concepts used in IPT are derived from organic templates. Below is information regarding the organization and how IPT is incorporated into their programs. To facilitate organic production, the IFOAM World Board has established the following official structures:

- The Norms Management Committee, which includes members of the Standards Committee and the Accreditation Criteria Committee
- The Development Forum, which works toward the development of organic agriculture in developing countries
- The Program Strategy Committee of the IFOAM Growing Organic Program
- The Africa Organic Service Center and the FAO Liaison Office
- Various working groups and temporary task forces
- IFOAM regional groups
- The Government Relations Committee, which works with governments worldwide to advance the interests of IFOAM

IFOAM member organizations have also established professional bodies such as the IFOAM Organic Trade Forum, the Organic Retailers Association, the IFOAM Aquaculture Group, the IFOAM Forum of Consultants, and initiatives like the Farmers' Group.

IFOAM's mission is to lead, unite, and assist the organic movement in its full diversity. Their worldwide goal is the adoption of sound ecological, social, and economical agricultural systems, which are all based on the principles of organic agriculture. In the rapidly growing environment of marketing and trade of products claiming to be organic, IFOAM and its standards provide a market guarantee of the integrity of organic claims. The OGS unifies organic producers through a common system of standards, verification, and market identity. It fosters equivalence among participating certifiers, paving the way for more orderly and reliable trade. The IFOAM OGS internationally assures organic integrity. In this way, IPT play a key role in providing authenticity for the organic claims.

The IFOAM OGS enables organic certifiers to become IFOAM-accredited and certified operators to label their products with the IFOAM seal next to the logo of their IFOAM-accredited certifier. More than 30 certifiers worldwide participate in IFOAM accreditation. IFOAM accreditation guarantees to buyers, government authorities, other control agencies, and the public that a product has been produced within a system that conforms to accepted international standards for organic production, processing, and certification. See Appendix G for a listing of IFOAM-accredited certification bodies.

The two pillars of the OGS are (1) the IFOAM Basic Standards for Organic Production and Processing (IBS) and the (2) IFOAM Accreditation Criteria for Certification of Organic Production and Processing (IAC). These two international documents are norms with which certifiers must comply when conducting organic certification.[*]

The IBS, whose seeds were sown in 1978 and came to real fruition in the mid 1980s, was guided by the work of a technical committee. From then until now, the IBS have undergone periodic revisions that have been approved by the IFOAM membership.[†]

The next phase included the continued development of the IAC for organic certification bodies. The IAC were at first developed from "best practices" along with ISO Guide 65 (1994), and later with even more reference to ISO guidelines (1998).[‡]

In 1997, IFOAM decided that the accreditation program was best administered by a third-party organization, and it founded the IOAS for this purpose. The IOAS is incorporated in the United States as a nonprofit independent organization registered in Delaware that offers international oversight of organic certification through a voluntary accreditation process for certification bodies active in the field of organic agriculture. In 2004, 29 certification bodies worldwide were IFOAM-accredited within the OGS. Supported by this system, these accredited certification bodies (ACBs) are developing more and more functional equivalence with one another to streamline trade for their clients. This is done formally through a MLA. The IOAS implements the IFOAM Accreditation Program, which is an industry-based global guarantee of organic integrity, unburdened by national barriers and implemented by one body that has no other interests.[§]

Codex's principles of organic agriculture are the roots from which their view of organic agriculture grows and develops. The principles express the contributions that organic agriculture can make to the world and a vision to improve all agriculture in a global context.[¶] Organic agriculture is based on the principles of:

- *Health:* Organic agriculture should sustain and enhance the health of soil, plant, animal, human, and planet as one and indivisible.

---

[*] IFOAM's Basic Standards and Accreditation Criteria are generally respected as the international guideline from which national standards and inspection systems may be built, and they have been used as a reference by standard-setters and legislators in national and international arenas. IFOAM Basic Standards have had a strong influence on the development of Codex guidelines for the production, labeling, and marketing of organically produced foods.

[†] In 1986, IFOAM launched the development of an evaluation program for certifiers, administered by IFOAM's technical committee. Evaluation included visits to certification bodies and the generation of reports, which were then shared among participating certification bodies. The purpose of the evaluation program was to enhance trust between certification bodies.

[‡] IFOAM additionally developed IAC to reflect the particular circumstances of certifying organic production and processing. IFOAM owns and develops these documents through further revisions that involve stakeholder participation.

[§] In 1999, the IFOAM accredited certifiers signed a MLA for mutual recognition and equivalency aimed at streamlining the approval of products that are traded among their clients. The agreement acknowledges the functional equivalence of these certification programs on the basis of the IFOAM Basic Standards and Accreditation Criteria. See http://www.ioas.org for more information. (accessed July 17, 2006).

[¶] The principles apply to agriculture in the broadest sense, including the way people tend soils, water, plants, and animals to produce, prepare, and distribute food and other goods. They concern the way people interact with living landscapes, relate to one another, and shape the legacy of future generations.

- *Ecology:* Organic agriculture should be based on living ecological systems and cycles, work with them, emulate them, and help sustain them.
- *Fairness:* Organic agriculture should build on relationships that ensure fairness with regard to the common environment and life opportunities.
- *Care:* Organic agriculture should be managed in a precautionary and responsible manner to protect the health and well being of current and future generations and the environment.

Throughout its history, IFOAM has consistently succeeded at fostering active debate; networking beyond the borders of class, gender, and region; continually improving organizational structure, policies, and standards; attracting volunteers; working with the diversity of organic movements; producing standards that provide a model for numerous major laws and voluntary standards, (Codex Alimentarius, EU, FAO); and integrating scientific expertise and business sense into the emotional realm of organic agriculture. IFOAM has observer status or is otherwise accredited by the following international institutions:

- ECOSOC status with the U.N. General Assembly
- FAO
- U.N. Conference on Trade and Development (UNCTAD)
- CAC (FAO and WHO)
- WTO
- U.N. Environment Program (UNEP)
- The Organization for Economic Cooperation and Development (OECD)
- International Labor Organization of the United Nations (ILO)

6.23   OCIA INTERNATIONAL

OCIA International
OCIA Research and Education, Inc.
1340 North Cotner Boulevard
Lincoln, NE 68505
Phone: 402-477-2323 ext. 326
Fax: 402-477-4325
E-mail: lschroedl@ocia.org

Available at http://www.ocia.org.

The OCIA was founded in 1985 and has been incorporated since 1988. It is a nonprofit, member-owned organization providing quality organic certification services and an entryway to global organic markets. OCIA provides organic certification services to thousands of organic farmers, processors, and handlers from over 20 countries in North, Central and South America, Europe, Africa, and Asia; more specifically, certifying crops, livestock, processing facilities, warehouses, importers, exporters, brokers, traders, community grower groups, and private labels. Its multiple verification programs and many certifications (e.g., its Transaction Certificate System) have helped OCIA become another well-recognized international organic organization.

6.23.1 Short History of OCIA

In the depression years of the dustbowl, farmers started meeting informally to share their mutual farming experience. Having no technical support to enhance the development of their profession, they formed the first "crop improvement associations." The principles were simple: farmers are the experts on their lands; have regular meetings as opportunities to share their experiences with such techniques and trials; and acquire the basics of adapted technology.

In the mid-1970s the notion of organic agriculture began circulating within a group of pioneers. A certain parallel was noted between the technological situation of the 1920s and the challenge of the new organic "movement." Work started on the idea of an "organic" crop improvement association, which was envisioned as farmers working together to facilitate the development and the transfer of technical expertise. In the early 1980s, certification guidelines were formulated that eventually formed the basis of OCIA's certification program. After a few years, a few farm groups (chapters) formed independently and assumed the leadership of a combined crop improvement/certification program.

In the fall of 1985, in Albany, New York, a group of farmers met and structured the concept of a "farmer-owned and farmer-controlled" association. During those early years, OCIA became well rooted in many farming communities in Canada and the United States. Two important events occurred in 1988. First, the OCIA program took on an international identity when a group of Peruvian farmers joined the organization attracted by the concepts of farmer-to-farmer networking and crop improvement. Second, OCIA International was incorporated as a nonprofit organization in the state of Pennsylvania. From then on, the program expanded throughout Latin America. In the early 1990s, membership from Europe and Asia added further dimensions to the international body.

In January 1997, OCIA moved the International Office to Lincoln, Nebraska, which provided access to the University of Nebraska–Lincoln's sustainable agriculture program. Today, OCIA is a key player and one of the world's largest organic certification agencies. OCIA has thousands of members in North, Central, and South America; Africa; Europe; and the Pacific Rim.

OCIA offers multiple certification and verification programs that provide access to the global organic marketplace and an opportunity to reach consumers who are willing to pay premium prices for certified organic products. OCIA's Transaction Certificate System offers participants and their customers a point-of-sale guarantee of organic integrity.

6.23.2 Steps to OCIA Certification

The steps to OCIA certification include the following:

- To request application information from a regional office, write to info@ ocia.org or download information from the OCIA website.
- Become familiar with organic requirements. Regulations may vary and depend on growing location and sales region. For instance, to sell a

product as organic in the United States, you must be certified with the NOP, whereas organic products sold in Québec, Canada, must be certified with the Conseil des Appellations Agroalimentaires du Québec (CAAQ). However, one common element of organic farming regulation is that no prohibited materials may have been used for 3 years prior to the first organic harvest.

- Submit completed Associate Licensing Agreement (ALA), applicable questionnaire(s), supporting documentation (field histories, maps, inventory sheets, logs, etc), and identify which OCIA program(s) [NOP, IFOAM, CAAQ, JAS, and the Costa Rica Ministry of Agriculture and Livestock (MAG)] and which product(s) are to be certified.
- After the paperwork has been received and reviewed, an OCIA-approved inspector will inspect the facility. Upon completion of inspection, the requester and OCIA International will receive a copy of the report.
- The OCIA Certification Decision Team will review the requester file to verify that the operation is in compliance with OCIA's certification requirements.
- If the operation is found to be in compliance, OCIA will send a Certificate of Organic Certification. This certificate will list the certified products and the specific certification program(s) that the products are certified under. A letter and checklist will accompany the certificate and will provide guidance on what can be improved. The certification will remain in effect until it is surrendered, suspended, or revoked. However, annual update forms and inspections are required to maintain OCIA organic certification.

Another IP certificate program that OCIA offers is called Transaction Certificate (TC), which tracks OCIA-certified products from the grower to the grocery shelf. A TC verifies the origin of the product and is a point-of-sale proof that the product purchased was grown in accordance with the standards for one of OCIA's international programs.

OCIA certifies the following:

- Crops
- Livestock
- Processing facilities
- Warehouses
- Importers/exporters
- Brokers/traders
- Community grower groups
- Private labels

OCIA maintains accreditation with the following organizations:

- *JAS:* OCIA has achieved Registered Foreign Certification Organization (RFCO) status from Japan's Ministry of Agriculture, Forestry and Fisheries (MAFF). This has expanded OCIA's members' marketing opportunities in

Japan. The JAS certification requires additional questionnaires and documents for application.

- *JAS Equivalency:* OCIA performs certification that lignin sulfonate, alkali-extracted humic acid, and potassium bicarbonate have not been used in the growing or manufacturing of the specified product. JAS Equivalency allows for NOP-certified product to be exported to Japan, providing the applicant will be shipping to a JAS-certified importer that is willing to affix the JAS seal in Japan, or be processed by a JAS-certified processor/manufacturer in the United States. This is required for issuance of a USDA AMS Certificate (AMS or TM-11 form) of export.

- *CAAQ:* OCIA's CAAQ accreditation program allows for product to be sold as organic in Quebec. It is based primarily on the OCIA standards with additional standards that must be observed to be compliant with this program. Products that are certified to U.S. NOP standards outside of Quebec may be brought into Quebec without additional review on the basis of CAAQ's agreement of equivalence with the USDA's NOP program.

- *IFOAM (see Section 6.21):* OCIA's IFOAM-accredited certification program is focused toward many European, Canadian, and other international markets. The OCIA international standards are accredited by IFOAM. Certification to these standards allows for the OCIA seal to be used. Certifying to OCIA international standards allows for products to be imported into many countries that do not have established organic certification programs. Although many European countries accept IFOAM-accredited certified product for import, several EU member nations are now requiring EU 2092/91 verification for import (see below).

- *EU 2092/91 Organic Regulations:* This regulation stipulates organic production standards for EU member states. EU 2092/91 is the equivalent to the NOP of the United States. However, EU 2092/91 only establishes a minimum standard for organic production, as in the United States and its individual states' organic regulations, individual EU member states may have higher standards. This is often the requirement for organic imports into the EU.

- *U.S. NOP (see Section 6.6):* OCIA has NOP-accredited organic certification services for members desiring to market organic product in the United States. OCIA is accredited by the USDA for this program.

- *Bio-Suisse:* Bio-Suisse (Switzerland-based) is a certifier similar to OCIA. They certify to several different program standards such as the Swiss Ordinance of Organic Farming and to their own set of standards. OCIA facilitates certification for their members to Bio-Suisse but does not make the final decision. An additional questionnaire must be completed to apply for Bio-Suisse.

- *Swiss Ordinance of Organic Farming:* This is a verification of the minimum standard for organic production in Switzerland and is also required for organic products imported into Switzerland.

Other OCIA accreditation includes ISO Guide 65; USDA ISO Guide 65, and MAG.

6.23.3 OCIA Fees

Below is a partial list of OCIA fees. See Table 6.5 for additional fees.

1. *Membership fees:* $ 95.00. Fees entitle associate to vote at the Annual General Membership Meeting and a 10% discount on OCIA training classes.
2. *Certification fees:* OCIA offers three certificates: OCIA, NOP, and CAAQ. It also offers four verification programs: EU 2092/2091, JAS Equivalency, Bio-Suisse, and Swiss Ordinance. One certification fee entitles a member to apply for all programs. Fees are nonrefundable and due at application. Fees are based on total organic sales of the previous year. First-year associates may use projected organic sales.

Projected Sales	Fee
$0–$24,999	$1,000
$25,000–$49,999	$1,200
$50,000–$99,000	$1,500
$100,000–$249,999	$1,800
$250,000–$499,999	$2,750
$500,000–$749,999	$3,750
$750,000–$999,999	$5,000
$1,000,000 and above	$6,250

3. *Re-application late fees:* There is a late fee of $100 for every month past the anniversary date of the previous year's inspection date.
4. *Inspection fees:* OCIA will charge exactly what the inspector charges OCIA. Inspection fees are due at application.

6.23.4 OCIA Research and Education Goals and Objectives

OCIA's research and educational goals and objectives include the following:

- To provide organic crop improvement through professional development of farmers, processors, and consumers, including technical assistance, education information, publications, and research
- To clarify and promote the image of organic products
- To identify the needs of organic farmers and producers
- To promote the general welfare and cooperation of organic farmers, organic consumers, organic agriculture, and the organic foods industry
- To support crop improvement and marketing with farmers, consumers, and growers in such a manner that their self-sufficiency is not destroyed in order to fulfill the needs of the global organic market
- To promote research into health, environmental, and socioeconomic benefits that pertain to the organization or industry and organic agriculture in general
- To educate those interested in the organic industry and others including producers, consumers, and decision-makers in the benefits of organic systems

TABLE 6.5

OCIA Additional Certification Fees (as of 2008)

Fee(s)	Application for JAS Only	Application for JAS in Addition to OCIA, NOP, or CAAQ
Application fees	$1,700	$300
Pre-inspection review fees	$100	$100
Inspection fee per day	Latin America: $150/day + $275 rpt Canada: $320/day United States: $450/day Quebec: $550/day	Same
Post-inspection review fees	$200	$200
Actual travel expense	As billed	As billed
Total fees	$2,000 + inspection + travel expenses	$600 + inspection + travel expenses
JAS verification inspection	Whole process	Partial process
Arrangement fee	$300	$100
Document review fee	$50	$50
Inspection fee per day	$100	$100
Travel days fee per day	$50	$50
Report writing fee	$300	$100
Review fee	$200	$100
Actual travel expense	As billed	As billed

JAS certification: OCIA-Japan is accredited by MAFF to administer JAS certification program.
Source: Information used with permission from OCIA.

- To create links or strategic alliances with research institutions, universities, and others to achieve common goals
- To increase the effectiveness and integrity of the organic system
- To develop and maintain a mechanism for identifying and facilitating the exchange of producer-based information and information needs

6.24 INTRODUCTION TO REGIONAL AND RELIGIOUS STANDARDS

This chapter points to other traits or credence attributes of interest desired by many individuals and cultures, namely geolocations, religion, and specific other qualities or characteristics. Section 6.23.1 illustrates the numerous entities and organizations available for various targeted groups of interest not discussed more fully within follow-up sections. These regional and religious systems include:

- Freshcare
- Woolworths Quality Assurance (WQA) standard
- NCS International

- British Retail Consortium (BRC)
- The Forum de l'Agriculture Raisonnée Respectueuse de l'Environment (Forum for Environment-Friendly Integrated Farming, or FARRE)
- Institut du Végétal (ARVALIS)
- Société pour l'Expansion des Ventes des Produits Agricoles et Alimentaires (Society for Expanding Sales of Agricultural and Food Products, or SOPEXA)
- The Federation Nationale des Centres d'Initiatives pour Valoriser l'Agriculture et le Milieu Rural (National Federation of the Centers of Initiatives to Develop Agriculture and the Rural Medium, or FNCIVAM)
- The Fédération Nationale d'Agriculture Biologique (National Federation of Organic Farming, or FNAB)
- The Organisation Générale des Consommateurs (Consumers' General Organization, or ORGECO)
- Groene Hoed, or Green Hat
- Buddhists
- Hare Krishnas
- Hindus (Lower/High castes)
- Mormons

Sections 6.24 through 6.27 provide greater detail regarding well established and recognized regional and religious programs.

- Safe Quality Foods (SQF), originally founded in Australia
- JAS and organic standards
- Halal rules
- Kosher rules

What follows are company/organizational/agency statements from their websites and naturally reflects their views.

6.24.1   OTHER REGIONAL AND RELIGIOUS STANDARDS

In addition to systems that will be expanded upon later in this chapter, other regional and religious programs and their short summaries are provided below. This is not a complete inventory of other systems or standards that require and incorporate IPT practices.

6.24.1.1   Other Regional Systems

6.24.1.1.1   Freshcare

Freshcare is Australia's national, on-farm food safety program for the fresh produce industry. Freshcare links on-farm food safety to the quality and food safety programs of the other members of the food supply chain. On the basis of HACCP principles, Freshcare provides independent verification that a recognized food safety program is followed by the certified enterprises. Freshcare was developed in response to requests from growers, wholesalers, packers, and processors for a food safety program that met the requirements of retailers and food safety legislation. The foundations of Freshcare are the Code of Practice and Certification Rules. The code describes the practices required on farms to provide assurance that fresh produce is safe to eat and has been prepared to customer specifications. The original Code of Practice and Certification Rules, developed in 2000, was reviewed and updated by the Freshcare Technical Steering Committee in 2004. Although the basic Freshcare Program addresses food safety issues, additional (optional) modules are being developed for the management of environmental practices and on-farm safety/welfare issues. This ultimately provides an option for EurepGAP equivalence for those members for whom EurepGAP compliance is an export market requirement.[*]

6.24.1.1.2   WQA

WQA is another Australian system to meet their public's demand for quality assurance. The recently released WQA Standard replaces Woolworths Vendor Quality Management Standard (WVQMS). WQA includes HACCP and specifically requires that at least one person from the business has attended formal HACCP training. It also requires formal training in internal auditing. The system requires procedures and records for other support programs, such as GMPs, pest control, cleaning, product identification, and traceability and product recall. The audit frequency is generally every 6 months.

6.24.1.1.3   NCS International

NCS International is the leading Asia-Pacific certification and auditing body for the food and agricultural industries. NCSI Food Division clients include small growers and packers, food processors, distributors, caterers, retailers, and the hospitality industry. They offer training and support to clients, and assistance with improving their businesses. As well as independent HACCP certification, NCS International audits to various commercial and international management and product certification standards designed to meet the needs of the food and agri-food industries as well as ISO 9001, ISO 14001, and AS 4801.[†]

6.24.1.2   Other Major IPT Trends in Europe[‡]

6.24.1.2.1   BRC Global Food Standard

The BRC is the lead trade association representing the whole range of retailers—from the large multiples and department stores to independents—selling a wide selection of products through metropolitan, suburban, rural, and virtual stores. Their aim is to bring about policy and regulatory changes that will ensure retailers can maintain their outstanding record on product innovation, job creation, and consumer choice. BRC requires food manufacturers to have in place a fully operational HACCP system; quality management system; factory environment standards; and product, process, and

---

[*] Freshcare Ltd., P.O. Box 247, Sydney Markets NSW 2129. Phone: +61-2-9764-3244, Fax: +61-2-9764-2776, E-mail: info@freshcare.com.au.

[†] See http://omega.jtlnet.com/~ncsi/ncsi.com.au/home.html for more information.

[‡] Excerpts and modified with permission from Northern Great Plains, 2003 Annual Report, "The European Study Tour: Rethinking the Role of the Producer." Available at http://www.ngplains.org/documents/2003%20Annual%20Report%20Final.pdf (accessed June 20, 2006).

personnel controls. It applies generally to food and beverage manufacturers supplying into British retail interests within the United Kingdom and overseas.[*]

6.24.1.2.2   FARRE

FARRE is known for integrated farming and certification with an emphasis on protecting the environment. They have a strong focus on environmentally sensitive agriculture production practices with acceptance of some use of chemicals and use third-party certification as ways in which farmers help protect the environment. FARRE is also perhaps the best organized effort to hold producers accountable for the environment by focusing on management, recordkeeping, labeling, and enforcement by the general public rather than government. Founded in France in 1993, FARRE has 400 active farm members in 53 regions of France working voluntarily to implement FARRE's program. FARRE's purpose is to promote *Agriculture raisonnée* (or integrated farming), a competitive form of farming that aims to satisfy three criteria: (1) the financial objectives of producers, (2) consumer demands and expectations, and (3) care for the environment. Members of the FARRE Farm Exchange Network are selected and approved by local committees and the National Executive Committee. All farmer members must sign the FARRE Charter, which details their commitment. They also agree to implement the environmental self-diagnosis process drawn up by FARRE's scientific advisory board. FARRE permits chemical use but emphasizes training of chemical applicators; extensive recordkeeping in the on-farm storage, use, and disposal of chemicals; and transparency of these records so that customers can know a farm's record of chemical use.[†] [‡]

6.24.1.2.3   ARVALIS

ARVALIS (*Institut du Végétal*) is known for producer support and research in support of environmental management and traceability (e.g., careful management of chemical inputs). This French organization takes a conventional approach to environmental protection by careful management of chemical inputs in food production. Their approach to protecting the environment has been to develop and use sophisticated technological controls to manage the amounts, combinations, dosages, and schedule of chemical fertilizer usage. As part of its focus on managing chemicals put into the environment by farming practices, ARVALIS has developed high technology software to help producers control their nitrogen dosage to achieve the twin goals of reducing costs to producers and reducing harm to the environment. Several ARVALIS-developed programs aim at giving producers the ability to control their use of chemicals.[§]

---

[*] See http://www.brc.org.uk/defaultnew.asp for more information.

[†] For more information see http://www.farre.org.

[‡] There are national associations similar to FARRE in six other European countries: Germany (FIP), United Kingdom (LEAF), Sweden (ODLING I BALANS), Spain (AGROFUTURO), Luxembourg (FILL), and Italy (L'Agricoltura che Vogliamo). These seven national associations are grouped together in the European Initiative for Sustainable Development in Agriculture (EISA).

[§] ARVALIS' protocols also address food quality and safety concerns. In 2001, the Technological Research Institute for the cereal-based food industry (IRTAC), the predecessor to ARVALIS, created the "Cereales de France" private label. This is a charter available to groups of ten or more producers who are committed to complying with the quality-assurance protocol rules established by

6.24.1.2.4 SOPEXA

SOPEXA is a French marketing company that emphasizes traceability as a way of guaranteeing safe and tasty food and pioneered the use of individual "passports" for meat products. SOPEXA has emphasis on detailed traceability as a way of guaranteeing safe and tasty foods. They heavily promote and market French food products.[*]

6.24.1.2.5 FNCIVAM

FNCIVAM is called "Durable Agriculture"—a network of local producers in opposition to industrial agriculture. They promote Durable Agriculture, with local farmers and consumers setting standards of taste and labeling as a guarantee of product conformity. The Durable Agriculture charter has three legs: (1) good for the consumer, (2) good for nature, and (3) good for the vitality of the countryside.

6.24.1.2.6 FNAB

FNAB was founded in 1975 and represents 70% of French organic farmers. France has about 12,000 organic farms, which comprise 1.25 million acres or 1.7% of French agriculture. They believe that organic farming is the best chance for a sustainable agriculture and emphasize rebuilding soil, direct contact with consumers, and a healthy environment. Some of their members use organic methods (such as green fertilizer, a compost-animal manure mix) on parts of their farms while reluctantly using some chemicals on traditional wheat, sugar beet, and barley acres. They receive some funding from the French Agriculture and Environment Departments.[†]

6.24.1.2.7 ORGECO

ORGECO defends consumers' interests in the food system, believes price is important, and lobbies for connecting public funds to providing public services.[‡]

6.24.1.2.8 Groene Hoed, or Green Hat

Green Hat emphasizes connecting city and countryside, establishing a personal connection between producer and customer, and selling locally or regionally. They believe that producers can play an important role in rural development. Their strategy is to reconnect the countryside and its nearby urban areas by using cooperative regional marketing and building Green Centers at the edge of cities.[§]

ARVALIS. The protocols are for wheat, durum wheat, malting barley, maize for forage, and sweet corn. Compliance with the protocols gives participating producers the right to use the private logo. ARVALIS requires producers to think through and record how their farms reconcile profitability targets with quality, protection of the environment, and traceability. For the 2003 season, nearly 30,000 French producers participated in the program, committing 1.2 million acres of wheat, durum, malting barley, and maize to assuring safe food by controlling the input of potentially harmful chemicals.

[*] For more information, see http://www.sopexa.co.uk.
[†] For more information, see http://fnab.org.
[‡] For more information, see http://www.orgeco.net.
[§] For more information, see http://www.groenehoed.nl.

As part of Green Hat's effort to reconnect rural producers with urban population centers, they are working out funding and other details to establish Green Centers as gateways between the city and the countryside. At these Green Centers

- Producers offer quality local products for sale.
- Professionals can be trained in nature, landscape management, and other green services.
- Groups can meet for planning retreats or conferences.
- There is a visitors and tourism center for individuals, families, schools, and groups.[*]
- Social service institutions can use farms as healing places for their clients.[†]

6.24.1.3 Other Religious Standards

Those with other religious standards include the following:

- *Buddhists:* Buddhist nutrition is generally veganic, with no bulb vegetables eaten (onions, garlic, etc.) and a ban of alcohol and caffeine.
- *Hare Krishnas:* Hare Krishnas practice a vegetarian diet with raw meals. Veganic nutrition is seldom followed. There is a ban of alcohol and caffeine.
- *Hindus:* The lower castes of Hindus have mixed nutrition with little meat, sheep, lamb, goats, pork, chicken, and fish (bovine and buffalo meat are not eaten). High castes (Brahmans) have a lacto-vegetal diet with exclusion of any kind of meat and fish and often an exclusion of eggs. Their diet avoids bulb vegetables (onions, garlic, and leek). Alcohol is forbidden.
- *Mormons:* Mormons are moderate in their nutrition. They generally eat fruits and vegetables and have a moderate consumption of meat. Such moderate nutrition is reduced in fat, albumin, cholesterol, and purines. Vitamins and dietary fibers are higher as found in normal nutrition.

6.24.2 OTHER PRIVATE SECTOR TRACEABILITY PROGRAMS

According to Hobbs (2003), voluntary labeling by firms, sometimes supplemented by third-party certification, is often used to identify credence attributes. If there is a market premium for "'safer" food, there is an incentive for firms producing products with enhanced levels of food safety to identify this attribute in a label. Irradiated meat products in the United States are a good example. A credible monitoring and enforcement mechanism is necessary to reduce the risk of cheating through mislabeling. A self-policing industry quality assurance or safety labeling program could be effective if those firms producing "high-quality" (or demonstrably safer) foods are able to censure those firms who free-ride on the certification program through false

---

[*] The Green Hat facility can be a learning center for nature education and information about regionally grown food, as well as a central point for arranging tours so that urban visitors can experience farms, the open peat bog meadow landscape, or water country.

[†] With proper organization, farms can become temporary healing places for those with psychiatric problems, the long-term unemployed, those with certain handicaps, etc.

or misleading labeling. In the absence of an effective self-policing mechanism, the market failure problem persists for products with negative quality or safety attributes. A firm will not voluntarily disclose low quality.

Private sector traceability initiatives in the livestock sector include individual supply chain initiatives and industry-wide programs. Supply chain partnerships delivering traceability have emerged in the U.K. beef industry, largely as a result of the loss in consumer confidence following the BSE crisis. One example is Tracesafe, a small farmer-owned company that has developed a network of cattle breeders and finishers who rear cattle to specific production guidelines. The production protocols specify the purchase of feed from a set of contracted feed mills and include an extensive system of on-farm recordkeeping. Tracesafe differentiates its beef on the basis of its ability to trace the history of individual meat cuts to the animal of origin, with an implied safety assurance. The beef is sold in specialist retail outlets and restaurants under the Tracesafe brand name (Hobbs 2003).

The meat-processing sector has also recognized the potential role of traceability in bolstering consumer confidence in food safety and as a product differentiation strategy. Michael McCain, President and Chief Executive Officer of Maple Leaf Foods, Inc. (a major Canadian pork and poultry processor) recently referred to traceability as the "holy grail of the food supply chain." Maple Leaf is currently funding the development of DNA identification technology to facilitate the traceback of meat to the farm of origin. Pressure from export markets, particularly the Japanese market, appears to be a significant driver for this development (Fearne 1998).

A voluntary grading system, Meat Standards Australia (MSA), uses a series of pre- and postslaughter measures to predict the eating quality of meat. Blood samples are taken from each carcass that qualifies for the MSA program while the carcass can still be identified with a seller. If a consumer complains of a bad eating experience from MSA-graded meat, a DNA sample from the meat and can be matched with the blood sample from the carcass. In this way, meat cuts can be traced through the supply chain and to the farm of origin. The traceback in the MSA system is focused primarily on quality rather than just food safety. It allows a direct link to be made between eating quality and production and processing methods. It can assist in identifying where improvements may be necessary or in identifying sellers who consistently misrepresent cattle on their National Vendor Declaration form (Fearne 1998).

6.25 SQF INSTITUTE (A DIVISION OF THE FOOD MARKETING INSTITUTE)

Food Marketing Institute (FMI)
655 Fifteenth Street, NW, Suite 700
Washington, DC 20005
Phone: 202-220-0635
Fax: 202-220-0876
E-mail: info@sqfi.com

Available at http://www.sqfi.com.

6.25.1 HISTORY OF SQF

The SQF programs have continued to evolve since 1995 in response to the needs of primary producers and food processors to consumers worldwide who are more frequently voicing their demand for safe food of consistent quality. The actual story below provides good reason for having an IPT program in place. During the years of *Escherichia coli* (*E. coli*) outbreak, there was no such system in Australia and innocent companies, such as Wintulichs Pty. Ltd. and the industry as a whole, paid a heavy price both in sales and brand names. What follows is modified and includes some excerpts from David Pointon's 1995 work.

6.25.1.1 *E. coli* Incident

In January 1995, one child died and many children and some adults were admitted to hospitals as a result of the presence of *E. coli* in some small goods products produced in South Australia, allegedly by Garibaldi Smallgoods Pty. Ltd. Unfortunately, a distinctly different manufacturer, Wintulichs Pty. Ltd., also lost sales because of *E. coli* and the industry's lack of IPT. Although Wintulichs Pty. Ltd. had a quality raw material purchasing and production process, this incident had a devastating impact on the company and the industry as a whole. The carrier, Metwurst, had sales fall to less than 10% of that before the *E. coli* incident. This incident illustrates that irrespective of how a company structures its marketing plan, how well established it is, or how well it complies with public health regulations, changing, uncertain, and unpredictable environmental factors can profoundly affect a company's performance (Pointon 1995).

E. coli is one of the predominant organisms found in the gut of all animals, including humans. It is usually harmless in its environment but certain strains can produce disease. In the process of disembowelment of cattle, poor standards in meat preparation can result in bowel rupture, allowing feces to contaminate the red meat. Inefficient production processes can fail to destroy the *E. coli* and therefore allow the affected meat to pass to humans (Pointon 1995).

Upon learning that children had suffered from food poisoning after consuming Metwurst, the general perception of consumers was that all Metwurst, irrespective of brand, was dangerous. The image of the product to the consumer was tarnished. Although the evidence was conflicting, it is believed that the general public in Queensland was not completely aware of the implications of the *E. coli* incident until a local company, concerned for the possible outcome on local sales, adopted a promotion strategy to persuade its consumers that it had nothing to do with the South Australian incident.

The publicity also provided the consumer with the details of how the product was made. Previously many consumers were unaware of the production process for Metwurst. The publicity provided information to the consumer that Metwurst was made of fermented meat and included the management of bacteria. As a result of the publicity, there was a lot of resistance on the part of retailers to promote or even stock the product. The company could no longer rely on retailers to "push" the product.

In response to the demands by the farming and small food manufacturing sectors in the early 1990s, the Western Australian Department of Agriculture began to search for a suitable system to implement. The systems that were available at that time did not meet the needs of those sectors that wanted a quality-assurance system that enabled their businesses to meet regulatory food safety and commercial food quality criteria. Because no suitable system could be identified, the department then set about developing the SQF 2000 Quality Code.

SQF means "Safe Quality Foods (Healthy Foods and of Quality)." The SQF Institute is a division established by FMI to manage the SQF Program.*

Information regarding the FMI:

- It is a nonprofit association conducting programs in research, education, food safety, industry relations, and public affairs.
- It has 2,300 members (food retailers and wholesalers).
- It includes 26,000 retail stores with an annual sales volume of $340 billion.
- It is also international, with 200 companies in over 50 countries.

6.25.2 SQF Institute

The SQF Institute comprises many entities; of importance is its Technical Committee, a team of food safety and quality specialists that reviews the SQF Program and makes recommendations on improvements to the codes; the training materials; and the implementation, audit, and certification requirements.

6.25.3 SQF Program

The SQF Program is recognized by the Global Food Safety Initiative (an organization representing over 70% of food retail revenue worldwide). It is based on the principles of HACCP, Codex, ISO, and QMS. The SQF Program provides a protocol or a frame for the implantation, administration, and verification of the HACCP, in agreement with the Codex Alimentarius and the standards of ISO.

SQF Program features:

- It is based on HACCP and quality management principles.
- It is an all-encompassing management system.
- It is aligned with International HACCP protocols (Codex Alimentarius).
- It is customer-focused and addresses the safety and quality of the food and its processes.
- Optional certifications modules are available.

SQF is recognized by the Global Food Safety Initiative as conforming to the highest international standards and utilizes protocols recognized by the IAF. Over 4,000 companies operating in Asia/Pacific, the Middle East, United States, Europe, and South America have implemented the SQF Program. Registered SQF experts and SQF auditors implement and audit SQF systems around the world.

---

* An FMI advisory board provides overall policy advice, guidance, and direction to the SQF Institute.

The SQF Program is regarded as rigorous and flexible, and it complements government programs and industry initiatives. It also attempts to avoid the duplication and confusion associated with the current array of industry sector programs. Simply put, the SQF certification provides and is

- A tool to build confidence and trust between retailers and suppliers
- The enabling tool for producers and manufacturers to demonstrate "due diligence" and compliance with regulatory and product traceability requirements
- An internationally recognized standard, suitable for all food suppliers operating in domestic and global markets
- A means to reduce the number and frequency of inconsistent and costly audits
- A proven way for suppliers to gain an advantage over their noncertified competitors and to increase profits by aligning their products to retailer/consumer requirements
- A management system that promotes cost efficiencies through waste reduction and "one system, one audit"

SQF certification provides an independent and external validation that a product, process, or service complies with international, regulatory, and other specified standard(s) and enables a food supplier to give assurances that food has been produced, prepared, and handled according to the highest possible standards.

6.25.4   WHY SQF IS OF VALUE

According to Noonan and McAlpine (2003) of Agri-Food Training Centre, the clear difference between SQF and many other quality-assurance programs is that it is outcome-focused and is not prescriptive. Although prescriptive schemes may have a strong appeal to some businesses, most have a desire to build dynamic management systems that have ownership and flexibility as the key components, which an SQF structure more readily enables. In addition, Noonan and McAlpine noted that it has often been said that implementation of HACCP at the farm level is too hard and that nothing could be further from the truth. There is clear evidence in the work being done in Western Australia by the Agri-Food Training Centre and QA Management Tek and others in Eastern Australia that it is possible and, for most primary producers not too hard, to implement a HACCP-compliant management system.

6.25.5   BENEFITS OF SQF

The ISO 9000 (and follow-up ISO 22000) series has had, and continues to have, a strong focus in business management. However, it has not been readily taken up by small businesses because of the high costs and overheads associated with implementation and maintenance.

HACCP is a tool, a methodology, and by itself has no constructs against which it can be audited other that the CAC or NSCMCF guidance documents. Herein lies the problem. The guidance documents for the application of the HACCP technique are not codified. As a consequence, it is difficult to get consistency of interpretation. HACCP does not address food quality issues per se (Noonan and McAlpine 2003).

The SQF 2000^{cm} and SQF 1000^{cm} codes, with the Victoria Meat Authority System, are the only HACCP-compliant or HACCP-based systems that have been endorsed by JAS-ANZ as third-party audited standards that comply with ISO guides 62 or 65.

These SQF codes have been developed so that they can account for factors, in addition to food safety, inclusive of, but not limited to

- Product quality hazards
- Environmental hazards
- Animal welfare hazards
- Production hazards
- Occupational health and safety hazards
- Regulatory hazards
- Ethical production
- GMO status

Of some importance is that an outcome-focused standard, which is what these SQF codes are, can lead to some ambiguity and lack of conformity in the audit process, which therefore requires a high level of training and understanding in the implementation and audit process. A leading and overriding factor is that the outcome-focused approach more readily enables a robust system of assurance. This is especially the case in varying production environments where hazards associated with producing and further processing of food and fiber can and do change (Noonan and McAlpine 2003).

6.25.6   SQF 1000 Code

The SQF 1000 code is designed specifically for primary producers. In addition to GAPs, a producer develops and maintains food safety and food quality plans to control those aspects of their operations that are critical to maintaining food safety and quality.

6.25.7   SQF 2000 Code

The SQF 200 code has wide appeal across the food manufacturing and distribution sectors. In addition to GMPs, a supplier develops and maintains food safety and food quality plans to control those aspects of their operations that are critical to maintaining food safety and quality.

6.25.8   SQF Certification

SQF certification provides an independent and an external validation that produce and other foods, their production and manufacturing processes, and related service complies with food regulations and other specified standards such as the SQF 1000 code or the SQF 2000 code. Certification enables a food supplier to give assurances that food they supply has been produced, prepared, and handled according to the highest possible standards. The certification of SQF systems is managed by internationally ACBs who are licensed by the SQF Institute. The ACBs oversee the activities of their SQF auditors, ensuring that they are qualified and provide a professional audit service. The ACB expert review panel reviews the results of an audit and a SQF certificate is then issued.[*]

6.25.9 SQF 1000 Quality Code[†]

SQF is divided into three levels:

1. *Level 1:* Food safety fundamentals
2. *Level 2:* Accredited HACCP food safety plans
3. *Level 3:* Comprehensive food safety and quality management systems development

Each level, which indicates the stage of development of a supplier's SQF system, builds on the previous steps, leading toward a comprehensive certification for food safety and quality. By dividing the implementation into more manageable and structured steps, the supplier can demonstrate continuous improvement while controlling costs and resources.

Only qualified SQF experts can implement SQF systems. Registered SQF auditors work with licensed and ACBs to provide SQF certification.[‡]

Below are illustrations of SQF 1000 and SQF 2000 codes with associated levels of development, and each code's strengths and weaknesses (see Figure 6.5).

SQF 1000 strong points include the following:

- It is a simpler system than SQF 2000 or ISO 9002.
- The system is designed for businesses supplying raw materials to SQF 2000 certified businesses or ISO-type businesses.
- Improvement of food safety and quality.
- Market place image enhanced.
- Development and strengthening of customer relationships.

SQF 1000 weaknesses include the following:

- SQF 1000 is a new standard and is not as widely recognized as ISO 9002 or even SQF 2000.
- Limited seafood HACCP practitioners available to assist with development.

---

[*] Certification is a statement that the supplier's SQF System has been implemented in accordance with the SQF guidelines and applicable regulatory requirements and that it is effective in managing food safety. It is also serves as a statement of the supplier's commitment to producing safe, quality food.

[†] Used with permission from Peter J. Bryar's "From Paddock to Plate—The First Step," published by AGWEST Trade and Development Principal, Innovative Horticulture, 80 Thompson Crescent, Research, Victoria, 3095, Australia, E-mail: pjbryar@netlink.com.au.

[‡] A supplier will be placed onto the SQF register (made available on the SQF website) after achieving Level 1, thereby immediately alerting their customers of their achievement and helping to raise customer confidence and support.

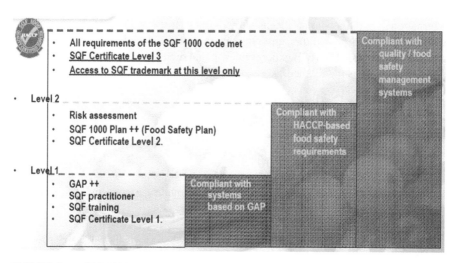

FIGURE 6.5   SQF 1000 illustration. (Reprinted with permission from SQF.)

- The code consists of the same six elements as SQF 2000 but is not as thorough in its application.

6.25.10   SQF 2000 QUALITY CODE

The SQF 2000 quality code is a network implementation program for small- and medium-sized businesses.* There are three levels of certification, similar to SQF 1000 code, but to SQF 2000 code level.

1. *Level 1:* Food safety fundamentals
2. *Level 2:* Accredited HACCP food safety plans
3. *Level 3:* Comprehensive SQF 2000 system implementation

The standard is utilized by organizations that produce, manufacture, or distribute food. It is relevant to fishing organizations that undertake simple processing or value-adding. The main feature of the SQF 2000 code is that it is a quality standard based on the Codex HACCP system. The code consists of six elements, including:

1. Commitment
2. Suppliers
3. Control of production
4. Inspection and testing
5. Document control and quality records
6. Product identification and traceability

---

* SQF 2000 quality code was developed by Agriculture Western Australia (AGWEST) as a practical alternative to JAS-NZS ISO 9001/2 for the small-to-medium food business (which includes primary producers) to meet growing demands by consumers (and retailers) for assurance on the quality and safety of foods that they consume.

See Figure 6.6 for an illustration of SQF 2000.

Primary producers who supply food that requires further processing through one or more steps use the SQF 2000 code. It is also used by small food businesses that are required to implement food safety programs specified as a requirement in the appropriate legislation of the country in which the food is processed or consumed.

SQF 2000 strong points include the following:

- It is a simpler system than ISO 9002.
- Improvement of food safety and quality.
- Market place image enhanced.
- Development and strengthening of customer relationships.
- Increased competitiveness.
- Staff responsibilities clearly defined.
- Reduction in waste and product rejects.
- Consistent supply of product to specification.
- Improving competitive advantage of the business.

The weakness of SQF 2000 is that it is a relatively new standard.

Several modules provided as voluntary options to suppliers whose markets require additional assurances for matters in addition to food safety and quality have also been developed to support the SQF program. They include social accountability, environment, animal welfare, organic, and bioterrorism modules.

System integrity is accomplished through

- Registered experts
- Registered auditors
- Licensed trainers
- Licensed and accredited internationally recognized certification bodies

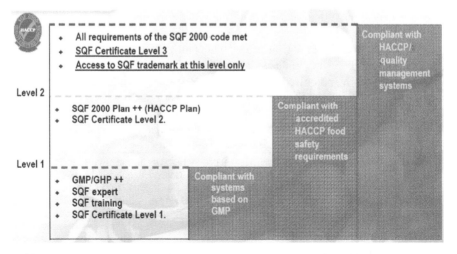

FIGURE 6.6 SQF 2000 illustration. (Reprinted with permission from SQF.)

6.25.11 SQF Accreditation—Qualification of Auditors

Accreditation of a certification body delivers confidence in the certificates and reports they issue. The international standard that a certification body must meet to be eligible to provide audit and certification services to the SQF Program is the ISO/IAF Guide 65 and other documents.[*] See Appendix H on SQF certification bodies for a listing of licensed SQF certification agents.

6.26 JAS

For the MAFF of Japan, see these websites: http://www.maff.go.jp/eindex.html, for labeling and standards see http://www.maff.go.jp/soshiki/syokuhin/hinshitu/e_label/index.htm.

6.26.1 Recent Events of JAS Importance

Since the enactment of the JAS law in 1950, the JAS system has contributed to improving the quality of agricultural and forestry products, facilitating simple and fair transactions of products, and providing consumers with information for informed choices. However, after deceptive labeling cases in 2002, the JAS system needed to be improved to assure reliable labeling and meet new social demands.

At the same time, the cabinet decided to review the government's involvement with the certification and inspection systems, including the JAS system, as a part of administrative reforms. On the basis of these situations, the committee to review the JAS system was established in October 2003. The committee consisted of stakeholders of the JAS, including consumers, industries, producers, distributors, and others.

After the intensive discussions at the nine sessions of the committee, the final report was published in October 2004, taking into account the public comments through the MAFF website and the opinions expressed at the public meetings. JAS has oversight of labeling for fresh foods, processed foods, and GM foods. They also have a certification system and regulate the process to import products with JAS marks into Japan.[†]

JAS is based on the Law Concerning Standardization and Proper Labeling of Agricultural and Forestry Products. It stipulates product information requirements for products such as processed food. The whole system is called the JAS System under the Law Concerning Standardization and Proper Labeling of Agricultural and Forestry Products (Law No.175, 1950), which governs all of the agricultural and forestry products, except for liquors, drugs, quasi-drugs, and cosmetics. The JAS

---

[*] These documents and additional requirements detailed by the SQF Institute in the document SQFI Guidance on the Application of ISO/IEC Guide 65:1996, General Requirements for Certification Bodies Offering Certification of SQF Systems, 4th Edition, April 2004. These documents address issues such as impartiality, competence, reliability of the audit, the certification service provided, and leads to confidence in the comparability of certificates and reports across national borders. Governments also have confidence in these testing and certification reports that support various regulatory requirements.

[†] Incorporating the recommendations of the committee, the laws and ordinances of JAS were revised and took effect on March 1, 2006.

System consists of the combination of the Quality Labeling Standard System and the JAS Standard System.

- *Quality Labeling Standard System:* This system requires all producers, distributors, and other parties to label in accordance with the Quality Labeling Standards established by the MAFF. All of the Quality Labeling Standards are mandatory so that all foods sold for consumers shall be labeled in accordance with them.
- *JAS Standard System:* The JAS Standard System refers to the certification system to attach the JAS marks to the products inspected in accordance with the JAS Standards established by the MAFF. The JAS Standards are voluntary, other than the JAS Standards for Organic Foods. Only certified business entities, such as producers and manufacturers, can attach JAS marks to the products.

According to *Food Traceability Report* (December 2006, p. 3), traceability in Japan is rising. Japan's MAFF reported that the rate of introduction of traceability systems rose last year across industries as a whole, but especially in the food retail sector. Use of traceability systems by Japan's food manufacturers rose by 3.5 percentage points over the previous year to 37.9% in 2005. For food wholesalers, it was 36.8% (a rise of 0.4 points compared to the previous year), whereas 35.8% of all Japanese food retailers had systems in place last year, a rise of 7.3% points compared with 2004.

Food producers were using identification systems to trace 76.3% of perishable foods and 71.5% of processed foods at the end of 2005. Rice farmers reported the highest level of recordkeeping for crops at 95.6% of all those surveyed. Fruit and vegetable farmers weighed in with 94.3% and 92.9%, respectfully, declaring they were keeping full records on cultivation and management.

6.26.2   JAS ORGANIC FOODS

Whereas JAS Standards are voluntary, the JAS Standards for Organic Foods are not voluntary. The JAS Standards for Organic Agricultural Products and Organic Agricultural Processed Foods were established in 2000 on the basis of the Guidelines for the Production, Processing, Labeling, and Marketing of Organically Produced Foods, which were adopted by the CAC.

The Organic JAS System has been further developed with the additions of the JAS Standards for Organic Livestock Products, Organic Livestock Processed Foods, and Organic Livestock Feeds, which took effect in November 2005.

The certified business entities are certified by registered Japanese certifying bodies or registered overseas certifying bodies that they produce or manufacture organic foods or feeds in accordance with the Organic JAS Standards for the products to be able to attach JAS marks to their products.

For more information regarding standards and criteria for JAS organic products, see http://www.maff.go.jp/soshiki/syokuhin/hinshitu/e_label/index.htm.

6.27 HALAL STANDARD*

The Codex general guidelines for use of the term "Halal" concerns specific process-based criteria for the use of the term "Halal" on food. As in the case with food origin's labeling, the only way to control that food delivered to the final consumer complies with the requirements for Halal is through an adequate paper-based traceability system. The same arguments also apply to the kosher code (see next section).

Many companies already have comprehensive traceability systems in place to facilitate effective safety and quality control. Such systems are also used in contractual agreements to guarantee companies further down the production and distribution line that products comply with legal and other quality requirements, thereby reducing the burden of testing of the contents of the food for subsequent operators. For such companies traceability is an essential risk management tool, enabling them to locate problems very quickly and cost effectively. A good traceability system increases the response capability of a company and reduces the risk of having to engage in extensive recalls, thus saving money and helping to maintain a reputable image. As is illustrated below, guidelines for Halal-prepared foods have well-established standards that preserve their identity and make traceability to their food sources easy and acceptable to consumers.

6.27.1 GENERAL GUIDELINES FOR USE OF THE TERM "HALAL"†

The CAC accepts that there may be minor differences in opinion in the interpretation of lawful and unlawful animals and in the slaughter act, according to the different Islamic schools of thought. As such, these general guidelines are subjected to the interpretation of the appropriate authorities of the importing countries. However, the certificates granted by the religious authorities of the exporting country should be accepted in principle by the importing country, except when the latter provides justification for other specific requirements.

The guidelines recommend measures to be taken on the use of Halal claims in food labeling.‡ By definition, "Halal" food means food permitted under the Islamic Law and should fulfill the following conditions:

- Does not consist of or contain anything that is considered to be unlawful according to Islamic Law.
- Has not been prepared, processed, transported, or stored using any appliance or facility that was not free from anything unlawful according to Islamic Law.

---

* From comments relating to the discussion paper on traceability (*Codex Circular Letter CL 2001/27 FBT*) by the European Commission, derived from http://ec.europa.eu/food/fs/ifsi/eupositions/tffbt/archives/tfbt_ec-comments_cl0127_en.pdf and www.codexalimentarius.net/download/standards/352/CXG_024e.pdf (accessed October 25, 2006).

† The Codex guidelines for the use of the term "Halal" were adopted by the CAC at its 22nd session in 1997.

‡ These guidelines apply to the use of the term "Halal" and equivalent terms in claims as defined in the General Standard for the Labeling of Prepackaged Foods and include its use in trademarks, brand names, and business names. The guidelines are intended to supplement the Codex General Guidelines on Claims and do not supersede any prohibition contained therein.

- Has not in the course of preparation, processing, transportation, or storage been in direct contact with any food that fails to satisfy both of the above.
- Halal food can be prepared, processed, or stored in different sections or lines within the same premises where non-Halal foods are produced, provided that necessary measures are taken to prevent any contact between Halal and non-Halal foods.
- Halal food can be prepared, processed, transported, or stored using facilities that have been previously used for non-Halal foods provided that proper cleaning procedures, according to Islamic requirements, have been observed.

6.27.2 CRITERIA FOR THE USE OF THE TERM "HALAL"

The term "Halal" may be used for foods that are considered lawful. Under Islamic Law, all sources of food are lawful *except* for the following sources, including their products and derivatives, which are also considered unlawful. Food of animal origin such as the following are prohibited:

- Pigs and boars
- Dogs, snakes, and monkeys
- Carnivorous animals with claws and fangs such as lions, bears, and other similar animals
- Birds of prey with claws such as eagles, vultures, and other similar birds
- Pests such as rats, centipedes, scorpions, and other similar animals
- Animals forbidden to be killed in Islam (i.e., ants, bees, and woodpecker birds)
- Animals that are considered repulsive like lice, maggots, and other similar animals
- Animals that live both on land and in water such as frogs, crocodiles, etc.
- Mules and domestic donkeys
- All poisonous and hazardous aquatic animals
- Any other animals not slaughtered according to Islamic Law
- Blood

Other prohibited items include:

- *Plants:* Intoxicating and hazardous plants, except where the toxin or hazard can be eliminated during processing.
- *Drink:* Alcoholic drinks, and all forms of intoxicating and hazardous drinks.

6.27.3 HALAL SLAUGHTERING

All lawful land animals should be slaughtered in compliance with the rules laid down in the Codex Recommended Code of Hygienic Practice for Fresh Meat and the following requirements:

- The person should be a Muslim who is mentally sound and knowledgeable of the Islamic slaughtering procedures.
- The animal to be slaughtered should be lawful according to Islamic Law.
- The animal to be slaughtered should be alive at the time of slaughtering.
- The phrase "Bismillah" (or "in the name of Allah") should be invoked immediately before the slaughter of each animal.
- The slaughtering tool should be sharp and should not be lifted off the animal during the slaughtering act, and the act should cut the trachea, esophagus, and main arteries and veins of the neck region.

6.27.4 HALAL PREPARATION, PROCESSING, PACKAGING, TRANSPORTATION, AND STORAGE

All Halal food should be prepared, processed, packaged, transported and stored in such a manner that it complies with the previous requirements and the Codex General Principles on Food Hygiene and other relevant Codex standards. In addition, in accordance with the Codex General Guidelines on Claims, claims on Halal should not be used in ways that could give rise to doubt about the safety of similar food or claims that Halal foods are nutritionally superior to or healthier than other foods.

6.28 KOSHER STANDARD*

"Kosher in the Mainstream" is an article from FoodProcessing.com and helps explain how kosher standards fit into IPT systems. The perception of higher quality is pushing "kosher," a Hebrew word meaning "connection," well into mainstream cooking and eating. Many perceive that kosher is of higher quality, possesses better taste and freshness, and includes a safety aspect by having a rabbi on scene for extra inspections. The notion that there are cleaner conditions in processing is a key component.

The acceptance of kosher food has spread, "Kosher quality has many non-Jewish customers eating kosher, that it makes them feel spiritual; it's food of faith," says Yossi Jacobson, Iowa's senior rabbi and the head of Chabad Lubavitch of Iowa. People believe kosher is better; it's like a "Good Housekeeping Seal of Approval." Consumers often feel they are largely on their own. The USDA cannot police everything, and there are 15,000 new food products released every year. There is reassurance associated with the fact that there is another set of eyes on kosher products.

Orthodox Jews represent a small segment of the overall kosher-buying population, which now comprises nearly 10.5 million consumers. Many kosher foods also meet the religious dictates of Seventh-Day Adventists, as well as Muslims who observe the tenets of Halal, the Islamic dietary laws. The fact that meat and dairy products are never mixed, necessitating clear labeling as well as innovative non-dairy recipe

---

* Excerpts and modified from Eric Schellhorn's *Kosher in the Mainstream*, 2005, with permission, as appeared in *Food Processing* magazine (http://www.FoodProcessing.com).

creation, makes kosher products attractive to lactovegetarians and the nation's 50 million lactose-intolerant consumers.

But even patronage by those groups does not fully account for kosher food's 15% annual growth rate and $175 million in 2003 U.S. sales. That year, a survey by Mintel Consumer Intelligence revealed that 28% of U.S. consumers had purchased kosher products. Of that group, 35% indicated they did so for "taste" or "flavor," whereas only 8% reported they kept kosher all year long. Christophe Hervieu, director of marketing for Osem USA, an Israeli kosher foods manufacturer owned by Nestlé, repeats the assessment: "Kosher products are looked at by non-Jewish people as being of a higher quality because there is a rabbi's supervision," he says. "If a product has kosher certification, they see there's been extra effort at the quality-control level."

Yaakov Luban, executive rabbinic coordinator at the Orthodox Union (OU—the largest of several hundred kosher-certification agencies), cautions that the "kosher-is-better" mindset invites some misconceptions. He stresses that the OU does not promote the idea that something kosher is always better or that the quality of kosher ingredients is always superior. The short answer is, kosher is the original "you are what you eat" model. The meaning of connection comes from the biblical implication that the foods we eat can enhance or detract from our connection to a higher power. Food that is kosher is "fit," or "proper," and is sourced, prepared, and served in compliance with laws derived primarily from the Torah (the first five books of the Bible) and the Talmud (the rabbinical interpretations and clarifications of the laws of Torah begun over 2,000 years ago), as well as the works of successive centuries of Jewish scholarship.

Pork is forbidden, as is meat from carnivorous animals, scavengers, and water-dwelling creatures without fins and scales. Kashrut (or Kosher certifier) also requires complete separation of meat and dairy products, including the utensils, equipment, containers, and surfaces used in preparation and packaging. Items that have come into contact with non-kosher food may not be used with kosher food. Wine and grape juice made by non-Jews may not be consumed.

Permitted animals must be slaughtered by a *shochet*, a ritual slaughterer, who slits the animal's throat with a special knife in a manner that minimizes suffering. The organs are inspected for flaws, such as adhesions on the lungs or a perforation of the brain, which could result in the animal being labeled *treif*, or unfit for consumption.

Sholom Rubashkin, whose Postville, Iowa-based firm Agri-processors slaughters most of the kosher meat produced in the United States, notes the emphasis placed on the intact brain. This factor raised the profile of kosher beef at a time when consumers are concerned about BSE, which attacks the brains of cattle and has been linked to Creutzfeldt-Jakob disease in humans.

"Consumers cite food safety issues as a reason to choose kosher foods," says Paul J. Albert, marketing communications manager for Empire Kosher, Inc.[*]

> At Empire, safety in all products is ensured by rigorous tests and temperature inspections throughout the process. Empire Kosher sells more kosher chicken and turkey than any other company, so we take extra care at every stage of growing, processing, selling, and

[*] http://www.empirekosher.com/index.htm.

distribution of our poultry to ensure the best quality and safety. We're the only kosher poultry processor to have two dedicated knife inspectors on the plant floor at all times.

Given the rigorous and complex nature of the kosher laws, it may seem surprising an estimated one-third of all food products, from crackers to corn syrup and club soda to caramels, are kosher-certified. Lubicom pegged the total number of kosher-certified products at 82,000 in 2003. Manufacturers are putting themselves through the paces just to ensure they can display a *hechsher*, a symbol attesting to a product's kosher status because the market opportunity is too great to ignore and the certification process is not as intrusive or cumbersome as to outweigh the benefit.

In kosher standards, the whole package is considered. Kosher extends to packaging as well. At first, the idea that containers, foils, or plastic wraps could be unkosher seems strange, until you delve a little deeper. Although the use of recycled cooking oils (in which unkosher foods could have been prepared) in food-grade lubricants is mostly a thing of the past, other contact ingredient issues still apply. According to Rabbi Evan Herbsman of the OU, some additives such as release agents or non-stick agents may in some instances be derived from animal products. Herbsman also points out there is a consumer-driven requirement that any item that comes in contact with food must be certified.

Another important checkpoint is with bulk ingredients. "Many times you can receive raw materials from a company which produces both kosher and non-kosher items," says Rabbi Levi Goldstein, a *mashgiach* (or kosher certifier) for the OU in Iowa. "An OU label will have its own special date code, so make sure you see that code." Today the OU certifies more than 660,000 products, making it the world's most recognized and the world's most trusted kosher symbol. The most controversial certification is the "K"—a plain letter K found on products asserted to be kosher. A letter of the alphabet cannot be trademarked, so any manufacturer can put a "K" on a product.

6.28.1 SEAL OF APPROVAL

There are hundreds of different hechshers (i.e., kosher seals) in use all over the world. Many states and even larger cities have their own *va'ad* (or kosher oversight group). Different organizations apply different levels of strictness in their adherence to kosher laws, so research is in order to find the kosher organization that best fits the needs.[*]

6.28.2 KASHRUT CERTIFICATION

The task of keeping kosher is greatly simplified by widespread kashrut certification. Products that have been certified as kosher are labeled with a mark called a hekhsher (from the same Hebrew root as the word "kosher"), which ordinarily identifies the

---

[*] Notion of kosher as a "style" of cooking—kosher is not a style of cooking and there is no such thing as "kosher-style" food. Chinese food can be kosher if prepared in accordance with Jewish law.

rabbi or organization that certified the product. Approximately three-quarters of all prepackaged foods have some kind of kosher certification, and most major brands have reliable Orthodox certification.

REFERENCES

Coreene, Catherine. "Data Track the Expansion of International and US Organic Farming." *Amber Waves* 5, no. 4, 2007: 36–37.

Delate, Kathleen. *Fundamentals of Organic Agriculture.* Ames, IA: Iowa State University Extension, Publication PM 1880, 2003, available at http://www.extension.iastate.edu/Publications/PM1880.pdf (accessed January 23, 2007).

Fearne, Andrew. "The Evolution of Partnerships in the Meat Supply Chain: Insights from the British Beef Industry." *Supply Chain Management: An International Journal* 3, no. 4, 1998: 214–231.

Glassheim, Eliot, Jerry Nagel, and Cees D. Roele. "The New Marketplace in European Agriculture: Environmental and Social Values within the Food Chain." *Plains Speaking* 8, no. 1, 2005: n.p.

Golan, Elise, Barry Krissoff, Fred Kuchler, Linda Clavin, Kenneth Nelson, and Gregory Price. *Traceability in the U.S. Food Supply: Economic Theory and Industry Studies.* Agricultural Report Number 830. Washington, DC: U.S. Department of Agriculture Economic Research Service, 2004.

Hobbs, Jill E. "Traceability and Country of Origin Labeling." Prepared for the Theme Day *Consumer Driven Agricultural Trade* at the International Agricultural Trade Research Consortium Annual Meeting, 2002 and the Policy Dispute Information Consortium 9th Agricultural and Food Policy Information Workshop, Montreal, Quebec, Canada, April 25, 2003.

Noonan, John, and Graham McAlpine. "Management of Value Chains through SQF: A Global Standard for Food Safety and Quality." Presented at the Muresk 75th Anniversary Conference, *From Farm to Fork: Linking Producers to Consumers through Value Chains*, 2003.

Pointon, David. "Wintulichs Pty. Ltd. and Marketing Environmental Disturbance 'Garibaldi Effect.'" *Agribusiness Review* 3, paper 6, 1995: ISSN 1442-6951.

Section III

Auditors and Laboratories

Part III includes two chapters that are critical to most identity preservation and traceability (IPT) systems. Auditors and laboratories are often required components of many IPT programs and are well documented for their detail and standards. Many auditing organizations and laboratories are national in scope, whereas others are global depending on their focus and accreditation.

7 Auditors

7.1 INTRODUCTION

According to Grande (2003) of IdentityPreserved.com,[*] for a successful identity preservation and traceability (IPT) program to exist, standards and verification of standards must be in place. Verification can take many forms and often extend to include laboratories (laboratories are discussed in Chapter 8). Although some verification is accomplished "in house," what is becoming much more important is the use of professional auditing firms to ensure compliance with IPT rules and regulations. Farms and firms use third-party auditors to prove compliance for numerous reasons, such as for the U.S. Department of Agriculture (USDA) National Organic Program (NOP) certification and seal. More often than not this includes review of a farm or firm's written IPT operating procedures or manual; review of the required documents that illustrate compliance, training records, and previous inspections; a walkthrough of the facility; and other requirements, which will depend upon what the purchaser requires, possibly included in a written contract and/or regulations and laws. Much of this introduction is condensed and modified from Grande's 2003 presentation for the Economic Research Service (ERS), USDA, and the Farm Foundation.

The typical approach toward a similar type of auditing has been conducted for the purposes of quality control. Manufacturing and processing firms have historically found benefit from improving quality control in production efficiency, fewer product defects, etc. Now, however, the focus has changed, or more precisely, taken on a much larger dimension to include not only quality and their subsystems such as chemical usage, but also more in intangible, less visible aspects such as in organic production, no trans fats, and country or region of origin or processing, to social aspects such as fair wages to laborers and animal living conditions. Many of these auditing firms have modified their audits to comply and ensure IPT standards and rules are met.

This chapter on auditors and the next chapter on auditing laboratories individually and in unison embrace the third-party responsibility for IPT systems to be credible. In his paper *Servicing IP Production and Marketing: A Third-Party Role*, Grande (2003) emphasizes how a well-defined IP programs assist producers in their capability to deliver differentiated products. These programs typically include specific services and tools that are customized to deliver "branded" crops to a predetermined end user(s) and/or market niches. Bovine spongioform encephalopathy (BSE, or mad cow disease), GMOs, StarLink, and Diamond Pet Food events have all brought global attention on the impact that agriculture practices and technologies can have on the resulting quality and safety of the finished product. Although each of these individual events has had different impacts on how the public perceives the

---

[*] http://www.identitypreserved.com

wholesomeness and safety of agriculture and the resulting finished products, they collectively set off a realization with some that a common trust between production agriculture and food companies is necessary to better understand and define particular roles and responsibilities in creating, producing, and delivering finished products to more sensitive customers.

This recognition has also advanced the role of a third party to validate the IPT of farming, processing, manufacturing, and marketing claims. Third-party involvement is now fixed in ongoing IPT discussion such as concerns with biotechnical applications in agriculture and food production, particularly with regard to consumers' "right to know," but also concerning environmental impacts, functional food, global population growth projections, malnutrition, sustainability, and trade. Third-party validation of raw material origin is becoming fundamental for any level of IP emphasis that might be placed on a differentiated food production system.

Agriculture responses toward IPT, in both production agriculture and food manufacturing, have gone beyond the well-established systems and protocols used by parent seed companies and animal genetics. Farmers and processors have been much more aware of not only seed inputs, but all other purchased and environmental inputs and have begun development or enhancement of their procurement functions for raw materials and key ingredients. The origin of grain-based food product factors (e.g., seed genetic makeup, production methods, ag-chem usage, etc.) and the ability to validate specified product attributes and/or process claims are becoming increasingly critical to the resulting finished product marketing strategies. A prime illustration of this is the USDA's NOP. Once again, the ability to maintain identity preservation (IP) of predetermined product information and/or production data will be critical to satisfactorily addressing customer and/or consumer expectations.

Production agriculture and food processing companies are improving their quality management system to address specific IPT goals and objectives. Often the "endorsement" aspect of successful program implementation comes from several certification bodies: private companies, industry associations, consumer groups, and government agencies. A secondary yet very important outcome from a certified quality management system is internal (e.g., production/processing efficiencies) and external (e.g., product differentiation in the marketplace) traceability.

As agriculture looks to the "value-added" returns promised by IPT, the numbers of market channels through which these products can flow are still limited in numbers. Handling, segregation, and transportation requirements for "high-value" crops do require additional planning in how they will be harvested, processed, stored, transferred, and eventually delivered to the customer. IPT systems are seen by some as being too costly and run counter to the current industry's commodity-based system. The cost structures of IPT products versus undifferentiated commodities also work against IPT systems' acceptance by farmers and processors.

Auditing firms provide specific services and tools designed to deliver "branded" crops to the marketplace. This is especially true given the challenging environment that IPT crops work in today: the notion "perception is reality" is well suited for defining why IPT product (and brand) authenticity is a necessary component of a successful sales and marketing strategy. And by authenticating how producers' conduct business (e.g., transparency, tracking, auditing), they will improve the perception of their brands and products.

The sampling of auditing firms cited below offers a glimpse of what is offered. These firms' services vary (some also include laboratory testing), and audits range from national and international to only foreign rules and regulations. The following companies have been selected for review:

- Caliso Consulting, LLC
- TÜV America, Inc.
- SGS SA (Société Générale de Surveillance)
- BRS Ltd.
- QMI—Management Systems Registration
- FoodTrust Certification
- BSI Americas & BSI Global
- Cert ID

What is included in this chapter are individual/company/organizational statements from their websites that naturally reflect their views.

7.2 CALISO CONSULTING, LLC

CALISO Consulting, LLC
1516 Oak Street, Suite 312
Alameda, CA 94501
Phone: 800-306-1366 (toll-free)
Fax: 510-217-6621

Available at http://www.caliso9000.com/index.shtml.

Caliso offers a full range of auditing and consulting services aimed toward helping organizations achieve competitiveness and develop market opportunities in the United States and overseas. Their notion is that via consulting, training, and auditing specific certification will help ensure IPT and quality systems compliance. During the past decade, increased emphasis has been on food safety and production agriculture. Caliso has relationships with certification bodies such as Underwriters Laboratories (UL) and RWTUV-USA of Germany and SGS SA of Switzerland. Caliso services include the following:

- ***Product and system certifications*** for organizations that want to enter or develop a particular market or sales opportunity. Compliance or certification to some of the following standards and regulations are achieved: International Organization of Standardization (ISO) Certification, ISO 9000, ISO 14000, Six Sigma, ISO 13485, ISO 16949, TL 9000, AS 9100, Good Manufacturing Practices (GMP), Occupational Health and Safety Standard (OHSAS) 18000, and Hazard Analysis and Critical Control Points (HACCP). Caliso coordinates with most recognized registrars to ensure certification has the proper national and international recognition.
- ***Operational and process improvement consulting*** focuses on improving an organization's effectiveness, efficiency, and profitability.

- *Market development and technology consulting*, which include market analysis, customer and vendor qualifications, and sales opportunity development.
- *Training:* Caliso also offers class training and online courses for ISO 9000, ISO/TS 16949, current Good Manufacturing Practice (cGMP), OHSAS 18001, ISO 13485, ISO 19011, HACCP, Six Sigma, and ISO 14000. (State funding California only: The State will fund training and implementation for ISO 9001, ISO 13485, OHSAS 18001, Six Sigma, GMP, ISO/TS 16949, or ISO 14000.)

7.2.1 Caliso Sample Price Schedule

ISO 9000:2000 Overview $79.95	ISO 13485:2003 Auditor $229.95
ISO 9000:2000 ....$149.95	ISO/TS 16949:2002 ....$189.95
ISO 9000:2000 Auditor $159.95	ISO/TS 16949:2002 Auditor .. $209.95
ISO 9000:2000 and Auditor $279.95	GMP: Medical devices $249.95
ISO 9000 in Spanish $149.95	GMP: Human food $179.95
ISO 9001 Business strategy $189.95	GMP: Pharmaceuticals $269.95
ISO 19011:2002 ....$99.95	Six Sigma course $109.95
ISO 14000:2004 ....$159.95	OHSAS 18001 kit $129.95
ISO 14000:2004 Auditor $169.95	HACCP $249.95
ISO 13485:2003 .... $199.95	

7.2.2 Funding to Pay for Certification

A federal fund is available through the Trade Adjustment Assistance (TAA), which will pay 50% of the consulting cost toward improving operations and/or achieving a certification if the organization was affected by import competition or delocalization. Funding for this program is sponsored by the U.S. Department of Commerce.

7.2.3 Sample of Consulting for ISO 9001, ISO 14001, ISO 13485, ISO 16949, and GMP Certification

The firm may already be certified to a standard and want to upgrade or implement an ISO 9001:2000, ISO 13485:2003, TL 9000, Canadian Medical Devices Conformity Assessment System (CMDCAS), ISO/TS 16949:2002, ISO 14001, GMP, or OHSAS 18000.

Caliso offers three options depending on the firm's resources toward the implementation of ISO 9001:2000 standards.

- *Turnkey certification consulting:* The option empowers Caliso to drive the implementation to comply with the chosen standard or any of the industry specific ISO standards, and follows a five-step methodology: GAP assessment, quality management system upgrade, training, internal audit, and certification audit. Caliso establishes all of the necessary ISO compliant processes and provides and generates all of the required documentation to meet the requirements of the standard. The implementation usually involves streamlining and simplification of operations to take full advantage of the benefits of the standard and has consultants in many states, particularly in California, Texas, and Illinois as well as Mexico, France, and North Africa.

- *Desk audit:* If the company has internal resources (management representative, quality-assurance (QA) manager/supervisor) and requires exact guidance on what needs to be done, Caliso offers desk audits that will assist in meeting standards.*
- *Online training and documentation templates:* If the company has internal resources (management representative, QA manager/supervisor) to conduct the implementation internally and does not need Caliso consultants, the company may need to train the staff to standards such as ISO 9001:2000, ISO 13485, TL 9000, ISO 13485, ISO 16949, ISO 14001, GMP, and auditing. Caliso offers online courses that come with documentation templates.

7.2.4 Example of Training Offered: Online HACCP Training for Meat and Poultry

The depth and breadth of training will depend on the particular employee's responsibilities within the establishment. Management or supervisory individuals need a deeper understanding of the HACCP process because they are responsible for proper plan implementation and routine monitoring of critical control points (CCPs) such as product cooking temperatures and cooling times.

- The cost ranges from $212.46 for group training to $249.95 for individual training.
- It provides detailed training on HACCP, which is a production control system for the food industry. HACCP is designed to prevent potential microbiological, chemical, and physical hazards, rather than catch them. The U.S. Food and Drug Administration (FDA) and the USDA use HACCP programs as an effective approach to food safety and protecting public health.
- It uses excerpts of the USDA Food Safety and Inspection Service (FSIS) 9CFR417 (Meat and Poultry HACCP) as well as the FDA/Center for Food Safety and Applied Nutrition (CFSAN) FDA 1999 Food Code.
- It uses a continuous evaluation method with ongoing quizzes to facilitate information retention.

---

* Caliso reviews all operational documentation to identify any requirements not being met. A copy of the QA manual or policies (Level 1), procedures (Level 2), and organizational chart for a thorough audit/review must be submitted. The review will be under nondisclosure agreement and will maintain full confidentiality. Caliso will also provide templates for the QA manual (Level 1), a continual improvement plan, and an internal audit checklist that can be customized to meet the new requirements of the standard. Once the modifications are made, you will need to implement the changes in accordance with the recommendations in the report.

7.3 TÜV America, Inc.

TÜV America, Inc.
5 Cherry Hill Drive
Danvers, MA 01923
Phone: 978-739-7000; 800-888-0123 (toll-free)
Fax: 978-762-7637

Available at http://www.tuvamerica.com/home.cfm.

The history of TÜV America and its parent organization, TÜV SÜD AG,[*] is as a technical service company that includes consulting, inspections, tests, expert opinions, certification, and training. Established in the 1870s as a steam boiler inspection association, TÜV SÜD (*Technischer Überwachungsverein,* or *Technical Inspection Association*) is globally active and represented internationally by more than 130 locations. Headquartered in Munich, Germany, TÜV SÜD is the largest of the German TÜVs with 2005 revenues of EUR 1.01 billion and over 11,000 employees. Globally, TÜV has issued over 190,000 product and 30,000 quality management system certifications.[†] During the late 1980s and early 1990s, deregulation, liberalization, and harmonization of trade practices in Europe allowed TÜV Bayern, whose activities were limited to Bavaria, to compete with other inspection agencies on both a national and international level. Globally, there are several TÜV organizations, including TÜV SÜD AG, TÜV America, Inc., and others.

TÜV America, Inc., a subsidiary of TÜV SÜD AG, is a business-to-business engineering services firm providing international safety testing and certification services. Founded in 1987, TÜV America has grown to more than 200 experts in over a dozen locations throughout the United States, Canada, and Mexico. Operating under the brand names of *Product Service, Management Service, Industry Service and Automotive,* TÜV America has partnered with more than 3,000 companies throughout the North American Free Trade Agreement (NAFTA) region, assuring product and management systems services and compliance in the global marketplace.

In the United States, TÜV America, Inc. is authorized by OSHA (Occupational Safety and Health Administration) as a nationally recognized testing laboratory (NRTL) capable of performing product safety testing to UL/American National Standards Institute (ANSI) standards. In Canada, TÜV Product Service is accredited by the Standards Council of Canada (SCC) as a certification body able to perform electrical safety testing to Canadian requirements.

---

[*] See http://www.tuev-sued.de for more information regarding the parent corporation.
[†] The original TÜV association in Bavaria, founded over 130 years ago, had 43 industry members and employed just 2 safety inspectors. With the advancement of technology, its presence and capabilities quickly expanded. In the 1900s, the group began working not only with electrically powered devices but also passenger elevators, diesel engines, sprinkler systems, and hydroelectric power plants. These inspection services further expanded into the transportation and motor vehicle industries and later to the nuclear energy industry. As late as the 1980s, the TÜV associations (TÜV Bayern being one of the largest) continued to operate independently in the federal states of Germany and their activities and name became synonymous with public safety, quality, and environmental protection.

7.3.1 About TÜV America, Inc. Management Service Division Accreditations

The testing laboratories and certification body of TÜV Product Service conform to the "General Requirements for the Accreditation of Testing Laboratories" (ISO/IEC Guide 17025 and 38; EN 45001 and EN 45002), "General Requirements for Accreditation of Certifying Bodies" [ISO/International Electrotechnical Commission (IEC) Guide 28, 40, and 48; EN 45011 and EN 45012] and in accordance with the provisions of CAN-P3: "General Requirements for Bodies Operation Product Certification Systems." TÜV can also provide a variety of services including online webinars, public training, private in-house seminars, supply chain management, retail supplier inspections, comprehensive training, product safety testing, and traceability software.

On May 1, 2006, TÜV America, Inc. announced that its Management Service division had certified the Colorado facility of Sparboe Farms, one of America's largest marketers and producers of shell eggs and egg products, to the Safe Quality Food (SQF) Program's SQF 2000 Code.[*] The scope of the certification is for the products of hen eggs and egg processing. Sparboe Farms' certification marks the first SQF certification awarded by TÜV since it received accreditation from the ANSI to the SQF program in late March 2006. SQF 2000 certification is a HACCP supplier assurance code for the food industry[†] (see Chapter 6 regarding SQF programs).

TÜV America certification provides various advantages.

- Proof from an accredited test and certification body that products meet all requirements of relevant European Union (EU) directives.
- Official statement of conformity accompanied by appropriate documentation as specified by EU regulations or customer requirements.
- Emphasis on special product properties such as safety, quality, durability, environmental compatibility, and conformity to standards.
- Protection against product liability claims.
- Up-to-date information on changes regarding technical regulations and developments in testing and certification. TÜV Product Service has fulfilled all necessary German and EU accreditation requirements and has received authorization to issue certification marks and certificates.
- The testing laboratories and certification body of TÜV Product Service conform to the "General Requirements for the Accreditation of Testing Laboratories."

---

[*] TÜV America, Inc. is one of only three registrars in North America accredited by ANSI to perform SQF 1000 and 2000 audits.
[†] The SQF program is a complete certification program for managing food safety and enhancing quality systems throughout the food chain, managed by the SQFI, an organization that is owned and operated by the Food Marketing Institute (FMI). The SQF program provides two standards on the basis of the type of supplier. The SQF 1000 code is designed for primary producers whereas the SQF 2000 code is designed for the manufacturing and distribution sectors—both codes are based on the HACCP method and principles.

7.4 SGS SA

SGS SA
1 Place des Alpes
P.O. Box 2152
1211 Geneva 1
Switzerland
Phone: +41-22-739-91-11
Fax: +41-22-739-98-86

Available at http://www.sgs.com.

SGS provides global inspection, verification, testing, and certification. With its 43,000 employees, SGS globally operates a network of nearly 1,000 offices and laboratories. Founded in 1878 in Rouen as a French grain shipment inspection house, the company was registered in Geneva as Société Générale de Surveillance in 1919. SGS is currently accredited by 32 national accreditation bodies and authorized to conduct certification audits under these accreditations in every country around the world. To date, more than 2,000 small, medium, and international companies use SGS as their certifying body to perform the audit of their Food Safety Management System against HACCP and HACCP-based Food Safety Management Systems (FSMS).

The core services offered by SGS can be divided into three categories.

1. *Inspection services:* SGS inspects and verifies the quantity, weight, and quality of traded goods. Inspection typically takes place at the manufacturer's/supplier's premises, at time of loading, or at destination during discharge/off-loading.
2. *Testing services:* SGS tests product quality and performance against various health, safety, and regulatory standards. SGS operates state of the art laboratories on or close to customers' premises.
3. *Certification services:* SGS certifies that products, systems, or services meet the requirements of standards set by governments, standardization bodies (e.g., ISO), or by SGS customers. SGS also develops and certifies its own standards.

7.4.1 IDENTITY PRESERVATION PROGRAMME AND TRACEABILITY GRAIN

SGS markets its Identity Preservation Programme (IPP) and Traceability Grain as transparent and logical. These services respond to clients' demand for information regarding the nature and origin of food products. Concerns over genetically modified organisms (GMOs) led SGS to put in place a program to trace the origins of soybeans (soya) and corn (maize) and to test goods at each critical point, from analyzing seeds for purity before they are planted through to storage, transport, and shipment. Non-GMO certificates provide evidence that every procedure has been taken for delivering authentic and untainted products. Traceability is a system process to retroactively detect where problems occurred in the supply chain. It entails keeping a record of relevant data for effectively tracing the commodity from the various production points to their destination.

7.4.2 ISO 22000 Certification

ISO 22000 is more than a set of standards. It is a business-building tool for organizations wishing to extend their reach, provide a more logical and structured approach to food safety management, or gain easier access to global markets. Throughout the food chain, from producers and suppliers to warehousing and grocery stores, an increasing number of food and safety standards are being certified by SGS to meet food safety needs such as HACCP, British Retail Consortium (BRC) food and packaging, International Food Standard (IFS), EurepGAP, and GMP. However, overshadowing them are the broader initiatives orchestrated by the Global Food Safety Initiative (GFSI) of the European Retailers. Rather than attempting to meet several or many of these standards, many organizations are now focusing on the ISO 22000 series.

ISO 22000 is a single standard published on September 1, 2005. It provides a single standard to encompass all of the needs of the marketplace. The key to this program is flexibility, and unlike some other schemes, ISO 22000 does not follow an exhaustive and prescriptive checklist approach. It instead allows an organization to develop its own food safety management system tailored to its particular suppliers, customers, and relevant parties.

7.4.3 SGS Helps Organizations Fulfill the Requirements of ISO 22000

As a first step, SGS helps organizations better understand the purpose and requirements of ISO 22000 as a tool for the continual improvement of a food and safety management system through a series of ISO 22000 training courses. SGS provides a visit to discuss the application of ISO 22000 to its business, determines critical steps to achieve certification, and indicates time scales and costs. SGS follows the initial visit with a preassessment (GAP analysis) to determine the state of readiness for meeting the requirements of ISO 22000 and to identify areas requiring attention prior to formal certification. Then SGS will issue a formal proposal for full certification to ISO 22000.

7.4.3.1 Example: SGS and Orthodox Union Kosher Food Safety Program

Consumers are becoming more discerning when it comes to their food. Many have turned to kosher products to provide them with the assurance that the food they eat is healthier and of better quality. To help ensure the safety of kosher food, the Orthodox Union (OU) and SGS have joined to provide a service that ensures that safety and quality management standards are met and that production steps and ingredients comply with the kosher requirements. By combining kosher certification with food safety certification, customers are provided a greater sense of trust in the food they eat (see Chapter 6 regarding kosher foods).

7.5   BRS LTD.

BRS Ltd.
Rim of the World Office
31977 Hilltop Boulevard, Suite D
P.O. Box 1020
Running Springs, CA 92382-1020
Phone: 888-285-5835 (toll-free)
E-mail: Globalnet@brsltd.org

Available at http://www.brsltd.org.

BRS Ltd., founded in 2003, was formed from BRS GlobalNet, which has operated since 1984. BRS, a wholly owned subsidiary and U.S.-based corporation, is an internationally registered body providing "adding-value" by focusing on competitiveness and reduction of risk management solutions. Members of the BRS GlobalNet issued certificates of ISO 22000 registration may participate in World Health Organization (WHO) programs. BRS is accredited as a QMS ISO 9001 and EMS ISO 14001 under partnership agreement with Raad voor Accreditatie (RvA), and provides ISO 22000, HACCP MS, OSHMS and ISMS management systems solutions.[*]

BRS, as an international registration body, provides management systems certification (registration) in accordance with International Management Systems Standards to organizations and governments in Asia Pacific (ASEAN), Europe, the Middle East, and other regions. BRS provides QMS ISO 9001, EMS ISO 14001, EMAS EU Regulation No. 761, FSMS ISO 22000, and other management schemes such as OSHMS, ISMS, and HACCP Management System (HACCP MS).

Effective September 1, 2005, the official release date of the international standard ISO 22000, BRS commenced offering certificate of registration against this international standard. The BRS ISO 22000:2005 certificates of registration combines ISO 9001:2000 (and Guidance ISO 15161) for organizations in the food supply chain wishing to meet contemporary requirements to best practices in FSMS fulfilling expectations for registration bodies.[† ‡]

As has been mentioned by other auditing firms, the introduction of FSMS ISO 22000 replaces the need for organizations to undergo multiple assessments. And

---

[*] Additionally, BRS offers Adding-Value-Assessment© (AVA©), an approach to help shift management paradigms, which they call I³ (Improve-Innovate-Invent).

[†] BRS meets the requirements and guidelines set forth by ISO/IEC 17024:2003, ISO 1901:2002, ISO/IEC 17012 (ISO 62 and ISO 66), and ISO/TS 22003. The requirements of these international standards for BRS certifying to ISO 22000:2005 are inclusive to the BRS operating management system.

[‡] Globally BRS provides services across the globe, including Asia/Pacific—Australia, New Zealand, Malaysia, Thailand, Taiwan, Korea, Philippines, China, Taiwan; North America—United States, Canada, Mexico; Latin America—NAFTA… Venezuela, Colombia, Cuba, Peru, Ecuador, Argentina, Chile… Central American Free Trade Agreement (CAFTA)… Puerto Rico, Dominican Republic, Costa Rica, El Salvador, Panama, Nicaragua, Honduras, Guatemala; Europe—Scotland, United Kingdom, Ireland, Western and Eastern Europe; and the Middle East—United Arab Emirates, Iraq, Syria, Egypt, Arabia, Qatar, Pakistan, Iran, Morocco, Libya, Qatar, Tunisia, Algiers, Bahrain.

with the large base and global food chain suppliers, this development has significant implications for the food industry. However, the effects of FSMS ISO 22000 are likely to impact many other organizations within the global base of suppliers because the food supply chain is more than foods and includes many others impacting the food sector (e.g., equipment, airlines, airports, tourism vessels, and utensils producers) that FSMS ISO 22000 will most likely require.

BRS provides HACCP (Codex) MS, HACCP combining ISO 9001, and ISO 22000:2005 certification/registration. ISO 22000:2005 provides for certificate of registration for FSMS.

Implementation of a HACCP MS provides the basis for reduction of risk, thereby improving trust and confidence in the processing and handling of safe food products. HACCP MS helps organizations advancing to ISO 22000:2005 management system certification. Organic farming certification is also achievable through ISO 22000 or HACCP MS.

Under ISO 22000:2005, organizations identify, control, and prevent effects and challenges brought about microbiology (e.g., *Escherichia coli*, *Salmonella*, and *Listeria*) in protecting the food supply chain. By applying HACCP's seven principles concurrent with ISO 9001 management fundamentals, it assists organizations to identify hazards; analyze; and pursue controls, effective actions, and preventative measures.

7.5.1 Certification of Individuals: Auditor and Assessors/Technical Experts

BRS provides QMS ISO 9000, FSMS ISO 22000, EMS ISO 14000, and other management systems auditor/assessor and technical expert certification. This certification of individuals operates under ISO 19011 and is aligned with BRS ISO/IEC 17024:2003 for certification of individuals. The certification programs include internal (first) and external (second) party auditors/assessors and technical experts. Certification of third-party auditors is exclusive to BRS assessors, auditors, and other registration bodies accepting BRS certification protocols.[*]

Fees	Auditor/Assessor	Technical Expert
Application fee (nonrefundable)	USD $35.00	USD $35.00
Certification fee	USD $215.00	USD $200.00

---

[*] The CE Mark (Conformité Européene), although not addressed here, yet is certifiable, is a product-distinctive mark that provides for conforming of specified requirements to the European Directives. The European Directives are a series of specific standards that relate to product safety and effectiveness for manufacturers that relate to a granted CE Mark. These Directives and granting of the CE Mark comes under a scheme set forth by the state countries within the boundaries of the EU and others including the Scandinavian region and Iceland (equally, other countries accept the Directives) and are generally applied toward maritime equipment/devices, medical devices/equipment (MDD), electrical/electronic devices/equipment, etc.

7.6 QMI—MANAGEMENT SYSTEMS REGISTRATION

QMI

8501 East Pleasant Valley Road

Cleveland, OH 44131-5575

Phone: 216-901-1911; 800-247-0802

(toll-free; United States and Canada only)

Fax: 216-520-8967

E-mail: clientservices@qmi.com

Available at http://www.qmi.com.

Founded in 1984, QMI was one of the first quality registrars in North America. They have registered more than 11,000 manufacturing and service firms spanning a broad spectrum of industry sectors that are serviced by more than 400 QMI audit professionals. QMI's registration services include the most important standards governing a wide range of businesses and industries. Available registrations include ISO 22000, Food Safety (HACCP), USDA-NOP, FDA audits, etc.[*]

7.6.1 ACCREDITATION BODIES THAT CERTIFY QMI

QMI certificates are recognized and accepted worldwide and are accredited by highly respected organizations such as the ANSI/ASQ National Accreditation Board of the United States (ANAB), the SCC, Entidada Mexicana de Acreditacion a.c. (EMA), and other important governing bodies. QMI's alliances with major registration organizations outside of North America through QMI's international partnership with the International Certification Network (IQNet) enable QMI to support world business markets.

ISO 22000 was launched in the fall of 2005 as a truly global FSMS standard. This international standard requires an organization to demonstrate its ability to manage food safety hazards and provide safe products that meet relevant regulations and the requirements of its customers. ISO 22000 goes beyond the prevalent "condition" audits of the past, where a snapshot look at the physical conditions and past paperwork was often all that was required. An ISO 22000-based FSMS audit looks at the organization as a whole and assesses its ability to satisfy its customers' needs. The key to ISO 22000 is that it requires the involvement and resources of the entire company to plan, design, and implement an effective FSMS that incorporates the use of safety measures and the HACCP methodology to ensure the delivery of safe food products to the consumer. Because of this, it applies to each and every aspect of the food chain, covering not just food manufacturers, but also the producers of ingredients, equipment, cleaning agents, packagers, transporters, distributors, and retailers.

---

[*] Other registration standards include: ISO 9001, ISO/TS 16949, QS-9000, TE Supplement, AS9100, AS9110, AS9120, ISO 14001, OHSAS 18001, RC 14001, RCMS, CSA Z809 SFM, SFI, Integrated Management Systems, TL 9000, ISO 13485 (CMDCAS), CE Marking, and more. They also specialize in automotive, aerospace, forestry, environmental, and food safety sectors.

7.6.2. How Registration Can Benefit an Organization

Adopting the ISO 22000 standard provides possible competitive efficiencies world-wide, such as

- A single, globally accepted standard
- Uniform food safety procedures
- Improved communication with trading partners
- Better understanding and implementation of HACCP principles
- A driver for continuous improvement
- Improved food safety hazard control
- A uniformly auditable standard

7.6.3 Key Components

The ISO 22000 FSMS is based on the ISO 9001:2000 quality management system model, and its requirements have been customized to address the specific needs of the food industry. Some of the key components of ISO 22000 are

- *Management responsibility:* This includes the policy, objectives, the food safety team, communication, emergency situations, defining organizational responsibility and authority, and the provision of resources and review of the FSMS. The standard is quite specific on the requirements for communication, external and internal, and includes the need for documented procedures for recalls and related notifications.
- *Product and process data:* This requires information to be documented on all of the materials and processes involved in producing the products, flow diagrams showing the sequence and interaction of all steps, descriptions of the steps, and other information that will provide the basis for the hazard analysis.
- *HACCP:* Built right into the ISO 22000 standard is the need and reinforcement of the HACCP system. Having a functional HACCP and prerequisite program (called "Supportive Safety Measures" in ISO 22000) is a cornerstone of an effective ISO 22000-based FSMS.
- *Measurement, analysis, and updating the FSMS:* This includes planning and implementing of all monitoring, measurement, inspection, verification, and related activities, including verification of the CCP plans and supportive safety measure (SSM) plans, as well as internal audits to confirm that the FSMS is effectively implemented. The requirements of ISO 22000 can be incorporated into any FSMS and can be applied at any stage or part of the food chain. It is not limited to feed producers, farmers, food producers, retailers etc., and includes suppliers of packaging materials, equipment, cleaning service providers, and others. It places the onus on the management of the business to fully understand and deliver the needs of their customers.

The demand for ISO 22000 is widespread across the food chain and there are a surprising number of interested parties that are not the typical food processor. As producers tighten requirements, their suppliers are increasingly drawn into the system. Ingredient suppliers, packaging suppliers, and the service providers to the food industry are keen to show that they support food safety and want to do all they can to improve confidence in the integrity of their processes and keep their customers happy. As the consuming public is bombarded with information about the safety of the foods they purchase, retailers see ISO 22000 as a means of demonstrating due diligence and controlling risk.

7.6.4 TRAINING COURSES AVAILABLE

QMI also offers a comprehensive and extensive range of learning products covering quality, environmental, occupational health and safety, and other management system standards to support organizational training needs that include a standards library, training courses, and interactive webinars that are hosted by certified product managers.

7.6.5 Organic Certification

QMI is an accredited organic certifying body. With trained inspectors and a certification committee of experts in the organic industry, QMI is accredited to certify crop, wild crop, livestock, and handling operations. QMI can certify operations, including:

- Farm and rangeland
- Forage
- Livestock
- Poultry
- Dairy
- Horticulture
- Seed cleaning
- Food processing and handling
- Retail operations

7.6.6 QMI Accredited Services

QMI accredited services include the following:

- Food safety
  - ISO 22000—FSMS
  - HAACP
  - Organic certification
  - ISO 9001 Quality Management Standard
  - ISO 14001 Environmental Management Specification
  - OHSAS 18001 Health and Safety Management System
  - Integrated Management Systems

- Environmental
  - ISO 14001—Environmental Management Specification
  - RC14001®—Responsible Care Management System
  - RCMS®—Responsible Care Management System
  - External Verification of Environmental Reports (EVER)
  - EVER Greenhouse Gas
  - Sunoco Contractor HES Management System

- Forestry
  - SFI—the Sustainable Forestry Initiative Program
  - PEFC—Program for the Endorsement of Forest Certification
  - PEFC—Chain of Custody and Labeling
  - CAN/CSAZ809:2002—Sustainable Forest Management System
  - CERTFOR Chile
  - ISO 9001 Quality Management Standard
  - ISO 14001 Environmental Management Specification
  - OHSAS 18001 Health and Safety Management System

7.7 FOODTRUST CERTIFICATION

FoodTrust
2806 Bernadette
Houston, TX 77043
Phone: 713-429-4092
Fax: 281-271-8112
E-mail: pwigginton@foodtrustcert.com

Available at http://www.foodtrustcert.com.

For over 35 years, the founders of FoodTrust Certification have been exclusively involved in independent third-party management system certification of quality, environmental, and safety management standards. Their experience includes developing and maintaining an accredited program, auditor training and qualification, audit planning, auditing, certification issuance, and continuous value-added client services.

FoodTrust Certification was established to specifically provide accredited third-party auditing and certification services to food producers, processors, transporters, retailers, and service suppliers to the food industry. FoodTrust Certification's program has been accredited by ANSI and selected as a pilot participant in ANSI's accreditation process for food safety programs. FoodTrust Certification has designed and operated its certification program while avoiding conflict of interest questions because they do not provide any consulting, implementation services, or private training as some auditing organizations do.

FoodTrust Certification management has been involved in a total of over 4,000 system certifications worldwide. In addition, the executive management personnel are HACCP and Safe Quality Food Institute (SQFI) trained.

7.7.1 SERVICES OFFERED BY FOODTRUST

FoodTrust offers the following services:

- SQF 2000 certification
- Level 1: Good Agricultural Practices (GAPs), GMPs, Good Distribution Practices (GDPs)
- Level 2: HACCP system
- Level 3: Comprehensive food safety and quality requirements
- HACCP certification based on Codex Alimentarius

7.7.2 SECOND- AND THIRD-PARTY AUDITS

FoodTrust Certification offers a comprehensive food safety management assessment program designed to accommodate growers, producers, processors, distributors, and warehousing and transportation organizations. The program has been designed to give public confidence that the organization's HACCP system is implemented.*

The application process includes evaluation of existing GAP, GMP, GDP credentials to assure that these requirements are part of an implemented FSMS before proceeding further. If no recognized credentials are held, FoodTrust Certification offers an additional assessment service to evaluate the implementation level of these requirements.

The Phase I assessment occurs on-site and includes evaluation of the documented system; verification that appropriate GAP, GMP, GDP are implemented; and a limited-scope audit covering a review of the HACCP system. The Phase I assessment results are discussed with the client and officially reported and plans are made for the Phase II assessment. The client will have the time necessary to address any identified deficiencies prior to the Phase II assessment. This second assessment is conducted as agreed upon between the client and FoodTrust Certification personnel.

The results of the Phase II assessment are discussed and officially reported to the client. The client then takes action to correct deficiencies, if necessary, and reports the corrective actions to FoodTrust Certification. FoodTrust Certification personnel review all documentation and make a certification decision. When the HACCP certificate is issued, it is valid for a 3-year period contingent upon the successful completion of the agreed surveillance audit program. At the end of the 3-year certification period, a recertification audit is conducted and the certificate is renewed if the system continues to meet requirements. The surveillance cycle then continues.

---

* FoodTrust Certification requires specific auditor qualifications on the basis of the ISO 19011 requirements, accepted worldwide as a comprehensive criteria set for third-party auditors. The FoodTrust Certification HACCP program has been designed to meet the criteria set forth in ISO Guide 65 and intends for this program to operate as an accredited third-party certification with accreditation by a national accreditation body that is a party to the International Accreditation Forum's Multilateral Agreement. FoodTrust Certification's application for accreditation is in process with the ANSI.

7.7.3 SQF 2000 CERTIFICATION PROGRAM

FoodTrust Certification offers a comprehensive food safety management assessment program designed to accommodate growers, processors, distributors, and warehousing and transportation organizations. The program has been designed to meet the criteria of the GFSI that was established by the Food Business Forum (CIES).

CIES is the independent global food business network. Membership in CIES is on a company basis and includes more than two-thirds of the world's largest food retailers and their suppliers. Representative members of CIES include Loblaw Companies Ltd. (Canada), the Coca Cola Company (United States), Kraft Foods (United States), Wal-Mart (United States), and Safeway (United Kingdom).

7.7.3.1 Certifying Bodies' Requirements

For certification bodies, the GFSI and SQFI (Safe Quality Food Institute) require specific auditor qualifications and training along with program requirements for the body itself, including accreditation by a national accreditation body that is a party to the International Accreditation Forum's Multi Lateral Agreement. FoodTrust Certification's application for Accreditation is in process with the American National Standards Institute (ANSI).

For the applicant, GFSI criteria include three key elements: FSMS; GAPs, GMPs, and GDPs; and HACCP. In the United States, the SQFI's SQF 2000 code is a GFSI-benchmarked standard and can be used as the base requirements for this program.

7.8 BSI AMERICAS AND BSI GLOBAL

United States—BSI, Inc.
12110 Sunset Hills Road, Suite 200
Reston, VA 20190-5902
Phone: 703-437-9000; 800-862-4977
    (toll-free)

**Canada—BSI Management Systems
    Company**
17 Four Seasons Place, Suite 102,
Toronto, Ontario M9B 6E6
Phone: 416-620-9991

Available at http://www.bsiamericas.com/index.xalter and http://www.bsi-global.com/index.xalter.

Founded in 1901, BSI Group is a business services provider of over 2,000 employees, in 86 countries, serving over 35,500 registered clients worldwide. Their services include: independent certification of management systems and products, product testing services, the development of private, national and international standards, performance management software solutions, management systems training, and information on standards and international trade.

The BSI Group consists of the following:

- **BSI British Standards** is the national standards body of the United Kingdom and develops standards and standardization solutions to meet the needs of business and society. They work with government, businesses, and consumers to represent U.K. interests and facilitate the production of British, European, and international standards. British Standards' products and services help organizations to successfully implement best practice, manage business critical decisions, and achieve excellence. This includes a wide range of published information and commissioned services delivered under the BSI Business Information brand.

- *BSI Management Systems* operates worldwide to provide organizations with independent third-party certification of their management systems, including ISO 9001:2000 (Quality), ISO 14001 (Environmental Management), OHSAS 18001 (Occupational Health and Safety), ISO/IEC 27001 (previously BS 7799 for Information Security), ISO/IEC 20000 (previously BS 15000 for IT Service Management) and FSMS, including ISO 22000. In addition, BSI Management Systems also offers a range of training services around management systems.

- *BSI Product Services* is best known for the Kitemark, the United Kingdom's first product quality mark. BSI Product Services exists to help industry develop new and better products and to make sure they meet current and future laws and regulations. It also provides third-party certification, specifically for CE marking, a legal requirement for certain categories of products to be sold within the EU.

- *BSI Entropy International* provides software solutions that enable organizations worldwide to improve environmental, social, and economic performance, thereby contributing to global sustainability. The Entropy System is a web-based application for enterprise-level risk and compliance management that helps businesses improve internal control and overall corporate governance.

7.8.1 BSI Food Safety Overview

Although much of our food supply is safe, several recent high-profile cases around the world underline the potential danger of food-borne illness to consumers, employees, and brand value. A few recent examples include BSE-infected beef and the *Salmonella* contamination of poultry and eggs. For these reasons and others, global retailers, distributors, food manufacturers, and food service companies are now concerned more about the safety of their food supply chain than ever before.

Organizations in the food sector must manage risk, demonstrate good corporate responsibility, meet legal requirements if they are to remain competitive, protect their reputation, and enhance their brand. An effective FSMS based on a proven standard helps organizations achieve their goals. Furthermore, assessment and certification of an organization's management system by an independent third party will optimize their food safety management.

BSI Management Systems is a leading registrar in the Americas, with offices in the United States, Canada, and Mexico. BSI employs full-time registrar auditors and is a provider of value-added auditing and training services for management systems. BSI provides auditing, certification, and training services to the food sector.

Numerous food sector businesses prefer to use BSI auditing for their FSMS over leading food safety standards such as the Dutch HACCP code, BRC Global Food, and ISO 22000:2005. BSI has achieved U.K. Accreditation Service (UKAS) accreditation to deliver ISO 22000:2005 as of May 30, 2006.

7.8.2 BSI MANAGEMENT SYSTEMS

BSI puts forward that good organizations have processes, procedures, and standards of performance to meet present and future challenges, but that great organizations will also have management systems registration. The implementation and registration of a management system helps an organization achieve continuous performance improvement. Use of a proven management system combined with ongoing external validation enables the organization to continually renew its mission, strategies, operations, and service levels.

Management systems registration means

* Verifying practice versus process
* Objective third-party validation
* Benchmarking

7.8.2.1 BSI Management Systems ISO 22000 Certification

On May 2006, BSI Management Systems was accredited to provide ISO 22000 certification worldwide. BSI Management Systems has strengthened its position in the global food safety certification market by being among the first organizations to be independently accredited to deliver certification against ISO 22000, the new international food safety standard. BSI's accreditation has been granted by UKAS, the globally recognized accreditation body.

ISO 22000, published in September 2005, specifies the requirements for a FSMS. The standard combines generally recognized key elements to ensure food safety along the entire food chain, including interactive communication, system management, control of food safety hazards through prerequisite programs and HACCP plans, and continual improvement and updating of the management system.

Organizations involved in the food supply chain are facing escalating demands to demonstrate that their management practices and procedures are of a consistently high standard across their business operations. Issues such as food safety scares, ethical trading pressures, and product quality and safety have put the accountability and transparency of the food sector under the spotlight. With supply chains now more diverse and internationally spread, bringing increased risk to those managing them, the support of an accredited certification body such as BSI is increasingly important.

By being audited and certified by BSI against the requirements of ISO 22000, organizations can demonstrate that they have the management and control systems in place to control food safety hazards and provide consistently safe end products that meet the requirements of all stakeholders. Ultimately, third-party ISO 22000 certification can independently demonstrate an organization's commitment to food safety.

7.8.2.2 BSI Recommendations for Implementing a Management System

Implementing a management system of any kind is a significant undertaking for an organization seeking business improvement. However, good planning and senior management support can significantly ease the process. For all management systems, there are some common tools to be used and a common processes that can be followed during implementation that include

1. *Understanding the host management system and its requirements:* All people involved in taking the decision to implement the management system need a basic understanding of what is involved.
2. *Implementation of the system:* Literature, consulting, and training must be effective.
3. *Register the management system:* Once the management system is in place, to ensure its long-term effectiveness, it is important to become registered by a third-party registration body.
4. *Promote and maintain the management system:* Promote the fact that a registered management system is established to customers and other stakeholders, with maintenance and continual improvement of the management system.

7.9   CERT ID LC

Cert ID LC
P.O. Box 1810
Fairfield, IA 52556-0031
Phone: 641-472-9979; 888-229-2011 (toll-free: United States and Canada only)
Fax: 641-472-9198
E-mail: info-na@cert-id.com

Available at http://www.proterra.at and http://www.cert-id.com.

Cert ID is the sister company, or spin off, of auditing laboratory Genetic ID (see Chapter 8 on auditing laboratories). Cert ID is a global company active in providing third-party certification programs to growers, agricultural processors, food ingredient producers, food and feed manufacturers, animal producers, and food retailers. Cert ID advertises itself as being born out of the requirement of the food manufacturing and retail industries to offer a non-genetically modified assurance to consumers. Cert ID is both the name of their company and the service it offers.

Historically, Cert ID's inception was during the advent of the commercialization of GMO products and was demanded from the agricultural and retail industries.[*] In 1999, Cert ID Ltd. was spun off from Genetic ID, in the United Kingdom, through a joint venture from the recommendations of the members of the BRC and other

---

[*] Cert ID began promoting the need of their services starting around the year 1990 when consumers in various parts of the world were confronted with news headlines informing them about many problems and even "scandals." The sheer number of catastrophes and near-catastrophes, many of them in Europe, including the dioxin scandal, BSE, foot and mouth disease (FMD), and the broadly expanding occurrence of food-related allergies, caused a rising wave of uncertainty among consumers, primarily in Europe, Japan, Korea, and lately also North America and Brazil.

European retailers and food manufacturers. Cert ID Ltd. was originally founded in the United Kingdom as a joint venture operation between Genetic ID, Inc., and a British laboratory. In 2000, Cert ID LC was founded in the United States.

Today, industries interested in Cert ID can use the full range of their programs. As an ancillary service, the company offers an extensive sourcing program of certified non-GMO raw materials and ingredients that enables the entire food production chain to go non-GMO. As an example, in the area of soy products, Cert ID clients range from suppliers of soybeans and soy meal to those offering lecithin and a wide range of protein products.

Cert ID services include the integration of state-of-the-art GMO testing, auditing, and recordkeeping, which helps to minimize expense to food producers and industry buyers and maximizes surety to consumers because they are a third party that is independent of any other industry. They assert to be in compliance with international standards for ethics, social responsibility, and environmental sustainability with many of their Cert ID standards such as Cert ID EU Regulatory Compliance Standards, Cert ID Non-GMO Standards, and Cert ID ProTerra Standards. These standards and tests are used in the application of agriculture production, storage, transport, and industrial processing of commodities.

Cert ID promotes themselves as being unique by enabling consumers to discern the Cert ID Non-GMO certification program, which goes beyond existing labeling laws such as those in the EU that require only labeling of food items if they contain more than 0.9% GMO ingredients. Cert ID is designed to help consumers identify at a glance whether a product is non-GMO, or, to be more precise, to tell whether a product is really made *practically without* genetically modified (GM) ingredients. Because of the complexity of food production and raw material sourcing, transportation, and storage, a food manufacturer or retailer cannot usually ensure that a product on a shelf is non-GM. Experienced and independent certification organizations such as Cert ID focus entirely on delivering this type of non-GMO assurance. Because Cert ID is a third party, independent of the agricultural, biotechnology, and food industry, it has earned recognition and credibility from the retail industry and consumer groups. Cert ID is an early pioneer in this type of certification. Products displaying its non-GMO seal are assured to contain a maximum of 0.1% GMOs. That is almost one-tenth of the current labeling threshold in the EU.[*]

7.9.1 Certification Process—Example

For a product to become Cert ID certified, each ingredient must be identity preserved throughout the production chain. In their company advertised summary, the process of certification focuses on one ingredient—soy lecithin—and the ways in which this ingredient is often used in the food production chain.[†] Lecithin is a natural lubricant and emulsifier that is used, for example, to keep the chocolate and cocoa butter in a candy bar from separating.

---

[*] The two main agricultural commodities—soybeans and corn (maize)—and their derivatives were the first raw material groups to be certified by Cert ID as non-GMO in 1999.

[†] Soy lecithin is a mixture of fatty substances that are derived from the processing of soybeans and it is often used in animal feed applications, pharmaceuticals, and protective coatings.

The certified lecithin begins as soybean seed and must be derived from non-GM seed stock. To reduce the chance of volunteer plantings (such as seeds from last year's harvest), the grower must take care to clean his/her equipment and sow anew on fields that have produced non-GM crops for at least two harvests. Also, the grower must take care during harvest to clean his/her equipment and not commingle the crop with that from other fields. A sample of the crop is tested for GMOs before being placed in storage or put into production. It is in the interest of the processor/manufacturer receiving raw material product that they obtain assurance that the shipment comes from a certified production system and that the shipment itself is product certified as non-GMO. Many of the requirements that Cert ID follows are similar to those used in USDA's NOP.

The processor receives raw product and turns it into ingredients. The processor may buy from a Cert ID non-GMO-certified supplier or he may choose to contract with growers to become non-GMO. Between facilities, all movement of the non-GM product must be tracked with the proper documentation to keep its IP status.

The verified non-GMO soy may now be processed into lecithin at facilities that were inspected by a qualified Cert ID inspector and shown to have processes in place to prevent inadvertent commingling of non-GM and GM materials. The final material can now be retested and certified as non-GMO lecithin.

Just as individual products may be Cert ID certified, so too may a supply chain be non-GMO certified. The supply chain, for nearly all food products, contains the following stages. All of them need to be inspected, audited, sampled, and tested for GMOs before non-GMO certification from Cert ID may be granted. Assessments of a farm/cooperative and its seed suppliers, the processors (seed crusher, processing plant, etc.), and of logistics [Transportation (e.g., trains, trucks, ships) and warehouses, silos, elevators, loading and unloading facilities, and ports] may be conducted by Cert ID.

7.9.2 Certification Methods and Tools

Below is an overview of the methods and tools applied in the process of Cert ID non-GMO certification. All Cert ID non-GMO certification clients are subject to these methods that include audits,* inspections, and unannounced audits. In addition, sampling, such as polymerase chain reaction (PCR) testing,† and certification of recordkeeping/record storage must certified by Cert ID.

---

* Standard definitions of the term "audit" give something like "an examination of records to check their accuracy." This comes rather close to what is done in a Cert ID audit. The inspector visits a client facility where he verifies whether "things" are in compliance with the respective module of the Cert ID Standard. Sometimes "things" may be the books and records of a trading company, but usually they are production or handling facilities and equipment. On their visits to client locations, Cert ID inspectors review all facilities for risks of GMO contamination. This is needed before a certification plan can determine how a particular client will be certified.The certification administrators decide if and when any unannounced audits should be conducted. In the end, this surprise tool gives consumers the assurance that a production facility does not just "groom" itself for the day an inspector has announced his visit.

† Any kind of certification for non-GM must have at its core a reliable testing method for the presence of GMOs. According to government regulations in many countries (e.g., the EU Lisbon protocol),

Cert ID has created a special database coined "Full Traceability Database" for storage of all certification records. The structure of this database enables certification administrators to relate all pertinent records. For Cert ID industry clients, this means that they are able to support their claim to have fully documented traceability for their certified product(s). They also meet legal traceability requirements where they exist, such as in the EU.* For Cert ID, it means that it is able to provide its clients with the complete set of records linked to their ingredients or raw material; in this way food manufacturers can stand behind the claim displayed on their packaging through the Cert ID seal.

Cert ID states that non-GMO certified and its "GMO-free" label ideally means that a product is devoid of any GMO material and, in reality, this is not scientifically verifiable with today's testing methods. Even in raw material and using the PCR method the limit of detection is approximately 0.01%. Although this is quite sensitive, it does not constitute material as GMO free. Therefore, any process guarantee (as opposed to a "content guarantee") given by a non-GM certification standard like Cert ID can only be a matter of definition. Cert ID's defined level is called "non-GMO." Another reason to guarantee process and not entirely content is one of statistics. Testing every last bean in a shipment of soybeans would mean that there is nothing left to process afterwards and is unrealistic. At the request of its European clients and with support

---

(Continued)

both protein-based and ELISA testing methods are ruled out. Cert ID accepts so-called strip tests as a screening method before crops are unloaded in processing plants (e.g., in the case of soybeans), but for the actual input and output tests rigid PCR testing is required. One of the purposes of certification is to be able to demonstrate good systems and procedures in compliance with legal regulations or, in the case of Cert ID, also with a certification standard. Such records typically contain audit and inspection reports, laboratory test results, shipping documents, photographs, maps, and other related documents. Although these records must be safeguarded, it would be quite cumbersome to save them all on paper in hard copy. At Cert ID, most records are recorded electronically in various file formats. This conversion is done soon after the record is received by the certification administration.

* Legislation of member states of the EU prescribes on an individual basis what positive claims in cases of non-GM labeling must look like. In some countries (e.g., in Germany) certain words (*ohne Gentechnik,* or "without genetic engineering") must also be displayed if a manufacturer wants to display the non-GMO seal of Cert ID (or any other positive statement regarding the absence of biotechnology) on their products. In other countries, (e.g., in Great Britain) it is sufficient that the positive claim is true and can be proven. Therefore, production certified by Cert ID should be devoid of GMO content being that a manufacturer must target his production at 0%. The 0.1% threshold is an acknowledgment of the possibility that an adventitious presence might occur during shipping, storage, handling, and transportation. The 0.1% threshold of non-GMO certification by Cert ID must not be confused with the thresholds of 0.9% (and 0.5%, respectively) that are stipulated in the EU regulations in force since mid-April 2004 regarding GMO labeling and traceability. The 0.1% threshold of Cert ID is about the permissibility of a positive claim (namely "without genetic engineering") whereas, in contrast to this, the legal threshold of 0.9% defines one legal consequence in case of an excessive GMO content. An important change of paradigms has occurred in the EU with the implementation of EU Regulations (EC) No. 1829/2003 and No. 1830/2003 (both of September 22, 2003): The application principle has replaced the detection principle that had been in place up until now. This means that any GMO labeling is not linked exclusively to detectability. Labeling is now necessary in all those cases where a food or feed product is made from GMOs, regardless of whether this can be detected in the final product or not. For the consumer, certification means "full" assurance of the non-GMO status of the product that he or she buys in a store. For the manufacturer, it means that the gap of credibility is bridged in this regard. Testing alone is only as valuable as the credibility the sampling process has to those who read a GMO testing laboratory's analysis report.

from consumer advocacy groups, Cert ID's assurance level for full non-GMO certification was set at 0.1%. The process guarantee underwritten by Cert ID for a fully certified product is, "This product has been produced without genetically modified ingredients, processing aids, additives, flavorings, colorings, or other inputs."

7.9.3 TRANSACTION CERTIFICATES OF COMPLIANCE

In the Cert ID non-GMO certification program, it describes the application of audits, inspections, and state-of- the-art GMO testing from the seed supplier all of the way to the food manufacturer of the consumer product, including the grower, storage and handling, shipping, processing, and the ingredient distribution, so that the paper trail of fully documented traceability is never interrupted. This paper trail is documented by Transaction Certificates of Compliance (TCCs) that are issued by Cert ID. The TCC documents accompany each and every shipment of product that is certified as non-GMO.

7.9.3.1 Testers and Inspectors

The testers are laboratories that meet the quality criteria set forth in the Cert ID standard. This does not necessarily mean it has to be a Genetic ID laboratory, but all members of Genetic ID's Global Laboratory Alliance operate according to the same analytical methods as the laboratories of Genetic ID, who, in turn, are accredited according to DIN EN ISO 17025. (ISO Certification Guide 65 forbids that a certain laboratory or method be specified exclusively.) The inspectors can either be Cert ID employees or professional members of the inspection industry specially trained by Cert ID on location for the purpose of rendering this kind of support to the program.

The Cert ID customers include all stages of the food and feed production chain, seed suppliers, farmers, producers, growers, cooperatives, trading companies, brokers, transport, shipping, storage, loading/unloading, processors, animal feed processors, food manufacturers, and distributors.

7.9.3.2 Example: Food Manufacturers

Many Cert ID customers request food products to be free of GMOs. Through certification, manufacturers are able to meet the demands of these consumers and are able to obtain premiums for their certified products. In light of liability concerns for allergic reactions to GMOs, manufacturers can be assured the certified status of their products will minimize their risk of using contaminated ingredients and thus also their liability. By far the most important reason for the food industry to have Cert ID certify their products as non-GMO is probably the reduced risk of brand damage and meeting regulatory compliance.

7.9.3.3 Costs to Implement Cert ID Non-GMO Certification*

Because of the differences between different production facilities, the cost of certification is calculated on a case-by-case basis. A reliable cost estimate can be issued

---

* Certification process: Steps toward Cert ID non-GMO certification: (1) application, (2) inspection, assessment (PA), (3) setting up a certification plan, (4) certification committee evaluation, and (5) licensing system certification.

to prospective clients interested in becoming certified. This is possible after Cert ID receives a completed System Assessment Worksheet form.

Cert ID Products (all downloadable)
Cert ID EU Regulatory Compliance Standard .... Controlled* $150.00
Cert ID EU Regulatory Compliance Standard .... Noncontrolled $140.00
Cert ID Non-GMO Standard .... Controlled $150.00
Cert ID Non-GMO Standard .... Noncontrolled $140.00
Cert ID ProTerra Standard .... Controlled $150.00
Cert ID ProTerra Standard .... Noncontrolled $140.00

7.9.4 Testing That Addresses Social and Environmental Concerns

The ProTerra Certification Program from Cert ID provides socially and environmentally responsible companies with the opportunity to obtain recognition of their practices and be confident that the materials they purchase have not been produced in a manner that contributes to social and environmental degradation. The program is also designed to help suppliers assure their buyers, and ultimately consumers, that their products have been produced in a sustainable manner. Cert ID argues that such assurances are vital as consumers become more and more ethically aware.[†] ProTerra considers various dimensions of social and environmental responsibility, including compliance with environmental protection laws, management of agronomic factors, preservation and restoration of fragile features of the ecosystem, and adherence to socially responsible practices.

REFERENCES

Grande, Bill. "Servicing IP Production and Marketing: A Third-party Role." Presented at the *Product Differentiation and Market Segmentation in Grains and Oilseeds: Implications for Industry in Transition Symposium.* Washington, DC: Economic Research Service, U.S. Department of Agriculture, and the Farm Foundation, 2003.

---

[*] Controlled version: This is if the customer wishes to automatically receive updates and revisions to the standard as they are released. Noncontrolled version: This is if the customer does not wish to automatically receive updates and revisions to the standard.

[†] According to *Pejling*, the magazine of the Swedish Dairy Association, 70% of Swedish consumers have a "personal blacklist" of products and companies that do not meet their personal standards for social and environmental responsibility. In addition, a survey by Market & Opinion Research International in the United Kingdom revealed that as many as one-fifth of the U.K. population boycott or select goods on social grounds. Similar statistics are found in many countries around the world.

8 Auditing Laboratories

8.1 INTRODUCTION

In addition to a firm having an internal or external verification of its identity preservation and traceability (IPT) system, it often is required by contract to have laboratory analysis of the crop or product being purchased at one or more points of the chain. This laboratory analysis may look at one or several aspects of the crop, again by contract, it may only matter that the crop is confirmed to not have any genetically modified organism (GMO) traits, it may also be analyzed for other specifics such as oil or protein content, or be analyzed to determine its origins of growth for country-of-origin-labeling (COOL).

Some of these laboratories also conduct field and processing auditing services, whereas some only provide laboratory services. Section 8.2 of this chapter will include a sampling of the various methods used to test sample crops and products. This is an important section because it helps to explain many of the methods used in greater detail than what will be shared within each laboratory firm's biography provided within each section. This nearly includes the full spectrum of analysis available, from enzyme-linked immunoabsorbant assay (ELISA) protein and polymerase chain reaction (PCR)-based testing to nuclear magnetic resonance (NMR) and atomic absorption spectrometry (AAS). Any repeating of test information within several of the organizations highlighted in this chapter is due to the organization's emphasis of the test. Allowing each organization to elaborate on their services may also help them to differentiate themselves from other like organizations.

The laboratories are public and private. Each of the following organizations will be reviewed for services that they offer and their prices, if available:

- Biogenetics Services
- California Seed & Plant Laboratory
- Canadian Grain Laboratory
- Genetic ID
- CII Lab Services
- EnvironLogix
- Eurofins GeneScan
- Mid-west Seed Services
- Neogen Corporation
- Protein Technologies
- Strategic Diagnostics

What follows are company/organizational statements from their respective websites and naturally reflect their views.

8.2 INTRODUCTION TO THIRD-PARTY CERTIFICATION/VALIDATION BY LABORATORIES[*]

From The Organic & Non-GMO Report (October 2006)[†] survey of non-GMO production testing, testing for GMOs has become almost standard procedure in non-genetically modified supply chains. Seventy-seven percent of their respondents reported that their products were tested. Testing was done by a variety of providers. The most common method was by sending samples to testing laboratories, most of which specialized in non-GMO testing. Third-party certifiers often conducted tests for GMOs as part of their non-GMO certification. Many of the respondents noted that they tested their own products, whereas 19% reported that the buyers tested their products before purchase. The most common form of testing was the rapid strip test, with 56% reporting that their products were tested using this method. In contrast, 43% reported that their products were tested using a PCR test. Finally, 30% reported product tests using ELISA tests. Many reported that their products underwent multiple tests using different testing methods.

The main point here is that laboratory testing is becoming much more important in food production to distinguish any number of important traits or attributes. Not only can laboratory testing confirm the presence of particular enzymes or proteins, but also what percentage of the volume are the enzymes or protein detected.

8.2.1 General

For food product species identification, methods based on protein, fatty acids, and deoxyribonucleic acid (DNA) allow a fast and unmistakable identification of animal species. Several methods accomplish this, such as protein-based methods,[‡] immunological methods,[§] and proteomics. They can be used to differentiate species, breeds, and varieties by their specific protein pattern. Infrared spectroscopy, both near-infrared (NIR) and mid-infrared (MIR) spectroscopy, can be used for analysis of the main components of foods as well as animal feeds, minerals, and vitamins.[¶]

---

[*] See reference works, Food Authenticity and Traceability, edited by Michèle Lees, for additional information regarding various laboratory methods for authentication and traceability food items; Schwägele, F. "Traceability from a European Perspective." Meat Science 71, 2005: 164–173. Available at http://www.sciencedirect.com.; and Auer, Carol. "Tracking Genes from Seed to Supermarket: Techniques and Trends." Trends in Plant Science, 8, no.12, 2003: 591–597.

[†] This report was formerly known as the The Non-GMO Report.

[‡] Proteins (enzymes, myoglobin, etc.) have been widely used as species markers. Applicable techniques include separation of water-soluble proteins by starch or polyacrylamide. Highly resolved water-soluble protein patterns can be used to genetically differentiate closely related species. The limit of detection of gel electrophoretical methods varies between 0.1 and 1% and depends on the visualization procedure of the proteins bands.

[§] For example, Western blotting and a specific type of enzyme immunoassay (EIA)—ELISA, which is performed on the solid surface of microplates—are using suitable target proteins for analysis. A qualitative detection of animal species is possible and the limit of detection depends upon their content in meat products (pork ≤1%; poultry and beef ≤2%; sheep ≤5%).

[¶] Gonzalez-Martin, Gonzalez-Perez, and Hernandez-Mendez (2002) successfully applied NIR to the determination of the concentrations of iron, calcium, sodium, and potassium in pork. Pires, Lemos, and Kessler (2001) demonstrated the potential of NIR to measure the concentration of 11 vitamin levels in poultry feeds. Garnsworthy, Wiseman, and Fegeros (2000) reported the application of NIR to the prediction of chemical, nutritive, and agronomic characteristics of wheat.

For traceability of production process and storage to determine the "history of meat and meat products" with respect to the production processes and changes occurring during storage, several technologies {DNA-based methods; electrophoresis, including capillary electrophoresis [CE]; immunological methods; high-performance liquid chromatography [HPLC, including HPLC with mass spectroscopy (MS)]; lipid-based methods [gas chromatography (GC), GC–MS]; infared [IR] and NMR spectroscopy; and electron microscopy} may be used. One of the significant challenges to identify irradiated food products is the different techniques necessary to cover the entire spectrum of products. Typical methods used include immunological methods, comet assay, photon-stimulated luminescence, and electron spin resonance. However, only a limited number of laboratories worldwide have the necessary capability for the reliable determination of food irradiation.

8.2.2   SHOULD FOOD PRODUCT LABELS SPECIFY THAT INGREDIENTS WERE DERIVED FROM GM CROPS?*

This question has been part of the international debate about agricultural biotechnology. Food labeling and traceability regulations are largely determined by economic, political, and social issues, leaving business operators and researchers to develop analytical methods that support compliance with the regulations within the existing system for crop production, international trade, and food processing.

With final approval from the 15 member states, European Union (EU) food processors and supermarkets are now required to label all food products containing approved genetically modified (GM) crops above a 0.9% threshold level for each ingredient. The establishment of a threshold acknowledges that conventional crops, such as bulk shipments of maize or soybeans, are never 100% pure and a low level of commingling with GM crops is expected. At present, only two transformation events [a herbicide-tolerant soybean and an insect-resistant Bacillus thuringiensis (Bt) maize] are authorized for human consumption in the EU. There is no acceptable threshold level for unauthorized GM crops, although GM crops that have received a favorable scientific assessment but are not authorized can be present below a 0.5% threshold. Labels will carry the words, "This product contains genetically modified organisms" or "Produced from genetically modified (name of organism)."

Analytical methods of tracking and testing to trace-back plant genes in the environment and the food chain are essential for environmental risk assessment, government regulation compliance, and production and trade of GM crops. Below are several laboratory methods used to track plant genes during pre-commercialization research on gene flow and post-commercialization detection, and identification and quantification of GM crops from seed to consumer or grocer. At present, DNA- and protein-based assays support both activities, but the demand for fast, inexpensive, sensitive methods is increasing. Part of the demand has been generated by stringent food labeling and traceability regulations for GM crops. The increase in GM crops, changes in GM crop design, evolution of government regulations, and adoption

---

* Excerpts from Auer, Carol. "Tracking Genes from Seed to Supermarket: Techniques and Trends." Trends in Plant Science, 8, no.12, 2003: 591–597. Reprinted with permission from Elsevier.

of risk-assessment frameworks will continue to drive development of analytical techniques.

The increased use of GM crops has created a demand for laboratory techniques that can track plant genes and transgenes in the environment and through the food chain. Two critical activities requiring laboratory analysis exist: precommercialization research on gene flow to support ecological risk assessments and post-commercialization detection, identification, and quantification of GM crops from seed to supermarket. The second type of analysis is most important to end users.*

For the first group, this has become critically important to the parent seed industry and the discovery of transgenes in corn (maize landraces) in Mexico during the past decade. Although crops and wild plants have exchanged genes throughout the history of agriculture, GM crops have raised concerns that gene flow will lead to negative environmental impacts in agricultural systems and/or natural areas.† The discovery of StarLink corn in human food and the resultant recall of hundreds of food products highlighted the difficulty of separating and tracking GM crops through the food chain. Because GM crops and the regulations pertaining to them are proliferating around the world, the current system of global agricultural trade demands laboratory techniques that support regulations, risk assessment frameworks, and contracts between trading partners.

8.2.3 HOW ANALYTICAL METHODS ARE DETERMINED

Analytical methods for tracking genes and transgenes are chosen based on contract, regulations, and a combination of other factors. The accuracy, precision, reproducibility, sensitivity, and specificity of the method must be understood in relation to the research question. Practical considerations include the cost and time per sample, the chemicals and equipment required, sample handling and processing, adaptability to field conditions, and technical expertise. For post-commercialization traceability and food labeling activities, methods must be practical for testing points at the farm, during transport, and in food processing. Regardless of the technique, appropriate experimental controls, reference materials, and information about parental crop lines and transgenes must be available.

The three most widely used laboratory methods are (1) DNA-based molecular techniques to characterize genetic markers, (2) isozyme analysis of protein profiles, and (3) marker genes that produce a selectable phenotype.

DNA-based molecular techniques to identify genetic markers and describe genetic relationships have become a powerful tool for crop breeding, population genetics,

---

* GM crops were grown on 58.7 million hectares in 2002, 99% of which was grown in the United States, Argentina, Canada, and China (http://www.isaaa.org). In most cases, this first generation of GM crops has been modified by the insertion of one or a few novel genes to produce valuable agronomic input traits such as herbicide tolerance or insect resistance.

† Environmental risk assessments generally require information about the probability of pollen movement from GM crops to fields of the same crop, closely related crops, wild ancestors in centers of biodiversity, intermediate weed species, and wild relatives in natural areas.

and studies on gene flow.* Molecular markers are advantageous because they are abundant in the plant genome, are not affected by the environment, and can be based on noncoding sequences that are selectively neutral and can provide a high level of resolution between closely related plants. Disadvantages of molecular markers include the requirements for expensive laboratory equipment, costly reagents, and technical expertise† (see Table 8.1 for comparisons).

Isozymes are related enzymes that catalyze the same reaction but have different structural, chemical, or immunological characteristics. Isozyme (allozyme) analysis uses the isozyme profile to distinguish between related plant species, an approach that has been documented for many crop species. Although laboratory equipment and costs are modest, isozyme variation will not always be sufficient to discriminate between species and might not be selectively neutral. Plant samples must be handled carefully to protect enzyme activity, and activity is affected by tissue type, developmental stage, and environmental conditions.

The most common selectable markers are antibiotic resistance and herbicide resistance, both of which are routinely used in the initial selection of transformed plant cells and plant propagation.‡ Visible markers or reporter genes can be inserted to study gene flow, including green fluorescent protein (GFP), ß-glucuronidase, and luciferase. The family of GFP genes provides the advantage of real-time, noninvasive identification of GM plants and pollen in the laboratory or field. For example, tobacco plants expressing GFP under the control of a promoter for anther and pollen expression demonstrated that a hand-held ultraviolet (UV) light can detect transgenic pollen carried by bees.§

8.3 TECHNIQUES FOR TRACKING GM CROPS FROM SEED TO SUPERMARKET

Post-commercialization activities conducted by industry, government agencies, and independent groups require fast, accurate, sensitive, and inexpensive methods to track transgenes from the planting of GM seed to the production of food products. Business operators use analytical methods to support seed certification, identity preservation (IP), traceability, and food labeling. Government agencies use laboratory tests for programs related to stewardship, seed quality, food safety, food labeling, environmental monitoring, and regulatory enforcement (see Table 8.2 for comparisons used to track transgenes from seed to supermarket).

---

* Many DNA sequence markers and assays have been developed for nuclear DNA or chloroplast and mitochondrial DNA that can trace maternal lines.
† The most useful molecular techniques to describe genetic relationships include amplified fragment length polymorphisms (AFLP), random amplified polymorphic DNA (RAPD), restriction fragment length polymorphism (RFLP), and microsatellite markers. AFLP and RAPD have an advantage in that they do not require prior information about DNA sequences or a large investment in primer/probe development.
‡ Other types of selectable markers include genes for resistance to cytotoxic agents, for auxotrophic markers to complement mutant's deficient in a growth factor, and for the use of mannose or xylose sugars.
§ GFP expression could support direct monitoring of pollen movement over different large distances and research on containment strategies. However, government approval would be required before unconfined release of the gene encoding GFP into the environment.

TABLE 8.1

Comparison of the Principal Laboratory Techniques (Conventional or GM Crop)

Laboratory Technique	Analyte	Useful for Conventional or GM Crop
Isozyme (allozyme) analysis	Enzyme profile	Conventional
ELISA	Novel phenotype	GM
Selectable marker gene, antibiotic resistance, herbicide resistance	Resistance phenotype	GM
Visible marker genes GUS, GFP, and Luc	Fluorescence or colored stain	GM
Molecular marker assays AFLP, RAPD, and RFLP	DNA sequences as genetic markers	Conventional
Microsatellite morpholgical character	Morphological trait or plant phenotype	Conventional
Flow cytometry	Nuclear DNA content	Conventional
Bioassay	Resistance phenotype	GM

AFLP, amplified restriction fragment polymorphism; GUS, ß-glucuronidase; Luc, luciferase; RAPD, random amplified polymorphic DNA.

Source: From Auer, C., *Trends in Plant Science,* 12, 591–597, 2003. With permission.

Post-commercialization tracking of GM crops requires three types of tests: (1) a rapid detection assay to determine whether a GM crop is present in a sample of raw ingredients or food products, (2) an identification assay to determine which GM crop is present, and (3) quantitative methods to measure the amount of GM material in

TABLE 8.2

Comparison of the Principal Laboratory Techniques (by Type of Measurement)

Laboratory Technique	Analyte	Type of Measurement
ELISA	Novel protein	Qualitative, semiquantitative, quantitative
Lateral-flow test strip	Novel protein	Qualitative
PCR-based methods: RTQ-PCR, QC-PCR, and multiplex PCR	Novel DNA sequence	Qualitative, semiquantitative, quantitative
DNA microarray	Novel DNA sequence	Qualitative, semiquantitative, quantitative
Spectroscopy and chromatography	Plant biochemical trait	Qualitative

QC-PCR, quantitative competitive PCR; RTQ-PCR, real-time quantitative PCR.

Source: From Auer, C., *Trends in Plant Science,* 12, 591–597, 2003. With permission.

the sample. The first stage can be accomplished by qualitative methods (presence or absence of transgene), whereas the third stage uses semiquantitative (above or below a threshold level) or quantitative (weight/weight % or genome/genome ratio) methods. Currently, the two most important approaches are immunological assays using antibodies that bind to the novel proteins and PCR-based methods using primers that recognize DNA sequences unique to the GM crop.

The two most common immunological assays are ELISA and immunochromatographic assays (lateral-flow strip tests). ELISA can produce qualitative, semiquantitative, and quantitative results in 1–4 hours of laboratory time. The lateral-flow strip tests produce qualitative results in 5–10 minutes in any location for less than $10 (U.S.). However, sufficient protein concentrations must be present for antibody detection, and protein levels can be affected by the plant's environment, tissue-specific patterns of transgene expression, protein extraction efficiency, matrix effects, and food processing techniques that degrade proteins (Sundstrom and Williams 2002).

The most powerful and versatile methods for tracking transgenes use PCR. PCR has many advantages but it requires DNA sequence information to design primers to identify a crop (e.g., Lec1 lectin gene for soybean), to detect a DNA sequence common to many GM crops (e.g., cauliflower mosaic virus 35S promoter), to detect a specific transgene, or to identify a specific transformation event using the unique transgene. Some sequence information can be found in biosafety databases, genome databases (e.g. GenBank), patent applications, and government documents. Theoretical detection limits for PCR have been calculated for various grain crops. Estimates for cost are $150–$1,050 (U.S.) and for time are from 4 hours to several days.

In addition to PCR and protein-based methods, chromatography, MS, and NIR spectroscopy can be used in some situations, such as for GM crops that have significant changes in chemical composition. However, these methods can fail when alterations in GM crop biochemistry are within the range of natural variation found in conventional crops.

The need to trace and identify GM crops has led to the suggestion that a universally accepted, noncoding DNA sequence be incorporated adjacent to the transgene to provide a unique identification tag. The identification tag sequence could contain information in an encrypted, artificial, triplet-based code and would not produce a protein or change plant fitness.

8.4 LABORATORY METHODS OF DETERMINING GEOGRAPHICAL ORIGINS OF FOOD[*]

According to Peres et al. (2003), modern analytical techniques can determine the plant or animal species present in food. However, to determine a food's origin is much more difficult. European regulation EU No. 178/2002 requires such information for its traceability of food. Physicochemical and microbiological analytical techniques provide, with a degree of exactness, processes to determine the origin of some foods. The choice of a technique depends on the level of study required. The region

---

[*] Reprinted excerpts Peres, Bruno, Nicolas Barlet, Gérard Loiseau, and Didier Montet. "Review of the Current Methods of Analytical Traceability Allowing Determination of the Origin of Foodstuffs." *Food Control* 18, no. 3, 2007: 228–235.

of food origin can develop from joint analytical techniques. The results can then be analyzed by mathematical/statistical methods to process the data.

Other notions that are encompassed concerning geographical origins include Protected Designation of Origin (PDO), which covers the term used to describe foodstuffs that are produced, processed, and prepared in a given geographical area using recognized methodology (e.g., Jamon de Teruel, Parma ham); Protected Geographical Indication (PGI) specifies geographic location must cover at least one of the stages of production, processing, or preparation [additionally, the product may benefit from a good reputation (e.g., Nürnberger Bratwürste]; and Certificate of Specific Character (CSC),[*] which means that a foodstuff possesses specific characteristics that distinguish it clearly from similar products in the same category (e.g., Münchner Weibwurst, Salami Milanese).

There are two types of methods: the physicochemical approach, which uses either the variation of the radioactive isotope content of the product, spectroscopy, or electronic nose; and the biological approach, which uses the analysis of total bacterial plant life through many procedures.[†] The goal or purpose of these analyses is to help in differentiating the origin of various cheeses or various wines, or of identifying the geographical origin of other foods like oysters, meats, fish, olive oils, teas, or fruit juices (e.g., a milk produced on a mountain from that produced on the plains).

Physicochemical methods use variation of radioactive isotopes. NMR coupled with MS of isotopic ratio (NMR/MS/IR) is used to detect variations in the nucleus of certain atoms; for example, there is a geographical effect on the polyunsaturated fatty acids. The microbial vegetation of the mountain pastures differed distinctly from that of the plains.[‡]

Another method is ion exchange chromatography/AAS. AAS permits the study of absorption of light by free atoms by the energy variation when one of the electrons passes from one electron orbit to another; for example, Emmental cheese, the type and quantity of which varies according to the geographical location of the cows. Strontium (Sr) is an artificial radioelement found everywhere in Europe, the presence of which is primarily due to the Chernobyl accident in 1986. Sr passes into dairy products by the water consumed by animals and can be used to distinguish Emmental cheese type produced in Brittany and Finland from those produced in the Alps (Switzerland, Savoy, Allgau, and Vorarlberg). The observed differences of this radioelement are explained by the geographical protective barriers against radioactive fallout and by the weather conditions just after the Chernobyl accident.

Other methods include site-specific natural isotope fractionation by NMR (SNIF-NMR),[§] electronic nose coupled with MS, Fourier transform MIR spectros-

---

[*] Recognized by EU member states.
[†] Such as denaturing gradient gel electrophoresis (DGGE) and denaturing HPLC (DHPLC), the polymorphism of conformation of the single strand DNA (SSCP), or DNA chips.
[‡] The mountain pastures are very rich in dicotyledonous and herbaceous nonleguminous plants, whereas the plain pastures are mainly composed by graminaceous and leguminous plants (Peres 2003).
[§] During food processing, isotopically accurate information is recorded about environmental conditions. Study of specific natural isotopic fractionation (SNIF-NMR) permits the association of a pure product or a component of a complex product with a particularly reliable identity card, and it is thus possible to know the geographical origin of a food.

copy (FT-MIRS),[*] MIRS and NIRS,[†] Fourier fluorescence spectroscopy,[‡] and other techniques such as Curie point pyrolysis coupled to MS (Cp–PyMS).

8.5 BIOGENETICS SERVICES, INC.

Biogenetic Services, Inc.
801 32nd Avenue
Brookings, SD 57006
Phone: 605-697-8500; 800-423-4163 (toll-free)
Fax: 605-697-8507
E-mail: info@biogeneticservices.com

Available at http://www.biogeneticservices.com.

Biogenetic Services, Inc., established in 1988, specializes in providing up-to-date protein and DNA analyses including isozyme purity tests and DNA genotypic profiling of plant and/or animal individuals, inbred lines, hybrids and breeding populations, and ELISA protein and PCR-based GMO/gene expression profiling (GEP) or transgenic (event) assays for seed companies, elevators, seed growers, other private individuals, plant and animal breeders, producers, ingredient suppliers, food companies, hatcheries, educational institutions, state and federal facilities, insurance companies, legal firms, and other associated industries.

Biogenetic is International Organization of Standardization (ISO)/International Electrotechnical Commission (IEC) 17025 accredited and acclaims being the first privately owned U.S.-based service company to receive ISO/IEC 17025 accreditation for protein and DNA genotyping, purity testing, pathogen diagnostics testing, and GMO/transgenic testing of plant (e.g., plant, seed, grain, feed, food ingredients, and food samples), and/or animal samples.

BGS testing services include inbred purity, single and multi-cross hybrid purity, and early generation analyses; genotyping (fingerprinting/profiling) of individual plants and/or animals;[§] and population (including multi-cross hybrids), fertility, and pest

---

[*] It is possible to differentiate *cis-* and *trans*-unsaturated fatty acids. It is reliable and more rapid than MIRS and traditional IR. This method can be coupled with other techniques to increase the accuracy of the results. For example, coupling FT-MIRS and GC made it possible to directly analyze complex mixtures such as flavors and fatty acid isomers in methyl ester form.

[†] NIR spectroscopy (and its applications in the animal feed industries) is a nondestructive analytical technique based on the principle of absorption of electromagnetic radiations by the matter. Spectra are analyzed and calculated by statistical methods such as discriminating factorial analysis (DFA) or by partial least squares (PLS). The DFA and PLS analysis of the spectra appeared to be powerful enough to authenticate the classification of wine produced from the same type of vine, but from three different French areas.

[‡] This method gives information on the presence of fluorophores and their environment in the sample. By using fluorescence properties of certain amino acids or extrinsic probes added to the medium, the structure of proteins alone or interacting with small hydrophobic molecules can be characterized. The data are analyzed by principal components analysis (PCA) and DFA. However, by those methods it is very difficult to geographically discriminate close regions.

[§] Genetic profiles frequently are included as part of the description of a newly developed line for plant variety protection (PVP) or patent purposes. Molecular marker loci such as isozyme loci, RFLP loci, and SSR loci are often used for marker-facilitated selection in plant and animal breeding programs.

diagnostic analyses. BGS is dedicated to providing comprehensive protein and DNA molecular marker screening services, including seed/grain/plant purity tests, GMO/non-GMO tests for presence or quantity of any GMO or for the presence of specific genetic events (e.g., Starlink Cry9C), and PCR diagnostics for organisms that cause damage in plants such as cyst nematodes (SCN) in soybeans and Mycobacterium avium subsp. Paratuberculosis (MAP), which causes Johne disease in ruminants.[*]

8.5.1 Plant Protein and DNA Tests

BGS uses isozyme (protein) assays for variety identification and seed purity and hybridity tests. ELISA protein assays are used for GMO event tests to determine the presence or quantity of protein of specific Bt or Roundup Ready (RR) events in a seed (plant) sample or a bulked seed or grain sample. BGS uses DNA technology, including restriction fragment length polymorphism (RFLP) and PCR-based technology, to assist in identification, protection, development of intellectual property, screening for resistance to pests, and marker-facilitated breeding (transgenic tests). In addition to standard PCR analysis with minimum detection levels of approximately 0.1% for any GMO in a non-GMO sample, BGS also routinely provides GMO testing services using real-time quantitative methods, which allow testing to the 0.01% detection level in a seed, grain, food ingredient, or finished food (or feed) sample. Event-specific testing using PCR-based methods looks for the gene conferring the trait of interest, such as the gene conferring the ability to produce Bt toxins. BGS utilizes protein (isozyme, ELISA) and DNA (RFLP, PCR, simple sequence repeat (SSR), single nucleotide polymorphism (SNP), and expressed sequence tag (EST) analyses, etc.) technology to provide information on sample purity (GMOs) and multilocus genotypes (genetic fingerprints) for organisms of any kind. In plants, emphasis has been on providing information on purity (hybridity and GMO presence and absence or quantity) and genotypes of inbred lines, single cross hybrids, and populations of corn, popcorn, sweet corn, cotton, sunflower, soybeans, common beans, potatoes, canola, wheat, oats, barley, hops, papaya, squash, and all types of other vegetable and fruit crops. DNA markers (RFLP and PCR) are also used to determine fertility in samples of maize inbreds and hybrids and are useful tools for marker-facilitated selection (e.g., backcrossing) in plant breeding programs.

Biogenetic customers include the following:

- Seed companies
- Plant breeders
- Animal breeders
- Ingredient suppliers
- Food companies
- Milling companies
- Insurance companies
- Sheriffs departments
- Judges
- Universities

---

[*] See http://www.cdc.gov/ncidod/EID/vol8no7/01-0388.htm for more information regarding this disease.

- USDA/Agricultural Research Service
- Brokers
- Organic growers
- Identity-preserved growers
- Fish and wildlife conservation facilities
- Attorneys
- Fish hatcheries
- Elevators
- Animal producers
- Wholesalers
- Seed growers
- Farmers

8.6   CALIFORNIA SEED & PLANT LABORATORY, INC.

California Seed & Plant Laboratory, Inc.
7877 Pleasant Grove Road
Elverta, CA 95626
Phone: 916-655-1581
Fax: 916-655-1582

Available at http://www.calspl.com/site/index.php and http://www.calspl.com.

California Seed & Plant Laboratory, Inc. (Cal-SPL) provides pathological and genetic testing to the vegetable seed, fruit tree, grapevine, and strawberry industries by approved or in-house improved methods at competitive prices with quick turn-around time, confidentiality, and real-time status reporting of pending orders.

Cal-SPL offers seed health tests and is accredited by the National Seed Health System (NSHS). Their tests include pathogen testing for seeds (vegetables, field crops, flower seeds, etc.). Standard or improved methods are used. For example, Cal-SPL uses liquid plating and Bio-PCR techniques for detecting bacterial pathogens in seeds. Viral pathogens are detected by ELISA. Several procedures are accredited by the NSHS and the California Crop Improvement Association (CCIA). Service includes fast turnaround time with real-time reporting of sample status via the Internet.

8.6.1   Sample Collection Guidelines

The guidelines for collection of seed samples is 10,000 seeds for testing at the 0.01% level. Fewer seeds can be sent if testing is needed at a higher threshold. For example, for testing at a 0.1% level, 1,000 seeds are enough.

- Processed food: 100 grams of any processed food
- Oil samples: 500 mL of crude oil

Types of plants include the following:

- Alfalfa
- Asparagus

- Beans
- Beets
- Brassica
- Carrot
- Celery
- Coriander
- Corn
- Cotton
- Cucurbits
- Grass
- Lettuce
- Onion
- Pea
- Pepper
- Potato
- Rice
- Soybean
- Spinach
- Tomato
- Valerianella

Types of tests include the following:

- Seed health
- Plant health
- Virus eradication
- Resistance screen
- Hybrid purity
- Variety ID
- GMO
- Germination
- Soil health

8.6.2 PLANT HEALTH

Cal-SPL provides rapid diagnosis of diseases of plants to help farm and nursery managers to take corrective steps on timely basis. For example, their turf disease program provides pathogen identity in 2–7 days to help golf course superintendents manage disease.

8.6.3 SEED HEALTH FOR SOYBEANS

Test ID	Description of Method	Quantity	Price
1184	Pseudomonas syringe pv glycinea liquid plating	10,000 seeds	$157.00
1185	Soybean mosaic virus ELISA (10 seeds/well)	100 seeds	$60.00

For example, for Soybean mosaic virus:

Test ID	Description of Method	Quantity	Price
1185	ELISA (10 seeds/well)	100 seeds	$60.00

Seed health for corn:

Test ID	Description of Method	Quantity	Price
1359	Pantoea stewartii ELISA (50 seeds/well)	400 seeds	$60.00

8.6.4   GMOs

Cal-SPL offers qualitative and quantitative PCR tests for GMO seed, processed food, and feed samples. Clients include seed companies, breeders, and producers of non-GMO food products. For example, Cal-SPL tests crude oil of canola, soybean, and corn. They offer real-time PCR tests for specific genetic events such as Roundup resistance in canola.

8.6.4.1   GMO—Corn

Test ID	Description of Method for 1000 Seeds	Price
1235	Bt11—cry1Ab gene PCR	$95.00
1236	E176—cry1A(b) gene (NaturGard, KnockOut, Maximizer) PCR	$95.00
1036	GA21—EPSPS gene (Roundup Ready) PCR	$95.00
1233	CBH351—cry9C gene (StarLink) PCR	$95.00
1133	NK603—EPSPS gene (Roundup Ready) PCR	$95.00
1446	T25—pat gene (Liberty Link) PCR	$95.00
1448	Mon810—cry1Ab gene (Yeildgard) PCR	$95.00
1449	35S (promotor)—CaMV 35S gene PCR	$95.00
1450	NOS (terminator)—nopaline synthase gene from A. tumefaciens PCR	$95.00
1447	Plant—zein gene PCR	$95.00
1453	5-event GMO panel (any five) PCR	$275.00
1451	10-event GMO panel (35S, NOS, Bt11, E176, GA21, NK603, Mon 810) PCR	$500.00

8.6.4.2   GMO—Soybean

Test ID	Description of Method for 1000 Seeds	Price
1455	35S (promotor)—CaMV 35S gene PCR	$95.00
1037	Roundup Ready soy 40-3-02—EPSPS gene PCR	$95.00
1456	Plant—Le1 lectin gene PCR	$95.00
1253	3-event GMO panel (35S, RR, Soy-specific) PCR 100 mL of soil	$185.00

8.7   CANADIAN GRAIN COMMISSION LABORATORY

Grain Research Laboratory
1404-303 Main Street
Winnipeg, Manitoba R3C 3G8
Phone: 204-983-2766

Fax: 204-983-0724
Peter Burnett, Director
Phone: 204-983-2764
E-mail: pburnett@grainscanada.gc.ca

Available at http://www.grainscanada.gc.ca/grl/grl-e.htm.

The Grain Research Laboratory (GRL) is an internationally recognized research center for research on grain quality. Its focus is to ensure that the processing quality of grain is maintained from cargo to cargo and from year to year. Analytical Services traces its origins to 1913 and the founding of the GRL. Analytical Services supports the quality assurance and market support programs of the Canadian Grain Commission (CGC).

8.7.1 ANALYTICAL SERVICES

From kernel to flour, GRL Analytical Services (AS) analyzes grains' functional components of quality. AS analyzes grain samples for breeders' line for variety registration, grain quality research projects, the annual harvest surveys, or cargo quality monitoring by providing a wide variety of analyses using advanced technology and standardized methods and procedures.

AS encompasses research, methods development, and testing through moisture determination, protein testing, and quality component analysis.* The Reference Protein Laboratory provides protein content determinations by combustion nitrogen analysis used to calibrate CGC operational protein testing instruments and for research and quality assurance programs. The analytical laboratory provides a full range of quality component tests, from flour ash to Zeleny sedimentation.

8.7.2 IMAGE ANALYSIS

The Image Analysis unit in the GRL is equipped to characterize, measure, and objectively assess the appearance of grain and grain products.The unit develops objective methods for grain quality assurance that enhance grain grading and inspection and to characterize the end-use quality of cereal grains, oilseeds, and pulses in the CGC's harvest survey and quality monitoring.

8.7.2.1 Variety Identification Research

The Image Analysis unit develops new methods for identifying varieties. Currently, the CGC performs protein electrophoresis and DNA fingerprinting on individual

---

* The moisture laboratory publishes moisture conversion tables for the Model 919 moisture meter, verifies the calibration tables annually, generates new calibrations tables as needed, and monitors the performance of Model 919 meters. CGC uses standardized technology and internationally recognized methods. Since 1996, CGC has used combustion nitrogen analysis, the emerging world standard for protein testing, as the reference protein method. The Model 919 moisture meter and conversion tables are checked against the appropriate reference air oven method with new crop samples each year. The analytical laboratory uses methods and procedures recognized by the American Association of Cereal Chemists and the International Association for Cereal Science and Technology.

kernels of grain. Many kernels must be analyzed to determine the variety composition of a sample. CGC's long-term goal is to develop a DNA-based method that will determine the variety composition of a ground sample of grain rather than multiple individual kernels. See Table 8.3 for fees and services.

The Variety Identification section supports the integrity of Canada's grain quality assurance system through variety testing and by researching and developing identification methods. The section has three programs:

TABLE 8.3

Sample of Services and Fees of the CGC

Fee Code	Name	Price ($)	Unit
	Sample of Services		
1201	Grading cert.—heat and corn (unofficial sample)	15.10	Sample
1203	Grading cert.—canola, rapeseed, mustard, (unofficial sample)	24.40	Sample
1601	Protein testing service	9.00	Analysis
1660	Vomotoxin (Don) testing by ELISA technology—batch run	50.00	Analysis
1672	Seed analysis—nondesignated crops	36.50	Analysis
1201	Grading cert.—wheat and corn (unofficial sample)	15.10	Sample
1203	Grading cert.—canola, rapeseed, mustard (unofficial sample)	24.40	Sample
1601	Protein testing service	9.00	Analysis
1660	Vomotoxin (Don) testing by ELISA technology—batch run	50.00	Analysis
1672	Seed analysis—nondesignated crops	36.50	Analysis
	Sampling Services		
Fee Code	**Name**	**Price ($)**	**Unit**
1310	Special services—where full inspection service not available	28.20	Hour
1510	Unsealed samples	35.00	Sample
1511	Sealed samples	41.00	Sample
1512	Samples—car/truck/container lot	2.50	Load
1651	Travel and living expenses (inspection)	Actual	
1694	Calibration samples for protein test equipment	31.00	Sample set
	Analytical Testing		
Fee Code	**Name**	**Price ($)**	**Unit**
1600	Test weight (by Schopper Chondrometer)	10.00	Analysis
1604	Analysis	23.50	First analysis
1668	Cut-off and cut-off post treatment samples	2.50	Sample
1669	Insect checks, car loading samples	5.00	Sample
9003	Moisture test by 919 meter	13.50	Analysis
9004	Falling number testing (Hagberg)	26.50	analysis

Source: Data adopted with permission from CGC.

- Variety identification monitoring
- Variety identification research
- GMO identification research

Through the work of the section, the CGC leads in the development of variety identification technology, the establishment of comprehensive variety fingerprint databases for wheat and barley, and in the implementation of these tools for the benefit of Canada's grain industry. The CGC is also committed to transferring variety identification technology to the private sector for use in commercial variety identification testing.

8.7.2.2 GMO Identification Research

The Image Analysis unit develops and evaluates PCR assays for detection, identification, and quantification of GMOs in grains and oilseeds. CGC also participates in GMO proficiency tests organized by AACC International, the Grain Inspection, Packers and Stockyards Administration (GIPSA), and the International Seed Testing Association.

8.7.3 METHODS AND STANDARDS

CGC uses polyacrylamide gel electrophoresis and HPLC for protein-based variety identification and microsatellite-based systems for DNA fingerprinting. See Tables 8.3 and 8.4 for CGC's services, fees, and GRL programs.

TABLE 8.4
GRL Programs

GRL Program	Website
AS	http://www.grainscanada.gc.ca/grl/analytical_serv/analytical_serv-e.htm
Applied Barley Research	http://www.grainscanada.gc.ca/grl/applied_barley/applied_barley-e.htm
Asian Products/Wheat Enzymes	http://www.grainscanada.gc.ca/grl/asian_end_pro/asian_end_pro-e.htm
Barley Research	http://www.grainscanada.gc.ca/grl/barley_research/barley_research-e.htm
Bread Wheat Studies and Baking Research	http://www.grainscanada.gc.ca/grl/baking/baking-e.htm
Durum Wheat Research	http://www.grainscanada.gc.ca/grl/durum/durum-e.htm
Grain Safety Assurance	http://www.grainscanada.gc.ca/grl/grain_safety/grain_safety-e.htm
Image Analysis	http://www.grainscanada.gc.ca/grl/image_analysis/image_analysis-e.htm
Milling Research	http://www.grainscanada.gc.ca/grl/milling/milling_research-e.htm
Mycology	http://www.grainscanada.gc.ca/grl/mycology/mycology-e.htm
Oilseeds Research	http://www.grainscanada.gc.ca/grl/Oilseeds/oilseeds_research-e.htm
Oilseeds Services	http://www.grainscanada.gc.ca/grl/Oilseeds/oilseeds_services-e.htm
Pulse Research	http://www.grainscanada.gc.ca/grl/pulses/pulses_research-e.htm
Variety Identification	http://www.grainscanada.gc.ca/grl/variety_id/variety_id-e.htm

Source: Information adopted with permission from CGC.

8.8   GENETIC ID, INC.

Genetic ID, Inc.
P.O. Box 1810
501 Dimick Drive
Fairfield, IA 52556-9030
Phone: 641-472-9979; 877-366-0798 (toll-free)
Fax: 641-472-9198
E-mail: info@genetic-id.com

Available at http://www.genetic-id.com.

Founded in 1996, Genetic ID, Inc. maintains its global headquarters in Fairfield, Iowa, and cites themselves as the first Genetic ID laboratory and first commercial GMO testing lab in the world. The company's GMO testing methods are used throughout the world by Genetic ID's Global Laboratory Alliance® members, including the company's Augsburg and Japan laboratories, as well as government and commercial laboratories in Brazil, China, Singapore, Taiwan, India, the United Kingdom, South Korea, the United States, and Italy.[*]

Historically, industries have undertaken steps toward consumer protection before most governments in the world enacted safety policies. Initially Genetic ID's laboratory was designed for the scientific analysis of agricultural and food item testing for GMO content. During 1997–1998, Genetic ID became a recognized non-GMO certification standard.

Genetic ID helps agricultural and food industry customers to grow and sustain their markets and exports by guiding them through various countries' government regulations and procedures concerning restricted ingredients such as GMOs. Genetic ID offers global reach and local support by providing laboratories in the United States, Japan, and Germany, plus more than 15 affiliated testing labs in the Global Laboratory Alliance, along with Genetic ID offices and representatives across 5 continents. They also offer problem resolution through third-party "defensibility" and a proprietary Rapid Response Protocol to save brands, costs, and recalls.

In recent years, regulatory requirements and market pressures around the world have prompted the food industry to address the questions: Does a product contain GMOs? If so, which GMOs? And how much is present? Genetic ID attempts to answer these questions via testing, such as using PCR, which is the technology of choice for detecting GMOs in a wide variety of food products. Sensitivity and specificity are two distinct advantages of GMO analysis via PCR testing over other methods (such as protein testing, including strip and ELISA methods). Capable of detecting genetically altered DNA content as low as 1 part in 10,000, PCR is considered at least 100 times more sensitive than protein tests. PCR is often more economical in practice than other testing methods because the much greater sensitivity of PCR means less testing is required. Other advantages of PCR testing include the capability to detect all, rather than some, GMOs and the capability to quantify GMO content in almost all food and feed products. The robust nature of the PCR method makes it possible to

---

[*] Genetic ID is the parent company of its spinoff, Cert ID, which is described in Chapter 7.

use PCR analysis to test for the presence of GM material at almost all points in the food chain, from the farmer's field to the retail shelf.

Key to U.S. export to the EU, in 2006, the U.K. Accreditation Service (UKAS) renewed its accreditation of Genetic ID's testing for detection of GM materials in raw foods, processed foods, and animal feed.* The UKAS accreditation renewal ensures that Genetic ID complies with ISO/IEC/EN 17025. Genetic ID's accredited tests provide cross-species, single-species, and variety-specific GMO detection, as well as detection of unapproved animal byproducts in animal feed. Genetic ID's DNA-based technology detects the presence of any and all commercialized GMOs to a 0.005% limit of detection and quantifies the amount of detected GMO material to a limit of quantification of 0.1% of the substance tested.

Many of Genetic ID methods are accredited to ISO 17025 laboratory standards by UKAS, which in turn helps eliminate false positives and false negatives on the basis of a wide array of safeguards. These safeguards include the use of statistically valid sample sizes to assure minimum risk of error (Genetic ID pioneered the use of the largest sample sizes in the industry); use of the proprietary Fast ID DNA extraction system, which eliminates DNA degradation and interference by PCR inhibitors maximizes yield; and running tests in duplicate from beginning to end to guard against operator or equipment error.

8.8.1 Genetic ID Products and Services

Genetic ID provides information and documents covering the full range of specialized services offered. These services are tailored to the operations and individual requirements of seed companies, growers, elevators, transporters, processors and manufacturers, retailers, and testing laboratories around the world, including:

- *Consulting:* Integrated quality-assurance (QA) systems planning and problem resolution
- *GMO testing:* Superior PCR detection technology
- *Varietal ID testing:* To detect specific GMO varieties
- *Animal feed testing:* To detect GMOs and regulated animal byproducts
- *Allergen testing:* Sensitive PCR testing for allergens in foods
- *Certified ID non-GMO certification consultants*
  - Third-party QA verification
  - IP system approval
  - Lot-by-lot certification
  - Full product line certification
  - Ingredient certification
  - Government regulatory verification
- *IP:* Compliance with traceability regulations
- *Global Laboratory Alliance:* Worldwide world-class services

---

* UKAS is a member of the European Co-operation for Accreditation (EA), the International Laboratory Accreditation Cooperation (ILAC), and the International Accreditation Forum (IAF). UKAS accreditation is accepted throughout the EU and in countries on five other continents where UKAS has bilateral and multilateral mutual reciprocity agreements.

Genetic ID offers a full spectrum of qualitative and quantitative testing options, such as:

- *GMO detection:* Genetic ID promotes their ability to reliably detect all commercialized GMOs. GMO testing is used to detect and quantify the presence of GMOs.
- *Varietal ID testing:* Varietal ID testing is used to detect the presence of specific GMO varieties. It is typically used when a food source must meet regulatory requirements for specific GMOs, such as the absence of StarLink or other unapproved varieties.
- *Animal feed testing:* Animal feed testing is used to detect the presence of animal-derived materials (e.g., meat and bone meal) in animal feed or its components, whether species-specific or for a general barnyard screen.

8.8.2   GMO DETECTING

Regulations requiring labeling of foods containing GMOs have now been adopted in a total of 36 countries throughout Europe and the Pacific Rim and are under development in other countries.

Genetic ID's focus is to offer services that assist in the following:

- Meeting regulations
- Delivering product to customer contract specifications involving threshold tolerances and unapproved varieties
- Optimizing sampling and testing programs to achieve efficiency in cost and operations
- Resolving conflicts in a rapid and cost-effective way

Genetic ID offers two fundamental types of Varietal ID for corn:

1. *Worldwide Varietal ID:* The Worldwide Varietal ID test can detect all of the GMO varieties of corn that have been approved by governments in North and South America. At this time, this list covers virtually all GMO corn varieties approved to date around the world. However, many of these varieties have not been approved for human consumption in other regions of the world. This test is particularly useful when a seller is not sure which country will be the final destination for the product and desires access to the widest possible market. This test can also be used to rule out particular markets (i.e., if the product is found to contain a variety unapproved GMOs in one particular nation or region then it can be sent elsewhere).
2. *Region-Specific Varietal ID:* The Region-Specific Varietal ID test can detect those GM varieties that are not approved in one specific nation or region. Using this more economical test, buyers and sellers can determine if a product is suitable for a particular target market.

8.8.3 ANIMAL FEED TESTING

To comply with domestic and international regulations on feed, comprehensive tests identify animal byproducts and species in animal feed. Genetic ID has responded to concerns over bovine spongiform encephalitis (BSE, or mad cow disease) by developing tests to detect specific animal tissue, bone, and blood byproducts in animal feed. U.S., EU, and Japanese regulations prohibit most animal products in feed. These tests can also be used for species identification of meat products and to detect adulteration of meat products with tissue from other species. Genetic ID has designed a wide range of primer sets and tests for PCR analyses of DNA isolated from animal feed samples. These tests include the Barnyard Test, which detects common barnyard species.[*]

- **Ruminant-specific test:** This test selectively detects members of the ruminant family by targeting a genetic sequence that is found only in this family, which includes cows, sheep, goats, deer, and elk.
- **Bovine-specific (beef), ovine-specific (mutton), porcine-specific (pork) tests:** These tests target a sequence unique to cattle and very closely related bovine species, sheep and very closely related ovine species, and pigs and very closely related porcine species, respectively.

8.9 CII LABORATORY SERVICES

CII Laboratory Services
10835 NW Ambassador Drive
Kansas City, MO 64153
Phone: 816-891-7337
Fax: 816-891-7450
E-mail: ciisvc@ciilab.com

Available at http://www.ciilab.com.

Since 1991, CII Laboratory Services has provided a full range of analytical services for the grain, milling, and baking industries. They conduct more than 100 different tests and analyses. They are also known for their Crop Quality Survey, a one of a kind since 1995, which involves 3 different classes of wheat samples (Hard Red Winter, Soft Red Winter, and Hard Red Spring) from 18 states for analyses and data. Tests and analyses offered include proximate analyses (moisture, ash, protein, etc.), physical dough testing, bake testing and product evaluation, grain and flour analyses, GMO testing, pesticide residues, and microbiology. CII Laboratory is a major laboratory in the United States for the baking, milling, and grain industries because of its extensive testing capabilities.[†] They also do consulting on control systems and

---

[*] This includes cattle, sheep, goat, horses, donkeys, pigs, chickens, turkeys, deer, and elk.
[†] The lab also performs testing for nutritional analysis, HPLC testing for mycotoxins and vitamins, mineral analysis by AAS, pesticides, microbiology, environmental monitoring, and sanitation for meat processors, food manufacturing, restaurants, and warehouses.

sanitation. They are an ISO 9001-2000 certified laboratory and follow standard AACC, AOAC, AOCS, USDA, and FDA-BAM methods.

CII Laboratory is the preferred supplier of analytical services for the American Institute of Baking (AIB) and supports their Bakers Seal and Gold Seal programs with the analytical services needed to comply with these programs. CII Laboratory is the only private laboratory in the country providing wheat and flour crop quality information through its annual Crop Quality Survey. This survey is considered "the Bible" of wheat grading.* The Crop Quality Survey is performed by CII Laboratory Services field personnel and follows each wheat harvest, picking up samples across 20 states and shipping them to their Kansas City laboratory where they are analyzed and the results are published daily on their website.

All testing is performed by approved methods and participates in numerous check sample and proficiency programs (AACC, API, industry collaboratives, and internal check programs). As part of CII Laboratory's ISO certification, they conduct internal audit procedures and have established internal quality review procedures.

8.9.1 EXAMPLE OF SERVICES PRICING

Procedures	Proximate Analyses		
	Sample Size (grams)	Turnaround Time	Pricing
Moisture	100	Next day	$11.00
Ash	100	Next day	$12.00
Protein (Combustion)	100	Next day	$15.00
Fat (Ether Extraction)	50	2 days	$20.00
Fat (Acid Hydrolysis)	50	2 days	$25.00

Other tests offered include:

- Fiber
- Physical dough testing
- Amylograph
- Glutomatic
- Grain and flour analyses
- Bake testing and product evaluation
- Physical tests
- Mycotoxins
- Pesticide residues
- Microbiology
- Minerals
- Lipid analyses
- Chemical
- Vitamin analysis
- GMO testing
- Nutritional labeling

---

* Available by subscription, the survey provides users with a timely look at the new crop wheat and flour qualities that can be expected from four wheat classes: Hard Red Winter, Soft Red Winter, Hard Red Spring, and Hard White Wheat.

8.9.2 PORT SERVICES

CII Laboratories also provides attendance at grain terminal facilities in the Gulf ports. These services include being in attendance as a vessel is loading, maintaining a log of the loading, receiving splits of Federal Grain Inspection Service (FGIS) sublots and composite samples for further testing such as mycotoxins, grades, proximate and physical dough testing, and many other quality attributes.

8.10 ENVIROLOGIX, INC.

EnviroLogix, Inc.
500 Riverside Industrial Parkway
Portland, ME 04103-1486
Phone: 207-797-0300; 866-408-4597 (toll-free)
Fax: 207-797-7533

Available at http://www.envirologix.com.

Founded in 1996 by a group of experienced immunoassay diagnostic test kit developers, EnviroLogix has built on its strong scientific foundation in the development and manufacture of immunoassay test kits for every link in the global food production chain, from seed to plant to grain handling and processing. EnviroLogix products are distributed in several countries outside of the United States by authorized distributors and in the United States and elsewhere worldwide directly by EnviroLogix.

EnviroLogix views itself as being innovative and a user-focused diagnostic tests manufacturer by monitoring and adapting to global issues in food production lifecycles, water quality, and environmental safety. Because of the various global issues facing the food industry, EnviroLogix has developed and modified immunoassay test kits in the rapidly changing field of GMOs. The immunoassay test kits allow for rapid, accurate, and easy-to-use diagnostics to identify transgenic markers in GMOs. To this end, they have introduced thier QuickStix™, QuickComb™, and QuickTox™ test strips.

In addition to these tests, they have utilized the technology in the field of plant pathogens and have products available in that category. At EnviroLogix, they manufacture and check all test kits on-site so they are able to monitor the quality control process from initial test development right through to manufacturing, packaging, and delivery to their customers. Their manufacturing facility is equipped for optimal manufacturing conditions and their management team has extensive experience in strict adherence to Good Manufacturing Practices (GMPs) and implementation of ISO standards and procedures.

In 2004, GIPSA approved EnviroLogix' QuickTox kit for aflatoxin detection, which was the first lateral-flow strip test for the detection of mycotoxins in grain.[*] In doing so, they also gained the USDA's Certificate of Performance. QuickTox for aflatoxin continues to offer a fast, reliable, and easy method for screening corn and cottonseed for this naturally occurring toxin. The test gives an accurate "yes/no"

---

[*] EnviroLogix also holds GIPSA COPs on many of their own GMO products.

result within 2–5 minutes, providing convenience where simplicity and speed are vital. This test has only three easy steps: extract the sample in methanol, dilute with water, and drop in a test strip. The results are read visually and costly equipment is unnecessary.*

8.10.1 OTHER TESTS KITS: GMO AND GRAIN MYCOTOXIN TEST KITS

Rapid and easy-to-use test kits are available to detect biotechnology-enhanced traits in plant tissue, single seeds, and bulk grain. Kit formats include QualiPlate and QuantiPlate microwell plates for laboratory analyses and QuickStix lateral-flow strips for on-site results in 2–5 minutes. Their Common Extraction™ method has enhanced testing for multiple genetic traits in corn.

8.10.2 PLANT PATHOGEN TEST KITS

These relatively low-cost kits provide accurate, rapid results and can identify the presence of various plant pathogens, including the first field test for soybean rust. State laboratories and crop scouts can more quickly and accurately identify or rule out these pathogens using these kits. Growers, producers, extension agents, and crop consultants may also use these simple tests to make informed decisions about treatment or remediation options.

8.10.3 MOLD AND MOLD TOXIN TEST KITS

The indoor air quality industry uses this fairly new analytical technology in different formats. QuickTox strips can rapidly and inexpensively identify Stachybotrys and Aspergillus niger on-site in 5 minutes. The QuantiTox plate kit can confirm mycotoxin-containing spore presence and quantitate the level of spore-borne trichothecene mycotoxins.

8.10.4 PESTICIDE RESIDUE TEST KITS

These low-cost kits with rapid results are used to determine pesticide levels in water and residues in foods. Water safety applications include drinking water monitoring, point-source testing, effluent monitoring, and runoff assessment and monitoring. The kits targeted for foods include several of the more commonly used fungicides and insecticides and the broad-screen cholinesterase and organochlorines assays.

8.10.5 ALGAL TOXIN TEST KITS

Blue-green algae or cyanobacteria produce natural toxins such as microcystins, which in high concentrations are toxic to humans and animals. The QuantiPlate and QuickTube™ kits detect microcystins at or below the World Health Organization (WHO) drinking water guideline of 1 ppb (one part microcystin per billion parts of water). See Table 8.5 for specific products offered.

---

* This is the first lateral-flow test to be certified by the GIPSA unit of the USDA, proven to meet the test performance claim of detecting aflatoxin contamination in bulk corn samples at 20 ppb and above.

TABLE 8.5

Sample List of Specific EnviroLogix Products Offered

Acetanilides plate kit

Aflatoxin QuickTox Kit

Bacterial Fruit Blotch QuickStix Kit

Bollgard/Roundup Ready QuickStix Combo Kit

Bollgard II QuickStix Combo Kit

Bollgard II + RR QuickStix Combo Kit

Cholinesterase plate kit

Cry1Ab/Cry1Ac QualiPlate Bulk Screening Kit

Cry1Ab Bulk Grain QuickStix Kit

Cry1Ab/Cry1Ac QuickStix Kit

Cry1Ab+Cry3B QuickStix Combo Kit

Cry1Ac+Cry2A QuickStix Combo Kit

Cry1Ac+CP4 EPSPS QuickStix Combo Kit

Cry1Ac+Cry2A+CP4 EPSPS QuickStix Kit

Cry1Ac+Cry2A+PAT/*bar* QuickStix Combo Kit

Cry1C plate kit

Cry1F QualiPlate Kit

Cry1F QuickStix Kit for bulk grain

Cry1F QuickStix Kit for leaf and seed

Cry2A QualiPlate kit

Cry2A QuickStix Kit

Cry34 QuickStix Kit, leaf & seed

Cry34 QuickStix Kit, bulk grain

Cry3Bb QualiPlate Kit

Cry3Bb QuickStix Kit, leaf & seed

Cry3Bb QuickStix Kit, bulk grain

Cry9C plate kit

Cry9C Bulk Grain QuickStix Kit

Cyclodienes plate kit

Herculex I QualiPlate Kit

Herculex I QuickStix Kit for bulk grain

Herculex I QuickStix Kit for leaf and seed

Herculex RW QuickStix Kit, leaf/seed

Imidacloprid plate kit

Imidacloprid plate kit for treated seeds

Isoproturon plate kit

LibertyLink (PAT/*bar*) QuickStix kit for seed

LibertyLink (PAT/*bar*) QuickStix kit for leaf

LibertyLink (PAT/*bar*) plate kit

LibertyLink (PAT/*pat*) plate kit

Metalaxyl plate kit

Methoprene acid plate kit

Microcystin QuantiPlate kit

Microcystin QualiTube Kit

Microcystin QuantiTube Kit

Modified Cry3A QualiPlate kit

Organochlorines plate test

Paraquat plate kit

PAT/*bar* QuickStix kit for seed

PAT/*bar* QuickStix kit for leaf

PAT/*bar* QualiPlate kit

PAT/*pat* QuickStix kit for bulk grain

PAT/*pat* QuickStix kit for leaf and seed

PAT/*pat* QualiPlate kit

QuickComb Kit for Bulk Grain testing multiple
   GM traits

RR, QuickStix Kit for plant tissue (corn and
   soybean)

Roundup Ready QuickStix Kit for bulk grain
   (corn)

Roundup Ready QuickStix Comb Kit for cotton
   seed

Roundup Ready QuickStix Kit for cotton leaf
   and seed

Roundup Ready QuickStix Kit for alfalfa hay

Roundup Ready QuickStix Kit for alfalfa leaf
   tissue

Roundup Ready, QualiPlate Kit

Roundup Ready, QuantiPlate Kit for soybean
   and soy flour

Roundup Ready, QuickStix Kit for canola leaf &
   seed

Soybean Rust QualiPlate Kit

Stachybotrys and *Aspergillus niger* QuickTox
   Kit (PRO 50)

Stachybotrys and *Aspergillus niger* QuickTox
   Kit (PRO 20)

Stachybotrys and *Aspergillus niger* QuickTox
   Kit (Homeowners')

StarLink plate kit

StarLink QuickStix Kit

Synthethic pyrethroids plate kit

Trichothecenes QuantiTox plate kit

YieldGard Corn Borer QualiPlate Kit

YieldGard Corn Borer QuickStix Kit, leaf and
   seed

YieldGard Corn Borer QuickStix Kit, bulk seed

YieldGard Rootworm QualiPlate Kit

YieldGard Rootworm QuickStix Kit, leaf and
   seed

YieldGard Rootworm QuickStix Kit, bulk seed

YieldGard Plus QuickStix Combo Kit

Source: Data adopted with permission from EnviroLogix.

8.11   EUROFINS GENESCAN, INC.

Eurofins GeneScan, Inc. (formerly known as GeneScan USA, Inc.)
2315 North Causeway Boulevard, Suite 200
Metairie, LA 70001
Phone: 504-297-4330; 866-535-2730 (toll-free)
Fax: 504-297-4335
E-mail: gmo@gmotesting.com

Available at http://www.eurofins.com, http://www.soyatech.com/bluebook/news/sponsor.ldml?a=35145, and http://www.EurofinsUS.com.

GeneScan USA is an ISO/IEC 17025 accredited commercial testing laboratory offering GMO testing by PCR and ELISA. GeneScan was recently acquired by Eurofins Scientific, a provider of bioanalytical support services to the food, feed, dietary supplement, animal health, biotechnical, and pet food industries. With this partnership, Eurofins and GeneScan provide over 50 service laboratories throughout the world serving the food and feed industry. The Eurofins U.S. laboratories specialize in GMO detection, quantitative PCR analysis, and IPT consulting. Traditional chemistry and microbiology is also offered, with special emphasis on residue testing and detection of acrylamides. They are also the exclusive licensee for Japanese Standard Method, accredited by New Zealand MAF, and audited by Fortune 500 Agro-Food Companies.

Eurofins GeneScan operates as an independent, third-party testing laboratory to maintain their position as a neutral arbiter of test results and will neither participate in nor provide support to special interest groups relative to genetic engineering. They use procedures that are proven in international collaborative studies and in check sample programs.[*] Eurofins' main U.S. laboratory is located in suburban New Orleans, Louisiana. In 1998, the company began offering GMO testing services to the U.S. market, utilizing the transplanted technical expertise developed in the GeneScan Research and Development hub in Freiburg, Germany. Since then, Eurofins GeneScan has quickly become a premier brand name for GMO testing in the United States.

Eurofins Scientific operates 40 laboratories, employs 2,000 employees, and performs more than 13 million assays per year to establish the safety, composition, authenticity, origin, traceability, and purity of food.[†] With over 10,000 reliable analytical methods, Eurofins Scientific is a major global provider of bioanalytical services.

A key emphasis of Eurofins GeneScan operates in the area of quality and identity control of food and animal feed. The focus of these activities is the detection of allergens and detection of GMOs in seeds, agricultural commodities, semi-finished, and finished products. They accomplish this by utilizing their laboratories in Europe, North and South America, and networks of strategic partners and licensees. They also offer the corresponding diagnostics kits to third-party laboratories. The group's portfolio also includes design, implementation, and certification of customized control programs and IP systems along the entire production chain.

---

[*] The Eurofins U.S. group includes laboratories located in Des Moines, Iowa; Memphis, Tennessee; New Orleans, Louisiana; and Petaluma, California.

[†] Eurofins has over 15 years experience in second-party auditing of industrial processes around the world.

8.11.1   Eurofins GeneScan IP Certification

Eurofins GeneScan recognizes that food and feed suppliers must comply with regulations on traceability and labeling of their products on parameters such as GMO content, allergens, country of origin, residues, etc. In addition, end users are demanding transparency throughout the entire supply chain. They understand that analytical testing is an important risk management tool, but it is not sufficient on its own to protect the value and integrity of raw materials or products. Further control, such as traceability and segregation along the supply chain, is necessary to fulfill legal requirements and reinforce consumer confidence. They promote their systematic approach by linking appropriate sampling and testing methods as well as efficient organizational measures (e.g., documentation, risk assessment, adverse event management and recalls) to be able to handle these complex situations at a reasonable cost.

Eurofins GeneScan provides solutions that incorporate current legislation/standards as well as best management practices. Their programs are custom tailored to meet the specific requirements of each individual client. Eurofins GeneScan Verified In-House Program (VIP) certification provides verification for customers who already have an operating control program. For those clients who want a complete traceability system, they offer the Eurofins GeneScan IP System. This approach allows their customers to achieve the level of product security and customer confidence appropriate to their situation.

8.11.2   Eurofins GeneScan IP Standard

The importance of IP is constantly growing in the foodstuff industry. This is based partly on consumers' refusal to accept GM foods and the industry's efforts to satisfy consumers' demands for GMO-free products. EU labeling regulations mandate that all products with a threshold of GM ingredients of 1% or more must be labeled. In every case, the supplier is obliged to provide evidence that due diligence has been taken to prevent his/her product from becoming contaminated with GM material. Even additives and flavors are subject to this labeling regulation if they contain detectable DNA or proteins. Every foodstuff supplier must therefore ensure that appropriate guidelines and measures have been integrated into his/her quality management system so as to comply with requirements outlined above.* Key elements for an IP system are (1) supplier's assurance, (2) segregation, (3) proofs of identity, (4) traceability through information systems, and (5) controls.†

8.11.3   Eurofins GeneScan's TRAC (Tracing Residues and Contaminants)

The TRAC system works well with other Eurofins systems of monitoring and testing. TRAC is a system to monitor application and residual amounts of pesticide and

---

* GeneScan Analytics GmbH, a company of the Eurofins group, developed the GeneScan General Standard as basis for the design and evaluation of programs to control the presence of G) material in food, feed, and seed production. This document is a catalogue of measures used as building blocks in individually tailored non-GM control programs in agreement with best management practices and relevant legislation.

† In addition to IP certification, Eurofins GeneScan is a global player in applied molecular biology, providing technical competence for GMO detection, allergen testing, and MBM detection.

herbicide compounds on food, animal feed, and grain products.* TRAC offers an independent, transparent source of confirmation concerning residue and contaminant levels at all points along a supply chain, from production to retail outlet. The system accomplishes this by providing a systematic collection of application records, sampling information, test results, and logistical tracking documentation. The documents and records that accumulate as the product moves through the supply production chain are compiled into a secure, web-based database offering various levels of access and review as determined by the customer. Customers may use TRAC to verify compliance with import-export regulations, protect brand name integrity, screen current or potential vendors, and identify and control hazards in their own operations to minimize insurance costs.†

8.11.4 Producer Audited Supply Systems

Producer Audited Supply Systems (PASS) is a comprehensive process management system focused on tracing movement of a product through a supply system. It optimizes the value of the delivered product by proactively managing each step in the supply chain and provides independent, third-party validation of production systems and practices.

For agricultural companies such as seed producers that are rapidly introducing a variety of specialty crop varieties and crop traits, PASS offers continued transparency along the food supply chain. PASS programs help the agriculture and seed industries document and audit their stewardship procedures for the commercial marketplace and for regulatory agencies and assist in systems to control the flow of crops into specific market channels. Eurofins GeneScan also offers the PASS program to help processors with reliable third-party verification from raw materials, product movement, and processing to plant-made pharmaceuticals (PMPs).

8.11.5 Examples of Eurofins GeneScan International Qualifications to Certify

8.11.5.1 Eurofins GeneScan's International Food Standard Certification

In 2002, German retailers developed a common standard called the International Food Standard (IFS) for food safety management systems. It has been designed as a uniform tool to ensure food safety and to monitor the quality level of producers of retailer-branded food products. The standard can apply for all steps of the processing of foods subsequent to their agricultural production. Eurofins GeneScan certification is accredited by the French institution COFRAC against the EN 45011 standard (ISO/IEC guide 65) for IFS. The certification committee, which involves food

---

* In 2006, a group of four companies joined together to offer a service that will provide systematic monitoring, tracking, and reporting of pesticide and herbicide residues on food, animal feed, and agricultural products. The TRAC system is the result of collaboration by RQA, Inc., Eurofins GeneScan, Inc., Illinois Crop Improvement Association, Inc., and Copesan, Inc.
† This assurance of system integrity assists all stakeholders in the value chain such as breeding, production, manufacturing, storage, logistics, and retail. This system information can be found at http://www.soyatech.com/bluebook/news/sponsor.ldml?a=35145.

industry and retailer representatives, strengthens the independency and recognition of the certificates issued.

8.11.5.2 Eurofins' British Retail Consortium Standard

In 1998, the British Retail Consortium (BRC), responding to industry needs, developed and introduced the BRC Food Technical Standard to be used to evaluate manufacturers of retailers' own brand food products. Most U.K. and Scandinavian retailers only consider business with suppliers who have certification to the appropriate BRC global standard.

Seed products/services and GMO testing product lines offered include the following:

- Certification
- Consultant
- Genetic Testing
- GMO Testing
- Overview of GMOS
- Quantitative and qualitative PCR
- Laboratory testing/equipment/services
- Non-GMO certification
- Seed testing services
- Testing laboratory
- Transgenic crops
- PCR methods
- ELISA methods
- Detection limits

8.12 MID-WEST SEED SERVICES, INC.

Mid-West Seed Services, Inc.
236 32nd Avenue
Brookings, SD 57006
Phone: 605-692-7611; 877-692-7611 (toll-free)
Fax: 605-692-7617
E-mail: info@mwseed.com

Available at http://www.mwseed.com.

Mid-West Seed Services, Inc. (MWSS) is a full-service seed testing laboratory co-owned and operated by Tim Gutormson, RST, President, and Sharon Hanson-Gutormson, RST, CGT, Vice President. The company has been in business since July of 1993 and works with over 1,500 seed company accounts from 43 states and several countries. MWSS employs more than 20 full-time seed analysts, six of which are Registered Seed Technologists (RSTs), three Registered Genetic Technologists (RGTs), and five Certified Genetic Technologists (CGTs). MWSS is an International

Seed Testing Association (ISTA)-accredited laboratory, as well as ISO 9001:2000 certified.

MWSS conducts germination, vigor, herbicide tolerance, physical and genetic purity, and GMO (ELISA/protein, DNA/PCR, bioassays) testing. They test hundreds of species every year, including corn, soybeans, alfalfa, canola, sorghum, sunflowers, cereals, grasses, native grasses and forbs, flowers, and vegetables. They also conduct workshops where seed technologists provide training sessions throughout the year. Seed Sampling, Canadian Graders, Pre-harvest, and Seed Quality and Seed Technologist training workshops are offered annually in Brookings, South Dakota, at the MWSS facility and throughout the country.

8.12.1 SAMPLE TRACK

MWSS promotes their Sample Track system of bar coding and scanning to track sample movement, PC Tablets to record results, and their database server and website to deliver real-time results to more than 300 customers that have chosen to stop receiving faxed reports. This software allows customers to electronically submit and retrieve information from MWSS' website. They see themselves as not only a seed testing laboratory, but also an information management company providing data management and software to its customers. Their website allows customers to retrieve data as far back as 5 years, which helps in audits and certifications. MWSS also uses this system to trace and track their own in-house processes from their computerized bar coding tracking system to monitor sample movement throughout their facility.

8.12.2 OTHER MWSS SERVICES OFFERED

In addition to corn and soybean seed testing services, MWSS offers PCR testing services (qualitative and quantitative) and adventitious presences (AP) testing services. Their AP testing service refers to GMO contamination caused by pollen drift/gene flow from one field to another. Their AP terminology has been accepted by the American Seed Trade Association. Recently, AP testing has become important to companies that supply organic, non-GMO, and low-GMO seed and grain to international customers. Testing for the absence of GMOs has become important in overseas sales of conventional planting seed, grain, commodities, and organic markets.

8.12.3 AP TESTING SERVICES

MWSS offers a variety of tests to determine the presence of AP material in corn, soybeans, cotton, rice, canola, sugarbeets, and other crops. These consist of herbicide, ELISA, and PCR assays. In the herbicide bioassay, nontransgenic seeds show distinct characteristics when placed on media that is moistened with the respective herbicide. ELISA detects a specific protein that has been captured by an antibody formed for the protein. ELISA is capable of quantifying amounts. PCR detects the presence of a certain DNA sequence and can be used to test all crops, tissue, or seed resulting in qualitative or quantitative results. AP testing differs from conventional seed testing by looking for the absence of a specific trait. Representative sampling of the lot is vital in attaining accurate results. The confidence level of the test becomes higher as the number of seeds tested increases. For this reason, MWSS pools samples in AP testing to increase the number of individual seeds tested.

8.12.3.1 AP—PCR

PCR has been used widely to detect the adventitious presence of GM materials for corn, soybean, canola, cotton, rice, and other economically important crops. PCR technology can determine the presence of GMO, called "qualitative PCR", or quantify the percentage of GMO present in the tested sample, referred to as "quantitative real-time PCR". The presence or quantity of overall GMO can be detected or measured by using DNA markers derived from promoters and terminators. Qualitative PCR can detect 1 GMO seed out of 10,000 conventional seeds, whereas quantitative PCR can quantify GMOs at a 0.01% level.[*]

MWSS also offers DNA-protein tests, real-time quantitative PCR tests, non-GMO certificates, ELISA testing [trait confirmation—Cry1Ab (Mon810, Bt11 and Event 176), Cry3Bb and Cry1F], mycotoxin testing, and other laboratory tests. In addition, MWSS offers consulting services, product development/research services, quality assurance, conditioning plant audits, and seed testing laboratory design.

8.13 NEOGEN CORPORATION

Neogen Corporation Headquarters
Food Safety Division, Acumedia
620 Lesher Place
Lansing, MI 48912
Phone: 517-372-9200; 800-234-5333 (toll-
    free; United States and Canada only)
Fax: 517-372-2006
E-mail: foodsafety@neogen.com

Neogen Europe, Ltd.
Cunningham Building
Auchincruive
Ayr, KA6 5HW
Scotland, United Kingdom
Phone: +44-1292-525-275
Fax: +44-1292-525-477
E-mail: info@neogeneurope.com

Neogen – Animal Safety Division
Life Sciences Division
944 Nandino Boulevard
Lexington, KY 40511
Phone: 859-254-1221; 800-477-8201 (toll-free; United States and Canada only)
Fax: 859-255-5532
E-mail: inform@neogen.com

Available at http://www.neogen.com.

Founded in 1982, Neogen Corporation has more than 350 employees at four U.S. and two international locations that develop, manufacture, and market a varied line of products dedicated to food and animal safety. According to Neogen, they offer tests that are easier to use and provide greater accuracy and speed than other diagnostic methods currently used.

---

[*] PCR can also identify the adventitious presence of individual GMO events such as Cry9C by using event-specific DNA markers. Specific event identification is available for corn, cotton, soybean, and canola.

Neogen prides themselves as pioneers in rapid diagnostic testing. The company has developed more than 200 diagnostic test kits, originally from complicated, expensive, off-site methods, to much easier, but no less precise and trusted, on-site test kits. Neogen's tests are quick and require minimal start-up costs and training. Their tests use built-in controls to provide added confidence in the tests' results. There is no guessing whether testers perform procedures correctly; if the controls perform as designed, sample results can be trusted. Neogen's test kits have gained worldwide use and acceptance and now serve as a "gold standard" for numerous domestic and international regulatory agencies and industries.[*]

Neogen's GeneQuence automated system is a fully automated four-plate processing system that is capable of simultaneously performing multiple assays. When combined with GeneQuence's genetics-based assays, the GeneQuence system quickly and very accurately detects pathogens in raw ingredients, finished food products, and environmental samples. GeneQuence is capable of performing up to 372 samples at a time and is available for Escherichia coli O157:H7, Salmonella, Listeria, and Listeria monocytogenes. Neogen advertises themselves as the one-stop food safety shop for rapid and easy-to-use diagnostic tests, "for food's journey every step of the way from field to fork." Neogen has solicited the help of several world-recognized business and scientific leaders to assist company management.

As an example, Neogen's Food Safety Division offers the following:

- Test kits
  - Natural toxins
  - Food-borne bacteria
  - Food allergens
  - Sanitation
  - GMOs
  - Ruminant byproducts
  - Sulfites
  - Pesticide residues
  - Drug residues
  - Centrus acquisition
  - Soleris products
  - Information on BSE
- Equipment
  - Starter kits
  - Readers and software
  - Filters, cylinders, and bottles
  - Pipettors
  - Training videos
  - ISO-GRID and NEO-GRID equipment

---

[*] Every year since 1994, the USDA's FGIS has awarded Neogen a contract for the exclusive use of its quantitative test for aflatoxin in grain commodities. Similarly, the USDA's Food Safety Inspection Service (FSIS) has used Neogen's rapid test for E. coli O157:H7 every year since 1994 to screen the nation's beef supply for the deadly pathogen. Neogen provides around-the-clock professional technical support should questions arise about one of its products.

Stringent research, development, and quality control practices have led to Neogen test kits' proven reliability and consistency. The accuracy and reproducibility of their products have inspired wide acceptance and use throughout the food industry. Their products have also earned official approvals and third-party validations, including:

- Association of Official Analytical Chemists (AOAC) International
- AOAC Research Institute
- USDA/FSIS
- International Union of Pure and Applied Chemistry (IUPAC)
- USDA/GIPSA (FGIS)

Finally, Neogen offers its customers the following:

- 24-hour technical support
- Hazard Analysis and Critical Control Point (HACCP) assistance
- Training programs
- Sample testing and commodity validation

8.14   PROTEIN TECHNOLOGIES INTERNATIONAL, LTD.

Protein Technologies International, Ltd.
16a Princewood Road
Earlstrees Industrial Estate
Corby NN17 4AP
Northamptonshire, United Kingdom
Phone: +44-01536-26732
Fax: +44-01536-261147

Available at http://www.farmindustrynews.com/mag/farming_grain_pays_premium_2/index.htmlandhttp://www.emeraldinsight.com/Insight/ViewContentServlet?Filename=/published/emeraldfulltextarticle/pdf/0170990603.pdf.

Protein Technologies International (PTI) has an IP system that ensures the delivery of non-GM soy protein to its customers. The system covers seeds, on-farm storage, planting, growing, harvesting, transportation, processing, and distribution along with independent third-party verification. It is, the company advertises, a way of ensuring that consumers can obtain the health benefits of soy protein consumption even if they are actively avoiding GM ingredients.

In investing in IP, one of PTI's key motivations has been to facilitate consumer choice. Although not anti-GMO, PTI provides non-GM soy protein to ensure that the benefits of soybeans can be delivered to consumers in a form that they find acceptable.

PTI requires use of a specific soybean seed known as STS1. This has been reviewed by the U.S. Food and Drug Administration, which has determined that it is not GM. Use of the STS1 soybean allows for postemergent treatment of the soybean plant with an herbicide (Synchrony1) that is hostile to the Roundup Ready GM soybean.

Synchrony1 will either kill or severely stunt the growth of Roundup Ready beans, thus guaranteeing that, at the time of harvest, the crop is 100% non-GM.

The added benefit of the use of STS1 seed with Synchrony1 herbicide is that the latter is an environmentally friendly herbicide, requiring lighter and less frequent application than that used with conventional commodity soybean crops.

A subsidiary of PTI, one of the world's largest producers of protein isolates, paid central Illinois, DuPont Ag Enterprise growers a $0.25/bu premium to grow and identity-preserve conventionally bred STS soybeans. Many of PTI's European customers are interested in non-GMO soybeans, especially with the EU's passage of a law requiring foods to be labeled as containing GMOs or as being GMO-free. Approximatley 50–60% of processed food in the United States and Europe contains soybeans. DuPont's STS soybeans are bred without GMO technologies to resist Synchrony herbicide, which is used (Synchrony) on fields to kill any GMO rogue beans.

8.15   STRATEGIC DIAGNOSTICS, INC.

Strategic Diagnostics, Inc.
111 Pencader Drive
Newark, DE 19702
Phone: 302-456-6789; 800-544-8881 (toll-free)
Fax: 302-456-6770

Available at http://www.sdix.com.

Strategic Diagnostics, Inc. (SDI), headquartered in Newark, Delaware, is a biotechnology-based diagnostic tests company that provides analytical food pathogen testing through immunotechnology. SDI is a leading developer and manufacturer of immunoassay-based test kits for both field testing and laboratory use. These products are used extensively for contaminated waste site assessment and remediation, water quality management, food labeling, and transgenic crop seed production. [*]

SDI is a major developer and producer of antibodies and immunoreagents for a broad range of applications. The company applies this extensive technical expertise to the rapidly growing agricultural markets for crop disease management as well as to medical and industrial problems. SDI specializes in developing immunoassay tests for the food industry. This technology provides users with fast, accurate, easy-to-use tests that are cost effective, require little space, and minimal capital investment. SDI is committed to using these technical capabilities to develop robust tests that provide value in real-world situations.

SDI advertises that whether it is the rapid analysis of GMOs in crops, mycotoxins in grains, or food pathogens in meat, dairy, or processed foods, they can help producers to reduce their risks, protect their brands, and gain confidence in the safety of their operation.

Three SDI test methods to detect meat and bone meal (MBM) in cattle feed include:

---

[*] Through its subsidiary, Strategic BioSolutions SDI also provides antibody and immunoreagent research and development services. In addition, in 2005, SDI announced a major distribution agreement with DuPont Qualicon to private label their products for sale outside of the United States.

1. *Microscopy:* Involves preparing a sample for microscopic examination by a qualified lab technician. This procedure must be done in a laboratory. The basis of the method requires the technician to visually identify specific components of the sample (e.g., bone or feathers). Because it is very labor-intensive, it is not generally thought of as a high-volume test method. It is estimated that a qualified technician may be able to evaluate fewer than ten samples per day.

2. *DNA testing or PCR methods:* DNA testing involves identifying DNA associated with MBM or specific tissues. This test must be performed in the laboratory by a highly trained scientist and requires sophisticated equipment and specialized, dedicated facilities. The automated nature of these tests would support processing of larger numbers of samples than microscopy, but the costs are high and the time to result is measured in days.

3. *Immunoassays:* These include ELISA and rapid lateral-flow devices (LFDs). ELISAs are multistep quantitative tests that utilize specific antibodies to react with the sample to detect the analyte of interest. They typically require several hours to perform and use specialized laboratory equipment. Some ELISAs may be automated and therefore are conducive to high sample throughput; however, they often require a labor-intensive sample preparation step. LFDs are inexpensive test strips that contain antibodies incorporated into the strip that react with the sample to form a color reaction. Crude sample preparations are typically used and, because the tests require only 10 minutes to conduct, they are also conducive to high sample throughput. If MBM or specific tissues are present in a feed sample, these strips develop a clear test signal indicating that the sample is positive. They are fast (10 minutes or less), do not require specialized laboratory equipment, can be run in the field, and require no special training. The SDI FeedChek™ MBM test kit is a lateral-flow device system.

8.15.1 EXAMPLE OF SDI FEED ASSURANCE: FEEDCHEK

FeedChek is a simple, highly sensitive lateral-flow test for the detection of MBM in feed and feed ingredients. Currently, the use of mammalian-derived MBM in cattle feed is prohibited or highly regulated in most countries because of its potential to spread BSE, otherwise known as mad cow disease. As a precautionary measure, some regions have restricted the use of MBM from any animal species in ruminant feeds. To accommodate user-specific requirements, the FeedChek test for MBM incorporates two tests into one test strip. One test line indicates the presence of any MBM (mammal and avian) in the sample and the second test line indicates only the presence of mammalian MBM in the sample. The test can detect less than 0.1% MBM in feed and other feedstuffs. The test that detects only mammalian MBM is directed against less prevalent muscle proteins that are mammal-specific. Because of their lower prevalence in MBM, the detection limit for the mammalian MBM test is 1% MBM in feed. FeedChek has a 15-second extraction process and provides results in 10 minutes. No laboratory equipment is needed to perform the test.

Other tests SDI offers include the following:

1. To identify genetic traits in seed, grains, and feed, SDI offers TraitChek and GMOChek tests to help producers to quickly meet regulatory and customer needs for genetic information.
2. With TraitChek test strips, customers receive on-site yes/no results at grain elevators, terminals, and barges in 5 minutes or less. SDI has a large selection of tests available and their simple, foolproof lateral-flow design provides hassle-free testing of seed, leaves, and grain.
3. SeedChek test strips and ELISA plates provide rapid, reliable, and cost-effective screening in the production and verification of purity of GMO and non-GMO seed.
4. GMOChek microwell plate test kits provide quantitative analysis with results in hours for semiprocessed ingredients such as flour, toasted meal, tofu, soymilk, grits, and more.
5. SDI tests are USDA-certified for traits in corn and soybean and have been validated by agencies in the United States, Europe, and Japan.
6. In 2006, SDI announced that the USDA's GIPSA had certified the performance of TraitChek LL rice test kit to detect unapproved LL601 rice variety.[*]

REFERENCES

Peres, Bruno, Nicolas Barlet, Gérard Loiseau, and Didier Montet. *"Review of the Current Methods of Analytical Traceability Allowing Determination of the Origin of Foodstuffs."* Food Control 18, no. 3, 2007: 228–235.

Sundstrom, Frederick J., Jack Williams, Allen Van Deynze, and Kent J. Bradford. "Identity Preservation of Agricultural Commodities." In *Agricultural Biotechnology in California Series,* Publication 8077, Davis, CA: University of California–Davis, 2002, pp 1–15. Available at http://anrcatalog.ucdavis.edu/pdf/8077.pdf.

---

[*] In addition, its Microtox® Bioassay technology was awarded the designation and certification as an "Approved Product for Homeland Security" under the Support Anti-Terrorism by Fostering Effective Technologies Act of 2002 (the SAFETY ACT), by the Department of Homeland Security (DHS). Accordingly, Microtox has been placed on the "Approved Products List for Homeland Security." Microtox is the water industry bioassay standard for rapid detection of toxins, because it detects toxicity over a broad spectrum of more than 2,000 biological and chemical toxins.

Section IV

Consultative and Service Contributors

Part IV, including Chapters 9 through 12, encompasses many entities that are essential for successful identity preservation and traceability (IPT) programs and their adaptation to changing demands. Chapter 9 provides a domestic and foreign sampling of policy and advisory organizations regarding IPT systems, policies, and standards ranging from local to global food systems. This chapter also addresses the various traits or credence attributes that IPT systems must adjust to. Following these chapters are Chapter 10 on software providers, Chapter 11 on process facilitators, and Chapter 12 on food recalls and insurance. These chapters include individual, company, and organizational statements from their websites and naturally reflect their views.

9 Policy and Advisory Organizations

9.1 INTRODUCTION

For identity preservation and traceability (IPT) to gain acceptance, from producer to consumer, costs of IPT are critical to consider within the creation of standards, rules, and regulations. Costs come in many forms and impact societies differently, as several of the follow-on sections will highlight. The key to Chapter 9 is in the sampling of various private organizations that, in conjunction with government agencies and industries mentioned in previous chapters, participate in the development of public/private standards, rules, and regulations. It is not unusual for external organizations like these to contribute—be it to oppose, agree, or promote—alternative ideas that may include social welfare, animal welfare, ecological issues, regional emphasis, and so forth. To illustrate the notion of cost versus benefits and hidden costs of IPT, studies by Kalaitzandonakes, Maltsbarger, and Barnes (2001) and Maltsbarger and Kalaitzandonakes (2000) lead off this chapter.

This chapter includes domestic observations from the Farm Foundation, Northern Great Plains, AgBioForum Journal, ATTRA—National Sustainable Agriculture Information Service, and the American Soybean Association.

The focus of this section is on the inputs or advice that domestic organizations provide to government and industry regarding policies, rules, standards, and practices. Each of these groups and organizations vary in size, scope, and mission. A key aspect of many, if not all, IPT programs or systems that are governed by governmental and/or industry rules and standards relates especially to direct costs associated with implementing, maintaining, and other related costs of overall IPT programs. The prevailing costs that many of the organizations in this section will discuss will be that of direct monetary costs rather than ancillary costs that may be incurred, i.e., social costs. This is not to imply that the latter are any less important or have less impact on society or the environment.

Kalaitzandonakes, Maltsbarger, and Barnes (2001) and Maltsbarger and Kalaitzandonakes (2000) begin this section highlighting the economic concerns that organizations have regarding the implementation and extent of IPT programs, standards, and rules. Their works point to the economics of IPT, which are very influential in the design and implementation of IPT policies and standards, be it for domestic or foreign IPT programs.

9.2   FARM FOUNDATION

Farm Foundation
1211 W. 22nd Street, Suite 216
Oak Brook, IL 60561
Phone: 630-571-9393
Fax: 630-571-9580

Available at http://www.farmfoundation.org and http://www.farmfoundation.org/projects/02-66/documents/ExecuSumFFtrace_lowres8-4-04.pdf.

For more than 70 years, Farm Foundation has worked to help private and public decision makers identify and understand the forces shaping the economic viability of agriculture and rural North America. Food traceability and assurance is one such issue, particularly because these protocols are more prevalent in several markets of the world, particularly the European Union (EU), than in the United States. The expanding volume of global agricultural production and trade, food safety concerns, genetically modified organisms (GMOs), and food industry biosecurity has focused attention on the viability of tracing food products from retail to farm and the need to ensure specific food ingredient attributes. Because food traceability and assurance represent a fundamental change in the relationships that exist among market participants, it is inevitable that important questions be raised about the motivations, constraints, and appropriate locations of responsibility in implementing these protocols in the United States.[*]

To accomplish the needed changes, the Farm Foundation brought together a panel of industry leaders from most segments of the grain and meat supply chains, as well as representatives from various agencies of U.S. Department of Agriculture (USDA). The charge to the panel was to define the forces, both pro and con, motivating the adoption of traceability and assurance protocols, and to explore the implications for the various sectors of the U.S. food system. This report is based on that dialogue. Farm Foundation's intent is for this report to aid informed decision making in both the public and private sector.

9.2.1   SUMMARY

Farm Foundation's Traceability and Assurance Panel debated many approaches to the challenges facing U.S. food and ingredient supply chains in dynamic global markets. On one issue, however, there was clear consensus: One size does not fit all.

Key issues identified by the panel include USDA agencies that have historically provided market facilitation and oversight through regulatory protocols, consistent with legislative authority, that do not recognize differences in firm size or

---

[*] Excerpts and modified from "Food Traceability and Assurance in the Global Food System" found in the *Farm Foundation's Traceability and Assurance Panel Report, July 2004,* with permission, http://www.farmfoundation.org/projects/02-66/documents/FINALFULLREPORTwCover8-5-04_000.pdf. The Farm Foundation thanks DeeVon Bailey of Utah State University and Eluned Jones of Texas A&M University for their leadership in coordinating this project.

strategic objectives, i.e., one size fits all. Thus, the difference between facilitation and constraint of markets may place the private and public sectors in opposition in a dynamically changing global market. IPT systems, which incorporate existing food safety and assurance elements such as Hazard Analysis and Critical Control Point (HACCP), International Organization for Standardization (ISO) series, total quality management (TQM), continuous improvement (CI), and application of electronic data interchange (EDI) in supply chain management (SCM) systems, have the potential to provide an umbrella framework for the diversity of public and private market demands. They may help address such issues as:

- Food safety contaminations
- Intentional biosecurity contamination
- Requirements established for market entry by country or firm
- Opportunities to address inefficiencies in the supply chain, such as non-safety contaminations that violate contractual specifications
- Opportunities to identify extrinsic characteristics such as animal welfare, environmental, and social responsibility
- Opportunities for gaining consumer, as well as internal supply chain customer, brand or private label equity through implied system integrity

The grain industry offers an example of where protocols tailored to meet specific food safety or quality assurance goals are preferable to blanket protocols. Grain and oilseed products are routinely tracked by lot number after initial processing but are typically commingled at the first assembly point at the country elevator. Segregation is used in the grain industry to ensure characteristics before processing, but this is not traceability per se because manufacturing and end-use product attributes are tracked rather than a chain of possession. Of greater concern with grains is the nonuniformity of recordkeeping systems across firms and whether protocols should be standardized to facilitate recalls. The direct economic benefits of using traceability to maintain the integrity of attributes within the chain are limited, except for high-value food chains such as soybeans for tofu products. Indirect benefits accruing to better management practices have been documented for several ISO-certified grain elevators in the Midwest. However, the cost-benefit relationship for a broader segment of the grains and oilseeds markets changed in the mid-1990s associated with the jeopardy of export market loss from rejected genetically modified (GM) grains and oilseeds ingredients.

The investment in implementing IPT protocols could be viewed as the option value, or premium, on ensuring future revenues from second-tier, demand-side GM products.*

---

* The second round of strategic positioning for competitive advantage can be illustrated by Cargill's Cerestar acquisition to take advantage of core competencies in cereal foods manufacturing; the alliance between Bunge Ltd. and DuPont to create Solae Company in 2003 to strategically benefit from DuPont's protein technology businesses and Bunge's ingredient assets to further develop consumer-oriented Solae-brand soy protein products; a joint venture between ADM and Kao Corporation to create the infrastructure necessary to meet Japan's strict standards of identity preservation and to build on ADM's NutriSoy brand of soy protein; and a venture between ADM and Volkswagon to further develop opportunities for innovation in biofuels; a joint venture between Tate & Lyle, PLC, and DuPont, DuPont Tate & Lyle BioProducts, LLC, that aligns proprietary fermentation and purification core competencies with DuPont's strategic objective of reducing dependency on petrochemicals in the production of textile fibers.

Whether for retail or for food service, the weakest segment in maintaining the integrity of the food chain is at the initial handling stages upstream. Thus, few developments have had greater potential for changing market infrastructure than the emergence of traceability and assurance protocols as a management tool. Processors and manufacturers in the middle of the supply chain, between retailers and producers, have traditionally dominated the U.S. food system. Because intermediate ingredient tracking quality was difficult, retailers and consumers relied on manufacturers and manufacturer brand names to signal quality in the system. Traceability and assurance protocols provide the ability for retailers to influence upstream management decisions through specifications that control all aspects of production and processing. The increasing influence of retailers in global markets, including the United States, and the emergence of consumer interest in extrinsic characteristics relating to production processes and inputs (e.g., animal welfare, environment impacts, social welfare of workers) will likely propel the need to document all management practices in a future driven primarily by retail and food service specifications. A recent example is McDonalds' requirement that larger cages be provided for laying hens in their egg supply system.

In the United States, consumers' willingness to pay for extrinsic characteristics has been at the center of most discussions regarding implementation of traceability and assurance protocols. Considerable debate has ensued in academia, government, and industry about whether firms should implement such protocols, and the scientific and economic justification for so doing. Most economic studies examining willingness to pay have revealed only small, positive premiums for traceability and assurance, indicating consumers perceive that many of the attributes studied are public goods or have insignificant additional value. In reality, it is almost certain that assuring traceability, source verification, and origination in U.S. markets cannot be justified solely on willingness to pay. However, characteristics related to nutrition and health could possibly generate premiums that would justify the costs of traceability and assurance.

The timeline of implementation of traceability and assurance protocols across global markets varies widely as a result of cultural differences, legislation that emphasizes protection of either the consumer or industry, and with experience of past food safety incidences. Substantial differences exist in the level of consumer trust in public oversight; the strongest example may be the market responses in the EU and the United States to their respective discoveries of bovine spongiform encephalopathy (BSE). U.S. market participants believe government regulation and industry compliance provide good control over the safety of the food system. In contrast, consumer confidence in the ability of government to effectively regulate food safety has been shaken across Western Europe by BSE incidences, dioxin contamination of poultry feed, and contamination of bottled beverages. The EU approach to new food introductions, such as GM ingredients and nutraceuticals, uses strict interpretation of the precautionary principle. In the United States, once the regulatory system designates a product as safe, it is considered to be so until proven otherwise.

Increasingly, market participants, rather than government agencies, are influencing the determination of acceptable levels of health and food safety. The leading global food retail chains, such as Tesco, establish acceptable thresholds based on their home nation's legal standards and cultural experience, as well as those pertaining to the country within which they operate. For example, Tesco responds to the consumer market of the United Kingdom (UK) and understands that one size does not fit all. The greatest challenge to implementing traceability and assurance systems may be adjusting a century-old public-private partnership that has been extremely successful using a "one size fits all" paradigm. Processors and manufacturers supplying retail chains must meet the public and private standards established for procurement, even though they may differ significantly from those prevailing in the country of origin. A significant question is whether U.S. multinational food corporations are adopting this model, and if such action diminishes or retains the public's role as a third-party certifier.

Globally, there is consensus that sound science should underlie oversight of food markets. However, increasing consumer awareness and knowledge of the limits and continual evolution of science increase the emotional response, rather than cognitive acceptance, to food products. This is particularly true in mature and emerging economies. It is the emotional response that activist minorities can sway, that corporate advertisers target in developing brand allegiance, and that retailers target to gain competitive advantage.

Both the public and private sectors use dramatic events to motivate paradigm changes. If, for example, government response to a life-threatening contamination of foods is a funded mandate to implement new oversight protocols, it is unlikely to be rejected by consuming taxpayers, demonstrating an indirect willingness to pay. Consequently, events dramatized in the media gain political support, even if the probability of a negative event is very low. In contrast, less dramatic but more probable negative events gain less political support but are no less critical to the overall integrity of the food system.

Implementation of traceability and assurance protocols may not eliminate the overall risk in a marketing chain because traceability does not guarantee that system breakdowns will not occur. However, these protocols provide an effective means of managing risk containment once a negative event is identified because the problem can be located efficiently and the impact can be minimized. For example, food recalls could be targeted and less market disruption would occur if a traceability and assurance system was in place rather than the conventional marketing chain. This suggests that a public interest exists in traceability and assurance systems; food safety breakdowns can be efficiently tracked and the consequences minimized. Use of these protocols can also help minimize damage to private brand equity, suggesting that private interests also benefit. For example, traceability can substantiate private standards used to determine whether there has been a breach of contract or other type of agreement.

9.3   NORTHERN GREAT PLAINS, INC.

Northern Great Plains Inc.
Valley Technology Park
510 County Road 71
P.O. Box 475
University of Minnesota-Crookston
Crookston, MN 56716
Phone: 218-281-8459

Available at http://www.ngplains.org.

Northern Great Plains Inc. (NGP) is a nonprofit research and demonstration organization working with a network of rural development, business, policy, and academic leaders to build a healthy economic and ecological future for the people
and communities on the Northern Great Plains of North America. They view
themselves as forward looking, creative, and directed toward ensuring that the
Northern Great Plains will be a place where today's families and future generations will want to live and work. Part of their research looks at farm policy,
in the larger context, as it applies toward subsidies, but even more than that,
traceability.*

According to Wagner and Glassheim (2003), NGP hopes not only to help mold
policy but also farmers' attitudes toward farming, which can be best summarized
by John Russnogle in *Soybean Digest*, "The farmer is no longer the customer for
the food industry. Rather they're one link in the food chain that strives to meet
the demands of the ultimate consumer of the product." So often, many farms let
government programs control their destiny. The welfare of individual farmers is
becoming less of an important consideration in the "New Agriculture." Budget
constraints, conservation, and public opinion have increased influence. Those
farmers who grow "bulk" commodities (i.e., corn, soybeans, and wheat) face the
greatest risk with reduced government subsidies and need to consider what alternatives are in their future. NGP puts forth that the best positioned to prosper without
government payments may be diversified crop and livestock operations. There are
many tools and strategies depending on a farmer's skills, location, resources, and
management capability. According to Moe Russell in a *Soybean Digest* article,
"One strategy is to go to an end user of your crops and find out the quality traits
they want, when, and in what quantity. Then you can put a plan together to meet
their needs."

In the 21st century, grain is no longer just grain. The generic wheat crops we have
dutifully grown year after year are, in some segments of the industry, on the verge
of disappearing. They are being replaced by function-specific, or identity-preserved
(IP), varieties. According to Jack Eberspacher, CEO, National Association of Wheat
Growers (NAWG), "There are some real opportunities, using plant breeding and
biotechnology, to develop wheat varieties with specific characteristics for specific
purposes."

When farmers switch from commodity crops to differentiated or value-added
crops, they will need a different set of skills, according to Mike Boehlje, Purdue
University agricultural economist. "With a commodity crop, farmers traditionally
added value to their crop by lowering costs through adapting new technology to
increase production and efficiency," he says. "That may be through product attributes, such as leaner pork, or it may be through the process you use to produce the
product, such as organic production."

Regardless of what crop a farmer decides to grow, it is the agriculture industry—
not farmers—that likely will dictate the changes. This is already happening in France,

---

* Excerpts and modified from "Traceability of Agricultural Products" by Gary L. Wagner and Eliot
  Glassheim. Published by Northern Great Plains Inc., May 2003, with permission.

where the four major food distributors have demanded that all major agricultural products have the ability to be traced from the farm to the consumer's table.

9.3.1 "TRACEABILITY" AND "IDENTITY PRESERVED" ACCORDING TO NGP

As the American farmer tries to understand this new concept in production agriculture, the terms "traceability" and "identity preserved" are frequently used interchangeably. Although the two terms are often blurred, they each refer to distinct ways in which products and information about the products will be handled in the food chain.

Traceability is a strict production and delivery method, with known procedures of observing, inspecting, sampling, and testing to ensure the presence (or absence) of certain traits, usually defined by consumer demand. Traceability is focused on food safety, consumer confidence, and a defined source. It often requires a certified "paper trail" so each step in the ownership of the product from farm to the final consumer is documented.

Identity preserved (IP) is a process by which producers contract with processors to deliver crops with traits that will increase processing quality and efficiency. Normally, the crop carries a premium price, and the processor is assured that the crop has maintained its unique identity from farm gate to processing plant. Another term coined by the National Corn Growers Association is channeling. *Channeling* is the act of keeping a crop separate after harvest as it is delivered to a specific market or end user. This term became necessary when the StarLink corn problem occurred in 2000. It was coined to protect the integrity of the IP process.

Specialty crops cost more to produce and, if they are to be IP, require increased special handling at the farm. Kansas State University agricultural economist Kevin Dhuyvetter stated, "Value-added grain buyers only have to pay enough to reward the better managers. IP farming will widen the gap between better and worse managers." A Kansas State Extension service report, "Economic Issues with Value-Enhanced Corn," projected premiums for high-oil corn were 21 cents and white and waxy corn 15 and 2 cents per bushel, respectively. These per-bushel premiums are needed to offset extra production costs and yield drag. However, these premiums did not consider extra costs for IP, storage, delivery, and segregation.

9.3.2 NGP PROPOSES A GENERIC FOOD TRACKING SYSTEM

Many models can be used to track food from the farm to the table. A sample generic system could implement the following steps to trace agri-food products:

1. Growers enter information about management practices, assurance schemes, and protocols into the standardized system.
2. When a crop is harvested, information about the growing process, such as seed type and chemicals applied, is recorded. In return, an "ID tag" is supplied that uniquely identifies this information and links to that already supplied. Additionally, there may be ID tags from many growers. As product is bulked up for processing, all the ID tags for the raw material are linked together.

3. When the crop is sent to the processor, the ID tag to the grower's information travels with it, either as an e-mail or barcode attached to the goods.
4. The processor is able to use the ID tags to access the grower's information and use automated tests to ensure compliance with the required standards.
5. Details of the production process, together with the ID tags from the raw materials, are recorded in the central computer database. Another unique ID tag is returned.
6. When the product is dispatched to the retailer, the ID tag to the processor's data (which includes ID tags to the grower's information) is passed on as before.
7. With access to all the ID tags relating to a final product, if there is a problem with the final product, a retailer is then able to trace the source of all the materials.
8. Information about products sourced anywhere in the world is treated exactly the same way using a centralized database.

Developing IP market opportunities provides farmers several risk management benefits, according to Cole Gustafson, an agribusiness professor at North Dakota State University:

- First, market premiums increase farm revenues and lower financial risks.
- Second, IP markets provide diversification opportunities, as these markets are less influenced by the supply and demand forces of traditional commodity markets.
- Third, adoption of IP crop production methods reduces food safety and market risks as purchasers are able to trace and verify sources.
- Finally, as farmers embrace IP market opportunities, human risks decrease.

The farm level IPT system may or should offer premium contracts. A traceability system helps to assure growers and their buyers of a crop's integrity and purity. Growers can prove the purity of the crop with an electronic or paper trail. This also provides an opportunity for farmers to consider becoming ISO certified (see Chapter 6 on ISO Standards).

At the farm level, a farmer would need to fill out forms recording everything done: where the seed came from, when and where it was planted, field rotations, genetically modified or not, type of equipment used, when sprayed and with what, and when equipment was cleaned.

In addition to the farm information collected, a complete system would require an independent auditor to conduct an on-site inspection, validate the information the farmer collected, and enter it into a central computer database. These auditors, who would need to be certified, might check fields two or three times a year and would need to record their findings into the same database used by the farmer. Any time during the growing season crop contractors then could log into this central database and check on each farm's crop to ensure they are meeting specifications.

On-farm IP systems will require more than separate storage structures. Producers will have to become their own managers of quality. At the time the crop is stored, tests will be needed to determine the level of purity after all instances of cross-pollination and mechanical mixing have occurred. Then, when the commodity is delivered to the purchaser, who will take samples of his or her own, the producer also may want to take samples to protect against any future claim that the commodity was genetically or environmentally contaminated.

The potential exists for some liability to be placed on the producer if contaminated food or a commingling violation was traced back through samples to a specific elevator. The retention of seed samples by producers could be essential insurance in case such a situation arose. A producer who retains a sample that is tested by a thorough testing system is in a stronger position to defend the farm operation against these kinds of liability claims; those who fail to manage their own quality may have no way of documenting that their production was not at fault.

The use of production contracts will increase. Both benefits and risks are associated with the use of production contracts. Producers will have to determine which contracts are best for them and which will increase the need for management education and additional market information. Currently, the demand for traceability is becoming more commonplace in food grade crops. Dry edible beans, food grade soybeans, confectionary sunflowers, potatoes, and sugar beets have contracts that require that, if asked, the grower provide documented proof of crop production.

Many farmers in the Northern Great Plains recognize that product traceability is needed even if at this time not all of their customers require the information. They also realize collecting this information requires an additional cost, which processors may not be willing to pay for.

One specialized use of traceability systems comes from the increasing use of GM crops. Producers will soon be required to grow any new-generation GM crops within strictly defined parameters. Examples of these parameters are mandatory buffer zones, regular use of certified seed, production contracts, and tolerance levels for commingling.

9.3.3 CROP CHARACTERISTICS

The following parameters are affected significantly by climate, soil properties, genetics, farming practices, and many other variables. Table 9.1 lists quality traits for wheat and assembled by Americrop.

9.4 ATTRA—NATIONAL SUSTAINABLE AGRICULTURE INFORMATION SERVICE

ATTRA—National Sustainable Agriculture Information Service
P.O. Box 3657
Fayetteville, AR 72702
Phone: 800-346-9140 (toll-free)
Fax: 479-442-9842

Available at http://attra.ncat.org.

TABLE 9.1
Crop Characteristics

Grading Data	Single Kernel Characteristics	Identity and Purity
Test weight (lbs/bu)/(kg/hl)	**Average and Standard Deviation**	Density (seeds/lb)
Damaged kernels (%)	Weight (mg)	Electrophoresis
Foreign material (%)	Diameter (mm)	Variety verification
Shrunken and broken (%)	Hardness	**Minerals**
Total defects (%)	Moisture (%)	Selenium (ppm)
Vitreous kernels (%)	**Flour Characteristics**	Calcium (ppm)
Contrasting classes (%)	Extraction (%)	Copper (ppm)
FGIS grade, class, subclass, special grade	Flour ash 14% MB (%)	Magnesium (ppm)
	Flour protein 14% MB (%)	Zinc (ppm)
Dockage (%)	Wet gluten 14% MB (%)	Phosphorous (ppm)
Wheat moisture (%)	Flour falling # 14% MB (s)	Iron (ppm)
Wheat protein dry (%)	Amylogram 100 g (B.U.)	Potassium (ppm)
Wheat protein 12% MB (%)	Amylogram 65 g (B.U.)	Total starch (%)
Other Wheat Data	**Baking Characteristics**	Amylose starch (% of starch)
1000 Kernel weight (g)	Absorption (%)	Polyphenyl oxidase
Ash 14% MB (%)	Dough handling (1–10)	
Falling number value (s)	Loaf volume (cc)	
Sedimentation (cc)	Grain and texture (1–10)	
	Crumb color (1–10)	
	Crust color (1–10)	

Source: Data adopted with permission from "Traceability of Agricultural Products" by Gary L. Wagner and Eliot Glassheim, 2003.

The National Center for Appropriate Technology (NCAT) launched the ATTRA project in 1987. ATTRA, or National Sustainable Agriculture Information Service, is funded under a grant from the USDA's Rural Business-Cooperative Service (RBS). ATTRA has often been cited as an example of a successful partnership between a private nonprofit (NCAT) and a public agency (USDA-RBS). ATTRA services are available to farmers, ranchers, market gardeners, extension agents, researchers, educators, farm organizations, and others involved in commercial agriculture, especially those who are economically disadvantaged or belong to traditionally underserved communities. The ATTRA project is staffed by more than 20 NCAT agricultural specialists with diverse backgrounds in livestock, horticulture, soils, organic farming, integrated pest management, and other sustainable agriculture specialties.

To promote more sustainable farm practices, ATTRA encourages and helps educate the importance of locally grown foods, how to transition to organic farming, environmentally friendly sources of energy (wind and solar power), animal identification systems, pest and water management, soils and compost, livestock, horticultural crops, etc. Their educational services include better understanding of the effects of GMOs and value-added products.[*] To this end, ATTRA offers several articles and websites regarding IP that include:

- "Adding Value to Farm Products: An Overview" by Holly Born and Janet Bachmann, NCAT Agriculture Specialists ©2006 NCAT-ATTRA Publication #IP141. http://www.attra.ncat.org/attra-pub/PDF/valueovr.pdf (accessed January 2, 2007).
- Website for International, National, and Regional Educational and Outreach Resources by Southern Organic Resource Guide http://attra.ncat.org/sorg/education_outreach.html and http://attra.ncat.org/sorg/downloads/sorg.pdf.
- "Seed Production and Variety Development for Organic Systems" by Katherine L. Adam, NCAT Agriculture Specialist ©2005 NCAT. http://www.attra.ncat.org/attra-pub/PDF/seed_variety.pdf.
- "Entertainment Farming and Agri-Tourism" by Katherine L. Adam, NCAT Agriculture Specialist, ©September 2004 NCAT. http://www.attra.ncat.org/attra-pub/PDF/entertn.pdf.
- "Organic Alternatives to Treated Lumber" by Lance E. Gegner, NCAT Agricultural Specialist, July 2002. http://www.attra.ncat.org/attra-pub/PDF/lumber.pdf.
- "Label Rouge: Pasture-Based Poultry Production in France" by Anne Fanatico and Holly Born, NCAT Agriculture Specialists, November 2002, ATTRA Publication #IP202. http://www.attra.ncat.org/attra-pub/labelrouge.html.
- "Transgenic Crops" by Jeff Schahczenski and Katherine Adam, NCAT Program Specialists ©2006. http://www.attra.ncat.org/attra-pub/PDF/geneticeng.pdf.

9.4.1 ATTRA's View of IPT and GMOs

The general and often misused term *biotechnology* refers to a broad spectrum of technologies, including conventional plant selection and breeding, in which humans intervene in biological processes of genetic alteration and improvement. The main concern of many when the term GMO is mentioned regards crop varieties created through a type of biotechnology commonly known as *recombinant DNA*, *genetic engineering (GE)*, *transgenic modification*, or *genetic modification (GM)*. The products of GE are often called *genetically modified organisms*, or *GMOs*. All these terms refer to methods of recombinant DNA technology by which biologists splice genes from one or more **differing** species into the DNA of crop species plants to transfer chosen genetic traits. This type of GE is also referred to as transgenic or transgenetic.[†]

---

[*] See ATTRA's website for additional information http://www.attra.org.
[†] Genes are segments of DNA that contain information that in part determines the structure and function of a living organism. Genetic engineers manipulate this information, typically by taking genes

With the advent of GE of plants around 1983, it appeared that this new biotechnology would benefit and even revolutionize agriculture. The transfer of desirable genetic traits across species barriers has shown promise for solving problems in the management of agricultural crops. This is often coined as the first realization of biotech innovation. Potential benefits include reduced toxic pesticide use, improved weed control resulting in less tillage and soil erosion, water conservation, and with increased yields, less time demanded in the field, and increased uniformity in crop. However, "emerging evidence suggests the promised environmental benefits remain small, uncertain or unrealized in the US, and some risks are real."

ATTRA's issues of concern focus on insufficiently answered questions about transgenic crops and their potential benefits, costs, and risks. The scope of concern is far reaching and involves ecological issues of gene flow to neighboring crops and to related wild species,* *pesticide resistance in insect pests,†* *antibiotic resistance,‡* *effects on beneficial organisms,§* *and reduced crop genetic diversity.*¶

(Continued)

from (more preciously) one species—an animal, plant, bacterium, or virus—and inserting them into another species, such as an agricultural crop.

* Ecological scientists have little doubt that *gene flow* from transgenic fields into conventional crops and related wild plants will occur. Gene flow from transgenic to conventional crops is of concern to farmers because of its potential to cause herbicide resistance in related conventional crops. Gene flow from transgenic crops to wild relatives creates a potential for wild plants or weeds to acquire traits that improve their fitness, turning them into "super weeds." For example, if jointed goatgrass, a weedy relative of wheat, acquires the herbicide-tolerant trait of Roundup Ready wheat, it will thrive in crop fields unless applications of other herbicides are made. Other traits that wild plants could acquire from transgenic plants that would increase their weediness are insect and virus resistance. Because of their experience with classically bred plants, few scientists doubt that genes will move from crops into the wild: seven of the world's thirteen most important crop weeds have been made weedier by genes acquired from classically bred crops. Because gene flow has the potential to affect farmers' crop and pest management, crop marketability, and liability, more research needs to be done to determine the conditions under which gene flow from transgenic plants is likely to be significant.

† Bt has been widely used as a microbial spray because it is toxic only to caterpillars. In fact, it is a pest management tool that organic farmers depend on, one of the few insecticides acceptable under organic rules. Unlike the commercial insecticide spray, the Bt engineered into crop plants is reproduced in all, or nearly all, the cells of every plant, not just applied on the plant surface for a temporary toxic effect. As a result, the possibility that transgenic Bt crops will accelerate insect pests' development of resistance to Bt is a serious concern. Pest resistance to Bt would remove this valuable and environmentally benign tool from farmers' and forest managers' pest control toolbox.

‡ The use of antibiotic-resistant marker genes for the delivery of a gene package into a recipient plant carries the danger of spreading antibiotic-resistant bacteria. The likely result will be human and animal health diseases resistant to treatment with available antibiotics. Research is needed on antibiotic resistance management in transgenic crops. Already the European Commission's new rules governing transgenic crops stipulate phasing out antibiotic-resistant marker genes by the end of 2004.

§ Evidence is increasing that transgenic crops, either directly or through practices linked to their production, are detrimental to beneficial organisms. New studies are finding that Bt crops exude Bt in concentrations high enough to be toxic to some beneficial soil organisms. The reason is that the beneficial rhizobium responsible for nitrogen fixation in soybeans is sensitive to Roundup. It also appears that disruption of beneficial soil organisms can interfere with plant uptake of phosphorus, an essential plant nutrient. Beneficial insects that prey on insect pests can be affected by insecticidal crops in two ways.

¶ As fewer and larger firms dominate the rapidly merging seed and biotechnology market, transgenic crops may continue the trend toward simplification of cropping systems by reducing the number and type of crops planted. In addition, seed-saving, which promotes genetic diversity, is restricted for transgenic crops.

9.4.2   ATTRA IPT Requirements

The need to separate transgenic crops from both conventional and organic crops opens farmers to liability for their product at every step from seed to table. Effective systems for segregation do not exist at present and will be costly to develop and put into place. Farmers may well end up bearing the added costs of crop segregation, traceability, and labeling.

In the meantime, farmers who grow transgenic varieties and, ironically, those who do not are liable for transgenic seeds ending up where they are not wanted: in their own nontransgenic crop fields, in neighbors' fields, in truckloads of grain arriving at the elevator, in processed food products on retail shelves, and in ships headed overseas.

Farmers who choose not to grow transgenic varieties risk finding transgenic plants in their fields anyway as a result of cross-pollination via wind, insects, and birds, which may bring pollen from transgenic crops planted miles away. Besides pollen, sources of contamination include contaminated seed and seed brought in by passing trucks or wildlife. Those farmers whose conventional or organic crops are contaminated, regardless of the route, risk lawsuits filed against them by the companies that own the proprietary rights to seed the farmer did not buy. Likewise, farmers who grow transgenic crops risk being sued by neighbors and buyers whose nontransgenic crops become contaminated.[*]

9.4.3   Regulation of Transgenic Crops[†]

Currently, three federal agencies regulate the release of transgenic food crops in the United States: the Department of Agriculture's Animal and Plant Health Inspection Service (USDA-APHIS),[‡] the U.S. Environmental Protection Agency (EPA),[§] and the U.S. Food and Drug Administration (FDA).[¶]

---

[*] Because contamination by transgenic material has become so prevalent in such a short time, all farmers in areas of transgenic crop production are at risk. Insurance, the most common recourse for minimizing potential losses because of liability, is not available to the nation's farmers for this risk because insurance companies do not have enough information to gauge the potential losses.

[†] Much of the controversy over transgenic crops, both internationally and in the United States, is in part a result of how the United States regulates transgenic crops. The federal government has determined that the commercial products of agricultural biotechnology are *"substantially equivalent"* to their conventional counterparts and that therefore no new regulatory process or structure is needed for their review and approval.

[‡] USDA-APHIS: The USDA looks at how a transgenic plant behaves in comparison with its unmodified counterpart. Is it as safe to grow? The data it uses are supplied largely by the companies seeking a permit for release of the new crop. Under "fast-track" approval, a process in place since 1997, companies introducing a crop similar to a previously approved version need give only 30 days advance notice before releasing it on the market. According to the Wallace Center report, APHIS staff estimate that by 2000, 95 to 98 percent of field tests were taking place under simple notification rules rather than through permitting.

[§] EPA: The EPA regulates the pesticides produced by transgenic crops, such as the Bt in Bt corn and cotton. It does not regulate the transgenic crops themselves. In contrast to its regulation of conventional pesticides, the EPA has set no tolerance limits for the amount of Bt that transgenic corn, cotton, and potatoes may contain.

[¶] FDA: The FDA focuses on the human health risks of transgenic crops. However, its rules do not require mandatory premarket safety testing or mandatory labeling of transgenic foods.

Central to the policy of substantial equivalence is the assumption that only the end *product* of transgenic technology is of concern, not the *process* of genetic modification. Canada has adopted a similar approach. Europe and other U.S. trading partners, however, have taken a more conservative approach. They focus on the *process* of genetic modification, the source of many of the environmental and human health risks of greatest concern.

How these different approaches play out in reality can be summed up simply: The United States and Canada assume a product is safe until it is proven to carry significant risk; the EU, which follows the "precautionary principle," assumes the same product may carry significant risk until it can be proven safe. The science used by the two approaches is not fundamentally different. The difference is in the level of risk the different societies and political systems are willing to accept.

9.5 AMERICAN SOYBEAN ASSOCIATION (ASA)

American Soybean Association
12125 Woodcrest Executive Drive, Suite 100
St. Louis, MO 63141-5009
Phone: 314-576-1770; 800-688-7692 (toll-free)
Fax: 314-576-2786

Available at http://www.soygrowers.com/step/darmstadt.htm.

The American Soybean Association (ASA), through its Soybean Trade Expansion Program, promotes the expansion of American farmer soybean sales. ASA advances the importance of U.S. soybean exports to Europe and a glimpse of how ASA views meeting EU requirements.

The EU is the largest regional market for US soybean exports. In 2001, the United States had record exports of just over 1 billion bushels (27.567 million metric tons) of soybeans. Of that amount, 253.5 million bushels (6.9 mmt) or 25 percent of total U.S. soybean exports were shipped to customers in the EU. In 2002, the United States exported nearly 15 percent more soybeans to the EU than were exported in the same period of the previous year. To ensure continued market access for U.S. soybeans in Europe, as well as other major markets, ASA holds firm to its policy of not commercializing unapproved-for-export soybean varieties and maintains an aggressive program to educate buyers and government officials about the safety of soybeans derived from biotech seed stock.

According to ASA, it is important to promote and market that biotech crops not only benefit farmers but also consumers by allowing farmers to use environmentally friendly farming practices that protect air, land, and water resources. Although ASA respects the rights of every nation to protect the safety of its food and feed ingredients and does not oppose science-based safety standards and regulations that serve the public interest, ASA asserts that products should be judged individually based on established scientific methods.

ASA promotes and understands that commodity soybean production differs greatly from IP production and that "non-GMO" soybeans cannot be guaranteed to be 100 percent free of biotech materials. A common problem being realized throughout the grain market, to include IPT products, is disconnect between what customers want, especially in Europe, and what they are willing to pay for. Public perception and understanding of commodities' production are weak. Regarding grain handling, the public does not appreciate that a single ton of commodity soybeans contains more than 7 million individual beans, which are collected and commingled at every point along a multistage handling and distribution system. A standard grain train car holds an average of 100 tons.

9.6   FOREIGN POLICY AND ADVISORY ORGANIZATIONS

This section includes foreign and international organizations such as the International Food Policy Research Institute (IFPRI), International Food and Agribusiness Management Association (IAMA), Food Standards Agency (UK), and International Seed Federation (ISF).

The IFPRI section provides details with regard to classifying of countries according to their approval and labeling regulations, and international institutions involved in the regulations of international trade of GM crops and GM foods.

IAMA is an international networking organization and acts as a facilitating intermediary between the agribusiness industry, researchers, educators, government, consumer groups, and non-governmental organizations (NGOs).

The Food Standards Agency (UK) provides advice and information to the UK public and government on food safety and protects consumer interests in relation to food safety and standards through effective food enforcement and monitoring.

ISF, a non-governmental, nonprofit organization represents the seed industry and serves as an international forum where issues of interest to the world seed industry are discussed.

What follows are individual/company/organizational/agency statements from their websites, and naturally reflect their views.

9.7   INTERNATIONAL FOOD POLICY RESEARCH INSTITUTE (IFPRI)

International Food Policy Research Institute
2033 K Street NW
Washington, DC 20006-1002
Phone: 202-862-5600
Fax: 202-467-4439
E-mail: ifpri@cgiar.org

Available at http://www.ifpri.org.

9.7.1   VISION

IFPRI's vision is a world free of hunger and malnutrition. It is a world where every person has secure access to sufficient and safe food to sustain a healthy and productive

life and where decisions related to food are made transparently and with the participation of consumers and producers.

9.7.2 Mission

IFPRI's mission is to provide policy solutions that reduce hunger and malnutrition. This mission flows from the Consultative Group on International Agricultural
Research (CGIAR) mission: "To achieve sustainable food security and reduce poverty in developing countries through scientific research and research-related activities in the fields of agriculture, livestock, forestry, fisheries, policy, and natural
resources management." Two key premises underlie IFPRI's mission. First, sound
and appropriate local, national, and international public policies are essential to
achieving sustainable food security and nutritional improvement. Second, research
and the dissemination of its results are critical inputs into the process of raising the
quality of the debate and formulating sound and appropriate food policies. Both of
these premises lend themselves to IPT policy development and systems advice by
IFPRI. Its mission entails a strong emphasis on research priorities and qualities that
facilitate change.

IFPRI is also committed to providing international food policy knowledge as a
global public good; that is, it provides knowledge relevant to decision makers both
inside and outside the countries where research is undertaken. New knowledge on
how to improve the food security of low-income people in developing countries is
expected to result in large social benefits, but in most instances the private sector
is unlikely to carry out research to generate such knowledge. IFPRI views public
organizations and the private sector in food systems both as objects of study and as
partners.

Given the large body of national and international food policy research, IFPRI's
added value derives from its own cutting-edge research linked with academic excellence in other institutions, such as other CGIAR centers, universities, and other
research institutes in the South and North, and from its application of this knowledge
to national and international food policy problems.

9.7.3 Consultative Group on International Agricultural Research (CGIAR)

IFPRI is one of 15 food and environmental research organizations supported by
CGIAR. The centers are located around the world and conduct research in partnership with farmers, scientists, and policy makers to help alleviate poverty and increase
food security while protecting the natural resource base. They are principally funded
through the 58 countries, private foundations, and regional and international organizations that make up CGIAR.

9.7.4 Labeling GM Foods

A more recent undertaking by IFPRI has been on labeling GM foods. Guillaume
P. Gruère's "An Analysis of Trade Related International Regulations of Genetically

Modified Food and Their Effects on Developing Countries" (2006) highlights the type of work IFPRI does to assist global policy regarding GMOs. This work provides a much clearer view of how particular countries' standards interact with labeling policy.[*]

Gruère's (2006) work reviews current trade-related regulations of GM food and discusses its effects on developing countries. There is a large variety of policies regarding import approval and marketing rules of GM food worldwide. At the international level, the coordination efforts are led by the Codex Alimentarius Commission, the Cartagena Protocol on Biosafety, and the World Trade Organization (WTO). Even within these groups, the regulatory process from approval to commercialization varies widely across individual countries. Figure 9.1 presents a schematic decision tree of countries according to their approval and marketing regulations. This diagram can be very helpful in better understanding the complexities involved with the designing of, not only labeling or GMOs, but also on how countries go about determining IPT policy and regulations. See Figure 9.1 of classification of countries.

Although internationally agreed on guidelines for safety approval have been finalized, there is no clear consensus on labeling regulations for GM food, and there is an increasing risk of conflicts among international agreements.

At the first level of division, at the top of the diagram after individual countries, countries may or may not have adopted any type of approval or marketing regulation on GM food. Then, among the ones with regulations (left side of figure), there are two main groups of countries: the ones that rely on a test of substantial equivalence (substantial equivalent products are exempt from specific requirements), and the others who generally do not and whose regulatory procedure depends on the production process (which means that any food produced with or derived from transgenic crop is subject to GM food regulations). Each country has also adopted its own set of safety approval and labeling policies with specific characteristics. *Key:* The specificities of labeling regulations are largely determined by the observable effects of regulations on international trade. More stringent regulations will generally require more costly procedures on behalf of exporters, and more comprehensive policies may have a more important trade effect. Alternatively, countries with no specific regulations (right side of figure) include those that are about to adopt approval or marketing regulations, the ones with no clear regulations, and the ones that have declared themselves GM-free.

In Figure 9.1, along the bottom edge of the division tree, are countries that are divided into eight categories or groups (defined by their eight bold-edged terminal boxes), according to their regulatory framework. Table 9.2 presents examples of countries in each of these eight groups. Organization for Economic Cooperation and Development (OECD) countries are represented in the first four categories (except Mexico and Turkey), and several countries with transition economies (such as Brazil or China) are also located in these four categories. All these countries have adopted specific regulatory framework for GM food and other products derived from GM

---

[*] Excerpts and reproduced with permission from the International Food Policy Research Institute www.ifpri.org. The paper from which this material comes can be found online at http://www.ifpri.org/divs/eptd/dp/eptdp147.asp. From Gruère, G. 2006. "An Analysis of Trade Related International Regulations of Genetically Modified Food and Their Effects on Developing Countries." EPTD Discussion Paper 147, pp. 5–8, Figure 1, Table 1. Washington, DC: International Food Policy Research Institute.

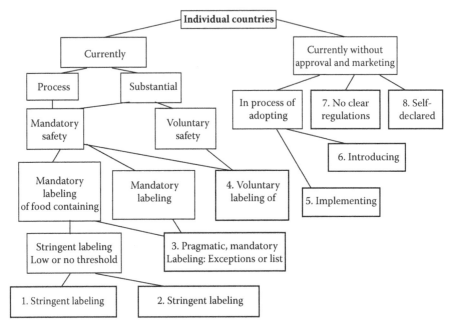

FIGURE 9.1 A classification of countries according to their approval and labeling regulations. (From Gruère, G.P., "An Analysis of Trade Related International Regulations of Genetically Modified Food and Their Effects on Developing Countries," International Food Policy Research Institute (IFPRI) EPT Discussion Paper 147, Environment and Production Technology Division, 2006.)

crops. In contrast, most developing countries are currently in Groups 5 to 8 because they are either without or in the process of adopting specific trade-related regulations of GM food. Table 9.2 provides characteristics and examples.

The large producers and exporters of GM crops have well-defined regulations, but most of them are in Group 4 (Canada, United States, Argentina, South Africa), with pragmatic regulations of GM food, whereas the last two are in Group 2 (Brazil and China), with stringent regulations. In contrast, large importers of these crops are in Groups 1 and 3 with relatively more stringent regulations. More specifically, Table 9.2 shows the level of stringency differentiating national regulations or approaches. Most groups of countries have adopted, are about to adopt, or intend to adopt mandatory safety approval regulations of GM food. The United States is a particular case; it has a voluntary safety consultation that is de facto considered a mandatory requirement because all companies comply with it for liability reasons. However, different groups have distinctive approaches on labeling of GM food; this reflects the level of success of international harmonization efforts: international convergence on specific requirements for safety approval and important divergences among countries with regulations on labeling and traceability of GM food.

To summarize, this overview of national regulations reveals that there is a large variation of specificity in regulations among countries, first in terms of development

TABLE 9.2

Characteristics of Group and Examples of Countries in Each Group

	Food Safety Approval Regulations	Labeling Regulations	Specificity	Countries
Group 1	Process-based mandatory	Stringent, mandatory, includes derived products	Traceability requirements, 0.9% threshold	EU, East Europe
Group 2	Process-based mandatory	Stringent, mandatory, includes derived products	No traceability, low threshold	Brazil, China, Russia, Switzerland, Norway
Group 3	Process-based mandatory	"Pragmatic," mandatory	Many labeling exceptions	Australia, Japan, Korea, Saudi Arabia, Thailand
Group 4	Substantial equivalence, mandatory (US: voluntary consultation)	Voluntary for substantial equivalent food	5% threshold level for labeling	US, Canada, Argentina, South Africa, Taiwan
Group 5	Mandatory (in place or pending)	Mandatory, introduced but not implemented	"Pragmatic" labeling requirements	Indonesia, Malaysia, Mexico, Philippines, Vietnam
Group 6	Mandatory (in place or pending)	Intention to require labeling	Slow regulatory process	India, Kenya
Group 7	Considering mandatory	No clear position	Wait and see approach	Bangladesh, most African countries
Group 8	No	No	GM free	A few African countries (Zimbabwe, Zambia)

Source: Reproduced with permission from the International Food Policy Research Institute (http://www.ifpri.org). The paper from which this table comes can be found online at http://www.ifpri.org/divs/eptd/dp/eptdp147.asp.

stages of regulatory framework, and second between countries with well-defined regulations. Developed countries differ in their general approach of regulations, with most GM producers and exporters in groups of pragmatic regulations, whereas importers tend to have more stringent marketing regulations for GM food and GM-derived products. Developing countries tend to have fewer regulations in place.

9.7.5   INTERNATIONAL EFFORTS FOR HARMONIZATION

There are six international organizations directly or indirectly involved in setting up harmonized rules, standards, and recommendations related to international trade in GM crops. Table 9.3 reviews these institutions' coverage, membership, and orientation.

TABLE 9.3
International Institutions Involved in the Regulations of International Trade of GM Crops and GM Food

Institution	Coverage	Member States*	Dispute Settlement Mechanism	Role
International Office of Epizootics (1924)	Infectious animal disease	167	Nonbinding; set WTO standards	Harmonizes trade regulations for animals and animal products
GATT/WTO (1947/1994)	Trade in goods and services	148	Binding	Sets rules transparency and dispute settlement
International Plant Protection Convention (1952)	Pests and pathogens of plants and plant products	136	Nonbinding; sets WTO Standards	Sets international standards for plants
OECD (1962)	Harmonization of international regulations, standards, and policies	30	None	Writes consensus documents and international data
Codex Alimentarius Commission (1972)	Food labeling and food safety standards	170	Nonbinding; sets WTO standards	Sets international standards and recommendations
Biosafety Protocol (2003)	Transboundary movements of GM organisms	120	None	Information sharing and biosafety measures

*As of May 2005.

Source: Data adopted from *Regulating the Liabilities of Agricultural Biotechnology*, Smyth et al. (2004), table by Buckingham and Phillips (2001). With permission from CABI Publishing.

9.7.6 UNITED NATIONS FOOD AND AGRICULTURE ORGANIZATION AND THE WORLD HEALTH ORGANIZATION (UN FAO/WHO) CODEX ALIMENTARIUS

The Codex Alimentarius is an intergovernmental organization managed jointly by the United Nations Food and Agriculture Organization and the World Health Organization (UN FAO/WHO). The Codex has two main purposes: (1) to protect the health of consumers and (2) to promote fair practices in international trade. The Codex provides international recommendations and standards based on a consensus among members. The Codex standards and recommendations are important for international traders because they are recognized as reference standards of food safety in the Sanitary and Phytosanitary Agreement of the WTO.

The Codex Commission has been working on finding a common terminology, a common food safety approval procedure, and a common position on the labeling of GM food since the beginning of the 1990s. The Codex Commission has published guidelines for the safety assessment of GM food, but it has failed thus far to reach any agreement on the issue of GM food labeling.

9.7.7   UN Cartagena Protocol on Biosafety

The Cartagena Protocol on Biosafety was introduced in January 2000 as part of the UN Convention on Biodiversity in an effort to set up a harmonized framework of risk assessment, risk management, and information sharing on the transboundary movements of living modified organisms (LMOs). The Protocol entered into force on September 11, 2003, 90 days after receipt of the 50th instrument of ratification, but the parties involved still have to decide a number of specific rules to implement it. GM organisms, GM seeds, and raw products from GM crops (used for food or feed) are considered LMOs.

9.7.8   World Trade Organization (WTO)

Unlike the two other international bodies presented in this section, the WTO does not have any mandate on GM food regulations. The WTO's role in the context of international trade and agriculture biotechnology is directly related to trade distorting regulations. There is no specific article of the WTO Agreement related to agricultural biotechnology; however, the general rules of the trade agreement are in question when biosafety and marketing regulations potentially act as barriers to trade. Many WTO country members have adopted different domestic regulations on the approval and the marketing of GM food, and in the absence of international consensus and standards the Dispute Settlement Body of the WTO can act as an arbitrator to resolve trade disputes among members.

Two WTO agreements are at the heart of the question of the legality of GM food regulations. First, the Agreement on the Applications of Sanitary and Phytosanitary Measures (SPS Agreement) provides rules related to safety regulations. Second, the Agreement on Technical Barrier to Trade (TBT Agreement) concerns domestic regulations that may be involved for other societal goals. In the case of GM food, the SPS Agreement would rule in a dispute related to the validity of GM food safety regulations (including bans) based on unproved risks of GM food. The TBT Agreement would rule if the importer raises technical standards or regulations (such as labeling) that are not directly related to safety or whose purpose is not related to safety but that still may be trade distorting.

The case of agricultural biotechnology presents new challenges to the application of the WTO trade agreement. First, the current WTO trade agreement does not provide clear guidance on the question of regulating products according to their process and production methods. Recent trade disputes have created precedents (tuna-dolphin and shrimp-turtle disputes), but there is a general lack of agreement, especially in the case of standards for non-product-related process and production methods (i.e., production attributes that cannot be verified in the product itself). At the same time, many national regulations covering GM foods are based on production process; for example, they do not apply to any product produced with conventional agriculture methods, even if this product is identical to a GM product. In other words, herbicide-resistant crops with the exact same property and characteristics as certain GM products but obtained through conventional breeding methods would not be subject to approval and marketing regulations in many countries. Moreover, a few countries (the EU, Brazil, and China) require labeling of GM ingredients even in highly processed products where there is no available precise method to quantify transgenic DNA or proteins synthesized by novel DNA. This raises the issue of regulation enforcement: if all final products are virtually unidentifiable, it is impossible to ensure that they were produced with GM or non-GM ingredients.

Second, the SPS Agreement bases safety standards on a scientific assessment of existing risks, which goes against the strict application precautionary principle supported by the EU (based on the presence of unknown risks). The SPS Agreement has two main objectives: (1) to recognize the right of nations to set up their own domestic regulations with respect to health, and (2) to ensure that these measures are not unnecessary barriers to trade. In particular, WTO members are not allowed to ban imports of products they consider risky for an extended period unless they are able to scientifically demonstrate the existence of significant risk or to prove that they are conducting a significant effort in scientific research to evaluate these risks. In other words, the SPS Agreement allows countries to use precautionary measures but only during a provisional period, and provided they show effort of evaluating the risk of the products. In the case of the Hormone-Beef WTO dispute, which was raised by the United States against a ban of beef by the EU based on unknown risk associated with the consumption of beef raised with growth hormones, the WTO settlement body ruled against the EU because the EU was unable to provide scientific evidence of the presence of risk to human health in a sufficiently timely manner.

Third, there is no clear rule for or against mandatory labeling but rather open rules under the TBT Agreement. The TBT Agreement includes two main clauses relevant to the case of mandatory labeling of GM foods. First, Article 2.1 restates the main principles of the General Agreement on Tariffs and Trade (GATT) agreement with regard to national preference treatment and most favored nation treatment. Imported products "shall be accorded treatment no less favorable than that accorded to like products of national origin and to like products originating in any other country." The main point of contention on this article relates to the definition of "like products," which could be based on end product differences (making GM food labeling a TBT illegal regulation only in some cases such as countries of Group 3) or on consumer preferences. Second, Article 2.2 of the TBT Agreement provides conditions under which a technical regulation is allowed for WTO members; it mainly requires two conditions: a broadly defined legitimate objective and the absence of any other less trade-distorting measures that could achieve the same objectives. For the case of labeling requirements, the interpretation would depend on the legitimacy of a specific labeling requirement, on its importance and visual effects to achieve the objective as compared with other measures (such as educational programs or voluntary labeling for the objective of information provision).

9.7.9 EU REGULATIONS

The EU regulatory approach is precautionary, process related, and includes mandatory labeling traceability requirements; it belongs to category 1. Requirements include food and feed crops, unprocessed or processed. Only non-food GM products (unseeded), such as textile or other industrial products, are not subject to any requirement. The EU regulatory system for GM foods has become increasingly more stringent. In 1990, the European Council adopted Directive 90/220 on the deliberate release of GMOs into the environment. The directive regulated approval of GM crops for field trials and cultivation, and it also governed the approval of GM food. This first regulation did not define any specific approval procedures or labeling regulations. In 1997, the EU Parliament and the EU Council adopted Regulation 258/97, entitled the Novel Foods Regulation. This regulation applied to new food products including GM foods, and it defined approval procedures requiring proof that any GM food is safe for human consumption. Later, the EU Commission and the EU Council published Regulations 1813/97 and 1139/98, which required the labeling of food products containing approved GM soybeans and GM corn. These regulations were augmented by Regulation 49/2000, introducing mandatory labeling of GM food and GM ingredients at the 1 percent level, and Regulation 50/2000, extending the labeling requirements to food ingredients containing GM additives and flavorings.

The EU's most recent laws on GM food authorization (Regulation 1829/2003 and Regulation 1830/2003) took effect April 18, 2004. These regulations established procedures for evaluating potential risks from GM food and laid down rules on labeling of GM food and feed. Approvals are now granted for a period of 10 years, renewable. There is a 0 percent threshold for unapproved GM crops. Labeling is extended to animal feed, food sold by caterers, and food derived from GM ingredients even if the end product has no significant traces of transgenic DNA or proteins. The threshold for labeling is 0.9 percent. One major addition is the traceability requirements for GM and non-GM food: any food potentially containing GM material has to be tracked all the way from the farm to the consumer. This requires food companies to keep track of all shipments and to conduct DNA or protein tests at different stages. There is no labeling requirement for products such as meat, milk, or eggs produced from animals fed with GM feed.

9.7.10 JAPAN REGULATIONS

Japan's regulations include mandatory safety assessment and mandatory labeling based on differences in products and with a number of exemptions. Labeling is based on the end products, which means that highly processed products are exempt from labeling. Japan can be considered in Group 3 of Table 9.2. In 2000, Japan introduced regulations defining the authorization procedure. The Ministry of Health, Labor, and Welfare (MHLW) is in charge of the approval procedure for GM food. All GM food, GM processing aids, and GM food additives are subject to premarketing safety assessment. The safety assessment includes information regarding the host, the vector, the inserted gene, the recombinants, and the toxicity levels. If the application to MHLW is complete, it is then submitted to the Expert Panel of the Biotechnology Subcommittee within the Food Sanitation Committee. The Panel reviews and makes recommendations to the Biotechnology Subcommittee, which then passes its judgment on to the Food Sanitation Committee. This committee makes a recommendation to MHLW's minister, and if approved the new variety is announced in the *Japanese Gazette*. It usually takes about one year to go through the regulatory process.

The MHLW enforces standards under the Food Sanitation Law (FSL), and it samples and tests imported foodstuffs at ports of entry. The testing focuses on GM foods approved abroad but not in Japan. There is a 0 percent tolerance for unapproved GM material. After the StarLink corn food scare, Japan increased the frequency of food safety inspections on corn from 5 to 50 percent of all cargoes.

The Ministry of Agriculture, Fisheries, and Forestry (MAFF) is responsible for environmental safety approval, feed safety assessment, and biotech labeling rules. The MAFF's environmental assessment is voluntary, but all companies comply. The MAFF's feed safety assessment is mandatory from April 1, 2003. All applications for feed approval are reviewed by the Feed Division of MAFF and then sent to the Expert Committee of the Agricultural Materials Council. There is a 1 percent tolerance level for the unintentional presence of GM feed that has been approved in other countries, under the condition that the exporting country's safety assessments are deemed equivalent to Japan's.

Japan's mandatory labeling scheme was introduced April 1, 2001 under the Law on Standardization and Proper Labeling of Agricultural and Forestry Products, which was introduced into the Japanese Agricultural Standards (JAS). Labeling is required for all GM food if DNA/protein can be detected in the finished food products and if the GM ingredient is one of the top three ingredients and accounts for more than 5 percent of the total weight. This 5 percent tolerance level is informal but currently applied. The MAFF list of products subject to mandatory labeling included 30 foods in 2003. Importantly, there are no labeling requirements for soy oil or corn oil, except if the oil has special properties (such as high oleic soy oil). The labeling regulations are enforced jointly by MAFF and MHLW under the JAS and the FSL, respectively. In addition to the mandatory GM labeling requirements, there is a voluntary labeling option for non-GM, subject to IP procedures.

Overall, the Japanese policy can be described as pragmatic, in the sense that it requires the labeling of GM food, but the regulations do not cover all products and the tolerance levels are higher than in other countries. Food processors and retailers in Japan have typically avoided products with GM labels. As in the EU, most GM products are used for animal feed, but unlike in the EU, many highly processed products derived from GM ingredients (e.g., soy oil) are sold without labels.

9.8   INTERNATIONAL FOOD AND AGRIBUSINESS MANAGEMENT ASSOCIATION (IAMA)

IAMA Business Office
333 Blocker Building
2124 TAMU
College Station, TX 77843-2124
Phone: 979-845-2118
Fax: 979-862-1487
E-mail: iama@tamu.edu

Available at http://www.ifama.org.

IAMA was formed in 1990 to encourage strategic assessment across the entire food chain. Today, IAMA serves as a valuable international networking organization and acts as a facilitating conduit between the agribusiness industry (e.g., farmers, elevators, and processors), researchers, educators, government, consumer groups, and NGOs. IAMA has more than 700 members in more than 50 countries.[*]

IAMA provides high-quality, value-added products and services to meet the needs of its members and addresses the many challenges and opportunities facing food chain participants through leadership and innovation. IAMA's members are stakeholders in the success of the organization through their involvement in volunteer networks and program activities.

IAMA advertises itself as a worldwide leadership forum, which brings together top food and agribusiness executives, academics, policy makers, and other concerned stakeholders to stimulate strategic thinking across the food chain. Focus area task groups are an integral part of IAMA's structure. Their responsibilities include identifying emerging issues, evaluating strategies and alternatives for managers, organizing education and knowledge transfer programs, etc. IAMA is dedicated to an efficient food system that is sensitive to the needs of consumers, is safe and environmentally responsive, and provides a high level of business integrity.[†]

Benefits of IAMA for:

Industry

- Unique opportunity to network with the world's foremost business, academic, government, and consumer representatives in an environment conducive to thoughtful and open exchange
- Exchange views, develop strategies, and evaluate the impact of changes taking place throughout the integrated food chain
- Opportunity to establish priorities in the development and direction of the global food system
- Interactive forum to evaluate the impact of modern technology and life sciences on business strategies in the food chain

Academia

- Access to the latest thinking on business issues and management strategies as articulated by the world's leading food industry executives
- Interaction and communication with academic, industry, and government colleagues in food and agribusiness programs throughout the world

---

[*] IAMA is incorporated as an international nonprofit educational organization; IAMA is financed by member dues and corporate sponsorships. The association is administered by an Executive Director, who reports to a multinational Board of Directors, which is representative of the membership. The IAMA Business Office is located in the Department of Agricultural Economics at Texas A&M University and is managed by a Business Manager.

[†] Program planning, development, and implementation within IAMA are accomplished through five Task Groups. These groups provide planning and development for the annual Forum and other special projects, set research and educational program priorities, and write articles that are shared with the IAMA membership through the *Chain Letter*. Membership in the Task Groups is voluntary, and communication is accomplished through e-mail, the IAMA website, and meetings at the annual Forum.

- Opportunity to influence the development and direction of the global food system through participation in conferences, task groups, executive development programs, and professional training programs
- Opportunity to publish articles in the *International Food and Agribusiness Management Review* (*IFAMR*), a premier publication outlet for food and agribusiness research, and the *Chain Letter*, IAMA's quarterly newsletter

Government

- Access to a neutral platform for discussion with industry and academic representatives
- Opportunity to test ideas and policies with industry and research experts and to obtain fresh ideas and information from the private sector

NGOs and Consumer Groups

- Opportunity to interact with academic, business, and government leaders and discuss important food and agribusiness issues
- Access to the most authoritative information on food quality, food production, and manufacturing practices, and the food industry's approach to connecting product values with social values
- Opportunity to interact with researchers on consumer and social studies related to the global food chain

Students

- Opportunity to interact with agribusiness executives, government officials, and leading academics
- Access to travel assistance (awarded on a competitive basis) to attend the World Food and Agribusiness Forum, Symposium, and Case Conference
- Source of relevant and timely subjects for research
- Opportunity to network with potential employers

9.8.1 ANNUAL MEMBERSHIP FEES

Advantages of corporate membership include:

1. Multiple individuals from an organization may take advantage of IAMA products and services
2. Invitation to additional networking events throughout the year
3. Opportunity to sponsor certain IAMA events

Industry Membership (Based on Annual Revenue)

President's Club (includes 10 complimentary individual memberships)........ $10,000

Corporate Large (greater than $500 million; includes five
complimentary individual memberships) $5000

Corporate Medium ($50 to $500 million; includes two
complimentary individual memberships) $2000
Corporate Small (less than $50 million; includes one
complimentary individual membership)..$500

University/NGO/Agency Membership
Institutional membership (includes one complimentary
individual membership) $500

Individual Professional Membership
1-, 2-, 3-Year..$125, $235, $350
Student membership (full-time students only)..$60

9.9 FOOD STANDARDS AGENCY (UK)

Food Standards Agency
Aviation House
125 Kingsway
London WC2B 6NH
United Kingdom
Traceability, Bill Drennan: bill.drennan@foodstandards.gsi.gov.uk
Labeling, Country of Origin; Derek Hampson: labelling@foodstandards.gsi.gov.uk
Food Authenticity (Research), Ruth Hodgson: ruth.hodgson@foodstandards.gsi.gov.uk
Genetic Modification, Labeling Legislation: gm.labelling@foodstandards.gsi.gov.uk
Novel Foods, Michelle Young: michelle.young@foodstandards.gsi.gov.uk
Organic Food, Richard Wood: richard.wood@foodstandards.gsi.gov.uk

Available at http://www.food.gov.uk and http://www.foodstandards.gov.uk.

The Food Standards Agency (FSA), under the Food Standards Act (1999), provides
advice and information to the UK public and government on food safety from farm
to fork, as well as nutrition and diet, to protect consumer interests in relation to food
safety and standards. It also protects consumers through effective food enforcement
and monitoring. Although the FSA is a government agency, it works at "arm's length"
from government because it does not report to a specific minister, and is free to publish any advice it issues. FSA is accountable to Parliament through Health Ministers
and to the devolved administrations in Scotland, Wales, and Northern Ireland for its
activities within their areas.

9.9.1 FSA STRATEGIC AIMS

The Agency's first strategic plan covered the years 2001 to 2006. In that time its
aims were to:

- Reduce food-borne illness by 20%
- Help people to eat more healthily
- Promote honest and informative labeling to help consumers

- Promote best practice within the food industry
- Improve the enforcement of food law
- Earn people's trust by what FSA does and how FSA does it

Their strategic plan 2005 to 2010 has as its key aims:

- To continue to reduce food-borne illness
- To reduce further the risks to consumers from chemical contamination, including radiological contamination of food
- To make it easier for all consumers to choose a healthy diet, and thereby improve quality of life by reducing diet-related disease to enable consumers to make informed choices

9.9.2 INTERNATIONAL RELATIONS

With the diverse range of foods from around the globe available to people in the UK and with free trade and markets within the EU, the FSA aims to ensure that imported foods meet the required UK standards to protect the safety and interests of the consumer. The most significant groups in which other countries participate and the FSA has a varied participation are:

- Joint FAO/WHO Codex Alimentarius Commission (Codex Alimentarius)
- World Health Organization (WHO)
- Food and Agriculture Organization of the United Nations (FAO)
- World Trade Organization (WTO)
- Office International des Epizooties/World Organization for Animal Health (OIE)

In particular, the FSA negotiates on behalf of the UK Government in the joint FAO/WHO body, Codex, which was created to develop food standards, guidelines, and related texts such as codes of practice. By active involvement in meetings and contributing to the EU's input to Codex, the FSA aims to influence the standards set for food traded globally and for better consumer involvement in the development of standards. The FSA also has links with food authorities around the world, including those in the United States, Canada, Australia, and New Zealand.

FSA focus areas:

- Nutrition
- Safety and hygiene
- BSE
- Labeling
- GM and novel foods
- Consultations
- Food industries
- Enforcement
- Science and research

9.9.3 FSA—GM Food and Feed, and Traceability and Labeling of GMOs

New rules concerning GMOs became legally binding across all EU Member States in 2004, one covering Traceability and Labeling of GMOs (EC No. 1830/2003) and the other, the GM Food and Feed Regulation (EC No. 1829/2003), dealing with authorization procedures and labeling issues. Under the food and feed regulation, labeling is required for all food and feed products derived from GM sources, regardless of the presence of detectable novel genetic material in the final product and regardless of the quantity of intentionally used GM ingredient present.

The European Community (EC) regulation concerning animal feeding stuffs, which includes pet food and feed for horses, farmed fish, and in limited cases wild animals, is harmonized throughout the EU and based on measures negotiated in Brussels by the Member States.

The regulation concerns the integrity of the feed chain and is primarily intended to safeguard animal and human health. Much of it concerns labeling and marketing to ensure both traceability throughout the feed chain and the provision of accurate information to purchasers and enforcement authorities.*

The GM Food and Feed Regulation includes two thresholds: a 0.9% threshold will apply for the accidental presence of approved GMOs in a non-GM source and a 0.5% threshold for those which have not yet been approved in the EU, but which have received a favorable assessment from an EC scientific committee. There is zero tolerance for any GM variety that is not approved and does not have a favorable assessment from an EC committee. Any GMOs that do not fall within the above categories cannot be imported into the EU. The intentional use of GM ingredients at any level will require a corresponding label.

The Traceability and Labeling of GMOs Regulation creates a regime for tracing and identifying GMOs and food and feed products derived from GMOs at all stages of their placing on the market. In addition, it will enable products to be withdrawn from the market if any unexpected adverse effects were to arise. The regulation requires business operators when using or handling GM products to transmit and retain information at each stage of the placing on the market. For example, where production starts with a GM crop, the company selling the crop for feed production would have to inform any purchaser that it is genetically modified. Information must be retained for five years.

These rules are applicable in the EU or on entry to the EU, and it is the responsibility of food manufacturers to ensure that any foods or food ingredients imported into the UK that are produced from GM crops are from approved varieties. New minimum traceability requirements, by virtue of the General Food Law Regulation 178/2002, were applied for the first time to all food and feed businesses since January 1, 2005. These do not, however, require "internal traceability," that is, the linking up of all inputs to outputs.

---

* "Quality" issues, such as the proportions of particular ingredients to be used in a feed, or their source, and the nutritional content (or "profile") of feeds are outside the scope of the legislation. These matters are generally covered by industry feed assurance schemes and other codes of practice, which have no statutory basis.

9.9.4   FSA—Update and Summary on Traceability in the Food Chain, October 2004

- New minimum traceability requirements—by virtue of the General Food Law Regulation 178/2002—have applied to all food and feed businesses from January 1, 2005.
- These *new legal requirements do not require internal traceability*, that is, a system that allows linkages to be made between the sale of individual products and the source of materials used to produce that product. Therefore, guidelines have been drawn up to cover this area. There are clear needs and benefits to be gained from such systems, specifically:
  - Improved consumer protection through better targeted and more rapid recalls and/or withdrawals
  - Greater efficiency within businesses, with more information to assist in process control and management
  - Provision of reliable information to consumers to support authenticity claims about products
  - Deterrence of fraud
  - Increased consumer confidence

- In recognition of the benefits that **internal traceability** systems can bring to consumers and industry, FSA has been developing **Traceability Guidelines** in conjunction with key stakeholders. These Guidelines are aimed at encouraging greater adoption of such systems. The Guidelines recognize, however, that the adoption of an internal traceability system remains a business decision.
- The Guidelines have already been subject to public consultation but were subsequently subject to certain significant revisions, in particular they now cover whole chain traceability and animal feed because these are integral parts of a "farm to fork" approach.

9.9.4.1   Introduction to Traceability (From FSA Traceability Guidelines—Annex A)

In practice, traceability systems often usefully include information about what has happened to the food or feed (its processing history), as well as from where it came and to whom it was sent. For example, specifications of ingredients and records of storage and processes applied to meet safety, quality, and legal requirements. This should include a link to records associated with implementing food safety management based on HACCP principles.

Since January 1, 2005, all food and feed businesses, regardless of size, have been required to implement basic traceability systems. The FSA wants to assist businesses to effectively meet these requirements but would also like to help businesses that want to implement more comprehensive systems. This is because FSA believes that there are significant benefits for businesses and consumers from the introduction of internal traceability systems. Again, however, the adoption of an internal traceability system remains a business decision.

9.9.4.2 Legal Requirements*

All food and feed businesses within the EU are required to:

- Identify the suppliers of food, feed, food-producing animals, and ingredients to their business
- Identify the businesses to which products have been supplied
- Maintain appropriate records and ensure that such records are to competent authorities

In addition, there is new EU Food Hygiene Legislation as of January 2006 that requires food businesses to be registered. Other traceability requirements of this legislation include:

- All food business operators, other than primary producers, to put in place food safety management procedures based on HACCP principles, including documentation procedures
- Primary producers to keep records proportionate to the size and nature of their business
- Food chain information relevant to food safety to accompany animals to the slaughterhouse
- Feed businesses, although not farmers, to follow the principles of HACCP, including documentation procedures
- Businesses to only source and use feed from establishments that are approved or registered
- Records to be kept showing the sources of raw materials and the customers of finished goods

9.9.4.3 FSA—Labeling Information

There are also sector-specific measures in place, such as the labeling of beef, fish, GMOs, and lot marking. Requirements for GMOs, for example, include indicating whether a product contains or consists of GMOs and providing a unique identifier with sale or supply. As legislation develops, sectoral measures may well increase.

The why of IPT: in addition to meeting regulatory demands, traceability systems have several key roles within business:

---

* In this context food means any substance or product intended to be or reasonably expected to be ingested by humans, whether as an unprocessed material or processed product. Therefore, food includes water and other drinks. It does not include medicinal or cosmetic products. In this context feed means any substance or product, including additives, whether processed, partially processed, or unprocessed, intended to be used for oral feeding to animals. Food and feed businesses include any undertaking, whether for profit or not and whether public or private, carrying out any of the activities related to any stage of the production, processing, and distribution of food or feed.

- To provide information within businesses to assist in process control and management, e.g., stock control, efficiency of material usage, and quality control
- To assist when problems arise. Traceability systems are important to support effective withdrawal or recall of products. They can also allow detection of the cause of a problem so that targeted action may be taken to prevent recurrence
- To help support claims about product and provide information to consumers. Traceability systems are important to authenticate marketing claims that cannot be supported by analysis, e.g., relating to origin or assurance status
- To help, if necessary, demonstrate legal compliance
- To prove organic standards compliance
- To provide information to customers with allergies
- To control processes, as well as protect name and brand, whether these are food products going to the supermarkets or byproducts going to the feed industry
- To provide assurance of the quality and composition of ingredients back to source if necessary

9.9.4.4 How a Traceability System Works

Traceability systems are in essence joined-up recordkeeping systems that bring together information collected at key stages in the production and supply process. The more stages at which information is gathered, the fuller will be the overall picture achieved. Key stages are:

- Deliveries from suppliers into the business
- Each of the steps in the processing or manufacturing of the ingredients into new intermediates and products
- Deliveries out of the business

Linking together this information can enable the path of a particular ingredient or unit of product to be established. Therefore, the accuracy of the records of ingredient usage, production, and dispatch are vital for achieving robust traceability. This is particularly important when a supply chain comprises a number of different businesses and when overall traceability is not specified by a single business.

It is important to remember that traceability systems need not be complicated or complex. The best traceability system in any business is one that fits into the normal working practice of the business and enables the right information to be collated and accessed quickly and easily. Thinking through a traceability system carefully can enable the most value from the information collected within the business.

9.9.4.5 Identification

Traceability relies on the clear identification of ingredients, intermediates, and products. Within a business, this identification can relate to production batches or lots, which may consist of a few kilos or many tons. However, in all cases their identification should provide a clear link to their production history. Consignments traded between businesses should be uniquely identified. There are a variety of identification systems available, from hand-written labels and accompanying documents to barcodes and radio frequency tags. No one identification system is right in all circumstances. Different types of identification might be used at different points in the same traceability system. However, there is a benefit in using standardized identifiers, such as European Article Numbering product barcodes, for labeling products traded between businesses.

9.9.4.6 Plus Information

In many traceability systems information about the product is recorded on data sheets that accompany each batch through all the stages of the production process. Increasingly these systems are being replaced by computer recording of data; in some cases the amount of data collected by a traceability system cannot be handled on paper. IT-enabled systems can provide automatic identification and data collection using equipment such as label printers, inkjet coding, laser coding, barcode readers, and radio frequency tags. These can bring increased accuracy and other operating efficiencies.

9.9.4.7 Key Steps in the Manufacturing Process

When goods are received: at this point records form a critical and complete traceability link in the food chain. Key records include:

- *From whom:* Name and address of supplier and/or transporter
- *When:* Keeping a note of the date and time on which goods were received can be important to help trace the path of goods through the food chain
- *What exactly is received:* Record the identity of food/feed supplied, the quantity, and any other information about the goods, entering them into the recording system
- *What happened to the goods received:* Added to Store A, mixed with Delivery B, etc. It is advantageous to consolidate and combine intake information with intake quality assurance records

Factors that may need to be addressed:

- New deliveries used to top up or top off a single store, e.g., a tank of oil or silo of flour
- Deliveries or collections when no one is on site
- Difficulties in getting the right information or poor information from suppliers; just because information is provided does not necessarily guarantee that it is correct
- Limitations on the information that can be obtained when basic raw materials are used. These might be produced by continuous extraction or produced and handled in very large batches (>1,000 tons)

9.9.4.8   Missing Link inside a Food or Feed Business

Can links between the products received and the goods or finished products sent out be made? The precision of a traceability system inside a food business is a business decision that requires careful thought. It depends on the balance between the difficulty and cost involved in being precise, i.e., operating with small unit sizes, and the commercial risks involved in being imprecise. The size of the unit chosen will affect the size of any withdrawal required; the larger the unit size, the more production will need to be withdrawn. The business determines the appropriate size of the unit, which may, for example, cover a single production unit or a period in a continuous process.

9.10   INTERNATIONAL SEED FEDERATION (ISF)

ISF Secretariat
Chemin du Reposoir 7
1260 Nyon, Switzerland
Phone: +41-22-365-44-20
Fax: +41-22-365-44-21

Available at http://www.worldseed.org.

The International Seed Federation (ISF) is a non-governmental, nonprofit organization representing the seed industry. With members spread over 70 developed and developing countries on all continents, ISF represents the mainstream of the world seed trade and plant breeders' community and serves as an international forum where issues of interest to the world seed industry are discussed.*

9.10.1   Mission

> *Represent the interests of their members at an international level:*

* Improve relationships between members
* Develop and facilitate the free movement of seed within the framework of fair and reasonable regulations, while serving the interests of farmers, growers, industry, and consumers
* Increase recognition of the importance and value of members' major contributions to world food security, genetic diversity, and sustainable agriculture, in particular through the development, production, and use of high-quality seed and modern technology

---

* ISF membership is of four kinds: *Ordinary* (national associations representing seed companies and enterprises within their respective countries), *Associate* (seed companies or enterprises), *Affiliate* (service providers to the seed industry) and *Tree & Shrub Seed Group.* It is also possible to have an *Observer* status within ISF. As a matter of policy, ISF encourages the formation of national seed trade associations and their application for ordinary membership in ISF.

- Promote the establishment and protection of intellectual property rights for seeds, plant varieties, and associated technologies, which follow from research investments in plant breeding, plant biotechnology, seed technology, and related subjects
- Facilitate the marketing of planting seeds and other reproductive materials by publishing rules for the trading of seed in international markets and for the licensing of technology
- Provide for the settlement of disputes through mediation, conciliation, and/or arbitration
- Encourage and support the development of national and regional seed associations
- Encourage and support the education and training of seedsmen throughout the world

ISF represents the seed and plant breeding industries at:

- UPOV (International Union for the Protection of New Varieties of Plants)
- OECD (Organization for Economic Cooperation and Development)
- ISTA (International Seed Testing Association)
- FAO (Food and Agriculture Organization of the UN)
- CBD (Convention on Biological Diversity)

ISF maintains regular official contacts with these bodies to promote the viewpoint and defend the general interests of its members, notably in improving the conditions of international seed trade and strengthening intellectual property rights worldwide.

ISF trade rules:

- Trade rules that clarify and standardize the contractual relations between buyers and sellers at the international level
- Procedure rules for dispute settlement for the trade in seeds for sowing purposes and for management of intellectual property

See Appendix I International Seed Federation for a listing of Network of Seed-Trade and Plant Breeders Associations members and e-mail addresses.

REFERENCES

Bender, Karen. 2003. "Product Differentiation and Identity Preservation: Implications for Market Development in U.S. Corn and Soybeans." Presented at the Symposium Product Differentiation and Market Segmentation in Grains and Oilseeds: Implications for Industry in Transition. Washington, DC.

Bullock, David, Marion Desquilbet, and Elisavet Nitsi. 2000. "The Economics of Non-GMO Segregation and Identity Preservation." Paper presented at the AAEA annual meetings, Tampa, FL.

Gruère, Guillaume P. 2006. "An Analysis of Trade Related International Regulations of Genetically Modified Food and Their Effects on Developing Countries." International Food Policy Research Institute (IFPRI) EPT Discussion Paper 147, Environment and Production Technology Division.

Kalaitzandonakes, Nicholas, Richard Maltsbarger, and James Barnes. 2001. "Global Identity Preservation Costs in Agricultural Supply Chains." *Canadian Journal of Agricultural Economics* 49:605–15.

Maltsbarger, Richard, and Nicholas Kalaitzandonakes. 2000. "Study Reveals Hidden Costs in IP Supply Chain." Economics & Management of Agrobiotechnology Center (EMAC), University of Missouri-Columbia.

Parcell, Joe. 2001. "An Initial Look at the Tokyo Grain Exchange Non-GMO Soybean Contract." *Journal of Agribusiness* 19:85–92.

Wagner, Gary L., and Eliot Glassheim. 2003. "Traceability of Agricultural Products." Published by Northern Great Plains Inc. Available at: http://www.ngplains.org/documents/traceability_report.pdf. Accessed January 24, 2007.

10 IPT Software Providers

10.1 INTRODUCTION

This chapter is not intended to be a comprehensive listing of providers, services, and costs for services. The information is to serve as another building block component toward creating a complete identity-preserved and traceable food chain system.

Many of these providers serve a variety of businesses, not only those found in the food chain. Most, if not all of them, work with established quality systems or programs. What follows are company/organizational statements from their websites, which naturally reflect their views.

This chapter is divided into three sections. The first section comprises smaller, yet agriculturally directed software firms that include IdentityPreserved.Com, Linnet, MapShots, PathTracer, Vertical Software, and AgVision. The second section consists of more formal providers that offer systems that work in a variety of disciplines and industries such as automotive or banking; these providers consist of Pacifica Research, Software America Enterprise Resource Planning (ERP), and CSB—Systems Enterprise. The last section highlights an electronic software system that is used nearly universally, namely, the GS1 and European Article Numbering Uniform Code Council (EAN.UCC) code systems.

The last section on GS1 represents a unique entity. Years ago, under differently named organizations, this entity began to establish an innovation of codes that were incorporated into the barcode system. The numbering system they established slowly evolved through acceptance as a quasistandard for merchandising and inventory control. With time, changes occurred that expanded their original scope to include its use as an identity preservation and traceability (IPT) tool. In this chapter they are viewed as a software provider, whereas GS1 is also intimately part of many organizations' IPT programs under larger global standards such as logistics control.

A key premise that ties software to the goals or objectives of any IPT system is the ability of technology's information and communication to interact in such a way to bring about positive, efficient results. For both the speed of tracing back to a source and communicating to both players in the food chain and the public, software compatible with the needs and goals of IPT systems is essential. The work of Pinto, Castro, and Vicente (2006) highlights that, "Both food industry and authorities need to be able to trace back and to authenticate food products and raw materials used for food production to comply with legislation and to meet the food safety and food quality requirements."[*]

During the past decade new, more focused, food safety and agricultural management policies have been implemented by governments and the agriculture and

---

[*] Reprinted excerpts from Pinto, D.B., I. Castro, and A. A. Vicente, *Food Research International* 39, 772–81, 2006. With permission.

food industries where food safety and quality assurance has become paramount. The public is demanding an unbroken accountability of not only food and feed but also accountability of all steps in the food production chain, namely, the supply of raw materials, food manufacturing, packaging, agricultural chemicals such as herbicides and pesticides, as well as nonfood agriculture products used in manufacturing and pharmaceuticals. For IPT software to accomplish its traceability and provide information to all participants, an interactive, transparent, yet flexible, efficient, and profitable system must be used. Software systems have evolved over time, just as the desktop personal computer (PC) has evolved.

Traditionally, the food chain has relied on a simple paper trail to perform rudimentary traceability. For those interested in identity preservation (IP), the early forms of IP were emphasized by brand name, region of purchase, or advertised trait, or quality of interest served as the manner to denote "identity preserved." The advent of several governments' regulations requiring a "one step back/one step forward" form of accountability led many software producers to cater to the growing IPT needs of the agri-food industry.

IPT systems can work well based on pen and paper versions. However, they are time and resource consuming, which makes them difficult to implement in small- and medium-sized companies, where the resources are scarce, and larger organizations, where established cultural processes are entrenched. Some larger organizations have instituted computerized logistic systems; however, most if not all of these systems can only accurately account for what is arriving on the loading dock by barcodes and bills of lading, and what leaves the dock by barcode or by nutritional food label. Nearly all software providers tie resources and supplies that enter or leave a facility, yet few if any can tie incoming resources from loading dock, through mixing, processing, packaging, and then to the outgoing pallet destined to the next facility.

For example, in the baking industry, the traceability process is very complex because of the multiplicity of raw materials used and the large number of different products that a single batch can produce. Moreover, there are several finishing raw materials used in the product that usually are not controlled or even traced back to the supplier. In light of these market realities, the development of specialized computer software applications, using user-friendly multiple interfaces, have been specially designed and incorporated for participants in the food chain.

10.2   IDENTITYPRESERVED.COM

IdentityPreserved.Com
21024 421st Avenue
Iroquois, SD 57353
Phone: 605-546-2299; 800-661-4117 (toll-free)
Fax: 605-546-2503

Available at http://www.identitypreserved.com.

IdentityPreserved.com is a business unit of Agricultural Information Technologies (AIT) (founded in 1991), which provides tools, software, systems, and services that help identify valuable crop attributes, document important processes, and communicate with all layers of the processing chain. IdentityPreserved.Com products offer the ability to have a single system to track input through the production process instantly and seamlessly, which streamlines the IP process and is of value to producers, contractors, and processors; their premier product for this is called IP Track. The notion is that value is inherent in the grain itself but is often hidden because of commoditized production, handling, and business practices. IdentityPreserved. Com helps capture this value by providing information, products, and services that streamline and enable IP.

To increase profits, crop producers must operate and manage the opposite of how they have traditionally farmed. A focus on specialized crops produced for specific customers provides the best opportunity for long-term profitability for the producer and for the food production chain. To be successful, the identity of the specialized crops must be preserved. Specific, valuable traits must be measured and tracked for value to be delivered. IdentityPreserved.Com advertises that their IP Track seamlessly combines a standardized information communication framework with a convenient communications interface for all aspects of the IP process. The system allows data to be input by any part of the production chain and then immediately made accessible to all authorized partners from any location.

IP Track is an online application that centralizes IP information. It is the centerpiece of IdentityPreserved.Com, a flexible and powerful nucleus for IP coordination, integrating data from many sources and dispersing it to widespread users. IP Track customizes information based on a user's profile, easing verification and streamlining operations. Access is secure, with information visible only to authorized users. By connecting participants and coordinating the tasks they perform, IP Track helps implement IP processes and maximize IP value.

IP Track operates in union with IdentityPreserved.Com's other products, Postmark signposts, CropTouch data collection technology,* SeedTag database, TraitCheck tests, and GMO Check tests. All of these products, services, and software focus on the identity-preserved production process, where IP Track provides the "golden thread of traceability from farm field to consumer's plate" by providing a standard framework for crop producers and independent field auditors to report the status of identity-preserved production. IP Track allows contractors to create and post protocol requirements to customization identity-preserved contracts. During the growing season, participants then use IP Track to record crop progress and protocol compliance. In the same way, independent auditors use IP Track to record protocol fulfillment observations. IP Track also automatically records and maintains a "pedigree" for each identity-preserved product it tracks, including how a product is managed and when and where it gets transferred from one location to another. IP Track's Pedigree Report provides buyers with complete source and management information about the products they buy.

---

* CropTouch system includes handheld computers, voice recordings and even paper forms. IP Track also supports many languages.

IdentityPreserved.Com views the current U.S. crop production system as having two important categories of IP, the first being the non-genetically modified organism (GMO) market* and the second the value-added crop market.†

Business-minded producers and processors have considered the costs as they evaluate the role for IP in their operations. A 1999 University of Illinois study identified the contract requirements and the related costs incurred in complying with identity-preserved contracts. The study showed that cost items included special seed to added transportation costs: in corn, yellow food grade corn costs were $1.61 per bushel, whereas tofu soybeans had the highest costs at $3.02 per bushel.

The specialty corn and soybean crops that are most frequently produced under contract are high oil corn, white corn, waxy corn, and tofu soybeans. For these crops, two-thirds or greater of the production was contracted in both 1998 and 1999. Organic or pesticide-free soybeans was the least contracted specialty crop, which is surprising given the high premiums and detailed quality control typically involved in this market. However, the small sample size for organic soybeans may not fully reflect the extent of contracting for this particular specialty crop.

The contract specification most frequently specified, regardless of crop, was delivery location. More than 80 percent of all specialty crop contracts included a requirement on specific delivery locations. Quality testing was a contract specification in at least half of all contracts for all specialty crops except tofu soybeans. Similarly, specific delivery dates were common contract requirements, included in at least half of the contracts for all crops except yellow food-grade corn and organic soybeans.

Partial Price List

Postmark Signpost, *The IP Foundation*.................$7.45–21.95, depending on model
Pocket Track, *The Pocket IP Companion*....................................Contact for pricing
TraitCheck, *On-site IP Tool for GMO Detection* ....$290 for 100 pack grain
GMO Check and GMO Quick Check,
Lab-based IP Tools for Quantifying GMOs $200–260
SeedTag Database, *For IP Certainty* .....$99 single copy, contact for licensing options

---

* USDA surveyed growers to determine what percentage had planted GMO varieties. In the June 2000 crop report USDA reports that 25 percent of corn and 55 percent of soybeans were GMO. Producers have found that some markets will pay premiums to get grain that is certified GMO-free. Market intelligence in grain trading channels tells us that sales of non-GMO grains will double in 2000 compared with 1999. The key for producers seeking these premiums is to confirm the identity through screening tests and preserve that identity as the grain travels to customers.

† Although this market is smaller in volume today, it holds the larger potential for the future. Agricultural economists agree that in the long term all grain production will be specialized to meet particular customer requirements. Today it is hard to find data on the amount of crop production that is produced with identity preservation in mind. Many of the production arrangements are contracts between local processors and local growers so they are not centrally reported. There is no CBOT for trading identity-preserved crops. Because this trend is growing, many groups are working to measure these market segments. One recent study in Illinois found that 25 percent of all country elevators and grain terminals handled some identity-preserved grain. These elevator managers were asked to estimate what percentage of their volume would be identity preserved by 2005. Answers range as high as 40 percent of corn and 45 percent of soybeans. Another study of Illinois crop producers showed that 18 percent of corn and soybean producers grew corn or soybeans under a value-added contract in the 1998 crop year. Thus, regardless of study or data, the perceived trend is that more crops will be grown with identity preservation as a trait of interest.

IP Track, *The Complete Software System* ....Varies with services requested
IP Audit, *Third-Party IP Inspections* ....Varies with services requested
IP Labs, *Integrating IP Testing Services* ....Varies with services requested

10.3 LINNET: CROPLANDS—THE SYSTEM (CTS)

Linnet
259 Portage Avenue, Suite 700
Winnipeg, Manitoba R3B 2A9
Canada
Phone: 204-957-7566

Croplands—The System (CTS)
1600 444 St. Mary Ave.
Winnipeg, Manitoba R3C 3T1
Canada
Phone: 204-957-7566

Available at http://linnet.com/app/product_solutions/croplands and http://www.croplandthesystem.com.

Linnet has been working in the field of food supply chain management software since 1998. They realize traceability of food products throughout the agricultural supply chain has gained intense worldwide attention in recent years and impacted every part of the food industry from governments and consumers to food retailers and restaurants right through to food processors and growers. Linnet's Traceability/Food Safety and supply chain management software "Croplands" manages the entire "front end" of the supply chain, providing complete traceability from the growers' fields and storage locations right through to the fresh pack and/or processing facility and finally integrating with the organizations' factory process control systems/ERPs to achieve traceability from the end product on the store shelf or at the restaurant back to the fields from where the product originated.

Linnet's Croplands—The System (CTS) software enables effective and efficient operational management of the raw material supply chain from the production of raw material (e.g., grain/pulses/oilseed, fruit, vegetable, livestock, aquaculture, or nutraceuticals) through to its delivery at the processing facility or distribution center and all points between. The impact spans from the field/point of origin to the end consumer.

A scaleable, spatially enabled Geographic Information System (GIS) enterprise land management solution suite of software modules integrates field production management, inventory management, product procurement, all quality tests including end product testing, and contracting/settlements. In addition, CTS can be easily integrated with other business systems, including those from back office accounting and manufacturing vendors such as SAP, JDE, i2, Wonderware, and integrated with scales and testing equipment to create a total business solution.

CTS is a software solution designed through intensive research and in-field development with industry leaders that can be deployed in either a GIS-enabled or

GIS-disabled environment. Croplands is built around an enterprise data model and spatial data warehouse; it covers the full range of agri-business operations, facilitating activities such as raw material contracting and inventory quality management to product procurement and traceability.

Croplands provides a field to fork solution in:

- Production contracting and settlement
- Automated scale integration
- Inventory management and raw product
- Product procurement and shipment planning
- Agronomic/crop or livestock production
- Planning (crop production planning)
- In-season (crop production)
- Logistics and field operations
- Quality assurance and testing

10.4   MAPSHOTS, INC.

MapShots, Inc.
4610 Ansley Lane
Cumming, GA 30040
Phone: 678-513-6093
Fax: 770-781-9471
E-mail: info@mapshots.com

Available at http://www.mapshots.com/default.asp and http://www.mapshots.com.

MapShots was created in 1999 and has corporate clients such as John Deere, Pioneer Hi-Bred, and Southern States Cooperative that use their software. MapShots provides the agricultural industry with the EASi Suite brand of crop management software such as the Windows-based EASi brand of grower software for crop recordkeeping that includes EASi Crops Professional series, EASi Planner, and EASi Map, which provides GIS capabilities. EASi Suite Farm Edition is used by growers to maximize agronomic management information. EASi Suite Professional is used by both crop consultants and crop input retailers to provide crop planning and nutrient management planning services to their farm clients.

10.4.1   FIELD OPERATIONS DATA MODEL (FODM USED FOR IPT)

MapShots believes that automating crop production records is essential to advancing industry's ability to efficiently manage agronomic practices: identity-preserved markets, nutrient management plans, manure management plans, watershed compliance, more sophisticated chemicals, and biological engineering, to name a few. MapShots has developed a significant a Field Operations Data Model (FODM) that was designed specifically to handle the wide range of data that can be captured from normal agricultural practices.

10.4.2 DESCRIBING THE FIELD OPERATION DATA MODEL

Two challenges are associated with describing a field operation. The first is determining the manner to summarize the information: by field, by product, or by combination of products. The second challenge is describing the data that are being recorded about a field process. To meet this challenge FODM uses a sensor-based metaphor for recording operating parameters, and it includes a detailed equipment

---

TABLE 10.1
Linnet: Croplands—The System (CTS) Products

EASi Suite (used for IPT) is a crop recordkeeping system that focuses on identity preservation, nutrient management plans, watershed protection, and GMO's traceability that emphasizes crop recordkeeping to minimize on-farm environmental risk, meet regulatory requirements, and at the same time manage crop production for maximum profitability.

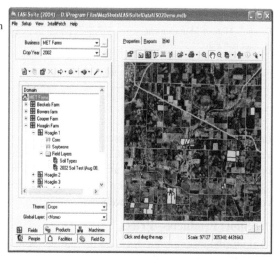

MapShots also offers more specific products such as **EASi Grain.** EASi Grain is a rather new software application specifically designed to help manage grain inventory. More powerful than a spreadsheet, EASi Grain allows:

- Tracking loads going into on-farm storage
- Tracking deliveries to elevators and processors
- Tracking in-transit grain
- Maintaining landlord inventory
- Recording bin cleanout events for IP documentation
- Performing landlord grain reconciliation

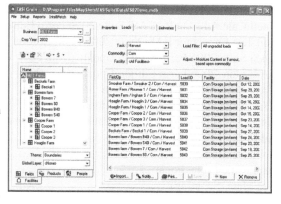

(Continued)

TABLE 10.1 *(Continued)*

IntelliCalc is also a new tool for merging multiple map layers, performing mathematical calculations, and generating new map layers. Useful in creating Nutrient Management Plans and preparing data for spatial analysis, IntelliCalc offers increased flexibility in the selection of source data and algorithms. Source data can be selected from external data such as shapefiles, internal layers such as soil type, or field operation data such as yield maps. IntelliCalc can even access external algorithms such as Purdue's Manure Management Planner software.

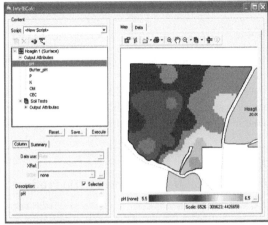

Source: Data adopted with permission from MapShots.

configuration metaphor for associating sensors with the pieces of equipment that are used in a field operation. Combining the metaphor with the sensor-based data collection metaphor produces an Operating Region (courtesy of VantagePoint). See Table 10.1 for more information regarding Linnet: CTS products. This information is derived directly from their website.

MapShots Products—Prices

EASi Grain* $495.00

EASi Suite Farm Edition

EASi Suite 2006 Farm Edition $995.00

IntelliCalc Add-On for EASi Suite 2006 Farm Edition $445.00

Soil Test Import for EASi Suite 2006 Farm Edition $200.00

EASi Suite 2006 Farm Edition Update (from 2005 version) $250.00

EASi Suite 2006 Farm Edition Update (from 2004 or earlier) $495.00

EASi Suite Professional Edition

EASi Suite 2006 Professional Edition $3,435.00

Annual Master Support Agreement for EASi Suite Professional..................$750.00

Additional Programs Available

Soil Test Manager $1,995.00

IntelliCalc Professional $995.00

Site Mate (Scouting)..$500.00

Site Mate (VRA)..$750.00

---

* The EASi software was developed by Charley Engelhardt who formed Engelhardt Agri-Services in 1982.

10.5 PATHTRACER

PathTracer
P.O. Box 1643
Denton, TX 76202
Phone: 940-498-9965; 888-398-3364 (toll-free)
Richard Ross, CEO
Phone: 785-218-7307
E-mail: richard.ross@pathtracer.net
Dan Brady, President, COO
Phone: 972-333-2444
E-mail: dan.brady@pathtracer.net

Available at http://www.pathtracer.net.

PathTracer is a patent-pending, internet-based, proprietary food safety software product that provides timely, accurate, and nearly effortlessness ingredient tracing. PathTracer offers themselves as software providers with solutions toward compliance with the law and to help meet company liability insurance requirements.* PathTracer advertises that they are the "only system" that provides the ability to capture the "missing middle part" of the equation, the specificity of linking the inbound ingredients, through processing by lot code including packaging, to a specific batch of the finished product.

PathTracer software system encompasses:

- Daily activity report by facility, what is unloaded and what is loaded out
- Links daily internal activities, data on blends, batches, and commingling
- Daily inventory, levels by product, i.e., moisture content, last fumigated, and last turned
- Inventory variances based on formula variations during production
- Improving purchasing decisions, right product to the right location at the right time
- Traces inventory by bin
- Links ingredient lot codes to specific feed batches
- Tracks trace elements
- Ties in specific lots of food contact packaging
- Speeds audit process from internal accounting aspects

PathTracer provides document proof for liability insurance carrier of compliance with the law. It is also used to demonstrate to customers that each product can

---

* Large companies must be compliant by December 9, 2005 with the Bioterrorism Act of 2005. Smaller companies, those with more than 11 employees, must be compliant by June 6, 2006. Records must be produced "as soon as possible, but no more than 24 hours" from time they were requested. Vertically integrated companies have one 24-hour period to produce records. Example: Country elevator transfers product to the terminal elevator that transfers product to the feed mill. If all three are under the same personhood (legal entity), the records from the first delivery, linked through all blends, commingles, transportation aspects, added ingredients, to the final feed batch, must be produced in one 24-hour period.

identify the source of all ingredients by lot code. PathTracer offers the ability to measure product usage, not just formula specifications, and can tie together accounting software and logistical recall software. PathTracer asserts that bulk food traceability cannot be achieved via accounting systems or formula/recipe stats but only by their PathTracer process.

Specifically, PathTracer offers applications for feed mills, feed lots, feed dealers, and farms engaged in feed manufacturing to bulk commodity elevators, brokers to aid with facility inventory, unloads, load outs, and blends. For bakeries and food processing/manufacturing PathTracer understands that these facilities face significant challenges, not with recall, but with back trace information and trace forward information, for which its system provides.

PathTracer helps coordinate:

- Immediate Previous Source (IPS), from whom ingredients came (bulk, bagged, liquid, or bulk)
- Who transported it, who delivered it (company, e-mail, address, rail or trailer number)
- What was received, as specific as possible, includes specific lot/date code of an ingredient, not just product description
- What bin it originally went into (bulk only)
- What subsequently occurred with each ingredient, blended, comingled, processed, etc.
- During production process, links each ingredient to specific batches including all food contact packaging
- What was shipped, as specific as possible, includes lot code information
- Who transported from facility, name of contractor that moves product from its present location (company, e-mail, address, rail or trailer number)
- Immediate Subsequent Recipient (ISR), who bought the product
- Records stored up to two years based on the shelf life of the ingredient or product

10.6   VERTICAL SOFTWARE, INC.

Vertical Software, Inc.
409 South Keller
Bartonville, IL 61607
Phone: 309-633-0700
Fax: 309-633-2328

Available at http://home.verticalsoftware.net.

Vertical Software began in 1981 and is in more than 30 states and Canada. The Vertical Point System is used for agri-business transactions, to make it easy to enter split invoices, to manage inventory by tracking product bookings, and to enter transactions from the scale directly into invoicing with no additional data entry. It is done by several products.

Vertical Software's primary IPT grain product is GrainTrac, which is used by river terminals, country elevators, grain processors, cereal makers, grain-trading houses, feedlots, and flourmills, from individual operations to companies with multiple locations. From contract, to delivery, to storage, through settlement, and to history, GrainTrac keeps track of grain data starting with ticket information, e.g., entered through ScaleTrac or GrainTrac's ticket entry.

Vertical Software offers GrainTrac Pass (Producer Accessible Secure System) for real-time secure software support.

Another system that Vertical Software offers includes BinTrac management system, which provides up-to-the-minute bin inventories from data received from GrainTrac, Grain Clerk, or ScaleTrac. The ScaleTrac system automatically enters scale ticket data into an accounting system by capturing weights electronically and calculating each ticket with computer accuracy. (ScaleTrac is National Type Evaluation Program [NTEP]-approved.) In addition to ScaleTrac there are other software tools that include ScalePoint, MixPoint, AgPoint, and TurningPoint, which integrates shipping scales with invoicing for products such as NH3, liquid nitrogen, water, and feeds.

Vertical Software also works closely with Archer Daniels Midland (ADM) and developed one of the grain industry's first electronic data interchange for contracts, farmer direct shipments, and elevator load-outs. One of these products is Vertical Software's SPEEDI product (by Electronic Data Interchange) in April 1995.

10.7 AgVISION/DMI COMPUTER TECHNOLOGIES

AgVision/DMI Computer Technologies
1601 North Ankeny Blvd.
Ankeny, IA 50023
Phone: 515-964-0708; 800-759-9492 (toll-free)
Fax: 515-964-0473
E-mail: dmi1@dmicomputer.com

Available at http://www.dmi-agvision.com.

AgVision portrays itself as an industry leader in designing powerful and intuitive software for agribusiness. Their customers, nearly 500 businesses nationwide, comprise grain elevators, seed processors, cooperatives, fertilizer retailers, feed stores, ethanol plants, and tree nut handlers/processors. AgVision software operates on stand-alone PCs and client-server configurations over both conventional, wireless, and Internet networks connecting single and multiple departments and locations.

10.7.1 AgVision Commodity Manager for Grain, Seed, and Tree Nuts

This system electronically tracks and manages handling, storage, and risk of commodities from the scale to the sale and shipment. This software also does advances, calculates discounts and storage, and handles multilocation inventories; automatically updates and provides client position and merchandising reports; and provides progress payment schedules to advance and deferred payments. Software can be used alone or integrated with the AgVision Financial Accounting System.

10.7.2   Scale Interface Software (NTEP-Certified)

AgVision's Scale Interface Software offers quick and efficient scale readings, creates on-line tickets, and transfers the information to selected AgVision Commodity Managers. Interfaces are available for a variety of scale models. The National Conference on Weights and Measures has issued a Certificate of Conformance for this interface.

Other interface systems and software suites include DICKEY-John's GAC2100 Moisture Tester Interface that automatically transfers moisture test data to the Commodity Manager; AgVision's Fertilizer Management Software that allows formulation, mix, bill, and keeps records of bulk fertilizer and ag-chemical transactions. (This Windows-based software tool is designed to be used with the AgVision Financials System.) AgVision's Financial Accounting System manages data from feed inventory and general ledger to degree days and deliveries/route scheduling.

As with many agriculturally focused software companies, AgVision has software that addresses Customer Sales History to record and track customers' accounts receivables and grain activities, Sales Commission Software to calculate sales force commissions, and Patron Equity Software to calculate and track agribusiness cooperatives' dividends. The software issues dividend checks and preferred stocks and includes a feature to enter, edit, and print all 1099 forms.

AgVision is a full-service computer and computer software company. In addition to software and hardware sales, it offers computer consulting, support, training, maintenance, and repair. It is an expert in designing networks, and it is a certified reseller of CISCO products.

10.8   CSB-SYSTEM

CSB-System AG – Headquarters
An Fürthenrode 9-15
D-52511 Geilenkirchen
Germany
Phone: +49-2451-625-0
E-mail: info@csb-system.com

CSB-System International, Inc. – Branch Office
2535 Camino del Rio South #350
San Diego, CA 92108
Phone: 619-640-0436; 800-852-9977 (toll-free)
Fax: 800-851-6299

Available at http://www.csb-system.com.

CSB-System AG employs nearly 260 personnel at its headquarters and 450 employees worldwide. CSB-System offers transparency of entire process chain spanning production, processing, and retail to meet organizational needs in accordance with statutory requirements, industry standards, and consumer demand. Streamlined and seamless capture of traceability data is efficient when all processes of material planning are fully integrated through CSB-System's industry-specific ERP systems specializing in innovative, complete information technology (IT) solutions for efficient management in batch and process-oriented industries, as well as the retail and logistics sector. This enables maintaining transparent documentation and proof requirements for Regulation (EC) No. 178/2002 and sales to Europe.

CSB-System provides clear proof of origin and safeguarded traceability for dairy, beverage, bakery and confectionery, and meat and fish segments of the food industry, and in accordance with all current international standards (including EC Regulation No. 178/2002, 1830/2003, Quality Assurance and Laboratory Information System [QLS/LIMS], EurepGAP, HACCP, ISO 9000, BRC, GLP, GMP, and GHP). Based on the cross-industry standard CSB has developed solutions that allow for flexible interchange of origin data between companies and organizations. With the help of this data interchange tool, client companies are ensured seamless farm-to-fork proof of origin for each batch that has entered the production process. The integrated laboratory information and management system QLS/LIMS extends the CSB-System to become a comprehensive ERP system.

For more than 25 years, CSB-System has partnered with companies in enterprise-wide business solutions consulting, as well as customer assistance and maintenance of soft and hardware in a one-stop solution.

The CSB-System encompasses all functions of a future-oriented ERP system. The enterprise-wide materials resource planning forms the basis for integrated information processing throughout the functional processes of resource management, procurement, inventory, production, sales, quality management paperless HACCP concept, and QLS/LIMS.*

10.9 AMERICANERP, LLC (ENTERPRISE RESOURCE PLANNING)

AmericanERP, LLC
6075 SW 124th
Beaverton, OR 97008
Phone: 503-924-4491
Fax: 503-924-4495

Available at http://www.americanerp.com/index.html, http://www.americanerp.com, http://www.marketwire.com/mw/release_html_b1?release_id=77402, and http://www.foodprocessing.com/vendors/products/2004/119.html.

---

* CSB also offers high-performance functions in the areas of Enterprise Resource Planning (ERP), Advanced Production Scheduling (APS), Automated Data Capture (ADC) and Mobile Data Capture (MDC), Computer Integrated Manufacturing (CIM) and Manufacturing Execution System (MES), integrated Customer Relationship Management (iCRM) and integrated Supply Chain Management (iSCM), or Management Information Systems (MIS) and Area Information Systems (AIS) for Management and Controlling round off the software portfolio of the one-stop system.

AmericanERP specializes in ERP automation for food plant strategies, bakery industry, baking industry, chemical manufacturers, pharmaceutical manufacturing, and cosmetic manufacturing. AmericanERP's food processing software clients include a wider range of beverage and food processors and include national brands in spice to sushi, as well as seafood processors to soup-processing plants. AmericanERP allows food/formula-based processors and manufacturers the technology to comply with the new, strict, and mandatory Food and Drug Administration product traceability requirements for all food- and formula-based manufacturers regardless of size.

AmericanERP, LLC is headquartered in Portland, Oregon. The company creates and markets intuitive manufacturing software designed specifically for formula-based (including chemical, cosmetic, and pharmaceutical), food, and other manufacturers.

AmericanERP software allows traceability for smaller processors, whereas larger retail chains have begun auditing their vendors to ensure they have in place HACCP and the corresponding "prerequisite" programs, including an effective recall program, or one up/one down product traceability. Lacking the technology for effective traceability has been a major stumbling block for small- to mid-size food processors, and AmericanERP offers a solution in the form of a new software package.

AmericanERP offers AFP (automated formula processing) Enterprise 2005 Edition and Automated Production Management (APM) as comprehensive ERP software systems that provide a "lot" search and traceability tool for small- to mid-size food processors and manufacturers, allowing them to compete and be accepted by the larger food chains. AmericanERP claims the system also significantly reduces processors' costs and increases profitability and can be used as a stand-alone ERP system or with any number of industry-standard accounting systems. *AFP software is:*

- Formula-driven with comprehensive Bill of Materials, AFP allows for any number of finished products from a single recipe/formula
- Inventory-trackable—provides the ability to monitor raw materials by lot number, location, and/or reorder points, providing an accurate view of inventory at any given time. For example, software solutions for food processors

AFP provides a comprehensive, integrated purchase order feature. Receipts automatically update inventory with assigned lot numbers for HACCP tracking. Complete purchase orders are then transferred to the accounting software. Purchase orders are created in the vendor section and then sent to the receiving area where there are received and assigned a location and expiration date.

AFP allows for detailed recipes/formulas and maintains an accurate inventory status and costing. AFP also allows for an infinite number of finished products from a single recipe/formula. Finished products can be customer specific for copyright formulas or produced and invoiced to several different customers. Product information includes complete costing information, with complete traceability for HACCP requirements. Batch templates allow repeat productions without the need to rekey product and customer information. A comprehensive Bill of Material is available showing yield loss and batch costing.

10.10 PACIFICA RESEARCH

Pacifica Software
202 E Street
Brawley, CA 92227
Phone: 760-344-1639; 800-536-5130 (toll-free)
Fax: 760-344-8952

Available at http://www.pacificaresearch.com and http://www.pacificaresearch.com/company.html.

Pacifica Research is a software publishing company, established in southern California, and offers a complete line of real-time accounting software in Windows for large and small business payroll, inventory control, accounts receivable, and accounts payable. Pacifica Research is a wholly owned subsidiary company of CShare Business Computers, which has manufactured innovative business management and accounting software since 1978.

Pacifica's Seed Inventory Control fully integrates seed inventory management within a financial system, with general ledger, payroll, payables, and receivables. Developed by seedsmen in 1978 and written for Windows, Pacifica Seed Inventory Software is a broad solution developed for the seed industry and addresses all the facets of producing, purchasing, selling, and costing seeds, and handles the unique challenges of seed companies.

Pacifica tracks sales and inventory by variety, lot, and sublot. It shows the quantities committed and available, the cost, price, treatment, current status, from where the seed came, and everywhere it has been sold or used. Pacifica software properly handles purchases, sales, and all adjustments in any unit of measure, including pounds, ounces, kilograms, grams, per seed, per M (thousand seeds), 10M or 100M, per acre or hectare, per bushel, each, or other units of measure, with automatic and transparent conversion to any other unit of measure.

Pacifica also offers Pesticide Use Management software, a system to supplement inventory management and invoicing, track pesticide and agrichemical applications, and print legal documents required by state and federal agencies and the Environmental Protection Agency.

10.11 GS1 AND EAN.UCC

GS1
Blue Tower
Avenue Louise, 326
BE 1050 Brussels
Belgium
Phone: +32-2-788-7800
Fax: +32-2-788-7899

GS1 US
Princeton Pike Corporate Center
1009 Lenox Drive, Suite 202
Lawrenceville, NJ 08648
Phone: 609-620-0200
Fax: 609-620-1200

Available at http://www.gs1.org, http://www.uc-council.org/home/tabid/36/default.aspx, http://www.eanucc.org, and http://www.uc-council.org.

For more than 30 years GS1 has been a leading global organization dedicated to the design and implementation of standards and solutions to improve the efficiency and visibility of global and sector supply and demand chains by offering a diverse range of products, services, and solutions. GS1 is a neutral, not-for-profit standards (and related services) organization. GS1 operates in more than 20 industry sectors ranging from retail, food, and fast-moving consumer goods to health care, logistics, and military defense.[*]

Safety, security, and traceability are currently at the forefront of both government regulations and industry concerns around the world. As a result, numerous incompatible track and trace solutions have been proposed to national, regional, and global supply chain participants. The cost of diverse government regulations, proprietary service offerings, and incompatible commercial solutions to consumers, companies, and the global supply chain called for defining traceability as a business process, which is supported by voluntary business standards that are accepted around the world.

Formed from the joining together of EAN International and the UCC, GS1 is truly global, with a presence in more than 150 countries driven by more than 1 million companies that execute more than 5 billion transactions each day using GS1 standards, solutions, and services. The GS1 Traceability Standard was developed by the GS1 Global Standards Management Process Team. This group was composed of 73 experts from 18 countries.[†]

GS1's portfolio of products ranges from GS1 BarCodes to GS1 eCom (electronic commerce tools) to next-generation technologies, such as GS1 EPCglobal (using radio-frequency identification), and solutions such as GS1 Global Data Synchronisation Network (GDSN) and GS1 Traceability.

---

[*] GS1 and GS1 US will be used interchangeably with EAN.UCC. Many texts still refer to EAN.UCC rather than GS1 and GS1 US. European Article Numbering—Uniform Code Council, more often know as EAN.UCC, was the former supply chain standards family name that included product barcodes that are printed on the great majority of products available in stores worldwide and electronic commerce standards. EAN International was the global office for the more than 100 member organizations around the world; in 2005 the organization changed its name to GS1. The Uniform Code Council (UCC) was the numbering organization in the United States to administer and manage the EAN.UCC System; in 2005 the UCC changed its name to GS1 US.

[†] This group included representatives of Allied Domecq, Albertsons, BASF, Carrefour, Casino, CIES, CPMA, Daymon, Dole, ECR Europe, FMI, General Mills, Glon, GMA, GS1, Imaje, John Deere Food Origins, Metler Toledo, mpXML, Nestlé, NTT Data Corp, P&G, Safeway, Syngenta, Target, TraceTracker, Tyson Foods, Verisign, Wal-Mart, and Wegmans, among others.

GS1's main activity is the development of the GS1 System, a series of standards designed to improve supply chain management. To accomplish their global mission GS1 interests are represented at meetings with official bodies (such as the United Nations and the European Commission), international associations, and other institutions. Member organizations (MOs) are usually national associations that provide tools and support that enable their own member companies to manage their supply chains and trade processes far more efficiently.

10.11.1 GS1 GLOBAL TRACEABILITY STANDARD

The global GS1 Traceability Standard has been developed to meet important business needs, including regulatory compliance. It addresses the entire supply chain and can be applied to any product. The GS1 Traceability Standard is based on current business practices used by a large majority of supply chain partners; this allows companies to leverage existing investments and more easily implement the Standard as part of broader product quality system. The GS1 Traceability Standard is one of many systems that, taken together, help companies continue and add to their ability to meet consumer expectations for safe, high-quality products.

The Standard maximizes the use of globally established and implemented GS1 System tools that uniquely identify any "traceable item," describe the creation of accurate records of transactions, and provide for fast data communication about the traceable item between trading partners. It meets the core legislative and business need to cost-effectively trace back (one step down) and track forward (one step up) at any point along the whole length of the supply chain, no matter how many trading partners and business process steps are involved and how many national borders have been crossed.

GS1 Traceability standards accomplish this by defining a shared minimum requirement and showing what action is required from trading partners. The GS1 Traceability Standard enables maximum interoperability between traceability systems across the whole supply chain at the same time as accommodating specific commercial, industry sector legislative requirements. It serves as a foundational standard for all GS1 members.

The GS1 System embodies an "open architecture" approach designed for modular expansion with minimal disruption to existing applications. The approach is made up of:

- *Open standards:* The goal is a single, open, business-led, integrated system of identification and information transfer technology standards that enable effective supply chain management in any company, in any industry, anywhere in the world.
- *Differentiation:* The system is founded on rules-based standards that, when followed, ensure globally unique and discrete identification of such things as products, handling units, assets, and locations. The system includes standard ways to transfer GS1 System identification numbers and relevant data related to these numbers.
- *Transparency:* GS1 System identification numbers must be relevant and applicable to any supply chain, independent of who assigns, receives, and

processes the standards. This should enable only one way to perform any given function. New features should only be introduced to the standard if they enable new applications or better ways to perform existing functions.

Because of its ability to provide globally unique identification of trade items, logistic units, parties, and locations, the GS1 System is particularly well suited to be used for these purposes. From an information management point of view, implementing a traceability system within a supply chain requires all parties involved to systematically associate the physical flow of materials, intermediate, and finished products with the flow of information about them. This requires a holistic view of the supply chain, which is best attained by deploying a common GS1 business language system. Its global reach and universal acceptance by consumers, businesses, and governments makes it uniquely positioned to provide the appropriate response to traceability system requirements.

The GS1 Traceability System focuses on parameters that affect the traceability of physical flows in whole supply chains between several distinct partners, i.e., on the interfaces, rather than the internal traceability procedures specific to each company and therefore strictly reliant on its transformation processes.

- *Multiple functions of traceability systems:* GS1 advertises that traceability is a tool intended for use in various predetermined objectives. It can be considered as one of several elements designed to improve security, control quality, combat fraud, and manage complex logistical chains. To effectively facilitate traceability both tracking and tracing capabilities must be in place.
- *Traceability principles:* Unique identification: any product that needs to be traced or tracked must be uniquely identified. The GS1 globally unique identifiers are the keys that enable access to all available data about the product's history, application, or location.
- *Identification of locations:* Unique identification of locations is ensured through the allocation of a GS1 Global Location Number (GLN) to each location and functional entity.
- *Identification of trade items:* Unique product identifications are ensured through the allocation of a GS1 Global Trade Item Number (GTIN) to each product (consumer unit). For traceability purposes, the GTIN has to be combined with a serial number or batch number to identify the particular item.
- *Identification of series:* Traceability of series is ensured through the allocation of a GS1 GTIN and serial number to each product (consumer unit).
- *Identification of lots/batches:* Traceability of lots/batches is ensured through the allocation of a GS1 GTIN and lot/batch number to each product.
- *Identification across product hierarchies:* A GTIN needs to be allocated to each of the three levels of the product hierarchy, namely, consumer unit, traded unit, and pallet. The latter should only be included if it is priced, ordered, or invoiced at any point in the supply chain; in other words, if the pallet is also considered to be a traded unit.

- *Identification of logistic units (pallets):* Identification and traceability of pallets are ensured through the allocation of a GS1 Serial Shipping Container Code (SSCC). Any pallet, independent of its type (mixed or uniform), needs to carry an SSCC allocated at source. A new SSCC must be allocated every time a new pallet (logistic unit) is created.

10.11.2 GS1 Focus on Identity Preservation and Traceability

Traceability requires associating the physical flow of products with the flow of information about them. To ensure the continuity of the information flow, each supply chain participant must communicate predefined traceability data to the next one, enabling the latter to apply traceability principles.

At present, GS1 has established several traceability programs and Traceability Standards (see their website for PDF format guidelines) for beef, fresh produce, fish, bananas, and wine.

10.11.2.1 Uses of GS1 Traceability

Increasingly, the ability to trace materials and products up and down the supply chain has become an integral part of doing business.

- To identify and locate unsafe foods or pharmaceuticals
- To validate the presence or absence of attributes important to consumers (e.g., organic foods, nonallergenic cosmetics)
- To fight product counterfeiting and protect brands
- A regulatory requirement to protect against bioterrorism

10.11.2.2 Basics of GS1 Traceability Standard

The GS1 Traceability Standard defines business rules and minimum requirements to be followed when designing and implementing a traceability system. They are clustered around a matrix of roles and responsibilities for each step of the traceability process. The following GS1 standards enable implementation of the GS1 Traceability Standard and may be considered tools within the GS1 standards program.

Benefits for using the GS1 Traceability Standard include:

- It is based on existing business practices, and there is no need to purchase, create, or integrate new systems.
- It uses a common language, the GS1 System of identification and barcoding, as well as GS1 EANCOM and GS1 XML messaging.
- It is broad-based: GS1 standards are used in more than 150 countries around the world.[*]
- It takes a global approach, addressing the supply chain as a whole rather than any particular individual partner.

---

[*] There are more than 1 million GS1 user companies.

- It is thorough, covering the fundamentals of traceability, identification, data capture and management, links management, and communication.
- It focuses on the interfaces of physical flow of materials and products, establishing an open, global relationship between independent partners.
- It is flexible, recognizing that circumstances vary within and between sectors, thus providing for tailored applications.
- It is not a standard for internal traceability, although it does show the inputs and outputs that must be linked by an internal traceability system.
- It is not a replacement for safety or quality programs. It complements them, such as the Comité International d'Entreprises à Succursales (CIES) Global Food Safety Initiative and quality programs such as EurepGAP.

10.11.2.3   Critical Points of a Traceability System

The risks of a traceability system are generally located at each point at which there is a change of partner and operation. Possible risks are break in the supply chain, break in traceability, loss of information, imprecise information, and human error.

10.11.2.4   Main Factors of Traceability

When implementing traceability systems four basic principles of traceability, regardless of sector, the country, or the tools, are involved.

- *Identification:* Traceability management involves the identification of all relevant entities of the transformation process, manufacturing batches, and logistic units, uniquely and nonambiguously.[*]
- *Data capture and recording:* Traceability management involves the predefinition of information to be able to record it throughout the entire supply chain.[†]
- *Links management:* Traceability involves managing the successive links between manufacturing batches and logistic units throughout the entire supply chain.[‡]

---

[*] To track and trace an entity it has to be unequivocally identified. The identifier is the key to follow its path and to access all available and related information. Most of the time, trade items are tracked and traced by the group of trade items that has undergone the same transformation, i.e., by batches (same production process) or logistics units (same transport conditions). Each time the unit is processed or transformed it should be assigned a new identifier. This may involve batches of raw materials, packaging, logistic and trade units, etc.

[†] The traced data cover variable elements in the transformation process (depending on the production line, time of manufacture, etc.). This information may be directly related to the batch or product group identifiers, or linked to the manufacturing order number, the time, or any other information that allows a link to be created with corresponding product batches. It has to be stored and archived in such a way that it can be available on request.

[‡] Within a company, the control of all of these links and accurate store accounting alone make it possible to make connections between what has been received and what has been produced and/or shipped (and vice versa).

- *Communication:* Traceability management involves the association of a flow of information with the physical flow of goods.[*]

10.11.3 PRODUCTS AND SOLUTIONS

The GS1 System is the foundation of a wide range of efficiency-building supply chain applications and solutions. Based on GS1's ID Keys, the GS1 System is composed of four important product areas: GS1 BarCodes, eCom, GDSN, and EPCglobal products.

See Appendix J for a summary of commonly used data carriers for GS1 Traceability and GS1 Methodology of Numbering and Identification Systems.

10.11.4 GS1 INTERNATIONAL STRATEGIC PARTNERS

10.11.4.1 International Standards Organization (ISO)

GS1 plays an active role in a number of ISO groups, with ISO/IEC/JTC1/SC31 being the most important. ISO/IEC JTC1/SC31 is focused on automatic identification and data capture (AIDC). The secretariat for ISO/IEC JTC1/SC31 is provided by GS1 US (through the American National Standard Institute), and many GS1 MOs take part in this process at all levels.

10.11.4.2 United Nations Directories for Electronic Data Interchange for Administration, Commerce, and Transport (UN/EDIFACT)

GS1 networks with all levels of the UN/EDIFACT organization. The objective is to ensure the EDIFACT development process considers the needs of GS1 user companies. GS1 eCom standards are closely linked with UN/EDIFACT.

10.11.4.3 Global Commerce Initiative (GCI)

GCI is a voluntary body created to improve the performance of the international supply chain for consumer goods through the collaborative development and endorsement of recommended standards and key business processes.[†]

REFERENCES

Pinto, D.B., I. Castro, and A. A. Vicente. 2006. "The use of TIC's as a managing tool for traceability in the food industry." *Food Research International* 39:772–81.

---

[*] To ensure the continuity of the information flow, each partner should pass on the traced batch or logistic unit identifiers to the next partner in the production chain, enabling the latter to apply basic traceability principles in turn. The link between the flow of information and the physical flow of goods is ensured by referring to the identifiers of both types of flow: shipment advice number, container serial code, shipment number, etc.

[†] Other partners included ISBN (International Standard Book Number) and AIM (Association for Automatic Identification and Mobility).

11 IPT Process Facilitators

11.1 INTRODUCTION

This chapter includes a sampling of service providers such as identity preservation and traceability (IPT) program design and developers; research on analytical methods for detecting genetically modified organisms (GMOs) and authenticity claims; research on biological containment methods and validation processes and systems that promote stable coexistence of biotech and non-biotech agriculture; operations, marketing, and training services; online services; general resource providers; and ingredient and nutritional labeling and testing providers.

These organizations provide a spectrum of services toward IPT that are not as readily provided for by sector organizations. These organizations provide some of the missing components that auditors and software providers cannot provide. This group includes FoodTracE, TRACE (**TRA**cing Food **C**ommodities in **E**urope), Co-Extra (**Co-Ex**istence and **Tra**ceability), Value Enhanced Grains (VEG) Solutions (website), Critereon Co. (training), Novecta (training and marketing), *The Organic & Non-GMO Report*, and the Food Consulting Company (labeling and analysis).

What follows are individual/company/organizational statements from their websites and naturally reflect their views.

11.2 FOODTRACE

FoodTracE Steering Committee
Project coordinator: Ian G. Smith
AIM Europe
The Old Vicarage
Haley Hill
Halifax
HX3 6DR, UK
Phone: +44-1422-368-368
Fax: +44-1422-355-604
E-mail: ian@aimuk.org

Mercedes Schulze
Centrale für Coorganisation GmbH
Maarweg 133
Koln
D-50825, Germany
Phone: +49-221-947-14-222
Fax: +49-221-947-14-291
E-mail: schulze@ccg.de

Noelle Vonthron
EUROCOMMERCE
Avenue des Nerviens 9/31
Brussels
B-1040, Belgium
Phone: +32-2-737-05-84
E-mail: vonthron@eurocommerce.be

Tony Furness
UCE Technology Innovation Centre
Millennium Point
Curzon Street
Birmingham
B4 7XG, UK
Phone: +44-121-331-7474
Fax: +44-121-331-5401
E-mail: tony.furness@tic.ac.uk

Dr. Ian Russell
CODEWAY
13 Telford Way
Colchester
CO4 9QP, UK
Phone: +44-1206-756-741
Fax: +44-1206-751-286
E-mail: ian.russell@codeway.com

Available at http://www.eufoodtrace.org/index.php.

FoodTracE is a European Union (EU)-initiated program that focuses on traceability. It affects all businesses and agencies involved with food.* FoodTracE relates to how the food system identifies every single item that passes through the supply chain. Traceability is both recognized, and the concept established, within the EU and legislation that came into force in 2005. The means of achieving full traceability has not been determined. FoodTracE seeks to find a common approach and deliver a standard framework based on a range of simple principles that will take existing systems into account and ensure smooth and efficient transfer of information through every stage of the chain. The basic premise of FoodTracE is that traceability data must adequately describe all products and processing in the supply chain. Processes include safety inspections and quality assurances. The general objectives of the project under traceability data are to define (1) standards for item identification and (2) methods for handling safety and quality data. The second of these objectives needs to be extended to (if not replaced by) the interchange of data between operators.

FoodTracE recognizes that "traceability" is a buzzword across Europe. Many Europeans believe that it is critical and urgent that a common approach is taken to ensure that current systems are compliant with each other. The primary objective of FoodTracE is to develop a practical framework for traceability of food and to develop the means to plan, model, validate, and implement it. The framework covers every aspect of traceability, with the "well-being" of the consumer of paramount importance. Its goal is to be pragmatic and worthwhile for businesses and the retail trade to implement, and be suitable for adoption by trading partners. The ultimate purpose of the framework is to support consumer enjoyment of a safe, diverse, and high-quality food supply. Although there is no mention of the environment and social welfare (wages, migrants), this system may include these aspects if enough consumers demand this information.

What is general to most, if not all, food is that it arrives on shelves in supermarkets and on plates in restaurants as the result of an overall process, comprising a set of stages.† How farmers, processors, and distributors perform their operations at

---

* Food sectors include animal feed, fish, meat, poultry, cereals, dairy, cheese, organic farming, processed foods, confectionery, wine, fresh produce, and supermarket supplies, and those covered national and international trade.
† FoodTracE's omission regarding farmers markets and road-side stands, it is assumed these entities are not within FoodTracE's scope of interest.

each stage has a cumulative effect on the condition of food reaching the consumer. Within the FoodTracE framework each participant in the food chain is described as a stage operator who hands over batches, or items, to the next stage operators in the chain. Suppliers of food-related materials and services, such as fertilizers, packaging, storage, and transport, also count as stage operators. A participant organization may be responsible for several stages.

To achieve batch/item traceability in practice, every stage operator must keep a set of records containing:

- The identity of each input batch/item and its supplier
- The attributes of each output batch/item, including the identities and characteristics of its input batches and details of its processing
- The identity of each output batch/item and its recipients, for example, transport companies

The other key element of the traceability framework is a set of mechanisms with protocols; for example, rules for the format and transmission of data, for stage operators and the authorities to exchange information efficiently.

These principles are a sufficient basis for traceability. Many food producers, processors, and retailers have adopted them in their own ways. FoodTracE aims to establish the common ground between the stakeholders and to establish the scope for further codification by industry bodies by the incorporation of specific identifiers. For this purpose, FoodTracE uses the European Article Numbering Uniform Code Council (EAN.UCC) (see Chapter 10, IPT Software Providers). EAN.UCC has established itself as the global standard for numbering and identification. Its standards are in general use for retail goods (consumer units), outer cases (logistic units), and pallets (transport units) in all sectors of industry. EAN.UCC allows identification of various product attributes and the need for differentiation by process changes that occur along the food chain. In addition, the EAN.UCC system allows for the relatively free flow of data between agreeing partners.

11.2.1 FoodTracE Food Attributes

A stage operator can point to the safety, origin, and quality of his or her product only by maintaining a record of its attributes, including details of its processing. Some of this information will depend on his or her suppliers' claims, which may need to be verified. The stage operator must decide what attributes, measurable or observable characteristics, mandatory (mandates) or voluntary (customer), to record.

11.2.2 Product Attributes

Classification of food composition and attributes is important for traceability when there are questions about a food item meeting its specifications and its supplier's claims. These include the composition tables published by national agencies, the proposed European food composition database, and the Global

Commerce Initiative (GCI) classification. Attributes may also extend to other various claims.*

11.2.3 BENEFITS OF COMMON PROTOCOLS

Efficient means of recording and exchanging information (attribute data) will eliminate duplication of effort by inspectors and reduce the stage operators' overhead costs. FoodTracE is looking to the Internet model for global electronic business for the protocols to support traceability. This is because traceability requires an analytic framework and will often depend on electronic communication. Extensible mark-up languages, like XML, enable trading partners and the authorities to store and exchange information in flexible yet defined ways. They enable the agencies, regulatory bodies, industry groups, and individual operators to set out their requirements on composite (electronic or paper) documents.

FoodTracE understands that although "one-up, one-down" is the legal requirement, essential elements of traceability in practice include identification of each entity or batch, verifiable records, and a flow of appropriate information at all stages in the food chain, through which only a barcode-like system may be most advantageous. FoodTracE has the concept of a stage operator as the basic link in the food chain. In this context, an operation is likely to be equivalent to the processing between critical points in Hazard Analysis and Critical Control Point (HACCP). It was strongly emphasized in FoodTracE meetings that, from the operators' point of view, traceability is largely a matter of identification (in a broad sense that goes well beyond existing barcodes on products and packaging) and the associated recordkeeping.†

11.2.4 MAIN PLAYERS IN THE FOODTRACE TRACEABILITY CHAIN

- The stakeholders include all the processors and the players that perform all the transfers of food products and supplies between them.

---

* For example: Additive Claim, Calorie Level, Concentrated, Cooking/Simmer Time, Country of Origin, Diabetic Claim, Edible State, Fair Trade Claim, Fat/Leanness Level, Food Quality/Food Assurance Claim, Form, Genetically Modified Claim, Gluten Claim, Health Claim, Instant, Intended Culinary Usage, Manufacturer's Treatment/Cooking Process, Marketed Recipe/Style, Method of Cooking/Reheating, Nut/Seed Content Claim, Opening/Closing/Application Device, Organic Claim, Salt/Sodium Level, Separate Ingredient, Spice Level, Sugar/Sweetener Level, Suitability for Home Freezing, Suitability for Vegetarians/Vegans, Texture, Variant, Vitamins & Minerals Claim, etc.

† The Tracefish concept, an electronic system of chain traceability, was developed under the patronage of the European Commission in its Concerted Action project QLK1- 2000-00164. As its starting point, the TraceFish team adopted the ISO definition of traceability and applied it to sea fish and farmed fish chains. A member of the team has since commented that "the ISO definition is far more powerful than that in the EU principles of food law, as it includes their constituents and processing history of products, what the food is made of and what has happened to it, not merely where it has been. This is crucial for food safety and for a number of other reasons such as labeling. An inevitable consequence of this is that an awful lot of information may be required, and it cannot all be carried with the item. However, the various definitions state that traceability is the "ability" to trace .... Therefore, the information only has to exist and be accessible when required for the purposes of traceability. This is not to deny the great value, in many instances, of being able to carry key information with the item."

- The traceability structure represents the traceability procedures and all its databases, software, communications infrastructure, and equipment.
- Farm suppliers represent all suppliers to the farmers, including feedstuffs, seeds, fertilizers, and crop treatment chemicals.
- The hauler represents all transport, storage, conveyance, and logistics operators.
- The processor represents all food processors, manufacturers, and packers together with the suppliers of additives, containers, and packaging that come into contact with food.
- The retailer represents all markets, supermarkets, shops, caterers, restaurants, wholesalers, and local delivery and other operations concerned with supplying food and drink to customers.

11.3 TRACE (TRACING FOOD COMMODITIES IN EUROPE)

TRACE

Project coordinator: Paul Brereton
Central Science Laboratory–CSL
Sand Hutton
York YO41 1LZ
UK
Phone: +44(0)1904-462700
Fax: +44(0)1904-462133
E-mail: trace.enquiries@trace.eu.org

Traceability systems issues: Petter Olsen
The Norwegian Institute of Fisheries and
Aquaculture–NIFA
Muninbakken 9-13, Breivika
N-9291 Tromsoe
Norway
Phone: +47-776-29231
Fax: +47-776-29100
E-mail: leadergroup2.TSG@trace.eu.org

Analytical tools issues : Michèle Lees
EUROFINS–EFS
rue Pierre Adolphe Bobierre, BP 42301
F-44323 Nantes Cedex 3
France
Phone: +33(0)2-51-83-21-07
E-mail: leadergroup1.ATG@trace.eu.org

Available at: http://www.trace.eu.org/index.php.

TRACE is a 5-year project sponsored by the European Commission (EC). Its mission is to provide EU consumers with added confidence in the authenticity of European food through complete traceability along entire fork to farm food chains. TRACE is also a forerunner in the development of cost-effective analytical methods integration within sector-specific and -generic traceability systems. This enables the determination and the objective verification of the origin of food. This project is funded by the EC through the Sixth Framework Programme under the Food Quality and Safety Priority. TRACE aims to improve the health and well-being of European citizens by delivering improved traceability of food products. This is the subsequent program developed following a similar format as TraceFish.

TRACE also assesses European consumer perceptions, attitudes, and expectations regarding food production systems and their ability to trace food products, together with consumer attitudes toward designated origin products, food authenticity, and food fraud. It also developed a "Good Traceability Practice" guide to food production systems, and its technology transfer activities will train industry, regulatory bodies, and analysts in the new systems and methods.

11.3.1 TRACE HIGHLIGHTS

TRACE involves 47 universities, research centers, and private companies (including small and medium enterprises [SMEs]) from all over Europe and one from China; is developing a cost-effective system that can identify where and how foodstuffs were produced; focuses mainly on products labeled "as of designated origin (DO)" or "organic," although it will have wider applicability to other foods and animal feed; and uses a combination of methods in geochemistry, analytical chemistry, molecular biology, consumer science, statistics, supply chain management, and information technology to create a cost-effective system to identify where and how foodstuffs were produced. The system may be extended to all food and animal feed.

The project has made major advances in meeting its objective "*to specify, develop and test a generic information infrastructure to ensure complete traceability along entire fork to farm food chains.*" In particular TraceCore, a generic XML request-response scheme, is a nonproprietary product that enables food businesses to more easily exchange information by using a common traceability language.[*]

TRACE's goals include:

- To specify, develop, and test a generic information infrastructure to ensure complete traceability along entire fork to farm food chains
- To correlate geochemical morphology and bioclimatic factors of locally grown food
- To develop rapid, robust, accurate, and cost-effective methods for determining species/varietal origin of food
- To develop rapid, cost-effective "fingerprint" methods that can typify food products
- To develop novel specifications from multivariate analytical data, which can be used for traceability and control purposes to characterize food products
- To develop an information platform mapping verifiable data to analytical methods, specifications, and thresholds
- To develop and exploit a communication and dissemination system that will be the focus of European information on food authenticity and traceability

---

[*] TraceCore XML is from TraceFish; for this reason, the TraceFish Technical Standard has been split into a core generic for electronic interchange of food traceability information in general, TraceCore. xsd and others. See specific WebPages for more detailed information.

- To assess European consumer perceptions, attitudes, and expectations regarding the ability to trace food products and food production systems and attitudes to DO products, food authenticity, and food fraud
- To develop Good Traceability Practice guides for the food industry
- To draft and demonstrate standardized XML "request-response" schemes

A key part of *TRACE* is the integration of new analytical parameters, relating to origin, into the traceability infrastructure. The resulting system has been demonstrated by industry within five food sectors: mineral water, honey, chicken, cereals, and meat. Traceability experts have conducted process mapping within a mineral water SME, assessed the company's present system, and have made preparations for installation of the new system (new process, XML). As a result, a Good Traceability Practice guide for the mineral water industry has been produced. In parallel to these activities a mineral water prediction model is being developed that will provide the traceability system with analytical specifications relating to the geographical origin of the product.*

Meanwhile, other teams are working on developing methods to determine the origin of honey, olive oil, cereals, and meat. To date, 800 samples have been taken to help build a similar model for those foods. In addition, spectroscopic and molecular biological methods are being developed for use in characterizing foods. These fingerprinting techniques will aim to differentiate between products based on their species/variety or the way that they were produced.

The TRACE goals include studying the relationship between tracers (isotopic and trace element data) found in the food with those in the local environment, i.e., geology and groundwater, by analyzing the soil, groundwater, plant, and animal tissue samples of certain geographical areas; further, using statistics, it will:

- Develop food maps indicating the specific characteristics a food should have when produced in a specific area.
- Develop generic, nonproprietary, and standardized solutions for transmitting the product information electronically, so each link in the supply chain will be able to provide and use this information. This will be achieved by incorporating all the information (e.g., origin, production, ingredients) on the product into the traceability system.
- Assess consumer perceptions and attitudes to DO and organic food products and their authenticity through a consumer behavior study.
- Particular attention has been given to consumer organizations and their input in the project. TRACE has a consumer non-governmental organization (NGO) (BEUC5) as a formal partner and another on the independent board that provide advice and comments on the project.
- Produce an information resource (http://www.trace.eu.org) that aims to become the central source of food authenticity and traceability in Europe. The website will be constantly updated in the course of the project and will offer an online reference tool for food authenticity and traceability, as well as providing information on the project.

---

* To this end 346 mineral water samples have been taken to date, in addition to 2,120 soil and groundwater samples. These samples are currently undergoing trace element and stable isotopic analysis.

TRACE's view of the state-of-art regarding traceability:

- According to EU Food Law, traceability systems have been operational since 2005. Food businesses are obliged to keep records for information related to products bought and products delivered/sold, i.e., to apply the so-called one-up, one-down rule.
- Current traceability systems address the logistic side of traceability of products in such a way that each link requires keeping records of preceding and succeeding links, but in general only the data from the previous link, which are deemed relevant, are stored.
- In many supply chains, this is current practice when dealing with traceability data, and obviously there are several limitations and weaknesses, for example, loss of data, no explicit link to the ingredients used, the potential of hundreds of identically marked units with inherently different properties, tracing back to origins, and difficulty effectuating a targeted recall.
- Numerous studies have shown that the information loss from one link in the chain to the next is large; in some industries losses are documented to be 80 to 95 percent.*
- There are a few companies that have gone to great lengths toward implementing the "Push"† or "Pull"‡ mechanism traceability model within their own chains, but much of the potential benefits are lost if the implementation is done in a proprietary and nonstandardized way. In addition, little work has been focused on developing systems that can be used for verifying the origin of food.

11.3.2 Analytical Lab for Food Authenticity

Labeling issues are of increasing concern as the European consumer becomes more discerning about food purchases. The information on a label, related to the claims made of that product, is generally limited to compositional and nutrition data. Many labeling claims that relate to perceived added value are rarely supported by analytical data, leaving regulators to rely solely on paper auditing procedures to monitor compliance. This is particularly important with the growth and promotion of "added value" regional foods, such as those produced under organic and DO labels. TRACE addresses this deficiency by supplying analytical specifications for labeling issues relating to food origin, especially in areas that currently rely mainly on a written specification, e.g., geographical origin and production origin. TRACE is developing generic low-cost analytical tools for use in the traceability infrastructure for

---

* The Nordic Council of Ministers project "Traceability and electronic transmission of qualitative data for fish products" studied material and information flow in cod and salmon chains, and concluded that only 5-20 percent of the number of properties that were known in one link of the chain were still accessible in the next link.

† Supplier sends data, electronically or manually, along with the batch.

‡ Supplier keeps data, but buyer is authorized to request more information.

verifying three types of origin: geographical origin, production origin, and species origin.

11.4 CO-EXTRA (CO-EXISTENCE AND TRACEABILITY)

Co-Extra
Yves Bertheau
Institut National de la Recherche Agronomique (INRA)
Phytopathologie et méthodologies de la détection
RD 10, Route de Saint CYR
F-78026 Versailles cedex
France
Phone: +33-1-30-83-32-04
Fax: +33-1-30-83-31-95
E-mail: bertheau@versailles.inra.fr

Available at http://www.coextra.org/default.html.

Co-Extra is a multiyear integrated program funded by the EC and coordinated by the National Institute for Agronomic Research (INRA) in France.* It studies the coexistence of biotech and conventional crops, and the detection of transgenic substances in the EU market. They also research biological containment methods and validate workable processes and systems that promote stable agricultural coexistence for Europeans. Of importance are their programs that not only design and integrate biotech detection tools but also help expand new techniques for cost-effective detection of as-yet unapproved or unexamined transgenic varieties.

The main drive behind Co-Extra goes back to consumer demand for freedom of choice when it comes to agricultural biotechnology and derived products. The issue is that many consumers are critical of genetically modified (GM) plants and products, whereas on the other hand, most of the experts in charge of GMO approvals do not see any real threats to health or the environment. Meanwhile, farmers growing GMOs in other countries report higher yields, greater profits, and seem to have cut back on pesticide use. The only way to solve this European challenge of offering both consumers and farmers the freedom to use or to reject GMOs is by implementing coexistence and traceability. Traceability has become expected in all European food and feed supply chains, but the traceability of GMOs adds the extra challenge of very strict legal thresholds for unwanted mixing.

To better understand this concept, the term coexistence must be defined. **Coexistence** means growing GM and non-GM crops side by side and keeping them segregated all along the food supply chain. With Europe's relatively small field sizes this becomes a difficult task. For European customers the general idea of traceability is to have producers preserve the identity of their goods, which

---

* This is accomplished through the Sixth Framework Program under the Food Quality and Safety Priority. Co-Extra involves over 200 scientists from 18 countries with a budget of €24 million.

ultimately allow consumers to select the agricultural system they wish to support. The end result of coexistence and traceability is having cost-effective ways of getting more information on the origins and safety of foods, which benefits more than just GMO foods.

Co-Extra, to make its point regarding the complexity the food supply chain, illustrates the challenges of frozen pizza, which combines a minimal legal threshold for unwanted mixing. In one small box of frozen pizza, a few dozen ingredients may come from several different countries, which may have changed hands several times along the way, and undergone numerous processes. Using this example, freedom of choice means knowing whether the bit of soy flour in the dough was made from any GM soybeans grown in Argentina, unloaded at a Dutch harbor, and processed in a French factory. Making all that information available in a reliable and cost-effective way touches on many different disciplines. That is why Co-Extra integrates the contributions from experts in agriculture, gene flow modeling, socioeconomics, logistics, and molecular biology. Co-Extra also involves legal experts for studying international legal regimes and solutions for liability and redress issues.

Co-Extra's main objectives are to be manifested on two levels: at the scientific technological level and at the industrial level:

- To achieve breakthroughs in the domains of biological containment, of horizontal (territory) and vertical (supply chain) organization, of supply chain economics, of detection methods targeting EU-approved and as-yet unexamined GMOs, and of control and validation strategies
- To integrate the results into user-friendly decision-support tools targeting all the stakeholders involved in the food and feed chains
- To provide analytical methods, decision-support tools, position papers, and guidelines to all stakeholders and enforcement bodies to implement, monitor, and control coexistence and traceability, from the technical and economic and legal points of view

11.4.1 Coexistence Aspect

The first point to be considered concerns the seed and crop productions. Co-Extra is surveying and developing novel biological methods and tools to prevent the contamination of conventional or organic seeds and crops by species containing GMO components. This includes identification of biological characteristics, breeds, and species likely to mitigate contamination, as well as interactions with farming practices and environmental features. The second point to be addressed is the organization of the supply chain in such a way as to prevent commingling of GM and non-GM products throughout their processing. Based on the different case studies, Co-Extra will model the different stages of the supply flow, then describe and assess the different phenomena occurring at each step, as well as their cumulative effect. In particular the territorial organization will be modeled to restrict the possible contaminations locations, while the socioeconomic and legal implications will be studied. The effect of imports in Europe and of third countries' practices to segregate supply chains on admixture possibilities will be also covered.

11.4.2 Traceability Aspect

For Co-Extra, traceability basically consists of ensuring the reliability of the information related to products all along the supply chains. Co-Extra is intended to address both analytical and documentary traceability. Regarding the first aspect, Co-Extra is establishing a state-of-art realm of onsite and laboratory GMO detection and assessment tools, integrating them into systems, and finally benchmarking selected systems in real conditions. At the same time, the project will continue to design and develop necessary tools that are currently missing. For example, although the relatively new EU regulations take into account many of the problems encountered by the analysis laboratories by providing them with several detection tools, the expected facilities in GMO detection and quantification do not solve all the questions raised by the application of European regulations.* The analytical traceability part of Co-Extra is overall facing problems of integrating or developing cost-effective and fit-for-purpose detection methods, as well as technical challenges. To be efficient, the developed methods need to be assessed with regard to internationally recognized performance criteria, including robustness, precision, sensitivity, and accuracy, and associated with control plans made up of sampling procedures and frequency schedules.

11.4.3 Current Co-Extra Innovations:

- Novel approaches to meet the technical and economic challenges raised by the increasing number of GMOs, stacked genes, and unapproved and as-yet unexamined GMOs to be detected; these approaches will consist in reliable multiplex—more than duplex—quantitative polymerase chain reaction (PCR), finger printing, and quantitative differential PCR
- Position papers and guidelines for routine analysis, for detection of stacked genes and unapproved or as-yet EC-unexamined GMOs
- Guidelines for validation of complex GMO detection methods such as those that combine different stages, for example, PCR approach followed by hybridization on microarrays
- Guidelines to reduce measurement uncertainty in quantitative GMO detection
- Evidence supporting the suitability of the European modular approach for validation of GMO detection methods
- Mathematical models for prediction of pollen distribution and impact over large distances and fragmented landscapes

---

* For instance, the detection methods provided by the client can use different reference genes to quantify the plant species (analytical translation of the labeling according the ingredient content of samples) whose compatibility is not ensured.

11.4.4   PROJECT STRUCTURE

The Co-Extra Project is structured in eight "work packages." The primary work package (WP) for traceability is WP8, with WP3, WP5, and WP7 providing ancillary information. A summary of each of these groups is listed below.

11.4.4.1   WP8

This WP is to develop the stakeholders' dialogue using an internet platform (Co-Extra website) and stakeholder workshops. The outcomes of Co-Extra will be disseminated to the different stakeholders. Links to user-friendly decision support tools for stakeholders will be provided. An editorial office as communication center is to provide a consumer-oriented multitarget website.

11.4.4.2   WP3

The WP assesses the internal and external costs and benefits generated by the implementation of coexistence and traceability.

11.4.4.3   WP5

The objective of this WP is to develop cost-effective and fit-for-purpose methods and tools for detection of GMO taxa and controls.*

11.4.4.4   WP7

This WP is to integrate the project outcomes to begin the initial development of decision support tools to stakeholders and policy makers; to define the most appropriate information structures, contents, and supports to ensure reliability and cost-efficiency of documentary traceability; and to assess the reliability of the coexistence and traceability systems from selected third countries (outside the EU).†

---

* **Workpackage 5:** EC Regulation 1829/2003 requires that creators of new GMOs provide sequence information characterizing the GMO, methods to detect and quantify the GMO, and appropriate reference materials, before authorization may be given. From this there is an urgent need for the improvement of existing methods (e.g. real-time PCR) and the investigation of new methods (hyperspectral NIR, loop-mediated amplification, ligase-mediated amplification) for the detection system to become more cost efficient.

† **Workpackage 7** is responsible for the looking into the legal, scientific, social, and ethical issues surrounding the co-existence and traceability of GM and non-GM supply chains. In effect WP7 has four major objectives. 1) an overview of the relevant guidance, national and EU legislation concerning GMOs, but especially in relation to co-existence and traceability will be produced. In addition, issues of intellectual property will also be addressed. The work of WP7 will ensure an understanding and compliance of all WPs with these regulations. 2) another objective of WP7 is the integration of results in preparation of (future) decision support systems for stakeholders of different supply chains. Formulation of recommendations for the set-up of effective management strategies for the selected food supply chains within the project as well as with a view on other types of food supply chains. 3) to study the compatibility of traceability and co-existence systems around the world. Lastly the legal, technical and political issues arising from coexistence, traceability and liability will be assessed. WP7 will be led by the Sheffield Institute of Biotechnological Law and Ethics (SIBLE), an inter-Faculty institute of the University of Sheffield, UK.

11.5   VALUE ENHANCED GRAINS (VEG) SOLUTIONS WEBSITE

US Grains Council
1400 K Street NW, Suite 1200
Washington, DC 20005
Phone: 202-789-0789
Fax: 202-898-0522
E-mail: grains@grains.org

Available at http://www.grains.org.

In the past, the Value Enhanced Grains (VEG) Solutions website advertised itself as offering objective information to buyers and processors of grain who are interested in efficiency gains and improved profitability. It is this service and information that are particularly helpful to those looking for a needed edge in a competitive market-place. VEG provides critical information on the types of value-enhanced grains, their specific uses and advantages, plus the quantitative results of objective tests.

Marketing of VEG marks a major break from the past. The U.S. grain production, marketing, merchandising, and export system historically has focused on volume and cost considerations to move as much grain as possible. Traditionally, corn from different farms is loaded together in the rail cars, hoppers, barges, and silos. To maintain their value VEG products must be kept segregated to retain their identity. Depending on the particular crop, farmers and elevators need additional grain bins for storage, upgraded combines for gentler harvesting, and upgraded drying equipment. Some products may be transported in containers instead of bulk. Widespread acceptance of specialty grains depends on the effectiveness of handling and transportation systems in delivering the crops to end users in consistent volumes, at consistent levels of quality, with consistent end-use characteristics, and at competitive prices.

Advances in VEG range from genetics and biotechnology to identity-preserved and value-enhanced marketing channels. Both these processes make it possible to deliver very specific commodity traits to the buyer. On the technology front, there is a new focus on "stacking" multiple end-use traits. In this regard, products with both animal-health and food-safety implications are on the horizon.

VEG traits can generally be classified into two categories:

- ***Compositional traits,*** or the qualities of the grain themselves. Examples include corn that has been bred or engineered to have high oil content, high levels of amylopectin starch, or white cob.
- ***Management and handling traits,*** which fit the end users needs for processing. Examples include products such as low stress crack corn, organic corn, or post-harvest pesticide-free corn.

The demand by grain buyers for these new products is evidenced by the five-fold increase over the past few years in land dedicated just to the production of corn higher in oil content. The VEG market is a fast-paced, ever-growing opportunity to provide new solutions, both economic and environmental, to the farm, feed, and food sectors. VEG marketing channels continue to develop, with attention focused now more than ever on the needs of the buyer. Feed manufacturers, corn refiners, and food processors, not to mention farmers and food consumers, all benefit from the development of VEG.

11.5.1 Example: Traits

Low-phytate corn helps address environmental concerns about livestock waste and helps nutritionists reduce supplemental phosphorus usage, decrease dietary total phosphorus, and lower dietary phytic acid content. In turn, the lower phosphorus content in the diet helps reduce phosphorus in animal waste, up to 22 percent for poultry and nearly 13 percent for swine. High oleic, high oil corn will offer livestock producers the ability to manage the fatty acid profile of lipid deposited in carcasses.

VEG products have many benefits for food and industrial processors, including:

- Increased yield such as starch content, purity, and quality
- Specialized physical attributes for the production of stabilizers and thickeners
- Uniform kernel/seed size, hardness, color, etc.
- Absence of stress or damage

VEG are grains with particular quality characteristics that may provide various users value. Examples of corn or grain sorghum types:

Corn		Grain Sorghum
Blue	Low phytate	Certified feed sorghum
Organic	Waxy/white	Certified food sorghum
High oil	Non-GMO corn	Certified specialty sorghum
High starch	Low stress cracks	Certified pet food sorghum
High amylose	Nutritionally dense	
High lysine/opaque	Low-temperature dried	
High oil/high oleic	Post-harvest pesticide free	
	Hard endosperm/food grade	
	Nutritionally enhanced (protein)	

In addition, VEG offers its customers Virtual Trade Show. This service provides a platform for buyers and sellers of VEG products to "meet" each other or to see what the industry has to offer and to become better acquainted with potential business partners. Buyers can see the types of products suppliers have to offer, and suppliers can see who can supply them with products that meet the needs of their operations. To use this "matchmaking" service, suppliers register the VEG-related products, goods, and services they offer. Buyers register their uses for corn products and any traits of interest that can be provided by VEGs. In addition to providing a searchable platform for both buyers and sellers, periodic e-mail updates are sent to registrants to alert them when potential business partners register or make inquiries matching their criteria.

See Appendix K for a listing of U.S. Grains Council offices located worldwide.

11.6　CRITEREON COMPANY, LLC (AUDITORS AND TRAINING)

Critereon Company, LLC
21024 421st Avenue
Iroquois, SD 57353
Phone: 605-546-2299; 800-661-4117 (toll-free)
Fax: 605-546-2503
E-mail: info@critereon.com

Available at http://www.critereon.com/index.html (accessed June 15, 2006)

The Critereon Company has more than a decade of experience in helping companies design, implement, and manage quality systems. With their consultative, customized business approach, Critereon has assisted corporations to proactively manage regulatory compliance records, research data, and distribution channel activities using unique, computer-based, and automated software and technology systems.

Critereon is a leading provider of procedural, quality, and compliance management tools to many influential global businesses. Authentix is their web-based authenticity management platform that allows manufacturers to manage the processes, protocols, and procedures involved in supply or distribution chains. Working in conjunction with Authentix is their handheld data collection tool Pocket Authentix. With the versatility and accuracy that Pocket Authentix offers, field staffs are able to quickly record process, product, and shipment authenticity at critical points along the supply and distribution channels.[*]

Critereon services deliver business services across the food industry that include:

- Animal handling and welfare
- Biotech compliance
- Counterfeit, diversion, and dilution
- Country-of-origin labeling (COOL)
- Franchise quality monitoring
- HACCP
- Identity-preserved production
- Non-GMO sourcing
- Organic production
- Plant-made pharmaceuticals
- Research trials
- USDA process verification
- Quality systems design

---

[*] To further assist companies in managing for enhanced results, Critereon established xPaper™ Technology. The xPaper™ Technology system works exactly like a common clipboard and ink pen. Forms are completed by hand in a normal fashion at remote sites or during plant operations using the uniquely designed pen. The digital pen records ink strokes and maps them on specially designed digital paper. When the recording duties are finished, the person doing the work simply inserts the pen into a "cradle," which is actually a mini-computer linked to the Internet. From the "cradle," all of the information taken down during an audit or inspection is digitally uploaded directly to administration computers. The very same forms used on the clipboard at the site, along with checked boxes, notes, and authorized signatures, are instantly reproduced and ready for viewing at any location in the world. Authentix and xPaper Technology enhance the process of capturing audit and inspection data, giving organizations the ability to quickly, efficiently, and cost-effectively gain control of regulatory, quality control and standard operating procedure forms and documents.

11.6.1 Authenticity Management

A parallel can be found between existing food safety programs and the new field of authenticity management. In much the same way, authenticity management monitors and manages protocols and procedures across the value chain to deliver the benefits of authenticity in production and processing systems. It is the best method to obtain these results because it is the most effective and efficient method. Authenticity management forges a bond between process and measurement to lower risks and build business value. Authenticity is all about attention to detail.

11.6.1.1 Example: Biotech Compliance

The problems facing agriculture are numerous. The development and application of genetically engineered products present new and increasing challenges for biotechnology companies. Researchers continue to discover new plant and animal genetic traits that promise to enhance a broad range of industries from manufacturing to petrochemical and from agriculture to pharmaceuticals. However, increased scrutiny from environmental and consumer groups, who have expressed concerns about possible disruption of natural ecosystems, unintended cross-pollination, compromised plant defense systems, and pharmaceutical-active contaminated food crops, has increased the demand for more rigid and mandatory regulatory and consumer testing. Moreover, international commercialization requirements for biotechnology products are becoming increasingly complex and stringent. In addition to environmental and international trade concerns, biotechnology companies also face internal challenges such as the financial risks in properly handling GMO products or the public relations liabilities that can occur with errors in the distribution channels.

To help solve these problems, Critereon works with organizations to:

- Collect the regulatory requirements that are important
- Analyze the current compliance process against the appropriate agency standards
- Bring the customer into direct contact with the regulatory agencies, providing them with enhanced understanding of compliance policies and international laws
- Develop procedures, processes, and protocols for exporters using hazardous and critical control point analysis
- Provide documentation management services for export shipments
- Thoroughly review the customer's compliance documents
- Provide a complete audit of the organization's compliance system

11.6.2 Country-of-Origin Labeling

COOL regulations require a covered commodity to have verifiable recordkeeping of its production and processing history to support accurate labeling. Critereon's compliance monitoring systems evolved from their experience in protecting the value of special crops and market animals. They design systems that address both regulatory compliance and identity preservation (IP) in the agri-food industry. Their compliance services support international market demands for traceability and specific production practices.

Every supply chain is unique and requires a tailored approach to keep costs low and to limit the potentially negative logistics effects of COOL. Authentix for COOL and its associated technologies provide proven and flexible tools to agri-food for these needs.

Benefits of Authentix for COOL:

- Authorized transparency to the supply chain
- Dynamic lot traceability
- Multidimensional protocol compliance monitoring
- Online posting and validation of affidavits
- In-field electronic inspections
- Integration with existing control systems
- Secure web application
- Multilevel access
- Multilingual capability

11.6.3 Quality Systems Design and Program Accreditation

Business customers and buyers today are interested in receiving comprehensive information about products and related services delivered by their supply-chain partners. Market forces often require specific information about the quality of products—who makes them, where, when, and under what conditions.

In markets where quality is no longer the sole focus on finished products and services, companies that supply products and services are adopting operating standards that help them carefully address selected processes and customer specifications that influence their operations. To meet customer demands for verifying that manufacturers are using quality systems in their operations, many companies are seeking third-party accreditation that provides a demonstrated commitment to excellence. To enhance an existing quality system or to establish an initial quality system program, it is critically important for companies to understand their ultimate goals and objectives; it is also indispensable for companies to enlist skilled guidance when they decide to commit themselves to building, implementing, and sustaining a quality system. In this light, Critereon is proficiently prepared to cost-effectively assist companies achieve quality management certification and to:

- Review and design a customized quality system that results in measurable operational efficiencies and verifiable process and/or product claims
- Align accreditation system functions with current quality control procedures
- Create documentation for achieving compliance with chosen quality system framework such as training and self-improvement
- Develop systems to ensure validation of successful program implementation
- Provide the necessary technology that will enable the organization to achieve quality management on an ongoing basis

- Present companies with the marketing advantage by clearly differentiating their products from the competitions' products

11.7 NOVECTA, LLC (FACILITATOR AND TRAINING)

Novecta, LLC
5505 NW 88th Suite 100
Johnston, IA 50131
Phone: 515-225-9242
E-mail: info@novecta.com
Brian Buckallew, Managing Director
E-mail: brianb@novecta.com
Gary DeLong, Project Manager
E-mail: garyd@novecta.com

Available at http://www.novecta.com/index.html.

Novecta was created in 2001, as agricultural biotechnology started to become an issue for farmers and consumers. It was created as a joint venture between Iowa and Illinois Corn Growers Associations and works closely with other organizations such as Iowa's Soybean Association, Farm Bureau, Coalition to Support Iowa Farmers, and others. Novecta has two sources of funding: an appropriation from the U.S. Department of Agriculture (USDA) and private/commercial sector funds. Novecta employees typically call on corn customers, such as processors, and find out what traits or qualities processors want, be it specified product traits or minimum oil or starch contents.

In their research it became evident that issues like quality, tracing components, quality assurance programs, and IP were becoming more critical to the public and processors. Many of Novecta's training programs are copied from the organic producer's game book. However, they are pro-biotechnology (and hybrids) and promotion of its wise use by educating farmers and the public. They hope to address processor concerns, cooperate in resolving problems, and to increase the speed of crop delivery.

They primarily focus on IP and quality assurance programs and training. They are also involved with USDA's process verification program and on International Organization for Standardization (ISO) 22002 and ISO 9000 projects. Novecta does little to promote commodity corn. Most of its focus is on value-added and differentiated corn products. They do not do much work with non-GMO or organic corn. However, they do promote ties to many industries. As mentioned, they use the template that organics use when they train on IPT for value-added corn products. In this way, for example, Novecta is sensitive to Japanese customers' demand for certification regarding chemical residuals or maximum residue limit (MRL).

Novecta realizes one obstacle for farmers is recordkeeping. Although many farmers are now becoming more familiar with computers, once they record information, they are reluctant to share their information with others. Novecta recognizes that there seems to be a natural distrust between grower, coop/elevator, and processor. For decades each group profited by not sharing market information with others. IPT requires information sharing and a different type of mentality toward agriculture.

They all can benefit by mutually sharing and when they better understand each other's problems and operational necessities.

Typically processors sponsor Novecta to train and motivate farmers to grow crops to certain minimum standards. The typical farmer that Novecta works with farms 800 to 1,000 acres. These farms are more adaptive than other size farms and can more easily modify production practices to take advantage of value-added crops and their traits. Once Novecta understands what the processor/customer wants and is willing to pay farmers to grow the crop, the processor then sponsors Novecta to train farmers on how to fulfill the processor's needs or contract. As Novecta's website suggests, they then help train farmers on various aspects of IPT. The key here is that Novecta looks at what the processor wants and makes sure that the farmers understand what is important for them to do to ensure compliance.

Novecta has a library of training programs for various levels of quality assurance and production practices. Novecta training and certification can be adapted to a variety of commodities or customer needs. By building on the ISO 9000 quality management system, Novecta offers programs to growers that allow them to initiate entry-level quality assurance programs and expand them as the market or needs dictate. This approach also allows growers to meet the requirements of various production contracts that may use ISO principles as a base.

Novecta also provides services to growers through the development of market opportunities for quality-assured commodities. By developing relationships throughout the supply chain, Novecta is able to build awareness for the capabilities of growers who have adopted quality assurance systems into their businesses. Novecta provides a service to both producers and buyers by facilitating trade between these two parties. The company works to develop both domestic and an export market for quality-assured production and maintains a close association with other commodity organizations and government agencies.

11.7.1 Objectives of Novecta's Identity Preservation System

It is primarily designed to provide assurances that the desired qualities or traits are present (or absent) in a product from the seed source, through all steps of production and delivery, to the end user. Typically, these assurances need to be documented in some manner from one party to the next throughout the entire chain.

In addition to initial seed purity, follow-on and continuous verification and documentation are paramount. More specifically, an IP system identifies and verifies that certain procedures were carried out during the growing, handling, transportation, and conditioning or processing of the crop. Sampling and testing may be part of an IP system, but the essence of the system is the procedures and the verification. Some examples of specifications and activities that may need documentation include seed breeders' statement of variety and breeding methods suitability of seed variety (variety release statement).

Novecta's program typically includes preplant planning (seed selection), planter preparation and documentation, planter clean-out documentation (planter items to check and clean, e.g., seed boxes), planting the field (field management), harvest, and its completion (when the storage bin has been filled or harvest of the IP crop is completed, ensure all records are complete and put with all other information regarding the contract), quality assurance testing (from storage bins samples must be tested and sent to a quality assurance laboratory), grain storage and handling (drying, etc.), and contract review (throughout the growing season, the IP grower should maintain a file on each IP contract as per each field's soil tests, purchase orders, etc.).

See Appendix L for a listing of Novecta's corn VEGs.

11.8   *THE ORGANIC & NON-GMO REPORT* (AND 2006 *NON-GMO SOURCEBOOK*)

The Organic & Non-GMO Report (**formerly known as** *The Non-GMO Report*)
P.O. Box 436
Fairfield, IA 52556
Phone: 641-472-1491; 800-854-0586 (toll-free; United States only)
Fax: 641-472-1487
E-mail: ken@non-gmoreport.com

Available at http://non-gmoreport.com.

The Organic & Non-GMO Report advertises itself as the only monthly newsletter that provides information needed to respond to the challenges of GM foods. This publisher is a prime example of using and promoting IPT systems. This group's focus is primarily on non-GMOs and organic products; however, the same principles may be used for other foods within the food chain.

11.8.1   Traceability Example of Food Safety Trend in EU

In the dispute over GM foods between the United States and the EU, one sticking point is Europe's requirement that all GM foods be labeled and traced back to their origin. Whereas the United States views traceability as a novel and, in some cases, a bad concept, the EU sees it as an essential element of food production and a way to assure consumers of food safety.

A good example is Tracemeal S.A., a company based in Geneva, Switzerland, that supplies soy meal to salmon breeders in northern Europe. Tracemeal buys soy meal from a Brazilian soy processor who contracts with farmers to grow certified non-GM soybeans. After processing, the soy meal, which is identity preserved at every stage, is shipped to a port in Denmark, which is dedicated to receiving only non-GM soy. The soy meal then goes to fish feed producers, where it is made into feed and given to salmon. Finally, the salmon are shipped to Japan where they are cooked and served fresh, just 72 hours after shipment from Europe. The entire chain, from the salmon dinner in Japan back to the soybean seed in Brazil, can be traced.

11.8.2   Traceability Laws

Although concern over GM foods is a factor, the demand for traceability extends beyond GMOs. Europe's main traceability efforts and regulations have focused on animal feed, which has been the source of several food scares, such as the bovine spongiform encephalopathy (BSE), "mad cow" crisis. At SGS, a non-GMO certification company based in the Netherlands, feed is a big issue (see Chapter 7, Auditors regarding SGS). The European Parliament has passed a series of regulations establishing traceability. In 2000, legislation was passed requiring traceability and labeling of beef products, and in February 2002, Parliament passed regulation Number 178/2002, which established the European Food Safety Authority and principles of food law.

11.8.3 TRACING GMOs

European consumers want GM food labeled and traced, and major food retailers, companies such as Tesco and ASDA (both based in the UK) and Carrefour (based in France), have eliminated GM ingredients from their house brand foods and require meat suppliers to raise animals on non-GM feed.

It is estimated that feed producers will pay a 5 to 10 percent premium over commodity prices for fully traced, identity-preserved, non-GM soy meal. In turn, feed producers that breed salmon can earn a 20 to 25 percent premium in Japan for salmon labeled as identity preserved and non-GMO.

According to Katrin Schröder, IP manager at GeneScan Analytics GmbH, traceability regulations are shifting the labeling criteria from detecting GMOs in the product to application of GMOs in the process or processing. "The European food industry is looking for avenues how to comply so they won't have to label their products," she says. As an example, the regulations require that all food and feed ingredients produced from GMOs be labeled even if GMOs cannot be detected in the final product. In addition, products exported to Europe without a label will be assumed to be non-GMO and be subject to PCR tests by authorities. Products that test positive for GMOs will prompt an investigation and may result in refused shipments. "Providing PCR test reports as proof that a product is non-GM won't be sufficient. Authorities will want to look at traceability documentation," says Richard Werran (a representative with Cert ID, based in the UK). "Exporters have to assume the worst possible case and have traceability in place."

11.8.4 *THE 2006 NON-GMO SOURCEBOOK* (EXCERPTS)

The Non-GMO publishers also published *The 2006 Non-GMO Sourcebook*, the fifth edition of the essential guide to the market for non-GMO seeds, grains, ingredients, foods, and related products and services.

As consumer concerns over GM foods continue throughout the world, the global market for non-GMO products continues to grow. *The 2006 Non-GMO Sourcebook* reflects this growth; this edition includes more than 560 suppliers of non-GMO products and related products and services. According to the Non-GMO publishers, their targeted readers are global and want healthier foods. Many of them see health and environmental risks with GM foods, which then fuels strong demand for non-GMO alternatives. This demand is strongest in the EU and Asia, particularly Japan and South Korea, where consumer opposition to GM foods is greatest. Demand for non-GMO is increasing in other nations, including the United States, with its growing natural and organic food industry. Another trend driving the demand for non-GMO foods is traceability. Consumers increasingly want to know the origin of their foods, and non-GMO food systems, such as IP and organic certification, meet this requirement.

11.8.5 "Farm-to-Fork" Products and Services

The primary GM crops grown in the world are canola, cotton, corn/maize, and soybeans. As a result, *The 2006 Non-GMO Sourcebook* focuses heavily on non-GMO alternatives to these GM crops, particularly soybeans, which are increasingly valued as an important protein source for both human food and animal feed.

The 2006 Non-GMO Sourcebook is global in scope, listing suppliers of non-GMO products and services not only in North America but also in Asia, Australia, Europe, the Middle East, and South America.

The *Sourcebook* provides suppliers of non-GMO products and services for:

- Seeds, including organic and food soybeans
- Non-GMO corn and soybeans and processed ingredients
- Specialty grains and oilseeds, such as flax, wheat, and sunflowers
- Minor ingredients and processing aids, such as vitamin E and enzymes
- Food products
- GMO testing, IP, organic certification, and other services that support non-GMO production

The 2006 Non-GMO Sourcebook lists many suppliers of organic products because organic food production, which prohibits GM products, is essentially non-GMO and because demand for organic is increasing worldwide.

11.8.6 *The Non-GMO Report* Recommendations for GMO Testing

The ability to detect GM material in seed, grains, and food has become critical to suppliers of non-GM products. GMO testing along with IP is essential to verify that seed, grain, or food products are non-GMO to meet regulatory requirements or a buyer's specifications. *The Non-GMO Report* recommendations for finding a GMO testing lab or test method:

1. ***Look for a lab that is accredited and participates in Grain Inspection Packers and Stockyard Administration (GIPSA)'s proficiency program:*** A lab should be accredited to ISO 17025 or United Kingdom Accreditation Service (UKAS). Accreditation requires that a lab provide evidence of good performance. It is also important to find a lab that participates in the USDA's GIPSA proficiency program, which tests the proficiency of GMO testing labs. Potential testing customers can look at GIPSA's website and see how laboratories perform.
2. ***Know the lab's capabilities:*** Ask questions about a GMO testing lab's capabilities. How long have they been performing GMO tests? Can they screen for all commercially available GMOs? How do they validate results? One should look for a method that has been proven over time.
3. ***Know the type of test needed:*** Enzyme-linked immunoabsorbent assay (ELISA) protein "strip" tests do a good job screening raw grains. For processed foods, PCR is recommended.

4. *If exporting, know what type of testing will be done at the destination country:* Learn as much as possible about the testing methods in the country one is selling to. This will be a guide in implementing a similar method.

5. *Avoid choosing a test based on price alone:* A few extra dollars up front are nothing compared with the costs of problems that can occur with inaccurate tests.

6. *Ask if the lab can test for specific GMO events:* Identifying specific GMO events is particularly important because one may not be approved in certain countries, which could cause major problems for an exporter.

7. *Get a representative sample:* Sampling is one of the most important aspects of testing. The sample must be statistically representative of the lot of material from where it came. If one does not have a representative sample, the validity of the result is in question no matter how good the method.

8. *Know the GMO threshold buyers will accept:* Do buyers need a qualitative, "yes or no," result about GM content or a quantitative test that determines the percentage of GM material present in a sample? Based on this tolerance, GMO testing labs and kit manufacturers will devise a sampling and testing protocol to meet the customer's and buyer's needs.

See Appendix M for National Laws for Labeling GM Foods.

11.9 FOOD CONSULTING COMPANY

Food Consulting Company
13724 Recuerdo Drive
Del Mar, CA 92014
Fax: 800-522-3545
Phone: 800-793-2844 (toll-free)
E-mail: info@foodlabels.com

Available at http://www.foodlabels.com/index.htm.

Food Consulting Company, founded in 1993, provides services for food labels, nutrition facts labels, nutritional analysis, and food label guidance to ensure Food and Drug Administration (FDA) regulatory compliance for small- and medium-sized food manufacturers, distributors, copackers, and importers. With the purchase of Nutrients Now in 1996 and Nutrition Labeling Services in 2000, Food Consulting Company is one of the largest contract providers of food labeling services with more than 1,000 clients in the United States and abroad.

Their mission:

- To become their customer's virtual food label department
- To provide expert food label solutions that position customer products well within U.S. and Canadian laws
- To make their customer's job of complying with the FDA food label regulations easy

- To guarantee 100 percent FDA regulation compliance

Specific services include:

- *Nutritional analysis:* To ensure accurate nutritional analysis and nutrition facts labels for products and recipe formulations, including both laboratory nutritional analysis and database nutritional analysis
- *Food labels:* To ensure full label compliance by providing development of nutrition facts labels and ingredient/allergen statements, product names, label claims, plus a review of final organizational label artwork
- *FDA regulatory support:* Food Consulting provides resources and answers to challenging regulatory questions to ensure food labels are FDA compliant

Food Consulting also creates Nutrition Facts and Ingredient Statements for private label manufacturers and food companies with new product introductions, which includes a Full Label Compliance Package that includes all required food label components, Ingredient Statements, label layout sketch and type-size for each component, advice on regulated nutrition label claims and National Organic Program

TABLE 11.1
Food Consulting

Food Labeling		FDA Regulatory Support	
Full label compliance Nutrition Analysis,* Nutrition Facts Panel Ingredient Statement, Allergen Compliance Product Naming and Label Claims Label Development Instructions Final Label Compliance Review	$795	**Annual regulatory support** Guidance for 10 questions/issues per year	$3,000
Nutrition facts panel Nutrition Analysis* Nutrition Facts Panel	$250	**One-time regulation support** Guidance for a single question/issue	$500
Ingredient statement Ingredient Statement Allergen Compliance	$250	**Client file review** Follow-up support (after free 90-day period)	$150
Laboratory nutrition analysis	Add $650	**Shelf life evaluation** Guidance for shelf-life issues, "best before" or "use by" dates	$475
Special requirements Bilingual Canadian and Bilingual US Diet Exchanges Child Nutrition Labeling Complex Formulas	Add $125 each		

* Includes database nutrition analysis or one's own lab analysis.
Source: Data adopted, with permission from Food Consulting.

requirements, and a Final Label Review. In addition, Food Consulting offers nutritional analysis and FDA-compliant food labels for beverages, baked items, snacks, condiments, dairy products, and more. Typical customers are manufacturers, ingredient suppliers, copackers, distributors, and marketers.

For food importers and brokers, Food Consulting can "Americanize" the food label content (U.S. FDA compliant) for the import of food products into the United States.

Ingredient suppliers receive 100-gram data (via Laboratory Nutrition Analysis or Database Nutrition Analysis) and Laboratory Microbiological Analysis for ingredient specification sheets.

Restaurateurs, food writers, and recipe publishers can receive ready-to-publish nutritional analysis of menu items, nutrition/allergen guides, recipes for meals, and for publishing in cookbooks, magazines, and websites, which ensure that nutrition claims (light, healthy, low fat) are valid and meet the FDA regulations in the Nutrition Labeling and Education Act (NLEA). See Table 11.1 for prices.

Also available:

- Dietary supplement labeling: available for products regulated as dietary supplements
- Glycemic testing and labeling: determines glycemic values and substantiates label claims
- EU Food Labeling: for products that will be exported to the EU

12 Food Recalls and Insurance

12.1 INTRODUCTION

In many ways this chapter should lead off this book as a motivator for the reasons why any participant within the food industry, with its interdependency on one another, should have a strong identity preservation and traceability (IPT) program well established. Often a firm's success or failure in the market is tied to its own actions, upstream suppliers of inputs, and downstream processors. The net results of failures of the food chain system include loss of brand name, public trust, and large expenses tied to recalls and legal liabilities.

This chapter is more eclectic than other chapters in that the resources range from the legal profession, private organizations, websites, to academia.

The first portion includes:

Legal and pragmatic view of recalls and insurance, by section:

12.2. Why food recall insurance is needed: a short history of food recall insurance
12.3. Product recall: disasters waiting to happen
12.4 Understanding the recall concept in the food industry

The second portion consists of ***informational resources and protocols:***

12.5. FoodTrack Inc. and FoodTrack incident report
12.6. ANZFA food industry recall protocol
12.7. OurFood.Com: food database

The last portion includes ***product insurances:***

12.8. Product recall insurance (type of insurance)
12.9. Seedsmen professional liability insurance (type of insurance)

What follows are company/organizational/individual statements from their websites and naturally reflect their views.

12.2 WHY FOOD RECALL INSURANCE IS NEEDED

12.2.1 Short History of Food Recall Insurance

According to *Food Recall Insurance: Why Your Company Should Take a Look* (2005), food recall insurance, which is sometimes included under the broader heading of product recall or recall insurance, began in 1980s as a result of the well-known and well-publicized Tylenol tampering incident. In that case, a number of Tylenol bottles were intentionally laced with cyanide. As a result, seven people died; the company spent more than $100 million dollars in remedial costs; and the Tylenol brand went from owning 35 percent of the nonprescription pain reliever market to around eight percent. After this incident, a few insurers began offering recall insurance; however, at that time, coverage only included malicious or intentional tampering. Nevertheless, and largely as a result of an increased number of recalls, accidental coverage began appearing in the early 1990s.[*]

Recall insurance has proven to be very popular in Europe, and with the passage of European Union (EU) Regulation 179, requiring that all food and beverage companies recall any product that violates the European Community (EC)'s food safety regulations. This type of coverage is also becoming increasingly more visible in the United States through some recent and well-publicized recalls. Although a number of insurers still do not offer this type of insurance, a handful of carriers do. Of course, as in any industry, companies that manufacture, sell, transport, or otherwise handle food come in all shapes and sizes. Food recall insurance can and should be specifically tailored to meet the needs of a particular company. For example, a local bakery whose distribution network extends to only a few counties or states would not need the same amount of coverage as a massive manufacturer who distributes globally.

A full-scale recall involving food products can be catastrophic to a food grower, processor, manufacturer, or retailer. Not only would a company have to pay for all the recalled products to be shipped to a suitable location and often destroyed, but also associated costs such as advertising the recall, public relations to rehabilitate a damaged reputation, and additional expenses to win back customer support will all be extremely costly. Perhaps more troubling is that these numbers represent out-of-pocket expenses that must be spent only to deal with a recall; once they factor in lost profits that result from their product no longer being sold, the outcome could close a company's doors for good.

Very few (if any) recall costs are covered by a general liability policy. General liability policies are designed to cover and protect a company from product-related tort lawsuits, not mishaps causing economic injury to food chain participants. To fully understand the economic consequences of a "recall," consider that food-related businesses generally operate with a two to five percent net profit margin. If a company has to order a recall that will cost hundreds of thousands or even millions of dollars, can they afford it? What about the cost to rehabilitate their name and win back the lost customers?

Food recalls generally result from identifiable contamination incidents. Contamination incidents are either accidental or malicious, with the latter consisting of intentional product tampering. Although contamination incidents have always

---

[*] From *Food Recall Insurance: Why Your Company Should Take a Look*, by Arthur S. Garrett with, permission from Keller and Heckman, 2005. Mr. Garrett joined Keller and Heckman in 1990 and practice focuses primarily on insurance coverage and insurance recovery; he has also worked on Superfund and toxic tort litigation.

been a concern for industry participants, the increased complexity and geographic reach of food distribution networks have dramatically increased the chances of accidental contamination. In addition, specific food contaminants such as *Listeria*, dioxin, lead, *Salmonella*, various underdeclared allergens, and now avian flu have been in the news almost nonstop. To get a true sense of the size and number of food (and drug) recalls in particular, one can visit the Food and Drug Administration (FDA)'s website where there is an active list of current and on-going food- and drug-related recalls in the United States.[*] All it takes is one positive test, one reported sickness, and a company could be facing a massive recall effort. In short, every company that deals with food or food products must be concerned with contamination in today's world.

Additionally, the public has become acutely aware of the possibility of malicious tampering since the 9/11 terrorist attacks. Although authorities have taken necessary measures to increase security, the possibility of a terrorist-related tampering cannot be ignored in today's environment. If this happens, the cost will be placed on the manufacturer, and the resulting publicity damage could be devastating.

Generally, recall insurance will cover most, if not all, of the costs associated with the recall. One prominent insurer provides a policy that covers both malicious and accidental contamination, as well as product extortion. As always, however, the value of the policy is in the details. Companies need to make sure that their company's policy covers shipping and destruction costs, media and public relations costs, and the amount they spend on replacing the recalled product in the market, as well as restoring its name with the public. In short, these are the logistical and reparative measures that are always associated with a recall. Without adequate "recall" insurance protection, these expenses will not be covered.

12.3 PRODUCT RECALL: DISASTERS WAITING TO HAPPEN[†]

Pan Pharmaceuticals Limited of Sydney is destined to become a benchmark case study of how not to handle a product recall. In January 2003, the Therapeutic Goods Administration (TGA) of Australia recalled the travel sickness product Travacalm, manufactured by Pan Pharmaceuticals. Travacalm was reported to have resulted in almost 100 people being seriously affected, including 19 who required hospitalization; more than 400 companies were involved and resulted in the recalling of more than 1,500 different products from the shelves of health food stores, pharmacies, and supermarkets. Pan Pharmaceuticals is just another company to get its corporate crisis management wrong. In other words, companies that ignore critical incident planning place their corporate reputations on the line.

The Pan Pharmaceuticals crisis has focused the spotlight on the dramatic and potentially disastrous impact of product recalls and the vital need for comprehensive,

---

[*] See http://www.fda.gov/opacom/7alerts.html for additional information.
[†] Excerpts modified from Patrick Weaver's "Product Recall: Disasters Waiting to Happen." Published in *Australian Business Report*, issue 3, vol. Jun/Jul 2003, an online business and current affairs magazine featuring commentary, issues, analysis, news, and information. Published by Fleishman-Hillard Stratcom for decision makers and opinion leaders in business, government, and the media. Available at http://www.fleishman.com.au/abr/jun03/product_recall.html.

professional critical incident management. As the *Australian Financial Review* observed, Pan Pharmaceuticals went "from market leader to industry pariah" in the proverbial blink of an eye. In years to come, the Pan Pharmaceuticals crisis will become a benchmark case study. However, the Pan Pharmaceuticals recall was not an isolated incident.[*]

According to Australian Treasury data, nearly 300 consumer products were recalled across Australia in the year to March 2003, costing companies many millions of dollars. In the United States, more than 1,000 products are recalled each year, and the bill soars to more than US$6 billion (2000 figures).The raw costs of a recall alone are damaging enough for a company's profit and loss figures. However, they pale into insignificance against the potentially disastrous impact on a brand's reputation and sales. And it can go even further. In the worst cases, the company faces liquidation, and the viability of the entire industry can be called into question. There are crucial points along the production and communication chains where critical incidents can be prevented or appropriately managed.

Failure to install effective quality assurance programs, failure to take the necessary withdrawal action at the right time, failure to create and follow a critical incident management plan, failure to implement a professional and comprehensive communications program, or simple ignorance of how to manage the situation all have the potential to escalate an issue into a commercial disaster.

12.3.1 GARIBALDI SMALLGOODS CASE STUDY[†]

The infamous Garibaldi Smallgoods crisis in the 1990s shows how a crisis management situation can spiral out of control, with disastrous and far-reaching consequences. It started with a contamination incident involving the South Australian company's metwurst in 1991. However, that was merely the forerunner of a fatal contamination in 1995.

In 1991, at the time of the initial crisis, Garibaldi was a category leader in South Australia's metwurst market. Its downfall began when a Port Pirie bride and others were struck down with food poisoning after eating Garibaldi salami at a wedding reception. The fatal crisis came with a major food poisoning outbreak in South Australia in January 1995, nearly 4 years later. Eventually, one child died and 24 people were hospitalized. The source of the poisoning was traced back to contaminated Garibaldi metwurst. Garibaldi was notified of the link with its product and

---

[*] The Therapeutic Goods Administration (TGA) of Australia has suspended the license held by Pan Pharmaceuticals Limited of Sydney to manufacture medicines for a period of six months with effect April 28, 2003 because of serious concerns about the quality and safety of products manufactured by the company. The suspension follows audits of the company's manufacturing premises, which revealed widespread and serious deficiencies and failures in the company's manufacturing and quality control procedures, including the systematic and deliberate manipulation of quality control test data. The license has been suspended to urgently address the safety and quality concerns posed by the multiple manufacturing breaches. Where the quality of a medicine cannot be certain, neither can the safety or effectiveness of that medicine. Available at http://www.tga.gov.au/recalls/pan.htm (accessed March 13, 2007).

[†] This event caused numerous regulatory changes in Australia and introduction of SQF.

immediately stopped all production of metwurst. "In a court of law, you're innocent until proved guilty. In the court of public opinion, you're guilty until proved innocent," warns Hayden Cock, Senior Vice President with communications consultants Fleishman-Hillard Stratcom. Cock says no product recall or other critical incident preparation is complete without a communications strategy and training. From this event, several programs were instituted by the government to protect the public from unsafe food.

12.4 UNDERSTANDING THE RECALL CONCEPT IN THE FOOD INDUSTRY*

Manufacturers strive to prevent a recall. Using Good Manufacturing Practices (GMP) and Hazard Analysis Critical Control Points (HACCP) plans are vital to preventing a recall. Even the best managed businesses can make occasional mistakes. It is important to be ready for a recall well before a problem occurs. Management must be part of an effective recall plan and team. The company management should not rely on product liability insurance in the event of a recall. Liability insurance might cover a portion of the losses resulting from a recall, but it will not cover the expense of product retrieval, and, most importantly, liability insurance will not help the company regain customer trust.

A food recall includes any corrective action by a company needed to protect consumers from potentially adverse effects of a contaminated, adulterated, or misbranded product. A recall is a voluntary action, and the recall decision is made by the company management. If the company does not initiate a recall, the government agency responsible for the particular product category may request that the company do so. Recalls are conducted by industry in cooperation with federal and state agencies.

Despite the undesirable nature of a recall event, it is in the best interest of the company to complete the recall quickly. Because the company is responsible for all of the costs involved in this process, it is critical to have a plan to cover recall expenses, to expedite the process without creating negative public opinion, and to prevent down time. When crisis hits, it is too late to work on the recall plan. Preplanning is vital to mitigate a crisis. Generally, recall events should be included in the Crisis Management and Emergency Contingency Program for a company.

Factors prompting a food recall include, but are not limited to, unsafe (toxin or diseased), contaminated, or mislabeled product, nonconformities to manufacturer's specifications, and missing allergen or other hazard warnings.

12.4.1 PURPOSE OF A RECALL

The basis of the recall concept depends on a company's food safety policies, ethical understanding, regulatory requirements, and financial constraints. A recall protects

---

* Excerpts modified from Gönül Kaletunç and Ferhan Özadali's "Understanding the Recall Concept in the Food Industry." Published by *Ohio State University Extension Fact Sheet* (nd). Available at http://ohioline.osu.edu/aex-fact/0251.html.

not only the consumer but also the company. A smooth recall process can save a company's name and prevent further damage resulting from negative publicity. Destroying, replacing, or altering the product are the three main corrective actions. A recall plan should strive to achieve the following goals:

- Protect consumer health
- Comply with existing rules and regulations
- Minimize the cost of the recall
- Regain and improve the company's reputation

12.4.2 Role of Government Agencies

Even though a recall is a company management decision, a government agency can force the company to recall potentially misleading and/or hazardous products from distribution and marketing. Two government agencies, the FDA and the U.S. Department of Agriculture Food Safety and Inspection Service (USDA FSIS), share regulatory responsibility for food product recalls. Although all recalls are voluntary, these agencies may ask the company to initiate a recall. To date, no company has ever refused a request from these government agencies to recall a potentially unsafe or hazardous product. However, if a company refuses to recall a product, the FDA and the USDA FSIS have legal authority to detain the product and to stop operations for good reason if the product constitutes a danger to public health. See Table 12.1 below for types of recalls.

12.4.3 Outline of a Successful Recall Process

- *Planning ahead:* A successful recall process depends on planning of the recall management well before a problem occurs.
- *Acting quickly:* Time is a vital factor in the recall process. The sooner harmful or misleading events are prevented, the faster the negative publicity and financial burden are eliminated.
- *Effective communication during a recall:* The firm should immediately provide recall instructions to everyone in the product distribution channels. Public notification about the recall through press releases and specialized media is also an integral part of the recall process.
- *Recall assessment:* Post-recall assessment is extremely important in determining the effectiveness of the recall plan to improve the efficacy of potential future recalls. The current recall plan also should be evaluated through simulated recalls.

12.4.4 Recall Overview

Planning ahead, rapid and well-coordinated action in the distribution channels, and truthful communication with the public are the most important elements for completion of a successful recall process and for regaining consumer confidence. The ultimate responsibility for removing the product from circulation before damage or

TABLE 12.1

Types of Recalls

Classification Definition Examples

Class I	This type of recall involves a health hazard where a reasonable probability exists that eating the food would cause **serious**, adverse health consequences or death.	Meat contaminated with *L. monocytogenes* in a ready-to-eat food product; *E. coli* O157:H7 in raw beef; allergens such as peanuts or eggs (not listed on the label).
Class II	This type of recall indicates a potential health hazard where a remote probability of adverse health consequences from eating the food exists, or if the resulting condition is **temporary** or medically reversible.	Presence of FD&C Yellow #5 dye in candy; presence of dry milk, a Class II allergen, as an ingredient in sausage without mention of the dry milk on the label.
Class III	This type of recall involves situations in which eating the food **will not or is not likely** to cause adverse health consequences.	A package containing fewer or lower weight products than shown on the package label or improperly labeled processed meat in which added water is not listed on the label as required by federal regulations.

Source: Data from the FDA.

injuries are caused belongs to the processor, manufacturer, etc. A recall requires manpower and financial resources. When a traceability system and a well-conceived recall plan are in place, the recall is likely to be successful and less expensive. Government regulatory agencies, FDA and USDA FSIS, are available to help companies with their hazard assessments.[*] If a company suspects a hazard, it should notify the Emergency Response Division (ERD), the Office of Public Health and Science (OPHS), or inform the nearest FDA or USDA FSIS office in the company's district so that the ERD office can be contacted as soon as possible.

- http://vm.cfsan.fda.gov/~lrd/recall2.html
- http://www.fda.gov/fdac/features/895_recalls.html
- http://www.fsis.usda.gov/OA/recalls/rec_intr.htm

See Appendix N for "Why Product Insurance Is Needed and What Is Offered," a discussion with Bernie Steves of brokerage house Insurance Brokers Services.

---

[*] The products under the jurisdiction of these two agencies differ. The FDA is responsible for domestic and imported foods. The USDA FSIS is responsible for meat and poultry. As an exception, responsibility for eggs is shared by the FDA and the USDA. USDA FSIS regulates pasteurized egg products (eggs that have been removed from their shells for further processing), and the FDA assumes responsibility for egg products after leaving the processing plant.

12.5   FOODTRACK, INC. AND FOODTRACK INCIDENT REPORT

FoodTrack, Inc.
111 Broadway
New York, NY 10006
Phone: 212-227-6460
Fax: 212-385-7870

FoodTrack Incident Report
838 Forest Glen Lane
Wellington, FL 33414-6328
Phone: 800-397-7202 (toll-free)
Fax: 561-333-7770

Available at www.foodtrack.net.

FoodTrack, Inc. is an international surveillance and food tracking service, which provides around-the-clock food incident surveillance and preemptive food event reporting on biosecurity issues, tampering incidents, terrorist events, product recalls, food-borne illness outbreaks, and similar product contamination events affecting food and beverage products and ingredients. FoodTrack views themselves as a strategic partner to leading food processors, distributors, supermarket chains, restaurant chains, wholesalers, and produce companies. FoodTrack offers real-time, mission-critical reporting before an incident becomes a corporate catastrophe.

A recent Lloyds of London study of the food and drink industry's corporate image found:

- A company's brand name is its most valuable asset.
- Product contamination is the most serious risk to its corporate reputation.

Fast detection and immediate corporate response can contain, even prevent, a crippling crisis. Failure to identify and quickly respond to an incident can lead to litigation, devastating financial loss, and irreparable damage to corporate reputation. Late notification from a supplier or none at all is a major risk for every food company, large and small.

Companies that thrive in the face of adversity are companies that prepare in advance and have crisis management teams that spring into action when a crisis occurs. However, even the best crisis management team cannot launch an effective response to a threatening event it does not know about.

The FoodTrack Incident Report focuses on:

Primary products: Food and beverage products, ingredients, raw materials, crops that are regulated by the FDA, USDA, Environmental Protection Agency (EPA), Health Protection Branch (Canada), and local, state, and provincial health authorities.

Events covered: Biosecurity issues, terrorist events, outbreaks of food-borne illness, product recalls, accidental contaminations, product tampering, product mislabeling, product adulteration and misbranding, product extortion, enforcement warning letters, and government agency warnings and alerts and similar food safety issues/incidents that are principally reported in the media and/or published on the Internet by government regulatory agencies.

12.5.1 PREEMPTIVE FOOD EVENT REPORTING

FoodTrack's Flash Product Alert Bulletins include:

- *Terrorist event alerts:* actual, threatened, or rumored attacks on the food, drug, or water supply
- *Biosecurity issues bulletins:* Periodic reports covering biosecurity issues and events
- *Product recall alerts:* Recalls and market withdrawals
- *Product tampering alerts:* Tampering and product extortion information
- *Outbreak alerts:* Outbreaks of food-borne illness that could affect business
- *Product contamination alerts:* Contamination events not otherwise classified
- *Heads-up alerts:* Situations that could lead to a recall or outbreak
- *Purchasing managers bulletin:* Bulletins covering product contamination incidents that may necessitate action by purchasing personnel[*]
- *TrendTrack reports:* Periodic reports on emerging outbreaks, contamination, tampering, and extortion trends or incidents
- *Foodtrack's warning letter bulletin:* Weekly summary of the most recent FDA warning letters, distributed on the date the information is made public by FDA
- FDA import alerts
- Weekly FDA enforcement reports
- Quarterly USDA/FSIS enforcement reports
- FedReg update

12.5.2 SURVEILLANCE: ELECTRONIC REAL-TIME MONITORING AND DATA FILTERING

FoodTrack uses a proprietary, state-of-the-art, information gathering and dissemination process for real-time monitoring (and manned full-text filtering) of more than 500 individual electronic news sources, including news wires (direct feeds), newspapers, e-magazines, select government websites, and television news transcripts. Leading news channels are also watched around the clock.

This information is filtered in real time through more than 150 unique "Tracking Profiles," developed by FoodTrack, consisting of customized code and queries resulting from years of research and development. This exclusive process yields prefiltered, up-to-the-minute, food safety and security news outputs for distribution.

---

[*] Recent examples include developments that unfolded subsequent to the StarLink corn contamination and disruption of supplies of baby back ribs resulting from the FMD outbreak in Europe.

For this type of reporting, 24 hours per day/7 days per week/365 days of the year, FoodTrack personnel are at work to provide clients the information they need to respond quickly and decisively when incidents occur. Other services and reports include:

Executive SnapShot summary: Highlights pertinent aspects of a breaking story, followed by the full text and a list of previous alert bulletins for cross-reference. Executive SnapShot summarizing is an additional level of filtering (reading and analysis) performed by the staff to provide clients the ability to quickly review key information in the bulletin, assess its relevance to their organization, and react immediately when needed.

Focused reporting/bulletin delivery: FoodTrack bulletins are distributed electronically, via e-mail, and via text messaging for immediate broadcast following an incident. This real-time information delivery enables clients to respond quickly and provides critical time needed to assess crisis situations that may threaten the integrity of their products, the safety of consumers, and the value of their corporate reputation.

FoodTrack safety and defense bulletins: FoodTrack offers customized reports that provide subscribers the information they need.

Worldwide events are covered selectively to accommodate clients with multinational operations and to identify food safety and security threats that may first emerge overseas and ultimately impact North American interests via imported products and ingredients, as well as to track threats, plots, methods, and tactics likely to be used by terrorists against food supply targets in North America.

FoodTrack's standard bulletins and incident alerts cover food and beverage products, ingredients, raw materials, livestock, crops, and seafood for which the information source states that the recalled or contaminated products were distributed.*

See Appendixes O and P for food recalls that pertains to Sudan 1 and other short case studies.

12.6   AUSTRALIA NEW ZEALAND FOOD AUTHORITY (ANZFA) FOOD INDUSTRY RECALL PROTOCOL

The Food Recall Coordinator
Food Standards of Australia and New Zealand
P.O. Box 7186
Canberra BC ACT 2610
Australia
Phone: +61(0)6271-2222
Fax: +61(02)6271-2278

Available at http://www.aar.com.au/pubs/prod/recalloct01.htm and http://www.food-standards.gov.au.

---

* Regardless of the selected zone, all recipients shall receive the Biosecurity Issues Bulletin, Terrorist Event Alerts, Weekly FDA Enforcement Report (Foods Version), Quarterly USDA/FSIS Enforcement Reports, Weekly FDA Warning Letter Bulletin, and selected TrendTrack Reports and Purchasing Managers Bulletins.

Every year in Australia a number of food manufacturers and distributors are faced with the prospect of having to conduct a food recall. According to the Australia New Zealand Food Authority (ANZFA) Australia ranked second only to the United States in terms of the number of food crises during the past 5 years. To address these concerns, the ANZFA Protocol was created. Excerpts from Kylie Giblett's work are highlighted in this section. The ANZFA Protocol contains a step-by-step guide to conducting a recall, including the following:

- Forward planning
- Convening a recall committee
- Conducting a hazard/risk assessment
- Determining the level of the recall
- Notification requirements
- Post-recall reporting
- Responsibilities of persons and companies at each level of the supply chain or network in the event of a recall

The ANZFA Protocol also contains up-to-date contact lists and sample documents relevant to each stage of the recall process.

12.6.1 Changes in the Protocol

The major change introduced by the revised Protocol is the new classification of different levels of recall. Under the previous edition of the Protocol, food recalls were classified wholesale, retail, and consumer recalls. The new edition of the Protocol has only two levels of product recall: trade recalls and consumer recalls. The Protocol describes these two levels of recall as follows:

- A trade recall involves the recovery of the product from distribution centers, wholesalers, major catering outlets (e.g., hospitals), and outlets that sell food manufactured for immediate consumption or food prepared on the premises.
- A consumer recall involves the recovery of the product from all points in the production and distribution chain or network, including recovery from consumers.

Obligation under the Australia New Zealand Food Standards, under Clause 12 of Standard 3.2.2 of the Code, all food businesses engaged in the wholesale supply, manufacture, or importation of food must:

- Have in place a system to ensure the recall of unsafe food
- Set out this system in a written document, and make this document available to an authorized officer on request
- Comply with this system when recalling unsafe food

12.6.2 Maintenance of Records and Contact Details for Distribution Networks

In a recall situation, it is vital that all products to be recalled can be located quickly and that the relevant people can be contacted to halt further distribution as soon as possible. The maintenance of up-to-date and easy to follow records is essential if a recall is to be carried out quickly and efficiently. The Protocol contains a number of suggestions as to what details should be included in records of product distribution.

12.6.3 INSURANCE

In preparing for a recall, it is important to consider who will be paying for the recall. Food businesses should review their current insurance cover to determine whether it includes the costs of a recall and any consequential loss.

Food Industry Recall Protocol also offers "A Guide to Writing a Food Recall Plan and Conducting a Food Recall" in their fifth edition, published June 2004.

Highlights include its executive summary: It is now a legal requirement under Chapter 3 of the Food Standards Code, Volume 2 (Food Standards Code) for manufacturers, wholesalers, distributors, and importers of food to have in place a written recall plan. It is noted that this legal requirement applies to Australia only and does not cover New Zealand. The purpose of a recall plan is to enable a food business to recall unsafe food from the market and consumers to protect public health and safety.

12.6.3.1 Product

- Product brand name and description, including package size and type
- Lot identification (batch or serial number)
- "Use by" date, "packed on" date, or "best before" date where relevant (may also be the lot identification)
- Australian sponsor and contact telephone number (including after-hours number)
- Quantity of the batch manufactured, and the date and the amount released
- Distribution within Australia
- Overseas distribution of any exported product

12.6.3.2 Other Relevant Information

- Name and telephone number of the person reporting the problem
- Date of the report
- Number of similar reports received (e.g., customer complaints)
- Availability for investigation of suspect sample or other samples
- Action proposed by the sponsor and proposed recall level

12.6.4 RESPONSIBILITIES OF MANUFACTURERS, WHOLESALERS, AND IMPORTERS

Sponsors who are manufacturers, wholesalers, or importers have the following general responsibilities in relation to food recalls:

- To maintain records and establish procedures that will facilitate a recall. Records should be in a form that can be quickly retrieved

- To have a written recall plan
- To initiate the action for implementing a recall
- Contact overseas supplier/manufacturer when initiating recall action

12.7   OURFOOD.COM—DATABASE OF FOOD
AND RELATED SCIENCES

OurFood.Com
Karl Heinz Wilm
Mühlenweg 5
D-26419 Schortens
Germany
E-mail: author@ourfood.com

Available at http://www.ourfood.com.

OurFood is a database containing information concerning food, related physiology, technology, analytical methods, bacteriology, and topics of general interest. This resource is more politically pointed or directed than others, and the author has strong feelings about how agricultural systems should work. His thoughts on health, industrialization, and globalization are played out below in his narrative on these subjects. However, the databases and resources made available online are very helpful and may be of use by specialists or for general inquiry.[*]

12.7.1   GENERAL INFORMATION REGARDING THE CREATION OF
OURFOOD.COM DATABASE

12.7.1.1   Health

No physician denies the truth that the most frequent causes of illness are based on wrong behavior related to food. More information about food is necessary to avoid an unhealthy lifestyle and to cut the cost of resulting medical care. In addition, often consumers cannot avoid contaminants and other dangers of modern food.

12.7.1.2   Industrialization

Food is increasingly industrialized. The health food (reform food), bio food, and alternative food are being commercialized. Because of a wide distribution, the shelf-life must be kept long. Vitamins and proteins lose their value.[†]

---

[*] This section's information is derived from the OurFood.com website, with permission. The author, Karl Heinz Wilm, is a biochemist, graduated from the Faculty of Pharmacy of the University of Belem do Para, Brazil. As a member of the Council of Pharmacy of Porto Alegre, he became director of the section of bacteriology of the Biochemical Laboratory Dr. Friedel in Sao Leopoldo, Rio Grande do Sul Brazil, and later chief chemist of the laboratory of food industry.

[†] The recent opening of the European Common Market adds further power to giant industries. Concentration on the retail sector has destroyed 60,000 full-time jobs in Germany. Mergers and acquisitions are the prime culprit. When a smaller company is taken over, a number of duplicated functions are amalgamated leading to low-overhead companies with smaller workforces.

12.7.1.3 Globalization of Trade and Industry

Globalization of multinational companies destroys the ecological isolated markets introducing the global business. Dumping prices from abroad destroys smaller industries, killing jobs. Economic and ecological isolated units like the habitation in the Amazon jungle as a self-feeding unit will be a picture of the past.*

To overcome the negative sides of dangerous foods, industrialization, and globalization, the author offers online databases like OurFood. These free databases provide information on how to avoid the menace of daily poisoning. According to Wilm, "Be careful not to fall into sectarian thinking—allow always arguments of the other side."

Topics of OurFood databases include:

- Introduction/about the author
- Genetic modification of food
- Parasites and pathogenic protozoa
- Radioactivity and food
- Future of global nutrition
- Foot and mouth disease

- Food-borne virus diseases
- Nutritional genomics

- Molds and yeasts
- Food, what is it?
- HACCP and ISO 9000
- Global food safety
- Food poisoning
- Hygiene monitoring
- Physiology

- Packaging

- Anthrax
- Phytopathology
- General bacteriology
- Ingredients
- Bioterrorism
- Dioxin

- Bovine spongiform encephalopathy
- Bibliography

12.7.2 SAMPLE: BIOTERRORISM SUBSECTIONS

Excerpts are illustrated below from various sections.

- Food and bioterrorism
  - The Bioterrorism Security Act
    - Dangerous agents
  - Food terrorism and sabotage
    - World Health Organization (WHO) food safety response to terrorist threats
  - Surveillance, Preparedness, and Response
  - WHO and food terrorism
    - International Health Regulations (IHR)

---

* The retail sector is also becoming global. Carrefour, a retail group with head in France, reports the opening of 10 new business fields in Brazil, China, the Czech Republic, Korea, Malaysia, Mexico, Spain, and Taiwan. The total number of Carrefour stores has grown to 345 in 20 different countries.

12.7.3   HIDDEN DANGERS IN INDUSTRIAL PROCESSING OF OUR FOOD: FOOD SAFETY AND CONTROL SYSTEM*

12.7.3.1   Simple System of Traceability—OurFood Recommendations

At the farm: if there is no official veterinary numbering system, a farm numbering system of the animals should be started:

- Tagging: animals should be tagged with an identification number
- Form: a form sheet for every animal should be created with the following data
  - Identification number
  - Date of birth
  - Species of the animal and other relevant information
  - Identification number of father and number of mother
  - Diseases during lifespan
- Agrochemicals and pesticides used on the farm
- Feedstuffs and its supplier; lot numbers with in and out date
- Veterinary chemicals used during lifespan with date of use
- Name of slaughterhouse or other enterprises taking over the animal

Note: Within this system there is no apparent information regarding the environmental concerns, nor about animal quality of life or farm management systems.

Under Database of Food and Related Sciences, food standards International Organization for Standardization (ISO) 15161:2001 and 22000:200x, OurFood.com notes that recently ISO published the standard ISO 22000 "Food Safety Management Systems—Requirements." This system is different than ISO 15161:2001. ISO 15161 has a wider scope dealing with all aspects of food quality and illustrates how the HACCP system can be integrated into a quality management system, whereas for IPT ISO 22000 concentrates exclusively on food safety and instructs food producers on how they can increase their food safety system.

12.7.3.1.1   ISO 22000 Food Safety Management Standard

- ISO 22000 aims to harmonizing the relevant national standards on the international level
- ISO 22000 will be international and will define the requirements of a food safety management system covering all organizations in the food chain from farmers to catering, including packaging

The standard has the following objectives:

- Comply with the Codex HACCP principles.
- Harmonize the voluntary international standards.
- Provide an auditable standard that can be used for internal audits, self-certification, or third-party certification.

---

* See website for expanded version: http://www.ourfood.com/download.html.

- The structure is aligned with ISO 9001:2000 and ISO 14001:1996.
- Provide communication of HACCP concepts internationally.

The ISO 22000 gives definitions on related terms and describes a food management system.

- It is a food safety management system.
- Can be used for verification, validation, and updating.
- There is correspondence between ISO 22000:200x and ISO 9001:2000.

12.7.3.1.2   Identification

An identification system using standardized identifiers, such as European Article Number/Uniform Code Council (EAN.UCC) product barcodes for labeling materials traded between businesses, may be very useful. Traceability is already a demand of ISO 9001:2000.

See Appendix Q for OurFood.com's Database of Food and Related Sciences table of contents.

12.8   PRODUCT RECALL INSURANCE

MarketScout.com
5420 LBJ Freeway, Suite 850
Dallas, TX 75240
Phone: 972-934-4299; 800-500-8720 (toll-free)
E-mail: nalberigo@marketscout.com

Available at http://www.productrecallins.com (accessed August 30, 2006).

MarketScout or MarketScout.com is an Internet company that offers product recall insurance through their website. Products Recall offers insurance protection in the event of a recall of an insured's product. This protection includes coverage for the insured's product recall expenses and liability to third parties for both finished and component goods. MarketScout emphasizes that a recall may involve numerous expenses, including:

- Costs associated with notifying customers
- Cost of shipping and disposal of the product
- Extra warehouse expenses
- Cost of extra personnel required to conduct the recall
- Cost to refund, repair, or replace and ship the product back to the customer

12.8.1   PRODUCT RECALL: INDUSTRY INFORMATION

The coverage has been around since the 1980s. The first type of product recall insurance was called malicious product tampering, which only responded to malicious incidents. The limits were $3 million, with six-figure premiums. Because it is catastrophic insurance in nature, when losses occur, they are generally major. Clients are not concerned with the smaller losses that they can handle financially in house. What they are looking for is protection from the large losses. Products Recall is designed to help the insured manage the crisis of such an occurrence and help protect against product degradation and third-party lawsuits.

Regarding which form of cover, Coverage A or B, is most advantageous for companies, MarketScout advises that many companies need both. Any company that sells finished goods under its own label has a greater exposure under Coverage A than Coverage B. They will handle the recall, in most cases directly incurring the recall expenses. However, if there is a third party between their company and the ultimate consumer, they also may have an exposure under Coverage B because that third party can claim loss of income or reputation as a result of the recall.

Similarly, any company that produces or processes a product that is ultimately sold under a third party's brand name, whether it is an ingredient or the finished product, has a greater exposure under Coverage B. The company may or may not be involved in the decision to recall or be involved in the recall itself but can still be held liable for damages by the ultimate seller. They also offer endorsements to extend coverage, including:

- Cost to refund, repair, or replace the insured's product
- Worldwide coverage
- Impaired property recall response is available

MarketScout can provide experience and knowledge of various recall coverages available and tailor a company's program to fit the requirements of an insured. In the event of a recall, claim expertise and a legal panel are available to guide an insured through government regulations and requirements.

12.8.2 POLICY FEATURES (SIMILAR TO OTHER RECALL INSURANCE)

Coverage A pays the first-party expenses associated with the recall, such as notification, shipping, warehousing, and additional personnel. Through attachment of an endorsement, Coverage A can be extended to include the cost of repair, replacement, or refund of the product.

Coverage B provides coverage for the claims by third parties seeking damages as a result of a product recall. Coverage B may be extended by endorsement to cover liability for impaired property. The optional impaired property endorsement provides coverage for the insured's products being incorporated into another company's product and causing it to not function properly.

12.8.3 TARGETED CLASSES

MarketScout's preferred market segments are accounts with annual sales of less than $700 million. Classes of business include food and beverage, medical, pharmaceutical, consumer, and industrial products. Target classes include:

- Meat
- Bakery
- Breweries
- Food flavoring
- Meat/poultry accounts
- Printing/packaging

- Computers
- Toys and games
- Can manufacture
- Bottle manufacture
- Firearms and ammunition
- Vitamins, furniture, fixtures

- Exercise equipment
- Household appliances
- Electronic components
- Medical/safety products

12.8.4 Product Features

Product recall covers expenses up to $10 million; product recall covers liability up to $10 million. The minimum premium is $25,000 per year.

In addition to Products Recall, MarketScout can provide a wide range of other risk management programs, such as Contaminated Products Insurance (CPI). This product is designed for food and beverage companies and covers losses associated with malicious product tampering and accidental product contamination. Coverage encompasses the far-ranging costs associated with these incidents, including the costs of the recall itself and related business interruption, business rehabilitation, and consultant expenses.

12.8.5 Classes of CPI Business

Classes of risks include a wide variety of manufacturers, processors, and retailers in the food supply chain, as well as pharmaceutical and cosmetics. The major eligibility factor is that the product must be ingestible or topical. Typical food risks include canneries of fruits and vegetables, manufacturers of grocery products such as breakfast cereals and boxed flour or grain products, condiments, baked goods and snack food, and supermarket chains.

Target classes of business include:

- Bakery
- Candy
- Dry ready-to-eat meals
- Spirits, wines, and breweries

- Chocolate
- Coffee/tea
- Cookies and crackers

- Spices
- Retail
- Supermarkets

Classes they DO NOT write:

- Nutraceuticals
- Basic grains and animal feeds
- Importers of food products
- Meat, poultry, slaughter, packing, and processing

- Bean sprouts
- Unpasteurized juices
- Restaurants, cafeterias, and buffets

12.9   SEEDSMEN PROFESSIONAL LIABILITY INSURANCE

Rattner Mackenzie Limited
37 Radio Circle Drive
P.O. Box 5000
Mount Kisco, NY 10549-5000
Phone: 914-242-7860
Fax: 914-241-8045

Available at http://www.rattnermackenzie.com.

Formed in 1988, Rattner Mackenzie has experience in insurance and reinsurance brokerage, and employs more than 70 employees located in London, New York, and Bermuda. In 1999, Rattner Mackenzie was purchased by HCC Insurance Holdings Inc., an international insurance holding company and a leading specialty insurance group established in 1974. Below is an example and excerpts of Rattner Mackenzie's Seedsmen's Professional Liability Insurance.

12.9.1   Seedsmen's Professional Liability Insurance[*]

Also known as "Seedsmen's Errors and Omissions Insurance" (E&O).

12.9.1.1   Insurance Program for International Seed Federation (ISF) Members

In recognition of the need for a global approach to risk management, the International Seed Federation[†] (ISF), in conjunction with Rattner Mackenzie Ltd. and Certain Underwriters at Lloyd's, has developed a tailor-made professional liability insurance product for the seed industry. Who needs seedsmen's professional liability insurance? The answer is everyone who is involved in growing, conditioning, or distributing seeds. At any stage of the seed business mistakes can occur in the selection, conditioning, packaging, or testing of the seed that can cause or contribute to the loss in whole, or in part, of the customer's crop. Even the most professional organizations may suffer from a lapse in standards by a distracted staff member, which can seriously impact the company's balance sheet if there is no applicable insurance. For those who produce parent seeds, product liability is not enough.[‡] Seedsmen's professional liability insurance should be carried in addition to products liability insurance because the coverage of the two insurances are entirely different but complement each other. Seedsmen's professional liability insurance covers claims against the seedsmen that may result from the failure of the seed sold to conform to the variety or other specified qualities or from the seed sold being unsuitable for the purpose

---

[*] Seedsman is another word for parent seed producer, breeder, or dealer in seeds.
[†] See http://www.worldseed.org for more information.
[‡] Products liability covers bodily injury and property damage, both of which are excluded by the seedsmen's professional liability insurance. Sometimes a products liability policy is enhanced to cover misdelivery, but that is not an adequate substitute for seedsmen's professional liability insurance.

specified as a result of an error, negligent act, or omission by the company or its employees.

12.9.2 TYPES OF CLAIMS MADE ON SEEDSMEN

There are six main categories of claims outlined below. In most cases the key factors are the adequacy of the seedsmen's quality control and seed-testing procedures, including the sampling procedures, and the depth of the plaintiff's distress at having lost all or part of his or her harvest and subsequent profit (sometimes blaming the seedsmen, instead of his or her own farming techniques).

1. *Mechanical error:* Such as errors in labeling, mixture of the wrong kinds of varieties of seed, inadequate labeling, or inadequate laboratory testing for germination.
2. *Overzealous distribution:* This includes verbal and catalogue warranties that may result from a salesperson or parties overrepresenting the seed product and are beyond the control of the seed producer. ISF recommends the use of a Standard Disclaimer of Warranty and Limitation of Liability, which is a protection against some of these claims.
3. *Germination deficiencies:* Careful grow-out testing and strenuous policing by official state and federal seed-testing laboratories can control this type of loss. Although claims may be less frequent, they tend to be particularly severe when they occur.
4. *Misapplication:* Claims resulting from seed failing to perform in a given area.
5. *Disease control problems:* Susceptibility to disease varies depending on the susceptibility or genetic resistance of the type of seed planted. Seedlings or plants may become infected with disease as a result of seed-borne organisms or may be infected by disease organisms in the soil or on plant residue. Damage can be reduced by disease control treatments and not planting in areas known to be infected with the disease.
6. *Miscellaneous problems:* Improper and inadequate pollination can produce substandard seed and consequential losses. Claims also arise from failure to carefully rogue undesirable plants and/or varieties from the seed field and, in particular, carelessness in harvesting the seed production fields.

12.9.2.1 Key Features

The policy wording covers claims made during the policy period, excluding claims/circumstances that are known at the inception of the policy. It defines "seed(s)" as including "seeds, bulbs, plants roots, tubers or other similar means of plant propagation." The policy covers worldwide sales of all crops, including genetically modified organisms. *Note:* There is no mention of pollen drift or other acts of nature or weather.

The limits and deductible are both inclusive of defense costs and expenses. This means that the insured must contribute to the defense of any claim, and the sum insured should be adequate to include these expenses. The deductible would be geared to an insured's turnover and to the insured's own claims record. Vegetable seeds would have a larger deductible than agricultural seeds.

The underwriters are willing to offer a catastrophe protection for those large risks that only require insurance to protect them against the unusually large claims. Capacity is available to provide limits up to US$10 million/UK£10 million or more.

In addition to seedsmen's professional liability insurance, they also offer:

- Allied health and nursing homes cover
- Directors and officers liability insurance
- Employment practices liability insurance
- Miscellaneous professional liability insurance
- Errors and omissions liability insurance
- Law firm professional liability insurance
- Medical and dental malpractice insurance
- Program business insurance

REFERENCES

Garrett, Arthur S. 2005. *Food Recall Insurance: Why Your Company Should Take a Look.* Keller and Heckman LLP.

Weaver, Patrick. 2003. "Product Recalls: Disasters Waiting to Happen." *Australian Business Report* June–July, issue 3. Fleishman-Hillard Stratcom.

Section V

Research Instruments

Part V's three chapters highlight how a scorecard matrix, cost-benefit spreadsheet, and cost-benefit questionnaire can assist in evaluation of identity preservation and traceability (IPT) system efficiency and purity cost-benefit comparisons, and improve understanding of IPT systems focusing on farm-level production (grain) data. These evaluation systems are based on auditing and toward a goal of International Organization for Standardization (ISO) 22000 compliance. The two spreadsheet analyses are related to one another; matrix represents effectiveness of a program, whereas the cost-benefit represents the efficiency of a program. Data for both spreadsheets were derived from the farmer survey questionnaire (see Chapter 15 for details) and used example data for analysis of the spreadsheets. Where survey data were not provided, data from agricultural literature were used.

Chapter 13's IP Scorecard Matrix provides an effectiveness comparison of a single-farm IP system, i.e., comparing the *standard* (specified—required documentation, procedures, and data) with what is *actually* accomplished. Three category areas are evaluated according to three criteria or objective characteristics. To help in the understanding of the scorecard matrix, a conceptual model is provided (Table 13.1). From the input data provided, calculations are compiled and are highlighted in Table 13.2 and output results are illustrated in Figure 13.1.

Chapter 14's IP Cost-Benefit Spreadsheet provides an extensive, but not exhaustive, spreadsheet that focuses on identity preservation costs and revenues generated as applied to varying purity levels of crop production. The chapter highlights the numerous cost components associated with grain production at various levels of purity. To better understand the spreadsheet, a conceptual model (Table 14.2) is provided. Table 14.2 provides a brief illustration of the spreadsheet used. See Appendix R, Cost-Benefit Spreadsheet—Complete, for the entire spreadsheet, which compares the various systems and associated costs. The appendix also provides individual costs on a per bushel basis for all levels of purity. Spreadsheet results are summarized in Figure 14.1.

Chapter 15's IP Cost-Benefit Questionnaire examines a farmer questionnaire that focuses on two critical periods of farming, namely, the two weeks that surround planting of the crop and the two weeks that surround harvesting of the crop. Because the owner/manager is most responsible for ensuring critical processes, inspections, etc., the questionnaire collects data regarding the time he or she spends on specific identity preservation tasks. The questionnaire also collects and compares standard and identity-preserved production data for comparing cost-benefit evaluations.

13 Identity-Preserved Scorecard Matrix

13.1 GOAL AND STRUCTURE OF THE SCORECARD MATRIX

The Scorecard Matrix provides an effectiveness evaluation of a single-farm identity-preserved (IP) system by evaluating three category areas: (1) the standard—required (i.e., purity levels, tolerances, etc.), (2) performance measurement entities/parameters (performed by farmer, buyer, and specified point items), and (3) communications—between farmer and buyer (transparency of nomenclature, measurements, software, etc.).

This evaluation system is based on auditing and toward a goal of International Organization for Standardization (ISO) 22000 compliance. This and the next chapter's spreadsheet are related to one another; the matrix represents effectiveness of a program, whereas the cost-benefit represents the efficiency of a program. Data for this spreadsheet were derived from the farmer survey questionnaire (see Chapter 15 for details) and used as example data for analysis of the spreadsheets. Where survey data were not provided, data from agricultural literature were used.

The effectiveness evaluation compares the standard or contractual specifications with what was performed or complied with. Each of these categories is evaluated by three criteria (objective characteristics). The three criteria follow along U.S. Department of Agriculture's (USDA) Golan et al. (2004) format and are used to evaluate identity preservation and traceability (IPT) systems, which look at breadth (amount of data recorded), depth (how far forward/backward data are recorded), and accuracy (the ability to measure standard tolerance to output measure).[*] To help in the understanding of the Scorecard, 13.1, Conceptual Model of Scorecard Matrix, is provided and highlights formulas used (with examples) and definitions for the criteria (objective characteristics). Table 13.2, Scorecard Matrix Spreadsheet, provides examples of the categories via input and output columns. The output results are summarized and highlighted in the far right (**Difference** columns), as well as in Figure 13.1, IPT Measurement Score graphic. Graphic criteria parameters (depth and breadth) are compared along the y-axis and are compared with what was recorded or accomplished. Accuracy, regarding harvested output purity, is also compared along the x-axis, as well as other types of IP systems. The other IP systems are illustrated (as examples) for comparison, comparing each standard's relative proportion of breadth, depth, and accuracy.

---

[*] In Elise Golan's work, she cites the objective characteristics as breadth, depth, and precision (precision meaning repeatability of testing). However, for this work, precision is replaced by accuracy to reflect laboratory and field tests accuracies, for comparison purposes.

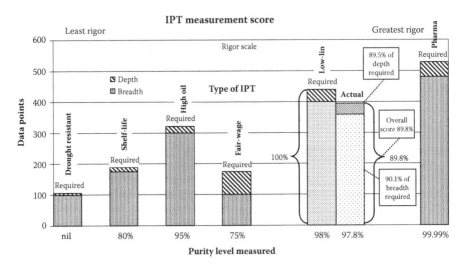

FIGURE 13.1   IPT measurement score. (Developed from research.)

13.2   SCORECARD MATRIX MODEL

The conceptual model of the IP Scorecard Matrix (Table 13.1) is divided into several sections. The top right portion of the page shows the weighted average models; along the left side are the two accuracy measurements; and at the bottom of the page are the definitions for each criterion (objective characteristic).

The goal of the Scorecard Matrix is to provide an IP program effectiveness evaluation based on the criteria of breadth, depth, and accuracy as applied to contract-specific categories. Breadth data standards are explicit checks, cleanouts, etc. that must be completed to specific criteria and recorded. Measured scores are proof or certified by a third-party auditor that the specific contract parameters were completed as per records and/or observations. The same type of auditable, third-party review is conducted for depth. In this case, the auditor confirms that previous stages of production records are included and appropriate, i.e., meet contract specifications. Accuracy data are composed of contract specifics, i.e., with laboratory and field test parameters, test dates/conditions, and associated paperwork results, as proof of testing parameters and output results.

The criteria (objective characteristics) of breadth and depth are each measured according to their weighted average score (or compliance). The reason that the weighted average is used is to now skew the value of one group of auditable parameters (e.g., few data points) over another (one with many data points). In other words, this is to provide appropriate value to a criterion that has two criteria points to one that has 200 criteria points. In this case, each of the latter 200 individual points has as much value as each derived from the two-point criteria category. For example, to determine the weighted average for breadth, several mathematic functions must take

TABLE 13.1

Conceptual Model of Scorecard Matrix (Developed from Research)

		Weighted Average (WA) Score or Weighted Average of Compliance
Accuracy measurements	Breadth WA =	$\Sigma\,[(1Ai^{Pr}_B * 1Ai^{Cr}_B),\ldots(3B^{Pr}_B * 3B^{Cr}_B)]\,/$ $\Sigma\,(1Ai^{Pr}_B,\ldots3B^{Pr}_B)$
Accuracy: Output Purity Level = Final test of crop purity	Depth WA =	$\Sigma\,[(1Ai^{Pr}_D * 1Ai^{Cr}_D),\ldots(3B^{Pr}_D * 3B^{Cr}_D)]\,/$ $\Sigma\,(1Ai^{Pr}_D,\ldots3B^{Pr}_D)$
Accuracy: Test Range = The minimum and maximum test scores from tests conducted	Example: $2Aii^{Pr}_B = 200$ $2Aii^{Pr}_B$ represents the **Points required (Pr)** [200 pts.] from Standard (required) column—Breadth (B), for category Performance Measurement Entity/ Parameters, subcategory Primary Entity (farmer)—Operations	Example: $3B^{Cr}_D = .93$ $3B^{Cr}_D$ represents the **Compliance rate (Cr)** [0.93] from Difference column Depth (D), for category Communications (Producer/ Buyer), subcategory Trait(s)/ Attribute(s)

Breadth describes the amount of information or data points collected (usually determined by agreement or contract). Breadth WA is the weighted average ratio (or percentage) of complied breadth points (actual) to mandated breadth points (required).

Depth is how far backward or forward the system tracks pertinent information (e.g., the total number of entities before or after the farm, including the farm). Depth WA is the weighted average ratio of complied depth points (actual) to mandated depth points (required). For example, an IPT system for fair wages would extend to harvest; for shade grown, to cultivation; and for nongenetically engineered, to the bean or seed.

Accuracy is the degree of *conformity* of a measured or calculated quantity *to its actual (true) value*.

Note: *Precision*, also called reproducibility or *repeatability*, is the degree to which further measurements or calculations show the same or similar results. *Precision* also reflects the degree of assurance with which the IPT system can pinpoint a particular food product's movement or characteristics. In some cases, the objectives of the system will dictate a precise system, whereas for other objectives a less precise system will suffice.

place. First, the compliance ratio (Cr) must be determined. This is done by dividing measured (actual) over standard (required) that produces a Cr value. Second, each category's (including subcategories) required data points or points required (Pr) are multiplied by its corresponding compliance rate (Cr) (Cr is the ratio difference between what was observed [actual] over the standard required [Pr]). All category product calculations are then added together into one overall (Pr × Cr) value. This Pr × Cr value is then divided by the total number of points required (Pr), derived from the total of all categories, which produces the weighted average score or compliance. The same process is done for the criteria of depth.

Accuracy criteria have two output measures. The first is the output purity level analyzed as the final test of crop purity. This is usually done just before or at the sale of the product (see Table 13.2, section 1Ai for seed purity and tolerance levels). This number is compared with the standard requirement to verify compliance. The second accuracy criteria measurement represents the minimum and maximum test scores derived from the performance measurement category for laboratories, field tests, etc.

The bottom portion of the model provides explanations for the terms used.

13.3 SCORECARD MATRIX SPREADSHEET

The spreadsheet has several sectional columns and a results row. Starting from the left: (1) definitions, (2) mathematic functions, (3) narrative of categories and subcategories to be measured, (4) the six vertical columns for data input (Standard—required and Measured—actual), and (5) the far right three columns (Differences), and along the bottom row are the results of various calculations.

Of importance is the Category narrative column (column three), which has three categories, each of which possesses its own subcategories that describe particular contractual points, or points required (Pr) to be measured. This area can be modified or tailored to meet other IP programs or contract specifications.

Regarding the input of data, under the three **Standards (required)** columns, the standard's number of auditable parameters, or point required (Pr) to be observed (contractually or by regulation), is inserted. These data can be derived from the contract or from whoever is conducting the survey, e.g., customer or auditor. It is envisioned that an ISO 22000 format will be used in the near future. This is the standard to which the bases of calculations are made. For example, if a farmer is to perform 200 recordable tasks or auditable parameters (the addition of all points, i.e., chemical data, storage, cleanouts, inspections, and prescribed tasks), then 200 is entered and represents 200 points required (Pr) data. Under the three **Measured (actual)** columns, the number of data points observed and/or measured, in accordance with the contract, is inserted. Typically this value or score will be verified and/or observed by a third-party auditor. A third-party auditor and laboratory should be used for credibility and transparency; often this is stipulated by contract.

The far right **Differences** columns calculate and portray the compliance ratio (Cr) from what was measured (actual) over the standard (required).* Breadth and depth **Weighted Average Scores** are calculated from breadth and depth **Differences** and **Standard (required)** columns. The **Weighted Average Score** represent the compliance level relative to the standard required. **Weighted Average Score** results are at the bottom of the **Difference** columns (found at the bottom right side of Table 13.2).

Accuracy is depicted in two forms (see Table 13.2), first at Output Purity (actual) (see section 1Ai) and second, the accuracy range, derived from the various laboratories, field tests, with the minimum and maximum results, found along the bottom row (Accuracy Range [Min, Max]).

---

* Accuracy, or in this case oval system accuracy, is not a function that is calculated in total (i.e., the mathematic formula include all tests conducted). It is measured individually for each subcategory. Results from the array of tests are provided as a Min. and Max, which shows the range of output results for given tests. Tested output purity level is deemed the best overall measurement of accuracy and compliance.

TABLE 13.2

Scorecard Matrix Spreadsheet

Scorecard Matrix

	Std (required)			Measured (actual)			Difference		
	Breadth	Depth	Accuracy	Breadth	Depth	Accuracy	Breadth	Depth	Accuracy
IPT Trait(s) / Attribute(s) Success Scorecard (e.g., organic product, fair-wage, pasture-fed, etc.) = Σ									
1) Controlling Std (contract/Regs.)									
A) Seed Purity (98%)									
(i) Output Purity ± 0.002-0.005	1	3	0.980	1	3	0.978	1.00	1.00	0.9980
(ii) Other purity data (pts.)	1	1		1	1		1.00	1.00	
B) Tolerance Level (pts.)									
(i) Other tolerance data	1	1		1	1		1.00	1.00	
B = Breadth (actual number of measurements and/or data points) = Σ									
2) Performance Measurement Entity/Parameters									
A) Primary Entity (farmer, etc.)									
(i) Inputs (pts.)	2	3		1.0	3.0		0.50	1.00	
(a) Seedpurity-98.0%				185.0	3.1		0.93	0.78	
(ii) Operations (pts.)	200	4							
(a) Chemicals data									
(b) Storage									
(c) Cleanouts									
(d) Inspections crop/field			0.98			0.9800			1.0000
(iii) Tests (pts.)	15	3		13.5	2.2		0.90	0.73	
(a) Field tests (A)			0.98			0.9600			0.9796
(b) Laboratory tests (A)			0.98			0.9750			0.9949
(iv) Administrative (pts.)	50	3		45.0	2.0		0.90	0.67	
(a) Training periods									
(b) Data collection									
(c) Inspection, records									
(v) Certification (pts.)	1	3		1.0	3.0		1.00	1.00	
(a) Organic									
(b) ISO									

D = Depth 1 = farmer 2 = farmer + 1 entity 3 = farmer + 2 entities

A = Accuracy (degree of conformity and/or measurement parameters; determined by tests, audits, etc.)

(Continued)

441

TABLE 13.2 (Continued)

	(pts.)		Tests (A)	Accuracy Range (Min, Max)			Weighted Average Score		
B) Buyer inspections									
(i) Operational (pts.)	8	4		4.8	3.2		0.60	0.80	
(ii) Administrative (pts.)	7	3		5.2	2.1		0.74	0.70	
(iii) Tests (A)			0.98			0.9700			0.9898
C) Third-Party inspections									
(i) Operational (pts.)	20	4		14.9	3.7		0.75	0.93	
(ii) Administrative (pts.)	15	3		13.0	2.0		0.87	0.67	
(iii) Tests (A)			0.98			0.9780			0.9980
D) Grader (pts.)	5	2		4.5	2.0		0.90	1.00	
= Σ 3) Communications (Producer/Buyer)									
A) Production Nomenclature (pts.)	25	3		22.0	2.4		0.88	0.80	
(i) Unit size									
(ii) Product									
(iii) Other inputs/Byproducts									
B) Attribute(s)/Trait(s) (pts.)	50	3		46.5	2.1		0.93	0.70	
(i) Data/process(s) of interest									
(ii) Measurements									
(iii) Test Methodology									
Accuracy Range (Min, Max)				0.960	0.9801		0.901	0.895	

Source: Developed from research.

442

13.4   IPT MEASUREMENT SCORE AND RESULTS

Figure 13.1 provides a visual comparison of actual versus required performance criteria. The Measurement Score, for illustrative comparison purposes, also includes alternative IPT program standards, i.e., fair wage, drought-resistant seeds, high oil content, etc., in regard to their breadth, depth, and accuracy standard or contractual requirements.

Along the left side is the number of data points, as a measurement reference. Along the bottom are the prescribed accuracy levels, or purity level standards, as desired by each IPT system. From left to right, the relative degree of requirement rigor increases as does each system increase in complexity. Each illustrated comparison score bar portrays the relative breadth, depth, and accuracy (purity) for that particular system. For example, **Fair Wage** standard (required) mandates the recording of 100 breadth, 75 depth data points, and with a targeted accuracy of 75 percent, for that particular attribute of interest.

Results show that for the trait of interest (low linolenic), the output purity level was 97.8 percent and within contractual limits. This is not to suggest that this particular IPT program would be sufficient for any other type of IPT system because each system is contractually different with regard to breadth, depth, and accuracy requirements. This low linolenic IPT system may be considered efficient because of it being within agreed on compliance specifications and tolerances. Still, depending on the exact specifications of the contract, a number of conclusions can be made. For example, if 89 percent is the agreed on cutoff between satisfactory and unsatisfactory IPT efficiency, the data and bar show that overall inspection point compliance (total) was 89.8 percent (satisfactory) and breadth criteria compliance was 90.1 percent (satisfactory). However, the 89.5 percent compliance for depth highlights that this particular criteria was in compliance (satisfactorily) but was the least in compliance. This could mean that if at a future time output purity level decreases or some other negative aspect arises, such as a recall of product, that some of the weaker points within the depth criteria should be looked at more closely.

After several years of the same farm scorecard measurements, data may show system trends, such as in increasing data loss during computer-to-computer (interface) communication transfer or by decreased buyer inspection points being recorded. Another aspect of this scorecard would be to compare the same crop, over several years, but for different purity levels. It would be interesting to note the difference and similarities. This would be true, especially if compliance, with a more rigorous and profitable IPT program, would not require many more steps. In total, the Scorecard Matrix can be a useful tool to evaluate an IPT system's efficiency. It represents a tool that can incorporate qualitative and quantitative measurements for evaluation.

REFERENCES

Golan, Elise, Barry Krissoff, Fred Kuchler, Linda Clavin, Kenneth Nelson, and Gregory Price. 2004. *Traceability in the U.S. Food Supply: Economic Theory and Industry Studies.* USDA Economic Research Service, Agricultural Report Number 830 (AER-830).

14 Identity-Preserved Cost-Benefit Spreadsheet

14.1 GOALS AND STRUCTURE OF THE SPREADSHEET

This spreadsheet brings together many of the components developed within this work focusing on farm level production. It also provides a statistical summation of identity preservation (IP) data, as it pertains to production purity levels, for comparative purposes. It is hoped that this methodology and data derived from this spreadsheet may help in decision making for crop selection (purity), which is best suited for the production skill level of the farmer and growing environment. This evaluation system is based on auditing and toward a goal of International Organization for Standardization (ISO) 22000 series compliance. This and the previous chapter's spreadsheet are related to one another; the matrix represents effectiveness of a program, whereas the cost-benefit represents the efficiency of a program. The spreadsheet was developed to provide production data regarding various IP programs, by both purity level and individual cost items, as compared with revenues received (benefit). A questionnaire, from which a sampling of data may be obtained, is found in the subsequent chapter—Chapter 15, which possesses a much shorter version of the questionnaire than what would be needed for this cost-benefit spreadsheet. Where survey data were not provided, data from agricultural literature were used.

What the spreadsheet provides: first level of inquiry—to discover the averages and boundaries of times and costs for specific IP events, given varying levels of purity, and to estimate costs versus profits to determine whether a particular IP crop and purity level (with its accompanied requirements) would be profitable and worthwhile to grow. Second level of inquiry seeks to discover strengths and weaknesses associated with various cost events, to determine more accurately (numerically) critical items, besides efficiencies, such as time/labor/cost items allotted to specific IP tasks, e.g., cleanouts, audits, lab tests, etc.

Follow-on cost-benefit questionnaires should more accurately account for the costs versus benefits tied to specific trait(s) and/or attribute(s) of interest and purity levels. At present, the evaluation of data is not as difficult as finding or creating specific on-farm IP data. As more surveys and data regarding on-farm identity preservation and traceability (IPT) practices become available, the distinctions between work (costs), market prices, and level of purity will be clearer.

This chapter provides an extensive, but not exhaustive, spreadsheet that focuses on IP costs and revenues generated as applied to varying levels of crop purity. To better understand the spreadsheet, a conceptual model is provided (Table 14.1).

TABLE 14.1
Cost-Benefit Model

Crop Production Financials (examples per purity level)

Std revenue	=	Number of standard bushels sold × sale price per standard bushel
IPT1 Total costs	=	$\Sigma\ C^{IPT\,1}_1, C^{IPT\,1}_2, \ldots$
IPT2 Profit	=	$R^{IPT\,2} - \Sigma\ C^{IPT\,2}_1, C^{IPT\,2}_2, \ldots$
IPT3 Revenue per bushel	=	$R^{IPT\,3}$ / number of IPT3 bushels sold
IPT4 Total cost per bushel	=	$\Sigma\ C^{IPT\,4}_1, C^{IPT\,4}_2, \ldots$ / number of IPT4 bushels sold
Standard profit per bushel	=	$[R^{Std} - \Sigma\ C^{Std}_1, C^{Std}_2, \ldots]$ / number of standard bushels sold
IPT1 Item cost per bushel	=	$C^{IPT\,1}_1$ / Number of IPT1 bushels sold

Example of $R^{Std}, R^{IPT\,1}, \ldots$	Example of $C^{Std}2, C^{IPT\,1}1, \ldots$
$R^{IPT\,1}$ = e.g., the revenue generated by the sale of identity preserved lot $R^{IPT\,1}$ (x bushels @ $x.xx/bushel). <u>This would include all premiums, bonuses, etc.</u>	C^{Std} = e.g., costs associated with the number of hours needed to complete standard planter cleanout, at $x.xx per hour. <u>IPT costs may often be in excess of standard practices, such as IP-specific required third-party audits and/or laboratory analysis.</u>

Source: Developed from research.

The goal of the spreadsheet is to offer a comparison of varying levels of purity with its prescribed costs to corresponding sale price of product. The chapter highlights the various cost components associated with grain production, comparing for purity levels: standard production (n/a purity level), IPT1 (5.0 percent), IPT2 (2.0 percent), IPT3 (1.0 percent), and IPT4 (0.1 percent). For example, a five percent purity level means that up to five percent may be of unknown composition or mixing. An example of the shortened single-page version of the spreadsheet can be found in Table 14.2. The entire spreadsheet can be found in Appendix R, Cost-Benefit Spreadsheet—Complete, which also provides individual costs on a per-bushel basis for all levels of purity. Spreadsheet results are summarized by Figure 14.1.

14.1.1 LIMITATIONS AND ASSUMPTIONS

Comparisons are based on contrasting data derived from the same or similar acreages; for example, a standard crop variety grown under typical management practices, such as Roundup Ready soybeans, are compared with an identity-preserved grown variety, such as ultralow linolenic soybeans.

This spreadsheet does not attempt to incorporate "other" social or environmental costs or benefits (to mean financial, loss/gain of jobs; social, loss/gain of businesses; and/or environmental, decrease/increase of water quality).

Assumptions for spreadsheet: interest rate, 0.08; units of measure, acre, bushel, and dollar; crop type, same species; crop cycle, same growing season.

TABLE 14.2

Cost-Benefit Spreadsheet—Abbreviated Single-Page Example

Background Information

Item	Measure Units	Standard	IPT1	IPT2	IPT3	IPT4
Personal information						
ID Number		1	2	3	4	5
Name		Bill Smith				
Address		Ames IA 50014				
Phone number		515-123-4567				
E-mail		isu@iastate.edu	=	=	=	=
General information						
Crop planted		Soybeans	UL Soybeans	UL Soybeans	UL Soybeans	UL Soybeans
Crop variety planted			DKB 2752			
Purity level (required)	%	**n/a**	**5.0%**	**2.0%**	**1.0%**	**0.1%**
Crop acres	Acres	200	200	200	200	200
Grain yield	bu./acre	55	55	55	55	55
Previously planted crop in field		Corn	Corn	Corn	Corn	Corn
Type of IP system		None	Non-GMO	Non-GMO	Non-GMO	Non-GMO
Trait(s) and/or attribute(s) of interest		None	Ultralow linolenic	Ultralow linolenic	Ultralow linolenic	Ultralow linolenic
Hourly wage information						
Management	$/h	$25.00	$25.00	$25.00	$25.00	$25.00
Labor	$/h	$15.00	$15.00	$15.00	$15.00	$15.00
Meeting, off season	$/h	$40.00	$40.00	$40.00	$40.00	$40.00
Contract or hired professional	$/h	$50.00	$50.00	$50.00	$50.00	$50.00
Operating assumptions						
Grain hauling, semi	$/mile	$0.250	$0.250	$0.250	$0.250	$0.250
Interest, carry-on operating money	%/y	8.00	8.00	8.00	8.00	8.00
Capital interest	%/y	6.00	6.00	6.00	6.00	6.00
Personal travel mileage	$/mile	$0.500	$0.500	$0.500	$0.500	$0.500
Personal travel meal expense	$/day	$50.00	$50.00	$50.00	$50.00	$50.00

(Continued)

TABLE 14.2 *(Continued)*

Item	Measure Units	Standard	IPT1	IPT2	IPT3	IPT4
Personal travel overnight expense	$/day	$100.00	$100.00	$100.00	$100.00	$100.00

Revenues

Production data/revenues						
Selling price	$/bu	$8.53	$10.35	$15.00	$18.00	$20.00
Bushels sold	Bu	11,000	11,000	11,000	11,000	11,000
Unit sale price	$	$93,830	$113,850	$165,000	$198,000	$220,000

Costs

Production data/costs						
Seed cost (total)	$/y	$40,000	$35,000	$37,500	$40,000	$55,000
Volume purchased	Lb	50,000	50,000	50,000	50,000	50,000
Seed $/bu @ 60 lb/bu	$/bu	$48.00	$42.00	$45.00	$48.00	$66.00

Source: Developed from research.

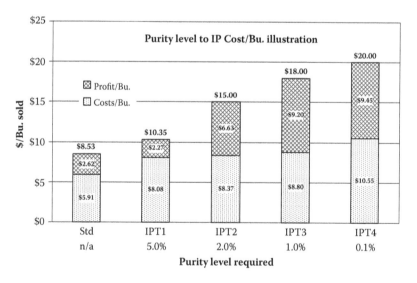

FIGURE 14.1 Purity level to IP cost/bushel illustration. (Developed from research.)

14.2 COST-BENEFIT MODEL AND SPREADSHEET

14.2.1 MODEL

The conceptual model (Table 14.1) was developed to help illustrate the much larger Cost-Benefit Spreadsheet and mathematic functions used. Along the model's left side, the various types of financial data desired, i.e., revenues, costs, and profits, are illustrated. It also delineates between each type of production purity level and output results in total and per-bushel values for revenues, costs, and profits. The right side provides examples of mathematic formulas for each type of financial inquiry. Along the bottom are two examples: (1) an example of revenues generated, and (2) an example of costs.

14.2.2 Spreadsheet

Although the spreadsheet is several pages in length, the computations used are very simple and forthright. Still, depending on the trait(s) and/or attribute(s) of interest, and especially purity level required, spreadsheet use comparing the various purity levels can be enlightening and helpful.

Spreadsheet results are calculated as an IP quotient or ratio; in this case, the output values are the overall (total) and individual per-bushel costs and per-bushel profit derived from each IPT program. IPT profit comparisons are derived from each system's revenues (output sale of product) less costs (accumulated costs associated with production), which provides a resultant profit or loss. The output can then be illustrated in graphic illustration (Figure 14.1).

For the purposes of brevity only the top portion of the spreadsheet is illustrated in the chapter, which includes all purity levels for comparison: standard (n/a), IPT1 (5.0 percent), IPT2 (2.0 percent), IPT3 (1.0 percent), and IPT4 (0.1 percent). Appendix R, Cost-Benefit Spreadsheet—Complete has the spreadsheet in its entity, including the per-bushel cost for each cost item.

The spreadsheet is divided into several columns: the first column contains the particular **Item** of interest (usually prescribed by contract); the next column provides for unit of measure or **Measure Units** (by percentage, acre, bushel per acre, dollar per hour, etc.); and the last columns are the input and output columns for the various purity levels tested. Rows are grouped by category and subcategory, i.e., Personal Information includes ID Number, Name, etc. The other major categories include General Information, Hourly Wage Information, Operating Assumptions, Revenues, Costs, etc. The Costs category is further subdivided by Production Data/Costs, Pest Management/Fertilizer Data/Costs, etc. The Summary Results provide additional data and per-bushel values for Costs, Revenues, and Profits for each purity level.

There are two sets of results: (1) is the purity level to IP cost per bushel as is illustrated in Figure 14.1, and (2) is found in the spreadsheet Summary Results portion (bottom half of Appendix R), which shows the individual cost line items per bushel costs.

The essence of the calculations is forthright: addition of all the various costs, subtracting the total costs from total revenues, and then dividing the results by the number of bushels sold, which provides an overall profit per bushel per IP purity system. Similar computations are done for per-bushel costs and per-bushel revenues.

14.3 PURITY LEVEL TO IP COST/BUSHEL ILLUSTRATION AND RESULTS

The graphic illustration summarizes the net results of the spreadsheet (Figure 14.1). It includes the various purity levels for comparison purposes, including standard (n/a purity level), IPT1 (5 percent), IPT2 (2 percent), IPT3 (1 percent), and IPT4 (0.1 percent).

On the left side of the graph is the dollar per bushel values. Along the bottom are the various purity levels reviewed. Each purity level has an associated bar above it that contains three stacked numbers. The lower number indicates the cost per bushel; the middle number indicates the profit per bushel; and the top number represents the per-bushel sale price of that particular system.

This graph illustrates that as purity level requirements increase, from left to right, per-bushel sale price increases. This also represents the buyers' willingness to pay for specific purity levels of production. It is the responsibility of the farmer to meet the specific purity level as dictated by contract and to contain cost expenses to ensure a reasonable profit. Each farmer must decide the risks of production and what is deemed the most reasonable and profitable approach. The graph also reveals that costs associated with each purity level do not increase uniformly. In Figure 14.1, IP costs quickly increase from the standard to IPT1 and at a lesser rate from IPT3 to IPT4. The increase from IPT1 to IPT3 shows a relatively slow increase in costs. Looking at profit, from standard to IPT1 there is a reduction in profit. This would indicate that changing production from standard to IPT1 would be less favorable unless that farmer intends to use an upper-level IP system in the following years. The farmer may be willing to make less profit this year for increased profits in the following years. Once the farmer had begun advancing within IP production management practices, from the graph and a pure profit to cost basis, it would be advised to produce at the IPT3 level of production. This would represent the greatest return for work (costs).

The spreadsheet offers the opportunity to manipulate input data and see the effects on production costs and profits. The spreadsheet has shown that various farm management practices have a tremendous impact on expenses and ability to meet specific purity levels. Most notable for IP-specific programs is the cost of increased original seed purity. As purity requirements increase, so too does the need for more pure seed, often from a parent seed company's foundation stock. Another notable expense has to do with storage and transportation. Usually the less rigorous systems require less storage and transportation expenses. As IP purity increases, it is not surprising to have much greater expenses for both storage and transportation.

This methodology and spreadsheet are examples of what can be surveyed and recorded. It would be ideal to have actual data from farmers for production at these purity levels, resulting in spreadsheet data. It is hoped that this research instrument will be of help for future research.

15 Identity-Preserved Cost-Benefit Questionnaire

15.1 INTRODUCTION OF THE QUESTIONNAIRE

The previous chapter portrayed an extensive spreadsheet. Unfortunately, the average questionnaire or survey mailed to participants is usually much shorter than desired to gain information. The ideal questionnaire fully reflects or asks participants for accurate data that fulfill all sections of the spreadsheet. Unfortunately, a questionnaire such as this would take pages of inquiry (questions with examples) and constitute many more hours to complete than the average grower(s) would provide. A truncated questionnaire has the best chance of being completed and returned. Thus, the following questionnaire is one that has been greatly shortened, from a much longer originally conceived questionnaire, and used to survey an Iowa organization comprising farmers growing identity-preserved crops such as ultra-low linolenic soybeans, both genetically modified organism (GMO) and non-GMO varieties. For some farmers this represents a double stack trait variety: the first being ultra-low linolenic and the second being that it is a non-GMO. So for some farmers in this group this could be considered double identity preservation (IP) for the two traits of interest (non-GMO and ultra-low linolenic traits).

This group welcomed the opportunity to participate. The total pool of participants comprised 42 growers. The organization's board of managers was personally visited and given a presentation regarding the purpose and scope of the questionnaire. No group meeting (to all of the participants) was given. The questionnaire with a single-page cover letter (see Figure 15.1) and return envelope* (preaddressed and stamped) was sent to each participant: 1 was returned because of a wrong address, and 11 participants mailed back the questionnaire (26.8%), of which 8 provided complete data (20%). The questionnaire was mailed out at the end of August 2007 and data stopped being recorded at the end of September 2007. The organization sent follow-up reminders via e-mail to each grower 2 and 4 weeks after the mailing questionnaire to encourage participation in its membership. Although the sampling is very small, the 20% return by respondents may be considered at or above the average in survey returns.

University rules require that the primary investigator (author) must complete the Iowa State University web-based training on the protection of human subjects in

---

* The questionnaire and return envelope were on yellow stationary to help increase the likelihood that they would be filled out and not be lost or misplaced.

May 2007

Dear Innovative Growers farmer,

Innovative Growers Board of Managers has agreed to participate in this ISU questionnaire project. Iowa State University is actively studying the IP processes that help retain the value of specialty traits. With regard to this new and evolving program, there has been much discussion concerning potential costs and benefits of IP as a management tool.

How well this process is adopted by farmers will have a great impact on agricultural customers and consumers, and most importantly, for farmers themselves, in how they pursue IP's use throughout all value-added farm production.

The questionnaire enclosed is focused upon you, the farm owner or manager, the person whose time is the most valuable of all labor inputs. Much has been studied regarding the time required to do cleanout of planters or combines. However, little data have been recorded regarding the time exerted by owners/managers during the critical times of planting and harvest, with regards to other IP requirements. Nor have studies tied these IP labor hours to other IP costs such as additional chemical usage, special storage, additional accounting needs. Accordingly, this questionnaire is looking at owner/manager hourly labor and other unique IP costs that are incurred during specific, time-critical periods within the farm management cycle.

We understand that this study comes shortly after planting, and many other tasks demand your time. The questions are directed in such a way as to compare costs of *standard* (e.g., Roundup Ready) soybean production to IP's (e.g., low linolenic) soybean production. The hopes are to gather as many owner/manager's costs that pertain to IP and then evaluate the data. You need not provide your name, so your anonymity is ensured. If you do wish to provide your name, no one at your organization will have access to your data. The numerical results of this questionnaire—without names—will be presented to your organization.

- The expected duration of this survey is 1–2 months.
- The procedures to follow: fill in the asked data questions, put the form in the enclosed envelope, and then put it in the mail.
- There are no foreseeable risks or discomforts to fill out the questionnaire.
- Regarding confidentiality of records: The forms will be coded so that participants' identity will be stored separately from the data. Data will be received and processed into the surveyor's (Greg Bennet) office computer hard disc. No other computer or person will have the information. The data form will be kept in a folder in the surveyor's locked office.
- It is understood that your participation is voluntary.
- Your refusal to participate will involve no penalty or loss of benefits.
- You may discontinue participation at any time without penalty or loss or benefits.
- A copy of the informed consent document will be provided to you.
- A signature line and date of participation is provided for at the end of this letter.

FIGURE 15.1 Letter to growers.

research (completed April 22, 2007) and have the questionnaire protocol approved by the Institutional Review Board (approved May 3, 2007—protocol ID Number 07-261).

15.2 QUESTIONNAIRE

Because a major concern in data collection is to receive as many returned questionnaires as possible, it is well understood that this is contingent upon the questionnaire being short—in this case one page of questions. In particular, this questionnaire concentrates on the basic cost differences between standard and IP crop production and on owner/manager IP-related task hours performed (see Figure 15.2).

Thank you for your participation. If you have any questions regarding this questionnaire or IP in general, please contact Greg Bennet at 515-294-6358 or by e-mail at gsbennet@iastate.edu. If you have any questions about the rights of survey participants or research-related injury, please contact Jan Canny, IRB Administrator, at 515-294-4566, e-mail IRB@iastate.edu; or Diane Ament, Director of Research Assurances, at 515-294-3155, e-mail dament@iastate.edu.

Respectfully,
Gregory S. Bennet
Ph.D. Student/Research Assistant
Dr. Charles R. Hurburgh, Jr.
Professor in Charge
Iowa Grain Quality Initiative
Iowa State University
515-294-6358

I _____ (print clearly) understand that my participation is voluntary.

Signature _____

Date _____

Please return this signed form with a completed questionnaire. Thanks again for your help.

FIGURE 15.1   *(Continued)*

The title addresses both the organization's name and (G)oal: To help determine IP costs and profit and (F)ocus: IP labor performed by the owner/manager during critical planting and harvest periods (within one growing season). The questionnaire is divided into two parts. The first provides basic data to compare standard crop production to identity-preserved crop production, as will be expanded upon in the next paragraph. The second part looks at IP-specific data and costs. From these two parts, interpretations were made and summarized.

The first part of the questionnaire asks for basic information and costs for standard and IP crop production; for example, the number of years growing IP crops followed by what the owner/manager's average hourly wage is (e.g., $25/hour). Next, the number of soybean acres and previous crop(s) grown in the IP field are asked for. The assumption is that like number of acres will be compared for costs, yields, etc. In this questionnaire, standard soybeans are considered Roundup Ready variety, whereas identity-preserved soybeans are the low linolenic/non-GMO (traits) variety. For comparison the questions include:

- Estimate of seed costs (total), $ per year
- Estimate of overall soybean pest management, chemical applications (fertilizer, etc.), costs $ per year
- Estimate of soybean total revenue from sales
- Estimate of soybean storage costs per year
- Estimate of total costs to transport crop to market
- Estimate of crop insurance cost per farm

The second part of the questionnaire asked the name of the particular variety of IP soybean being grown, if the grower is International Organization of Standardization (ISO) compliant or certified, and, if so, by whom. Next, emphasis is made toward estimating the time and labor that the owner/manager performs during the critical

■■■ LLC, place, & date	ISU IRB # 1 07-261
	Approved date: 17 May 2007

Speciality soybean identity preservation (IP) questionnaire

Goal: To help determine identity preservation (IP) costs and profit

Focus: IP labor performed by owner/manager during critical planting and harvest periods

Years growing IP crop: _____ Owner/manager, average hourly wage? _____ (e.g. $25/hr)

IP soybeans acres: _____ Previous crop(s) growth in IP field: _____ Please remember, data evaluation is based upon comparing **equal** number of standard acres to IP acres that you grow or have grown.	Type of SB prodution	
	(For comparison)	
	Standard	IP
	(e.g. Roundup ready)	(e.g. Low linolenic)
Est. of seed costs (total) $/yr		
Est. of overall SB pest mgmt/chemical (fertilizer, etc.) costs $/yr		
Est. of SB total revenue from sales		
Est. of SB storage cost/yr (costs may vary)		
Est. of total costs to transport crop to market (distances may vary)		
Est. of crop insurance cost $/farm (costs may vary)		

IP Soybean production information

IP Soybean variety planted _____

ISO-**Complaint**: Y N   ISO-**Certified**: Y N   Certified by _____

Est. of IP time and labor during ***critical*** planting and harvest periods

FIGURE 15.2   Identity preservation questionnaire.

planting and harvest periods. Specifically, the 2–3 weeks encompassing the planting and harvest, for a total of a 4- to 6-week period. These questions include:

- IP preparation hours (e.g., coordinating separate storage)
- Documentation hours spent on IP (i.e., field recordkeeping, logs of grain movement)
- Planter cleanout (e.g., hours beyond standard cleanout, number of hours)
- Management hours spent on fields and/or facilities beyond the standard
- Inspection hours spent related to fields and/or facilities beyond the standard
- Combine cleanout (e.g., hours beyond standard cleanout)
- Handling and separate storage hours spent beyond standard practices

Specifically the two to three weeks that involves both planting and harvest, for a total of four to six weeks.

IP preparation hours, e.g., coordinating separate storage _____ hours

Documentation hours spent on IP; i.e., field record keeping, logs of grain movement _____ hours

Planter cleanout—hours *beyond* standard cleanout? _____ hours

Management hours spent on fields and/or facilities *beyond* the standard? _____ hours

Inspection hours spent related to fields and/or facilities *beyond* the standard? _____ hours

Combine cleanout—hours *beyond* standard cleanout? _____ hours

Handling and separate storage hours spent *beyond* standard practices? _____ hours

Managerial/ operational hours spent on any other tasks related to the IP systems? _____ hours

IP related meeting **hours**_____ travel **miles**_____ **overnights** _____days

Mileage cost per mile, (e.g. 0.365) $_____ Average meals/lodging per day? (e.g. $125) $ _____

Other IP systems costs associated with any other IP production tasks; i.e., <u>inspectors</u>, <u>auditors</u>? $ _____

Thank you for sharing this information——which will be kept confidential.

FIGURE 15.2 *(Continued)*

- Managerial/operational hours spent on any other tasks related to the IP systems
- IP meeting hours
- IP overnights
- IP travel miles
- Mileage cost per mile, (e.g., $0.362)
- Average meals/lodging per day? (e.g., $125)
- Other IP systems costs associated with any other IP production tasks (i.e., inspectors, auditors)

See Figures 15.1 and 15.2 and Appendix S for a full summary of spreadsheet data.

15.3 INTERPRETATION OF DATA

Because of the lack of surveys returned and completed, the interpretation of data weighs more heavily as a descriptive narrative than statistical analysis. Some of the data lend to inference by comparison, whereas some response outliers may be attributed to human input due to misinterpreting a question. For the participants with more complete data, their analysis will serve as a type of mini-case study. Within the conclusion, participant statements will be used to reflect individual views. This questionnaire used the same spreadsheet as described in the previous chapter; however, for the sake of space, only the items of interest have been focused upon.

This section includes number of years growing IP crop, wage, IP acres, and standard versus IP crop production comparisons. Participants #1 and #3 provided no data and #7 only provided partial data. Comparisons will generally be between standard and IP crop production by individual participant. When feasible, comparisons between participants will be provided as a narrative rather than a statistical analysis. A few of the participants provided short narratives that will be included at the end. Overall some general data were observed

- 75% of the participants (6 of 8) had grown IP crops for 4–12 years.
- 75% estimated their wages were $20–$30 per hour.
- 72% had 199–400 acres of IP crop being grown.
- The previous crop grown in all reported cases was corn.

Note: The hourly wage of Participant #2 hourly wage was $55, outside of the average $25–$27 per hour rate; this rate may skew Participant #2's overall IP costs. It was noted that the question could have been worded better by asking the participant what they would be willing to pay a manager to complete IP tasks.

No data were provided citing the number of acres growing IP or standard crops, thus a more accurate overall cost of IP production divided by total IP acres is incomplete. Data provided by participants varied in units measured. This also presented a problem for comparison from farmer to farmer. Sometimes the data could be converted for a participant because of additional information being provided, such as yields per acre. Otherwise, exact comparisons such as mean, minimum, maximum, and standard deviation within categories were not always accurate or possible. Examination of the survey forms, and its raw data, shows the inconsistencies of data collected. This, in turn, could corrupt follow-up formulas, ranges, and other vital survey measurement tools used for possible interpretation. In any case, efforts were put forth to best summarize data collected for useful interpretations by comparisons and mini-case studies.

15.4   STANDARD/IDENTITY-PRESERVED COMPARISON DATA—BY CATEGORY*

Standard and IP seed costs varied from a low of $9 per acre (this appears to outside of the norm—possibly the farmer was using his/her own seed from the year before) to $36 and $37 per acre, for standard and IP, respectfully. Sixty-six percent of the participants had IP seed cost being less than standard seed cost; $19 less per acre was the largest difference, with three other participants IP seed cost being $7.50–$9.50 less than the standard seed cost. Two participants had IP seed costs of $5 and $11 higher than standard seed cost. It is unclear whether these prices were current prices or from the year before. One farmer had IP seed costs 54% higher than standard seed cost, whereas another farmer had standard seed costs 53% higher than IP seed costs.

---

* Participants are referred to as Participant #1 through #10 or #1, #2, etc.

There was insufficient data regarding the varieties bought to determine these cost differences or if technology fees were actually included.

Pest costs were higher for IP crops by 80% of the participants—$10–$25 more per acre than standard production. Only one participant had higher standard pest crop costs than IP pest crop costs, and two respondents had identical pest costs. The percentages of pest cost increase for IP crop over standard crop were from a low of 22%, to a high of 265%, or more than 2.5 times the cost of standard pest costs. Assuming that most standard production is Roundup Ready and that the IP production was a non-GMO variety helps to explain the cost difference.

Farmers providing data regarding total revenue indicated that one-third of the IP crops generated greater revenue than standard production, one by as much as $50 per acre. Conversely, two-thirds of the standard production generated greater revenues than IP production, by $15–$40 per acre. Percentage wise, the range of difference between the two systems, as provided by the farmers, was [(4% $13), (−6% -$26), (7% $26), (11% $40), (−13% -$50)]. Positive numbers indicate that IP production produced greater revenues as provided by the participants, whereas negative numbers indicate that IP crops created less revenue than standard crops. The survey did not provide the participant the opportunity to explain how the total revenue was determined (i.e., premiums paid, etc.).

Two participants had IP storage charges higher than standard storage (50% and 183%), two producers had standard storage charges higher than IP storage (15% and 16%), and two producers had identical storage costs. It is surmised that transport distance, crop price, availability of storage, and other considerations play an important role in determining the necessity of using storage, its cost, and the wide variation of responses.

Transportation costs were higher for all IP crop participants except one, which had identical costs. For example, one grower who had the largest change had transport costs go from 0.05/bu (standard) to 0.25/bu (IP). In nearly all of the cases the costs for IP transport were substantially higher than those for standard production. The increases by percentage were from a low of 20–79% to highs of 300–500% increases over standard transport costs. It is probable that the transport costs were much less for standard cost because of the nearby availability of coop or elevator, than for the IP because of IP's inherent contract specifications of designation. There were no data for premiums being paid to compensate for transport distance traveled. Also, there was no provision made for waiting time at the unloading destination.

Insurance costs were identical for both crops, although the price paid from farmer to farmer varied. There appears to be no premium reduction because of IP accountability for IP crops at this time.

15.5 SUMMATION OF STANDARD/IP COMPARISON DATA

IP seed costs were overall within ±$8 of standard seed costs. No exact reasons could be inferred for these differences. IP pest costs were nearly always greater for the IP crop production, that average being nearly 20% higher. Considering the non-GMO aspect of the IP crop, the higher pest costs were expected.

Total revenue from sales had standard and IP production averages of $369.57/acre and $370.20/acre, respectfully. The range for standard production was larger ($293–$439/acre) than IP production ($306–$413/acre), but these differences were not significant. The question openly asked the participant to estimate total soybean revenues from standard and IP sales. This could infer what the participant truly believed or calculated the revenues were from each system. The values provided were not accompanied by justifying accounting data.

Storage costs varied greatly between the two systems. As mentioned before, it is surmised that transport distance, crop price, availability of storage, and other considerations play an important role in determining the necessity of using storage, its cost, and the wide variation of responses for both standard and IP production.

Transportation costs, by the largest degree, were much higher for IP than for standard production. This was not a surprise because of contract requirements and limited locations to which IP production is delivered. As mentioned, the degree of extra IP expense was as high as 3–5 times more expensive as standard production. Aside from solely IP-related managerial tasks (discussed below), transportation costs stand out as a major IP cost obstacle. Some organizations have offered premiums to offset the varying distances that producers must travel to deliver their crop. This may be reflected in the total revenue from sales cited above, but is not conclusive. Insurance cost comparison between each system appears to have a neutral effect—or no cost adjustment.

15.6 DATA REGARDING VARIETY AND ISO COMPLIANCE/CERTIFICATION

Five of eight participants provided variety information. Regarding ISO compliance, one was in compliance (16.6%) and five were not in compliance; for certification, two were certified (28.5%) and five were not certified.

15.6.1 CRITICAL PLANTING AND HARVESTING PERIOD'S DATA[*]

IP preparation hours, coordinating separate storage, etc., ranged (75% of surveyed) from 1 to 5 hours with 3 hours being the average. Two of the respondents had IP preparation hours of 10 and 49 hours. Although these latter two may not be unreasonable, given the generality of the question, they were well outside of the average of the respondents.

Documentation hours spent on IP, field recordkeeping, logs of grain movement, etc. ranged (75% of surveyed) from 2 to 4 hours with 3.2 hours being the average. Again, as in the IP preparation hours above, two respondents had their times of 1 and 20 hours, which may not be totally unreasonable given the parameters (period) of the survey. Some of this work may have been done earlier or later than the time surveyed.

Planter cleanout and hours beyond standard cleanout ranged (75% of surveyed) from 1 to 3 hours beyond the standard cleanout with 1.9 hours being the average. Two of the respondents cited 8 and 12 hours of additional cleanout time beyond the

---

[*] Specifically, the 2–3 weeks that encompasses planting and harvest, for a total of 4–6 weeks.

standard time required. No accounting for the extended times for cleanout other than the planter had previously planted non-IP soybean seeds followed by IP soybean seeds. The shorter times for cleanout by the majority could have been due to having the previous seeds being corn rather than soybeans.

Management hours spent on fields and/or facilities beyond the standard ranged (62.5% of surveyed) from 2 to 10 hours beyond the standard production, with 5.4 hours being the average. Three of the respondents cited 0, 1, and 21 hours of additional management time required. There were no indications why these three respondents were outside of the main group. It could not be determined if these times were due to longer years or fewer years growing IP crops. In other words, the experience or inexperience level of the respondent dictated the time required.

Inspection hours spent related to fields and/or facilities beyond the standard ranged (75% of surveyed) from 1 to 6 hours beyond the standard time required, with 3.8 hours being the average. Two of the respondents cited 0 and 24 hours. The 0 hours seems unlikely, whereas the 24 hours appears well beyond the group's range. It is possible that this high number of hours spent on inspection is by the same participant that cites him/herself with wages of $55/hr and also has several hours spent toward IP in excess of the other participants surveyed.

Combine cleanout and hours beyond standard cleanout ranged (75% of surveyed) from 1 to 4 hours beyond the standard with 2.8 hours being the average. Similar to the planter cleanout, two of the respondents cited 12 and 24 hours of additional cleanout time beyond the standard time required. No accounting for the extended times for cleanout other than the combine had previously planted non-IP soybean crop followed by IP soybean crop. The shorter times for cleanout by the majority could have been due to having the previous crop being corn rather than soybeans.

Handling and separate storage hours spent beyond standard practices ranged (75% of surveyed) from 1 to 6 hours beyond the standard, with 3.5 hours being the average. Two of the respondents cited 10 and 40 hours of additional handling and storage activity time beyond the standard time required. No accounting for the extended times were claimed. The higher additional time could have been tied to building additional storage, but this is only speculative.

Managerial/operational hours spent on any other tasks related to the IP system ranged (75% of surveyed) from 1 to 6 hours, with 3.3 hours being the average. Two of the respondents cited 0 and 10 hours of additional handling and storage activity time beyond the standard time required. Neither of these extremes could be explained.

Total hours spent on critical planting and harvesting period's data ranged (75% of surveyed) from 21 to 60 hours, with 37.9 hours being the average. Two of the respondents cited 10 and 161 hours of additional IP activity time. Again, neither of these totals could be explained.

15.7 SUMMATION OF CRITICAL PLANTING AND HARVESTING PERIOD'S DATA

Regarding the above section, there appears to be an incremental increase of 1–5 additional hours to perform IP tasks, with the average increase being 3–4 hours. The only exceptions were for cleanouts of the planter and combine, which averaged 1.9 and 2.8 additional hours, respectfully.

The totals of the data above (ranges and averages) ranged (75% of surveyed) from $530 to $1,800, with $961 being the average. Two of the respondents cited a low of $100 and high of $8,855 of additional IP activity costs. Again, neither of these hourly totals could be fully explained. However, regarding the overall costs, the hourly rate of Participant #2 helped to push his/her costs to $8,855.

Although academic studies may contend that IP systems take more time than illustrated here, in addition to individuals cutting corners, experience and farmer knowledge of his/her farm may greatly aid in reducing time for IP-related tasks. This in turn would reduce overhead costs of IP production. Much more research and data collection need to be done to determine the effects of not only experience, but even more important, the level or degree that purity has upon time, costs, willingness of the farmer to participate, and premiums that buyers are willing to provide.

Time and costs associated with IP meetings include:

• IP-related meeting hours, ranging from 3 to 12 hours, with the average being 6.8 hours
• Travel miles ranged from 110 to 300 miles, with the average being 181 miles
• Number of overnights and mileage cost per mile were negligible
• Average meals/lodging per day ranged from $25 to $200
• Other IP systems costs associated with any other IP production tasks (i.e., inspectors, auditors) ranged from $50 to $500

15.8 MINI-CASE STUDIES

Four participants that provided the most complete data will be divided into two groups for these mini-case studies. The division is based on the format of the data; similar formatting participants were grouped together. As mentioned before, the questionnaire did not ask for yields or selling price per bushel, thus the data provided varied by unit measure.

Participants #4 and #9 provided all of their standard and IP production comparisons in unit measure per acres. Seed costs of Participants #4 and #9 were very close to the U.S. soy crop statistical average for Iowa soybean seeds,* whereas the IP seeds on average were $10 less per acre. Pest costs were greater for IP production; this is mostly due to the seeds not being Roundup Ready. However, Participant #4's overall pest management costs for both systems were much more than any other participant (no reason was given). Total revenues between both participants and systems were nearly even. Participant #4's standard production produced $400 per acre whereas the IP production produced less, $350 per acre, a difference of $50. Participant #9's

---

* See U.S. soy crop statistics at http://www.soystats.com/2007/page_12.htm (accessed September 10, 2007).

standard production produced $366 per acre whereas the IP production produced more, $392 per acre, a difference of $26. These numbers, without any accounting for costs, show that for these two producers the revenues varied greatly. Producers did not expand on how they came to their data conclusions.

Storage, transportation, and insurance costs between standard and IP production were minimal for each producer. For several of these costs there was no difference between standard and IP production costs.

Regarding time and labor during critical planting and harvest periods—in excess of standard production time requirements—in nearly all categories Participant #4 performed more IP-related hours than Participant #9. There were no reasons provided; however, Participant #4 had nearly 4 times the number of years growing IP crops. No explanation was given to explain the wide variation of time needed to clean-out the planters. According to Darren Jarboe of Iowa State University's Center for Crop Utilization Research (CCUR), planter cleanout, much like combine cleanout, may vary greatly depending on the type and size (number of rows) of the machine. Compounding the difficulty in determining the time needed to perform an adequate cleanout is the variability between farmers to determine time needed to cleanout to purity desired by the customer. During this period, Participant #4 spent an average of 10 additional hours toward IP production requirements whereas Participant #9 spent 4.5 additional hours during the same period. The research data did not provide enough information to form conclusions why a much more experienced farmer spent more time on IP than the less experienced farmer. However, this may indirectly explain why Participant #9 had the higher total revenue from IP sales over standard sales. This could indicate how closely related many of these operations are to one another (i.e., standard to IP production).

Participants #2 and #8 provided their standard and IP production comparison unit measures in either per acres or per bushels; this was one of the primary reasons that these participants were grouped together. Participant #8 seed costs was very close to the U.S. soy crop statistical average for Iowa soybean seeds,[*] whereas Participant #2's costs were nearly $4 less per acre. The IP seed costs diverged, with Participant #8's costing $37 (no explanation) and Participant #2's, more in line with other participants, being nearly $10 less per acre. Again, pest costs were somewhat higher for IP production. IP pest costs for Participant #8 were $10 per acre greater than for standard production. The pest costs for Participant #2 were over 2.5 times higher than standard production. Total revenues between both participants and systems were nearly even. Participant #2's standard production produced $439 per acre whereas the IP production produced less—$413 per acre—a difference of $26. Participant #8's standard production produced $350 per acre whereas the IP production produced more—$390 per acre—a difference of $40. Again, these numbers, without any accounting for costs, show that for these two producers, revenues varied greatly. Producers did not expand on how they came to their data conclusions. Storage and transportation costs for Participant #2 were 350% and 80%, respectfully, more than standard production. Although Participant #8 had no change in storage costs, the

---

[*] See U.S. soy crop statistics at http://www.soystats.com/2007/page_12.htm (accessed September 10, 2007).

transportation costs increased by 500%. Standard and IP production insurance costs were identical for each producer.

Regarding time and labor during critical planting and harvest periods—in excess of standard production time requirements—in nearly all categories Participant #2 performed more IP-related hours than Participant #8. Participant #2 also had more total hours relating to IP than any other participant; for example, Participant #2 had 49 (preparation), 24 (inspection), 24 (combine cleanout), and 40 (handling and separation) hours of IP in excess of standard production. Participant #2's data still held that they receive less for IP, yet the amount was higher than many other participants, although Participant #8 (possibly reflecting greater experience) spent nearly 110 fewer hours than Participant #2 toward IP and still showed increased IP revenue over standard production, albeit, at an amount less than Participant #2.

15.9   OVERALL SUMMARY

No organizational standards were shared with the surveyor to compare questionnaire data (e.g., standard operating procedures (SOPs), purity standards, etc.) with soybean quality standards for production or purity level. The precision of participant inputs cannot be accurately measured to determine their validity (e.g., the data that they used to determine questionnaire answers).

The notion of ever-increasing transportation expense, over which producers have little control, looms much larger in cost importance than many of the other actual IP hands-on activities and management practices. A direct correlation between transportation and storage could be better made if the terms of the contract were known, as well as the status of on-farm storage availability. These two items could be directly linked if more data were presented.

Much like transportation costs, pest costs were tied to the type of production, in this case a non-Roundup Ready product, thus incurring a higher pest control cost.

Regarding insurance, at this time larger processors and elements further along the food chain seem to benefit more from lower insurance premium costs when the underwriting insurance agent confirms that the entity has an established traceability program to minimize liability and/or increase quality/purity (value-added traits) to reduce recall costs or product rejection costs. It seems advantageous for processors, wishing to extend the safety net further and reduce liability exposure more, to include their input ingredient (raw materials) producers, such as farmers, into this type of program.

15.10   NARRATIVES FROM PARTICIPANTS

Quit raising IP crops; cost of production too high, compared to premiums for IP crops. Buyer call contracts are bad for producers, will never do them again. Premiums are just too low to justify raising IP crops. Yields are always less for IP crops, never had any that yielded the same or better than regular crops.

—Participant #2

I grew no beans for I.G. [Innovative Growers] last 2 years because it takes more work and there is no payback. Vistives take very little extra work, seed beans more on cleanout, but the rest is similar to Vistives.

—Participant #5

Although more studies are needed to help equate time needed to meet various IP purity levels (i.e., organic, non-GMO, etc.), from these two farmers' perspectives the participants were/are willing to look and try alternative farm processes such as IP; however, as stated, payback is essential. See Table 15.1 for a summary of spreadsheet data.

TABLE 15.1

Summary of Spreadsheet Data

	Participant							
	#2	#4	#5	#6	#7	#8	#9	#10
Basic data								
Years growing	7	15	7	2	10	12	4	5
Wages	$55	$30	$30	$25	$22.5	$23.50	$20	$10
Acres	1	400	199	100	600	300	287	340
Previous crop		Corn	Corn	Corn	Corn	Corn	Corn	Corn
	Standard and identity-preserved comparison data							
Standard								
Seed cost	26.00	36.10	20.42	25.00		32.00	28.75	9.00
Pest cost	14.90	96.00	37.44	25.00		45.00	15.68	50.00
Total revenue	439.00	400.00	46.00			350.00	365.85	293.00
Storage	10.01	5.00	0.00	14.00		0.35	3.48	10.00
Transportation	0.14	12.50	0.03			0.05	6.97	3.00
Insurance	100.00	15.00	Same			15.00	6.27	7.00
IP								
Seed cost	18.50	17.00	31.50	17.00		37.00	19.16	9.00
Pest cost	39.45	87.00	37.44	40.00		55.00	41.11	50.00
Total revenue	413.00	350.00	61.00			390.00	392.00	306.00
Storage	28.32	4.25	7.08	11.80		0.35	5.23	10.00
Transportation	0.25	12.50	0.06			0.25	8.36	9.00
Insurance	100.00	15.00	Same			15.00	6.27	7.00
	Output for standard and identity-preserved comparison data							
	(– means IP data were less)							
Differences standard/IP								
Seed cost	(7.50)	(19.10)	11.08	(8.00)		5.00	(9.58)	0.00
Pest cost	24.55	(9.00)	0.00	15.00		10.00	25.44	0.00
Total revenue	(26.00)	(50.00)	15.00			40.00	26.13	13.00

(Continued)

TABLE 15.1 *(Continued)*

Storage	18.31	(0.75)	7.08	(2.20)		0.00	1.74	0.00
Transportation	0.11	0.00	0.03			0.20	1.39	6.00
Insurance	0.00	0.00				0.00	0.00	0.00
Differences by percentage								
Seed cost	−29%	−53%	54%	−32%		16%	−33%	0%
Pest cost	265%	−9%	0%	60%		22%	162%	0%
Total revenue	−6%	−13%	33%			11%	7%	4%
Storage	183%	−15%	-	−16%		-	50%	-
Transportation	179%	0%	200%			500%	120%	300%
Insurance	0%	0%				0%	0%	0%
IP time and labor during critical planting and harvest period (hours)								
Category								
Preparation	49	10	1	2	4	5	4	2
Documentation	4	4	1	3	20	4	2	2
Planter	8	12	2	2	3	2	1.5	1
Management	7	10	4		2	21	4	1
Inspection	24	5	5	6	2	1	4	0
Combine	24	4	3	3	4	12	1	2
Handling	40	10	3	6	3	4	4	1
Managerial	5	5	2		10	1	6	1
Total hours	161	60	21	22	48	50	26.50	10
x wages =	$8,855	1,800	630	550	1,080	1,175	530	100
IP meeting data								
Meeting hours	12	8	3	8		4	6	
Miles	150	300	140	300		100	100	
Overnights	0	1	0				0	
Mileage rate	$0.33	$0.33	$0.35	$0.40		$0.35	$0.40	
Lodging/meals	$60	$200	$25	$-			$-	
Other	$125	$-	$-	$-		$50	$500	
Mile costs	$50	$99	$49	$120		$35	$40	
	$235	$299	$74	$120		$85	$540	
Total other IP $	$9,090	$2,099	$704	$670	$1,080	$1,260	$1,070	$100

Source: Developed from research.

16 State of the Science—Interpretation

16.1 INTRODUCTION

Identity preservation and traceability's (IPT) state-of-the-science, thus far, has been the collection of the various components, entities, and participants involved with IPT. This interpretation chapter will give definition to the present state and also attempt to suggest what we may expect in the future. The state of the art could have included more, or at least different, entities and components on which to base the interpretation. This interpretation, imperfect as it may be, provides another way of looking at agricultural production as it applies to our present needs. Other studies could easily dedicate their whole scope of research to any individual component. The interpretation is also not a quantitative piece. Thus, it was deemed critical to provide whatever anecdotal, ancillary, and spherical qualitative information (from entities, components, and participants) that was most readily available and thus provide greater information and interpretation in as complete a work as possible.

16.1.1 FUZZY SCIENCE

This is a good point to make mention of experimental and "fuzzy science" as it applies to IPT research. Measuring the effects of IPT is not always an exact science. The challenge when attempting to measure and evaluate IPT solutions is that no one discipline or group of scientists can claim the upper ground of knowledge or domain. IPT incorporates new notions and combines many subjects that have traditionally been discrete entities of focus. Our computer age environment leads us to believe that exact statistics and laboratory results will provide overwhelming evidence and solutions. Still, there are many questions that have areas of gray; what is right for one group may not be right for another. This is especially true for scientists researching outside their area of specialty. Conventional mathematics joined by fuzzy science may help guide us toward systems and possible solutions that society seeks. Often the distinction between conventional science and fuzzy science is unclear. However, what is clear is that transparency of information and programs should bring credibility and confidence toward the systems that compose IPT.

16.1.2 THE "STATE OF THE SCIENCE"

So what does the "state of the science" show us? IPT is here to stay and is a well-established system within agriculture's framework. Traceability programs, mandated by industry, national laws, etc., will continue to be integrated within all facets of food production. Changes and improvements, regarding traceability, will usually be measured and based on the most minimum of requirements. For the marketplace, business logistics will help promote traceability programs. However, competition and greed may also attempt to cloud transparency and established processes.

The most dramatic occurrences already happened when many developed nations and international organizations (e.g., United States, European Union [EU], Canada, Japan, Codex, International Organization for Standardization [ISO], etc.) instituted traceability rules and regulations governing agriculture and trade, whereas identity-preserved systems generally continue to be dictated to by the marketplace. Systems and programs that address both traceability and identity-preserved needs, which result in increased efficiencies and are favored by customers, should be seen as long-term winners for agriculture and consumers.

16.2   GENERAL

Our food supply and agriculture face many challenges, not only to provide enough food to feed a growing population but also to preserve the environment in the process. Supply and demand for agricultural products are becoming more critical, and lands that were once deemed unsuitable for agriculture are now being put into production, usually at the cost of the environment. Energy needs are pushing agricultural lands into bioenergy production where human food and animal feed production once occurred. Global production and trade have increased the incident of contaminated food and substitution of inferior ingredients. Fortunately there are solutions. Traceability and identity preservation (IP), fairly new concepts, are being incorporated into such areas as quality assurance and use within advanced software/hardware to help provide solutions. IP and traceability are, as mentioned throughout this text, both independent systems yet may also be tied together depending on the agricultural system or product being discussed. Both are affected by many factors. Presently, and for the foreseeable future, energy issues, global trade, and food safety issues are all expected to be the major contributors toward governing IPT systems, as will be discussed within this chapter.

Traceability systems are expected to show steady, incremental refinements along the entire agriculture food chain. Traceability's framework is built on two pillars, usually sovereign safety regulations and business's desire for improved logistics and reduced liability exposure. Changes in traceability regulations have been driven primarily by nationalistic attitudes as a result of perceived dangers derived from unregulated food production methods, environment abuses, fears of technology run amuck, etc. However, a very close second reason for traceability systems' advancements has been their ties to logistics and supply management efficiencies. Traceability-type systems should see further expansion into all facets of agricultural production, as required by law. However, most of traceability's innovative advances should be expected to come from industry. Business will continue to be motivated to meet legal requirements and still show increasing profit. In addition, these changes should address an array of needs besides logistics, from expanding value-added products offered to reduction of recall/product liability insurance expenses. Industry is expected to provide incremental changes based on research and market trends.

Regulations are not expected to change greatly unless food safety issues and recalls become alarmingly more prolific and, if so, may act as a catalyst for further regulation modifications.

Identity-preserved systems are expected to divide into two groups: (1) the group of products/processes that survive the willingness to pay for product market test, and (2) those that customers are generally unwilling to pay for. This is not to suggest that niche products, special geolocations of production or processing, fair wage, etc. will not still have a place in the market. Rather, these types of traits or attributes may be greatly diminished in the overall market place. Identity-preserved product survivability will be determined by IP systems economically providing a product that the consumer desires and believes is factually true, often provided by independent parties and laboratory analysis.

Just as globalized trade increased the exchange of goods/commodities, demand for identity-preserved products will be determined by the degree of willingness to pay and/or marketing motivations of processors/manufacturers. The interactions of traceability requirements and identity-preserved requirements, i.e., the cost of implementing and processing these programs, should be reduced over time as these processes become more imbedded within corporate processing/manufacturing standard operating procedures (SOP). However, global demand for fuel/energy and the displacement of agricultural products from food/feed to fuel/energy may interrupt the cost/benefit structure that now exists for many products such as organics. If basic commodity prices increase, so too will the prices for IP products because within the IP consumer price will be the farmer's premium to grow IP products. The question will be how high can commodity and IP products go before consumers will be unwilling to pay. It is projected that as food prices increase, IP production will decline overall in number and selection regardless of reason. Agricultural economists are better suited to explain the reasons behind agricultural supply and demand effects and associated elasticity of demand for various agricultural products and are beyond the scope of this work.

From another view, as tools for IPT become more plentiful and less expensive, a general shift (within agricultural products) may occur from the more basic or minimal requirements of traceability (only) to the more detailed identity-preserved production process. Again, this change will depend on a combination of customer/consumer willingness to pay and processor/manufacturer-perceived marketing benefits.

16.3 IPT THEORY, DESIGN, AND COMPONENTS

Traceability theory, design, and components have improved overall and provide greater understanding of growers and processors to customers and end users. However, there is a long way to go. Regarding theory, most advances are to the result of logistical considerations (software integration) and less toward nontangibles, non–computer-based solutions, i.e., social issues. Most newcomers to traceability concepts are still struggling to meet minimum legal requirements, whereas those with well-established logistics systems have expanded their software's technological scope to address traceability's legal requirements and more. Traceability design, as will be discussed in more detail, has been primarily tied to logistics software, minimal laboratory testing, and very little consulting (outside the firm or external). This is not to say it will not change. The shift in increased commodity values—be it good or bad—as a result of increased fuel/energy prices, food safety, accountability, and global trade should favor enhanced traceability especially for commodity production. Traceability components, be it items within a farm or along the food chain, are expected to become more easily identified and understood as the adoption of traceability increases. IPT components have also dramatically improved because their importance within the agricultural chain has been put on notice and highlighted as a result of regulations.

Identity-preserved theory, design, and components are very much at their infancy stage. With few exceptions, such as organic production, general identity-preserved production has much potential but still much farther to travel to become recognized and acceptable to the general public. It makes sense that identity-preserved theory, design, and components have more potential to achieve because IP is much more encompassing (holistic) and has additional demands beyond basic traceability requirements. In a different vein than traceability, most identity-preserved theory has evolved from the notion of trait(s) and/or attribute(s) of interest desired by customers or consumers' willingness to pay. Again, the effects of increased global fuel/energy needs, food safety, and international trade will come into play. It is unclear how the effects just mentioned will be realized on IP products. One scenario is that as the value (costs-prices) of commodity products increases, traceability for these products will be improved. Unfortunately, for identity-preserved products, for farmers to be willing to produce enough IP products there would also need to be a corresponding increase in both the premium paid to farmers and resultant cost increase that customers and consumers would need to pay. The unclear part is how high (traced) commodity prices will go, with corresponding increase of identity-preserved prices, and when customers/consumers willingness to pay for IP products will start to decrease.

IP design is expected to follow on the coattails of traceability design because insofar as software goes, both systems share many common logistical requirements. Then again, because IP production takes accountability often many steps beyond basic traceability, IP design features will tailor themselves for general or specific traits/attributes of interest. An ideal IP design would include an IP standard (contractual), which continues through and addresses the entire food chain, typically within the purview of the parties that are willing to pay for such traits/attributes. At this time, other than some conventions for organic production, for which an agreement between nations is still far off (but being worked on), most IP design is predicated on basic requirements of general traceability (one up, one down) and specifications, usually by contract, that prescribes audits, inspections, documentation, and specifics of laboratory analysis. It is in the contractual specifications of audits and laboratory analysis that more players have entered the marketplace. As it will be described, third-party auditors and laboratory analysis have greatly expanded to meet the growing number of traits/attributes for which parties/customers are willing to pay. It is assumed that these auditors and laboratories will expand to accommodate further market needs. As such, the cost of auditors and laboratory tests for the more standard practices, e.g., audits of fields, bins, equipment, documentation, and laboratory analysis of well-established tests, should remain stable or possibly decrease.

IP components should develop akin to traceability components but with even greater emphasis on contractual development needs. This should include aspects of enhanced computer linkages between parties (i.e., software compatibility), common unit sizes, training, greater standardization of auditing practices that promotes increased transparency, improved standardized laboratory analysis, and on-site or field-strip testing. Regarding IP traits, the scope of what may be of interest appears to be continuously growing and tied to the economics of customers/consumers' willingness to pay. Government actions, via regulatory requirements, could change the notion of IP being tied to only customers' willingness to pay, e.g., organic versus natural food production. Through regulation, the government could encourage market changes by way of incentives and taxation to promote long-term social and economic goals. It should be noted that for whatever reason a trait/attribute is desired need not be on a global scale (market). Small producers may still provide for niche markets. This too offers unique challenges such as for those interested in products derived from, or processed at, specific geolocations. This has been important for the EU customers. The auditing and laboratory analysis techniques for these attributes are often as unique as the geolocation itself. This would be another example of IP theory, design, and components accommodating distinctive traits/attributes of interest.

16.4 PROGRAMS AND STANDARDS

Seed agencies are expected to continue in their practices, striving for increasing quality standards and auditing for all types of seed produced under their responsibility. These organizations already have traceability programs in place (one up, one down) and are expected to promote increased purity and quality of seed product. However, little work has been done toward qualifying and connecting specific identity-preserved production to targeted traits of interest, although some seed agencies include in their publications' narrative that particular IP seeds claim to produce high or low "x, y, z" traits. Seed agencies thus far do not test to confirm the accuracy of the claims, nor do these agencies, regarding identity-preserved production, certify a farm's practices, e.g., if they claim to be organically produced. Agencies traditionally certify, for example, if a particular foundation seed was from genetically modified organism (GMO) or non-GMO original sources and foundation seed's field conditions during routine inspections. Much could be done through these agencies to expand their scope regarding IP production, but they are unlikely to do so because of the higher costs needed to perform additional testing, costs to be borne by farmers, and, at this point, general lack of interest from its farmer customers.

16.4.1 INDUSTRIAL PROGRAMS

Most, if not all, industrial programs have well-established traceability programs. From parent seed producers to food processors, traceability has generally been an outgrowth of their quality control and logistics programs. Regarding identity-preserved products, it is in their IP programs that many of their premium products show tremendous promise and sales. Highlighted by the success of TraceFish, many commodity producers, especially those that produce organic products, have excelled in market growth and consumer acceptance. Many industry IP programs include or offer consulting options, market information, and more. They are also on the forefront of IP innovation in both systems approach software suites and laboratory/ field analysis and testing. The software approach has greatly assisted in streamlining inter- and intrabusiness communications, standard units of measure, and—very essential—smoother flow and transparency of data from one entity to another.

Many IP industry programs have been well established before governments implemented traceability regulations. Some industry programs offer additional services ranging from satellite imagery and educational classes to broader, nonspecific IP practices. Industrial programs show a tremendous opportunity for IP growth and acceptance. However, it is estimated that IP growth and acceptance will primarily be derived from, or the result of, customer demand, especially from end-use customers' willingness to pay. In time, a gradual split in IP products demanded may occur. For example, because of energy costs, environmental concerns, etc., governments, industry, or consumers' advocacy groups may cause a shift in production requirements. This usually causes an increase in production costs, possibly without an appreciable increase in product value, as perceived by a lack of consumers' willingness to pay. Other examples include products produced at locations especially known for their quality, or any new trait or attribute on the horizon, that consumers would flock to and be willing to buy.

Overall, much of the success of industrial IP programs will be predicated on the present or near-present market demand for a new or current product, efficiencies of production to include IP-specific accounting practices, improved laboratory analysis and field testing, and greater adoption and familiarity of IP production processes from one end of the food chain to the other. It is on this last point that uncertainty in IP adoption is very high because it is understood that more farms will be adhering to traceability practices because of regulations, with an overall increase in cost of production reflecting this change. Unfortunately, farmers must usually absorb these additional production costs, even if there becomes an oversupply of, or low demand for, their product. If, however, energy demands take product from food/feed production to bioenergy crop production, there appears to be an inherent lower supply (and relative higher demand because of scarcity) for particular commodity food/ feed products. Again, this may help many commodity (non-IP) producers but may also push out some IP producers. The increased premium needed to produce the IP product may cause them to price themselves out of the market or customers' willingness to pay.

In this regard, industrial programs must be very sensitive to not only the demands of the market at present but also to other external events that may be influenced by energy costs, the environment, special needs, etc. Product differences resulting from various production requirements, i.e., traceability versus identity preserved, should become much clearer as external factors pressure the cost of production and the market's willingness to pay.

It is projected that nearly all of the external influences mentioned will have an incremental impact on production. For example, a projected increase in demand for corn ethanol should initially represent a shift from food production toward energy production, one that may result in higher food prices but hopefully still provide enough food production (at least of U.S. customers). Behind the scenes, non-end-use customers (e.g., processors) are also expected to increase their demand for specific processing IP traits of interest. For processors these traits may include extended shelf life, benefits to packaging, and other qualities. At present there is a strong demand for non-GMO products around which entire food chains and corporations are built. If, however, a new production process is developed, which has been noted in several journals, that does not use transgenetic engineering to obtain the same benefits as GMO products, the notion of GMO versus non-GMO may become mute. However, this may be some distance into the future. Who knows what changes may occur that will promote some new process, trait, or attribute of interest?

16.4.2 STANDARDS—UNITED STATES

IPT standards have primarily been motivated by bioterrorism laws and tailored for marketplace implementation. In practicality, U.S. IPT regulations are primarily designed for food safety recalls (traceability) and specialized niche production, such as organic food production (identity preserved). U.S. traceability regulatory trends are expected to be refined with time because of ruling modification needs for dealing with complex, changing food safety issues. Most traceability regulations are focused on accountability in the case of food recall. In the United States the assumption is that unless otherwise noted, food products are safe to eat and that testing and approvals have already been done by the producer or industry to ensure that it is safe. This is much different than in the EU, which requires many more types of approvals and testing. In the United States, industry has a greater say and control of what is produced and offered to the market. This may be because of the public's trust in both the products being produced and in government's ability to provide oversight. There will always be some disconnect between the amount of government regulation needed (and enforcement) to provide enough food safety for the public and yet allow enough freedom to innovate (and profit) for producers. What appears most evident regarding standards throughout the globe is that most regional regulations enacted are agreed on by that particular group (public) as understood or perceived as appropriate for their community.

In the United States, identity-preserved product sales should steadily increase based on successes of improved IP product lines, efficiencies of IP processing, and expanded global market. At this point, however, traceability progress with regard to some production groups (e.g., cattlemen, vegetable producers, etc.) has been slow because of particular industry-specific traceability needs and challenges. Often the reason is the result of management and process challenges tied to reluctance to change traditional practices and its associated cost of production changes. Identity-preserved products face an additional hurdle of how to finance or pay for increased identity-preserved processing costs. This is usually from customers (processors) or consumers. If costs of agricultural products continue to climb substantially, the willingness to pay for a particular IP trait may not be realized. In other words, external forces may shift production requirements, which can greatly impact domestic and global food prices, and therefore affect the purchasing trends of commodity and IP products.

Indirectly, future changes to U.S. IPT standards may come from energy and farm-ing bill implementation. Both these bills can greatly determine what types of crops are grown and where they are grown. This then would affect the supply and demand for various crops and either help promote some products (high demand/lower prices) or curtail other products (low supply/higher prices). As mentioned earlier, this can affect the sales of products tied only to traceability systems and those grown under identity-preserved programs.

16.4.3 STANDARDS—EU

EU traceability standards are nearly akin to U.S. traceability standards. However, generally speaking, EU regulations are approached and designed to better pro-tect the public from danger or perceived danger. The assumption is that produc-ers must show that the food is safe to consume. This is especially true for the EU's notion of GMOs and GMO contamination. EU standards specify labeling requirements that highlight GMO ingredients in raw and processed foods and animal feeds. A possible loophole in EU labeling regulations allows approved GMO animal feed, appropriately labeled, to be fed to animals (livestock), which are later processed and sold for human consumption without a GMO labeling. This may be because of the EU's need to economically import enough feed to supply its livestock industry while not curtailing its domestic meat industry sale of products. This could change with further EU regulations. Also, changes in technology may open doors for U.S. products into the EU, especially if techno-logical advances mitigate the need of transgenetic gene use or GMO products, thus removing the stigma of GMO labeling. Newer techniques, improved hybrid-ization, etc. may overshadow the challenges that now face GMO products when attempting to introduce them into the EU marketplace. Again, the perception of the EU citizenry and their numerous consumer advocacy groups toward any new technique or technology requires that it will need to be proven safe and appeal to their notion of what is right or appropriate.

EU regulations, compared with the United States, are much more prescriptive and detailed, describing the how and why rather than mere quality or tolerances of the end product. EurepGap is a good example of an on-farm prescribed management and documentation program. In some cases, this prescriptive type of program assists in protecting the environment, people, or culture, whereas at other times it may sty-mie innovation and commerce. EU regulations have also had the effect of influenc-ing peoples' purchasing, agriculture, and eating habits of other nations, i.e., in what they import/export from other countries. In one case, a country that had planned to import GMO corn (from the United States) for domestic use faced possible trade sanctions by the EU as a result of the GMO importation. The EU threatened to ban agricultural imports from that country if it imported GMO commodity products. This type of practice is especially troublesome for developing nations regarding imports and exports. Additionally, the understanding of each country's particular IPT regulations becomes especially critical to know and is not always harmonious with others. This is not to suggest that progress is not being made. The United States, EU, Canada, Japan, and several other developed and nearly developed countries have agreed on many facets of traceability. The same can be said for identity-preserved organic products, although much work still needs to be done.

16.4.4 STANDARDS—OTHER

IPT rules and regulations, whether promoted by industry or the state, have grown and advanced tremendously. Some standards such as Codex, ISO 22000, and Hazard Analysis and Critical Control Point (HACCP) are internationally recognized and have helped streamline and refine the many regional and international regulations into focused, yet much more acceptable international standards. Although some of these regulations appear more detailed than most U.S. companies are used to, they are relaxed enough to offer opportunities toward innovation and competition. As mentioned previously, several national and international organic organizations are working toward greater rule harmonization for accreditation and certification. This may further assist in expanding the sale of organic products in the global marketplace.

It is expected that developing nations exporting to nations with IPT regulations will meet the requirements of the importing country; this should also help the exporting country by increasing the standards of its domestic farmers, processors, etc. Additionally, developing countries that import products from countries with established IPT regulations should not only benefit from the exporting countries' standards but also this may help improve the quality, if not the expectations, of the importing country's populace that safer and more accountable food quality is possible and available. Unfortunately, there are still many obstacles for developing countries to overcome to achieve the many benefits of IPT programs. Often this lies within the countries' own infrastructures. It will be interesting to observe which developing countries will take advantage of the opportunities IPT has for not only exports but also to improve its food quality issues for its own populace. Within the category of developing countries there appears to be several layers or distinctions between these types of countries. Some developing nations such as India and China appear much closer to being considered developed countries. However, this chapter does not cover global economics and individual nations' development status in detail.

Additionally, developing countries may also be affected by the influences of foreign entities wishing to expand internationally, from large farm cooperatives, processors, and retailers to non-governmental organizations (NGOs). We already see some of the influences of NGOs regarding identity-preserved traits highlighting fair wage and fair trade, and several other traits and attributes of interest. However, the link between NGOs and the implementation of regulations (that are enforced) for many developing countries is still tenuous. Fortunately, in some countries, it appears progress is being made.

It is unclear how other regional and religious standards, as well as their governed products, will be affected by the larger global influences that affect agricultural food, feed, and biofuel production. There are numerous smaller organizations vying for attention to persuade government officials to enact rules and regulations that favor these particular groups. Often these groups are of the grassroots type and propose the benefits of "buy local"; preserve our economy, our countryside, our culture; and an assortment of other mandates that are important to them. Some, depending on the nature of the proposal, such as highlighting the origins of the product (geolocation of production and/or processing), have been brought into legislation within the EU and several other countries. Still other standards, such as SQF guidelines and rules, have expanded from a region/province (Australia) into new markets, countries, and production industries and are expected to continue to do so. The kosher standard is also expected to expand not only in both the types of food prepared (not of Jewish or Hebrew heritage) but also in pure volume prepared under kosher guidelines. This is not because of a substantial increase in the Jewish or Hebrew population but rather the perceived cleanliness and safety of kosher-prepared foods by the public.

16.5 AUDITORS AND LABORATORIES

Auditors and laboratories are expected to increase in number and in scope because of greater market demands caused by new laws, innovations, and technology tied to both traceability and IP systems.

16.5.1 AUDITORS

Auditors have traditionally been used to validate company claims, especially relating to quality control procedures and logistics management. Besides standard management and production audits, some auditing firms perform unannounced mock recalls as part of their suite of services. Auditors provide verification of traceability requirements typically prescribed through both paper trail and electronic data collection accounting, which is expected to continue. Auditors and their services are also expected to be demanded by more agricultural customers and businesses associated along the entire agriculture food chain. This increase is expected to enlarge traceability data requirements regarding the proof of recordkeeping, training, etc. by food chain participants. It should be expected that auditing firms that already audit a wide variety of industries will expand their repertoire of services to meet the additional needs of agriculture traceability more easily than smaller firms, which may offer less flexibility in services. Auditors are greatly expanding their services into areas well outside the traditional expectation of one-up, one-down traceability toward the more dynamic requirements tied to identity-preserved contracts and regulations.

This is especially true as auditors venture further away from, by comparison, the more rote requirements of basic traceability and move more toward the value-added products associated with identity-preserved production. The most common example of this is of organic food production. Auditors in this case not only check the required records but also on-site management, procedures, etc. typically for all organizations that wish to market officially recognized organic foods. Thus, auditors, as needed or required, have and are expanding their services. Just as organic production has been growing at a remarkable rate, so too have other identity-preserved traits and attributes. Specialized organizations such as Rainforest Alliance[*] and Cata[†] act as

---

[*] The Rainforest Alliance has information at http://www.rainforest-alliance.org (accessed January 25, 2008).

[†] Cata (El Comité de Apoyo a Los Trabajadores Agricolas—The Farmworker Support Committee): its information can be found at http://www.cata-farmworkers.org (accessed January 25, 2008).

quasi-auditors that verify specific identity-preserved traits and attributes well beyond traditional IP requirements. Other nonstandard auditable IP items include fair trade, fair wages, sustainable agricultural practices, non-GMO products, geolocation of production/processing, etc. It is in these areas where auditing is showing great growth and tremendous promise. Again, success for specific traits and attributes will be directly tied to market effects from external influences.

Although standards from country to country are becoming more in line with one another, there are still large differences between them. Auditors also vary in their ability to accommodate various rules and regulations, in quality, scope, in services they provide, etc., just as any other industry. Auditors who are themselves audited or accredited, tested, and reviewed, and who have a good track record, should be used for their auditing services. Some of the larger auditing firms provide services for many types of industries such as aerospace and medical industries. The needs of specialization in auditing agricultural products, production, processing, etc. are expected to increase as more traits/attributes become desired. As such, it is also projected that larger auditing firms will buy out smaller, more specialized auditing firms, especially as the value of identity-preserved contracts grows in value and volume. This may be because of larger auditing firm's greater ability to incorporate auditing with laboratory analysis and field testing services.

It is also expected that as agricultural biofuel production becomes more critical and essential that some identity-preserved traits/attributes, as a result of shifts in production, will fall to the wayside, whereas others, such as identity-preserved distillers dried grains (DDGs) may become more important because of the very large quantities of DDGs produced and used as livestock feed—all of which will require various levels of traceability and IP auditing.

16.5.2 Auditing Laboratories

Both auditing laboratory analysis and associated field tests used for IPT production are very similar in design. However, each may differ in scope; that is, to the breadth (amount of information analyzed or tested) and depth (how far back/forward within the supply chain) for which they are being contracted to conduct.

Traceability requirements for auditing laboratory analysis and field testing, as has been illustrated, are usually narrower in scope, with fewer demands or conditions tied to them, as prescribed by regulations or contract. Typically traceability laboratory analysis and field testing have specific purposes, for example, to confirm analysis/tests substantiate management data/records and claims (oil or protein content). These analysis and tests monitor specific desired features and/or quality at one or many particular points of production or process. Often analysis/tests may record a commodity's oil content or confirm a non-GMO quality at a particular stage or stages of production.

It is expected that improvements will be made regarding analysis/testing techniques required to confirm current and future traits/attributes of interest. In addition, the statistical modeling used for analysis and testing is expected to advance to meet tighter measurement standards. It is unclear exactly how the scope of analysis and testing will change as demand for traceability matures. However, what is evident is that there is still deep division between nations and organizations regarding the parameters to determine quantitative measures and tests to be conducted to prove claims. Auditing laboratories and manufacturers of field test kits are attempting to tackle which analysis/tests will provide the appropriate level of confidence at reasonable costs. For example, debates include the use of nuclear isotope testing to determine geolocation of a product's origin. Although experts in laboratory and field testing may agree or disagree about the merits of various tests and modeling used for interpretation, so too do the politicians, as they interject often other, less scientific aspects to be considered for instituting regulations. On a larger scale, official international committees and organizations are joining to discover and determine which tests are most reasonable, accurate, and cost efficient to promote specific standards that will be acceptable to industry and trading members.

Regarding identity-preserved analysis and tests, laboratories and field test kit manufacturers are expected to continue to work closely with corporate and research facilities' laboratories and engineers to be better able to meet the changing analysis/ testing needs as new IP products enter the marketplace. Because of market instability derived from the current fuel production situation domestically, global trade, and food safety, the trends of identity-preserved products for the future are most difficult to imagine. It will be interesting to see how smaller, specialized laboratories and field test kit providers will perform if larger laboratories absorb these smaller entities as consolidation occurs. Auditors and laboratories that have international connections and shared resources should be better able to provide more up-to-date analysis/ testing that conforms to required standards.

16.6   CONSULTATIVE AND SERVICE CONTRIBUTORS

Both *domestic and foreign policy and advisory organizations* provide opportunities to modify system structures and manipulate our environment and economic marketplace by recommending and influencing industry policies and government regulations. Business, consumer, environmental, and other advocacy organizations bring specific views and concerns from their particular fields. Sometimes work is on an international level, whereas at other times the focus may be only at the local level. Predominantly the focus has been on traceability, with less but growing interest regarding identity-preserved products and production.

The EU's primary disposition has been, and is expected to continue to be, anti-GMO, as it pertains to perceived consumer safety and environmental issues. They have also expanded rules in regard to growing, processing, claims, and testing to EU food's geo-origins/processing. At present, the major catalysts for many of the EU's present and expected future regulations changes are the result of continued food safety scares and activists or advocacy groups. EU public debate has a much larger cross-section of vocal participants, and this is expected to continue. Compared with the United States, the EU allows greater weight of discussion or influence from non-scientific attributes or values. In other words, the EU regulatory foundation is more culturally based and influenced rather than being solely market-based (industry-driven) or scientifically based for regulatory development. Much of the EU's developing food regulations appear to be more driven to maintain the cultural status quo, with safety a primary component. This is not to suggest that scientific and industry are not heard or involved with the development of regulations, only that the EU's public sense of appropriate balance toward regulation development is different from the United States. Already there is much concern within the EU regarding the environmental effects of the global energy situation and the production of ethanol. EU legislators are considering rules that govern not only the sources of ethanol but also the production processes and appropriate lands from which it is derived. The United States, with its interest in exporting agricultural products to the EU, should consider it better business and politics to meet EU rules and regulations rather than challenging them (barring obvious international trade violations). The United States should consider providing product (its export surplus) in the manner that the EU wants (importer regulations).

The United States continues to be protechnology with continual research with GMOs. The United States appears to be more scientifically focused in their approach toward food safety, the environment, and community. This is expected to continue. Their acceptance of scientific and market approaches, from buy local and environmental concerns to global sales and fair trade, are becoming more predominant within U.S. culture. Food safety, the environment, and food prices are expected to direct and influence agricultural production, the direction of research, and market demand. Two items may greatly influence future production, like the EU, the global energy situation and the prospect of a new technology, one that is non-GMO in nature, for development of commodity products. Although this may be many years in the future, the notion of not having the issues of GMOs or GMO ingredients being a barrier to trade, or safety concerns, may greatly ease many food concerns. This is not to say that this type of new technology will not require appropriate safety testing and acceptance throughout the globe. Cultural issues may still be of grave concern.

Other developing or nearly developed nations face trade issues with regard to food imports and exports. Most developed or nearly developed nations are involved with and interested in agricultural imports and exports. Of particular interest for imports/exports are traceability and its incorporation of labeling, appropriate tests to confirm claims, enforcement tools, jurisdiction, etc. At present, most traceability concerns are focused on food safety (e.g., EU—GMOs and United States—bioterrorism). Developing countries encounter challenges different in scale and ability to manage as compared with the EU and United States. This is especially true with regard to the influences of international or multinational corporations and NGOs, as they address their developing countries' issues as they pertain to environmental and food safety to financial and economic well-being. It is expected to continue that the more powerful developed nations (i.e., United States or EU) will attempt to influence what agricultural imports or exports developing nations will trade. For example, it is well documented that EU countries in the recent past would not import from African countries that had imported GMO commodities or GMO seeds. This type of influence can greatly influence trade for developing countries.

As organizations around the globe become better acquainted with the challenges of traceability and share their views and approaches, it is expected that more unified approaches and consensus will develop. Most regulations are, at present, mostly directed toward traceability with regard to food safety issues. IP has been addressed in a backdoor manner as regulation enactment has occurred, i.e., with regard to ingredient labeling and country–of-origins labeling (COOL). Still, the primary concrete identity-preserved regulation, for many nations at present, deals with organic food production. Although a consensus of compatible national regulations is still far off, work is being done to harmonize various aspects, such as organic accreditation and certification. It is also expected in the not too distant future that regulations will be developed that govern geo-origin and geo-processing, associated social issues such as fair wage and fair trade, and the influence of agricultural production on the environment. These issues are expected to guide both traceability and identity-preserved regulations for many years to come.

16.6.1 Software Providers

Traceability software has been and is well established as a result of corporate logistics and quality assurance interests. Providers are expected to continue to improve their products and accompanied services in efficiencies, ease of use, transparency, and transferability of information. This is an area where the supply of providers, from specialized (discrete) to general (complete food supply chain), is growing and expected to continue. The primary challenge for software providers will be in determining the challenge(s) that customers face, be it basic traceability requirements to refined niche dictates prescribed identity-preserved attributes, and then incorporating the appropriate tolerance level(s) and methods of evaluation or testing within their software programs and systems. Selection of a software package or software firm can then be determined from the strengths and weaknesses of their services. Although this sounds straightforward, it is not. Software and services considered for purchase should include the ability of expansion/growth of the product line and modifications of regulations. As newer external laboratory analysis and field tests evolve, so too is it expected that associated software will incorporate these changes. In addition, as tests and regulations are introduced, so must the advances of modeling tools of greater power be included for analysis purposes.

Software providers that provide identity-preserved products are much more limited and fewer in number. This is most probably because of IP's more fluid or custom requirements, which are usually dictated contractually. Most IP software is derived from traceability software, with additional accessories such as digital satellite imagery, geographic information systems (GIS), or global positioning system (GPS) software. Nearly all IP software is tailored for specific product lines or chains. Many of the original IP software providers started by designing products for non-GMO production and then expanded to include non-GMO processing and accompanied non-GMO food chain participants, e.g., contracting, inventory/logistics management, quality assurance and testing, network security, and standards integration. IP software, much like traceability software, is limited to the input devices that can easily interface with specific software. Many traceability or quality control systems are automated; unfortunately, many identity-preserved items of interest, i.e., organic production, require onsite third-party inspection or auditing. Identity-preserved software can greatly contribute by providing specific data such as dates, measurements, standard requirements, outcome projections, and who performed what task, etc. This is where it needs to be understood that software, in and of itself, will not solve all identity-preserved challenges, although software is continuingly improving. Who knows, it may be possible for software and associated systems to monitor growing conditions or wages paid in foreign lands. However, many situations still require third parties to visually check and interview to validate claims.

Most, if not all, IPT software integration includes GS1 and European Article Number/Uniform Code Council (EAN.UCC)—barcode standards and operating nomenclature. Software development is expected to increase, especially regarding the ability to integrate tests/analysis from various input sources and overall systems assimilation with other organizations/companies, all while decreasing in price.

16.6.2 Process Facilitators

This is an area of tremendous growth and expansion for IPT systems. The demands of mandated industry and government regulations have affected an array of organizations such as TRACE and FoodTracE, specialized websites, IPT training/marketing organizations, and labeling, analysis, and media groups. These organizations fall within a wide spectrum of services, from traceability or IP only to a blending of the two, all dependent on the organization type.

Traceability, with its associated supply chain components, incorporated many complementing participants tied to food safety, trade, etc. issues. As a result of this incorporation, traceability process facilitators are expected to grow in numbers to meet expansion. Developing nations that wish to export may find traceability processor facilitators especially well equipped to handle exporting needs on a number of levels, such as training personnel on food handling requirements to customs declarations. This should help in the smoother flow of commodities. As specialized commodities become more apparent, it will be these types of facilitators that should help accelerate customer acceptance, lower overhead costs, etc.

Identity-preserved process facilitators have aligned more with particular IP traits and attributes of interest, such as non-GMO and organic production. In addition, IP process facilitators have also focused on new or young farmers, those looking beyond standard commodity production and desire to learn how to transfer from traditional to nontraditional farming systems. IP process facilitators are expected to provide information, education, and analysis as market demand necessitates. Opportunities in this area of instruction will be predicated on market demand of IP products.

16.6.3 Food Recalls and Insurance

16.6.3.1 General Food Recalls

Many factors suggest that there will be an increase in food recalls, especially for products grown in one or more countries, processed in a third, and sold within a fourth, all of which can be a recipe for disaster. Additionally, as more agricultural production is shifted to biofuel production, the resultant decrease of supply in food production without a corresponding decrease of demand in food should pressure both agricultural commodity prices to new heights and marginal lands (lands not traditionally farmed or ecologically sensitive lands) to be put into agricultural production.

Record high commodity prices are expected to encourage cheating, substitution of inferior cheaper products, with a resultant increase in food recalls as a result of safety issues, contamination, etc.

It is unclear whether the likelihood of recall is greater for traceability products than for identity-preserved products. One could suggest that the volume of food produced under traceability rules (larger volume) would necessitate a larger number of recalls than identity preserved (smaller, niche volume). Although the proportion or percentage of recalls is purely speculative, it should make sense that IP products, being more tightly governed, should have fewer instances of recall. Studies would need to be conducted to detect a trend and comparisons.

A major theme in the use of traceability and IP systems is a firm's real or perceived exposure to recall and its financial resources. Generally, firms that have well-established traceability and identity-preserved systems in place regard the expense of these systems as costing less than the cost of recall, lost brand name value, etc. Some even use this built-in management instrument as a marketing tool to enhance sales. Not all food chains have the same degree of negative exposure to recall. For example, within the meat industry, which has had many recalls (although usually regional in scope), traditionally processors or meat packers have been responsible and thus conduct recalls. The cattlemen, as individuals, were not directly responsible for conducting the recall or direct expense. At present, it falls onto processors, especially those with more recognized brand names such as Tyson Foods Inc., to take effective action. The meat industry as a whole is grappling with the notion of animal identifications (or animal IDs). At present the conflict for them is not only whether animal IDs are desired but also who is to pay for such a system. Other developed countries, such as Australia, are incorporating animal IDs for nearly all livestock animals.

16.6.3.2   Resource Protocols

With the passage of time more corporate, government, and academic resources are being drawn on to address food safety and food recall issues. Protocols are greatly expanding to improve the need and speed of recalls (speed of detection, notification, testing, etc.) and to isolate the recall item(s) to a specific company (location, date, batch, or bin number). Regardless, IPT systems' ability to quickly and accurately provide information to determine the likelihood of particular batches or lots being affected will greatly mitigate recall costs and brand name damage. For example, in recent years protocols for the detection of GMOs have been devised to determine whether a product lot had above a specific threshold of particular GMO traits. Unfortunately, there appears to always be a race within the market, especially for new products, with the possibility of some level of nonconforming ingredient being introduced to the public. Then it is the ability of government and auditors to be able to detect the nonconformity and where well-honed IPT programs can accomplish this cost effectively.

In this way, there may always be this cat-and-mouse game, especially as the stakes become higher. In addition, for individual companies and industry as a whole, the ability to isolate which products are and, very importantly, are not in violation or needing to be recalled will be and is essential. Improved public education regarding food safety, traceability, recalls, and IP by all organizations involved with the food chain could greatly increase food safety and reduce unnecessary food product recalls.

16.6.3.3   Product/Recall Insurance

This is an area that deserves much attention for its pivotal role and potential to influence and expand not only traceability rigor but also especially identity-preserved product accountability. All too often the role of product/recall insurance has been to shield and protect processors and major brand-name product lines. Typically insurance underwriters provided insurance premiums at less expense when a firm illustrated competencies with improved or enhanced risk management practices, confirmed by mock recalls and written SOP. Product/recall insurance usually works backward, from retailer (with an abundant of product/recall insurance), back to the processor (with somewhat less product/recall insurance because of less exposure and greater opportunity to control risk). For example, commodity or specialized grain producers do not typically have the option of purchasing product/recall insurance. At most they may purchase income insurance, which is tied to the crop's Chicago Board of Trade (CBOT) commodity futures market value. Specialized grain producers are only covered for the CBOT commodity grain value of their IP crop in case of loss. In the meat industry, if a recall is deemed necessary, the specific processor's lots are recalled, and insurance company coverage expenses stop at the processor, covering specific *processor* expenses. This is not to say that the USDA will not further investigate the source of any outbreak to the farm level if need be. Unfortunately, although a processor under recall may be covered by product/recall insurance, the damage to the meat industry, even temporarily, may affect not only the cattlemen/feedlots of the affected processing plant but also all cattlemen, affiliated businesses partners, and corollary service establishments not directly involved with the recall. For cattlemen there appears to be no generic product/recall insurance options at this time in the United States. In time it is expected that U.S. cattlemen will participate in an animal ID program that will assist in recalls and possibly product/recall insurance. Presently the details of this type of program are still being determined by meat industry participants.

It is unfortunate that no product/recall insurance tools or instruments are currently available at the commodity and livestock production levels. It is true that traceability has provided better accountability with the use of barcodes, ear tags, software, auditing, etc. However, typically an insurance underwriter will only provide coverage if the organization or industry can provide proof (SOPs, records, etc.) of active risk reduction, like IPT programs foster, and in large enough participant numbers to provide the insurance for profit. For processors and manufacturers, product/recall insurance is available and used as an economically efficient business tool. It is expected that many farmers and cattlemen would not voluntarily seek out this type of insurance, especially a new type of insurance, which would be very expensive initially. A candidate farmer or cattleman would incur an additional cost without any realizable or perceived benefit, for at least the present time, because most recalls do not seek liability damages from individual farmers or cattlemen. Regulatory changes would need to change traceability accounting transparency of all involved, which could greatly change the map of liability and cost issues. In other words, farmers and cattlemen who had been shielded from being noticed or identified for inferior product production would be exposed and open to litigation.

As agricultural vertical integration and consolidation increases, it is expected that various insurance tools will be made available, especially to contracted growers. This only makes sense for processors desiring to minimize risk of substandard ingredients, to reduce liability exposure, and to save money. Additionally, processors contractually would also benefit by achieving year in/year out consistent ingredients, derived from farmers who had been providing the company product for years. It is likely that a few forward-looking processors could extend nontraditional insurance benefits (contractually) to their growers to insure product consistency and long-term production. As new types of insurance become available to farmers and ranchers, such as product/recall insurance for specialized production, it would not be far-fetched to envision other insurance options not yet conceived to be offered or developed. Insurance companies, via processors, may offer alternative insurance options that could be made available at the producer production level. Processors, by way of contracting with their growers, could bundle into the contract other items such as health and life insurance. However, this is way off in the future. At present, the extension of product/recall insurance to the growers would seem to be advantageous, but the market still needs to be made aware of its feasibility and benefits. This is a case where possible government assistance (pluralism), to encourage this type of insurance with its many players, could accelerate its development and use for society's benefit.

16.7 IPT MEASURING AND QUESTIONNAIRE ANALYSIS

The Scorecard Matrix, Cost-Benefit Spreadsheet, and Cost-Benefit Questionnaire research instrument tools help provide analysis of IPT systems. Not unlike other data research collection tools, continual changes and evolution of the topic will help modify and sharpen spreadsheet and questionnaire usage. As such, the gathering of enough—and the correct—data to produce data of statistical significance will always remain a challenge. There are numerous websites, consulting services, and software providers that offer spreadsheet tools that can be used for IPT applications. Research institutions and industry alike are continuously attempting to gather data from questionnaires and surveys for new and upcoming opportunities on which to capitalize, such as organics, sustainable agriculture, etc. Unfortunately, farmers and cattlemen are particularly difficult to get enough detailed data from for many reasons. It is hoped that these research measuring tools will assist in further studies and research. The Scorecard Matrix should be of assistance in how it works toward evaluating the efficiency of an IPT system, a more qualitative approach. The Cost-Benefit Spreadsheet helps provide comparisons between a variety of production purity levels by providing costs per bushel, profit per bushel, etc. to help in a system's evaluation. The Cost-Benefit Questionnaire seeks to clarify costs associated with IPT production done by the farm owner/manager during the critical times during planting and harvest. In total, these research instruments can assist by providing an evaluation of how well an IPT system is performing or for comparison purposes.

16.7.1 SCORECARD MATRIX

The Scorecard provides a more qualitative approach toward evaluating an IPT system. Most evaluations, much like the next cost-benefit spreadsheet, offer purely statistical data to substantiate claims. Often this is enough; however, it is not always the case, especially when other less data-driven or less quantitative inputs must be considered. This is where the Scorecard Matrix can help. It provides an approach toward the efficiency of the infrastructure of testing an IPT program. It is understood that many of the concepts tied to traceability and IP are new and still being explored. Tests and evaluation protocols for IPT are in their infancy stages of development. Often the questions needing to be asked are still unknown. This is where the Scorecard Matrix makes some basic checks and comparisons to evaluate an IPT system. What was found was that it can evaluate what should be accomplished against what was accomplished, in accordance with agreed-on specifications. Categories known for the IPT system's weaknesses were focused on, and criteria, specified by USDA and very much measureable, were observed for compliance. The output data were provided in usable weighted average of compliance for breadth and depth of data required. Accuracy was measured by output tests and by the range of test results as they were taken during production. In total, this type of measurement can provide a view of system health with some statistical evidence. Further expansion of this type of evaluation may greatly assist the less common traits or attributes of interest, such as fair wage or substantiation of geolocation of production. The weighted-average approach provides a different avenue to evolution and systems testing. As accuracy measurements become more standardized and recognized by industry, the ability of accuracy and precision of measurement will become more common and better able to refine system processes.

16.7.2 SPREADSHEET

A large challenge in the development of an IPT spreadsheet is that many of the quantitative questions have traditionally not been asked or measured before, i.e., time to clean out type "x, y, z" combine to specific accuracy (e.g., 99%). For many farmers and cattlemen IPT poses many unforeseen challenges, such as how to quantify actions or processes that had not been previously calculated or measured before. Then, if that data can be recorded, the question arises, what do the spreadsheet data tell us? Often an IPT spreadsheet is one that typically compares a traditional crop (and its production management practices, inputs, etc.) with an IPT crop (with its unique production management practices, inputs, etc.) to help determine which system is more profitable. Other IP spreadsheets help to compare costs per acre, bushel, etc. and revenues generated (what is also known as an IP premium) with those of traditionally grown crops. Many spreadsheets have originated from other industries and are modified for IP use. The unfortunate part of IP spreadsheets, aside from the normal ambiguity associated with spreadsheet data, is the attempt to quantify

IP traits/attributes that may not be quantifiable, i.e., social impact attributes or data beyond the scope of farmers' or cattlemen's knowledge. Another problem is that many times a spreadsheet may have a line item to be filled in, for example, where the question asks for input regarding the time to perform a specific task, when in fact the farmer may be doing several tasks simultaneously in the same chore. This can often skew data and interpretations and must be taken into consideration.

As mentioned, websites and other entities provide samples of spreadsheets for particular usage; often these spreadsheets highlight, under close analysis, the challenge to IPT in gathering enough pertinent and detailed data. The number of adequate questions, which produce clear, concise answers, are usually very time consuming to gather from a large enough body of willing participants (observations)—especially when you consider that similar type farms are needed to be surveyed and over several years' duration. However, it is hoped that this type of spreadsheet analysis will mature and be refined with time and innovation.

Another great challenge is that farmers/cattlemen are traditionally very independently minded and guard their operations, especially financial data, closely. Typically, data have been difficult to obtain from these sources, except for the very basic information. Often questions asked by parent seed companies, cooperatives, feed lots, etc. have been answerable and pertinent to the author but much narrower in scope than what typical IP accounting necessitates. With time, and many more spreadsheets from which to learn, studies may provide enough statistically significant data that will help to bring a better understanding of IP production, products, inputs significant, programs, etc. The significance of this spreadsheet is that it offers concise measures for prescribed tasks (work) for the various purity levels considered. It can greatly assist in determining what purity level will be most advantageous, given particular information. The close examination of costs per system can also greatly aid in determining strategies used to reduce management and labor expenses. This approach can be used by various food production industries, such as vegetables to livestock industries, to better determine costs associated to IP production. Companies that are more vertically integrated should be able to extend the use of the spreadsheet from farm field to final warehouse or point-of-sale counter. As such, evaluations of IPT can help improve overall cost reduction, diminish liability exposure, etc. The benefits of using this spreadsheet, with a Scorecard Matrix, can only help improve the understanding and financial implications of a company's IPT program.

16.7.3 QUESTIONNAIRE

IPT questionnaires, like their recipient spreadsheets, have many challenges. Typically traceability-directed questionnaires are more focused on the lines of logistics (one up, one down) prescription as predicated by law. A key issue for both traceability and IP questionnaires, regarding farming, revolves around the key IPT issues of trait(s) and/or attribute(s) of interested (contractual), associated tolerances (auditing and laboratory details to substantiate claims), and agreed-on nomenclature of bin size, lot numbering, etc. All too often questionnaire units of measure are not articulated well, i.e., given in per bushel, per acre, per year, etc. If the units of measure are articulated, it is not guaranteed that respondents will fill in the appropriate unit measure data, but rather put in their estimates for the unit of measure that comes most easily to them. Often missing essential data must be extrapolated from several other provided answers. If possible, it may be advantageous to have these types of data gathered directly from observations by a member of the questionnaire survey researchers. However, this would be time consuming and possibly too expensive to conduct for a large number of farms. An area that can greatly assist the formation of conclusions and future questionnaires is to offer open-ended questions or an area for any dialogue they wish to express. This allows respondents to put forth new ideas and suggestions. In some cases respondents offer notions and conclusions well outside the academic's purview.

The challenges of questionnaires and associated spreadsheets are well known. Much more needs to be done in this area to provide enough useful data to support arguments for trends derived from IPT production questionnaires.

16.8 CONCLUSION/SUMMARY OF INTERPRETATION

IPT systems are here to stay. Although traceability has a longer, more consistent history tied to logistic systems and food safety, its new sibling, IP, has emerged at a time when the issues of globalization, GMOs, bioterrorism, and food contamination issues are much more prolific, newsworthy, and affect larger portions of the global community. Steps are being taken along all fronts to answer these challenges, from government regulations, industry standards to policy and advisory organizations—the list is lengthy. Although many hands are involved with changes and implementation regarding IPT, I believe that the influence of government regulations and market forces, especially legal issues surrounding product/recall insurance, will greatly accelerate the use, profits, and better understanding of IPT systems.

Conclusion

This research is an attempt to further this relatively new field of study and to shed more light on its fundamentals, interactions, and interdependence. This work helps to define identity preservation and traceability (IPT); expands on its various subsystems; highlights the rules under which it functions; elaborates on IPT's primary, supportive, and ancillary components, which ultimately affect food safety and the market; and provides an interpretation of the science at present and near future possibilities.

Part I. General introduction to IPT, history, theory, design, and components is important because it sets the foundation for understanding IPT. As has been documented, numerous studies have attempted to dissect specific portions of the food system in the endeavor to simplify the complex into discrete parts. Many of these works do not truly portray the importance of interactions and interdependence within the food system. Academics have traditionally, and by training, researched and studied discrete parts and events to better understand a system or phenomena. The results of these works, many of which have helped to explain part of the picture, often omit large or essential parts of the landscape. For the most part, these types of studies do not provide a holistic approach to the problem, but they do provide incremental solutions. The commercial market also has contributed solutions that range from computerized machines, software, and consultants to newly discovered or created biological "answers" to society's hunger problems—again, often by offering discrete solutions without understanding the whole picture. Part I helps to consolidate and at least help to define the environment that IPT works within.

Part II. Programs and standards: official seed agencies, industrial programs, and country, regional, and religious standards establish many of the essential rules that govern much of the developed world's food system. Further, this part expands to illustrate how various rules have resulted in industry programs, such as within the parent seed industry. However, many of our more modern technology systems are still very fragmented, distinct, and uncomplementary in regard to integrating individual corporate IPT systems with one another within the food chain. In addition, disassociated training of personnel and fragmented system accounting/recordkeeping need to be more transparent, linked, and standardized to improve IPT interactions. One can begin to see the evolution of IPT rules and resultant changes by industries, nations, and regions that are in the forefront of food safety. For example, as more countries better understand the dynamics of the food system, the procedures for verifying and testing of grains and livestock become more routine and second nature. The notion of third-party or government oversight and testing is becoming the norm. As more of the food system becomes involved in global trade, the need for standardization and transparency of data is much more critical. Those countries and regions that take the lead in providing what the food system customers desire, safe food, with the traits and/or attributes of interest, and at a reasonable price, should benefit the most. Harmonizing of rules and standards in such a manner that provides adequate transparency should provide the global customer sufficient information to make informed decisions.

Part III. Auditors and laboratories provide the current method to verify and test both products and processes for IPT. This part plays an intricate role in better understanding the overall concepts and challenges of IPT with regard to quality control and verification of claims. Traditionally first or second parties were the sole judges of quality and safety. As the challenges of safety, bioterrorism, and other liabilities to the food system have emerged, the need for third-party impartiality and certification of auditing and laboratory methodology has become essential to help reestablish public confidence in the food system. Some of these third-party firms also offer consulting services. Questions have been raised and further research is needed regarding certifying auditors. For example, how unbiased are organic certi- fiers when many themselves may grow organic products and be "pro" organics. This type of situation may not be at arm's length, which is typically the desired form of auditing. Although this part highlights the benefits of auditors and laboratory tests, more study should be considered to compare consistency within each of these verifi- cation systems. For example, it has long been understood that laboratory test results of same-bin samples may vary greatly from lab to lab. The precision of auditors and laboratories is essential to establish baseline requirements, which further establishes credibility of the food supply. It is expected that these third-party auditors and labo- ratories will continue to improve their proficiency and accuracy as experience and technology increase.

Part IV. Consultants, policy and advisory organizations, software providers, process facilitators, and food recall and insurance issues are primary contributors for changes within the food system. Many of these facilitators and enablers of change include independent, industry and non-industry participants, who advocate various positions such as fair wages, fair trade, the environment, animal living conditions, low income groups, regional processing, etc. Others represent add-on tools or instru- ments such as software systems and training for management and employees. Each entity's goal varies depending on its focus. For example, many software companies claim that their products help meet government regulations while reducing overall costs. Still others suggest that their product or service will mitigate liabilities or expo- sure to undesired recalls. The growth of these providers has increased dramatically during the past decade. Unfortunately, diverse and ever-changing government regu- lations, expensive proprietary service products, combined with incompatible com- mercial solutions results in additional cost to consumers and companies throughout the global supply chain. It is hoped that as these organizations better understand the dynamics of IPT and how it integrates within their society and region that more clear regulations and processes are established.

Part V. Research instruments: The Scorecard Matrix, Cost-Benefit Spreadsheet, and Cost-Benefit Questionnaire offer methodology for analyzing and interpreting research information and data.

The scorecard, spreadsheet, and questionnaire illustrated in this work focus on farm-level aspects of identity-preserved production. This chapter does not cover, but further studies should include, other aspects of IPT as it relates to processors, warehousing, up to and including grocery stores. Additionally, an area of increasing importance that has had insufficient statistical research involves the less common but becoming more popular traits and/or attributes of interest such as fair wages, fair trade, animal health concerns, environment, pollution, etc. One reason that this type of research is important is to better quantify corporate costs versus revenues and to see how it plays out as societal output products or ancillary byproducts, which may result in increased unpriced societal costs. For example, the organizational costs and benefits of IPT may not be the same as the social costs and benefits, so that the private and government supply of IPT may fall below society's desirable levels. If the case is made that societal costs are too high, such as farm nitrate and phosphorous runoff that creates dead zones in the Gulf of Mexico and kills fish, then government regulation may be the appropriate tool to reverse this negative trend and social cost. In this way we can see how spreadsheet data can be used for varying purposes. In the same way, questionnaires and surveys must be designed to capture the appropriate information and data. Questionnaire and survey data may always suffer because of their dependence on the whims of participants' answers. Much work and study have been done to minimize the noise of unclear questions and choice of possible answers that may be provided. Unfortunately, questionnaires and surveys may only act as models of real-life situations. Many more studies and duplication of study results are needed to advance any theorem and change.

Future trends for IPT are unknown because many of these concepts are still in their infancy. It is unclear how they will develop. They could mature as individual entities, like two offspring born of the same system, distinct and completely independent, or as a type of combined entity, like conjoined twins, reacting according to each's independent thought, yet tied together because of shared internal operations, e.g., software, auditing, etc. Much like humans, internal DNA (specific system's makeup and dynamics) and external environmental (marketplace) will shape how IPT will react and evolve in reaction to the various food supply chain's customers, legal aspects, laboratory and field tests, auditors, rules and regulations, and so forth.

We do know that our food and medical supplies, air and water qualities are coming under increased scrutiny. Traceability should continue to provide logistical support and act as an instrument to remove unsafe products from the market through established accounting systems, which are auditable and verified. Increased tightening of rules and monitoring are causing this evolution. Using many of the same tools as traceability, identity preservation keeps track of physical products or ingredients, which should help improve product claims and therefore increase customer satisfaction. The future should see the food system accommodating a spectrum of foods and consumer tastes, with appropriate levels of oversight and auditing. Where the marketplace does not provide adequate mechanisms to provide what consumers want with sufficient safety, the government should provide needed guidance through its regulatory tools. There may always be a disconnect between the marketplace and government, especially in how they react to the changing sea of consumer wants. IPT then becomes the middle ground that attempts to accommodate each side.

As a final thought, *The Food Traceability Report* (March 2007) illustrated one of many directions IPT is headed. In their *Adventitious Traces* section, in an article titled "Frequent Flier Penalty," The Soil Association, which sets UK organic standards, is considering denying the organic label to food products imported by air transportation. At its 2007 annual meeting, Executive Director Patrick Holden said, "There is growing demand to reduce the carbon footprint of food distribution and we in The Soil Association take that very seriously." This is especially interesting when you consider that the UK is an island nation and that organic foods may be considered perishable and a time-sensitive product.

This is one way that IPT concepts may be affected by governmental agencies or associations, which in turn will affect industry and consumer choice. In the same way government provides influence, new industry products, processes, and systems will shape output products and the manner in which consumers perceive products. Regardless, IPT has set its footprint on the food system.

Appendix A

Identity Preservation and Traceability Systems at Seed Production, Processing, and Retail Stages

A.1 SEED PRODUCTION STAGE

Historically, this has been the starting point for crop supply chains as seed development firms commercialize new crop varieties and market the benefits to agricultural producers. This push version of supply chains has had difficulty adapting to consumer demands for a pull supply chain (Smyth and Phillips 2002).

Identity-preserved production systems are developed voluntarily by private firms to ensure that all stakeholders in the supply chain for a specific product capture a share of the value from specialty traits. Private firms may use technical use agreements (TUAs) to protect the intellectual property of the specialty traits, or they may use production contracts that have specific conditions that producers must meet to receive relevant premiums. These systems are common for niche markets and are typified by small acreage and low volumes. There is presently some debate as to whether long-run premiums for producers are sustainable because they may be bid away through competition[*] (Smyth and Phillips 2002).

A.2 PROCESSING STAGE FEATURES

Processing stage features are those of firms involved in the manufacturing of food products. Most of these features contain aspects of quality assurance and industry-developed standards.

---

[*] Traceability is very fragmented at the farmer-producer stage. Production arrangement is accomplished largely through membership in the organization (cooperatives), which was established to create and manage the industry. Production control is accomplished through industry standards and stringent recordkeeping. The cost of initially becoming involved in a traceability system results in short-term premiums being available to attract producers. Long-term benefits are not evident because the premiums evaporate when the desired number of producers become involved (Smyth and Phillips 2002).

The processing stage is very important for identity preservation and traceability (IPT) systems because this is the stage in the supply chain where tracking and tracing systems begin to be rigorously applied. Enforcement of standards is valued in these systems because of the nature of focusing on increased food safety. The lack of high standards and careful enforcement of the standards results in costly recalls of products, therefore the enforcement of standards is done collectively. Quality is focused on the production processes to ensure that the highest standards possible are maintained at all times. Tolerance levels exist for food safety reasons because no product can be entirely free of potentially harmful effects, therefore tolerance levels are established at levels that ensure safe consumption. When tolerance levels are exceeded, a risk of harm to consumers develops; these products must then be recalled from the marketplace. The costs of recall are substantial. Not only does the firm incur the cost of gathering and disposing of the product in question, it may also incur a loss of consumer trust in its brand name that will require aggressive marketing campaigns to overcome. Third parties do testing and auditing of traceability systems.

A.3   TRACEABILITY IN THE MANUFACTURING STAGE

Although traceability in food processing systems is important, some data are essential to fulfill ethical and legal responsibilities of food manufacturers to customers and consumers. Other data are less crucial but also relevant (e.g., for consumer information, price setting, optimal processing, etc.). The desired degree of detail of information (number of subdescriptors, size of traceable resource units) varies according to the purpose. The processing step is the step in the chain that may be interested in the highest degree of detail of information. Hence, the number of subdescriptors laid into a chain traceability system may be significantly fewer than the number of subdescriptors used in an internal traceability system. This is a problem area. Finally, it may not always be possible to establish the ideal traceability system with traces unbroken. Where loss of traceability of a product is unavoidable, effective alternative methods of control should be ensured* (Moe 1998).

A.4   RETAIL STAGE

The final stage of the supply chain is the retail stage. The features in this category apply to those firms that are involved with selling food products to consumers. This is the stage of the pull supply chain that is now seen as driving many modern supply chains (Smyth and Phillips 2002).

---

* Traceability components: In principle, there are two main ways of managing information in the chain where full traceability is required. (1) Older information is stored locally in each of the steps in the chain sending only product identification information along with the product. Thereby the product and its subdescriptors can be traced by going backward in the chain one step at a time. (2) Newer information follows the product all off the way through the chain. The latter is necessary if it is desired to bring information from early steps in the chain to the consumer or to advertise and market special features of a product (e.g., organically grown, free of genetically manipulated materials, freshness from a certain area caught yesterday, special slaughtering method used, etc.) (Moe 1998).

Identity-preserved systems may play a large role in the introduction of new GM food products. New GM products may be introduced without complete international market acceptance, and identity preservation systems can be used to ensure continued market access. An IPT system is able to provide consumer information on the uniqueness of the branded product. For an IPT system to function properly and ensure that all stakeholders remain committed to the process, final market price premiums must be available. If this premium is not available for the retailer, an incentive is created for the retailer to no longer carry the product. Products of IPT systems will need to be labeled because if the consumer has no means of identifying the value of the product, the consumer will not pay a premium to purchase it (Smyth and Phillips 2002).

Appendix B

Farm Identity Preservation and Traceability Program and Its Components

B.1 EXAMPLE OF A FARM IDENTITY PRESERVATION AND TRACEABILITY PROGRAM AND ITS COMPONENTS (GENERAL)[*]

Identity preservation and traceability (IPT) systems do not begin with testing of the end product. Rather, IPT is a system of standards, records, and auditing that must be in place throughout the entire crop production, harvesting, handling, and marketing process. Figure B.1 represents identity preservation (IP) processes and factors to consider at various steps, including testing and auditing points.

Seed certification is an example of a successful IPT program. Seed certification was introduced in the 1920s–1930s as a mechanism to maintain the genetic purity of publicly released crop varieties. These programs have been highly successful in maintaining the integrity of crop varieties and in providing farmers with seeds of known pedigree with high purity and quality. Because IPT programs are developed for agricultural commodities, they often follow principles similar to those used in seed certification. Thus, in describing the components of IPT programs, seed certification is often used as the model.

In grain production, IPT programs and processes are designed to keep lots of grains or oilseeds with special qualities separate from the bulk commodity. For this special identity to be maintained, IP systems must be in place throughout the supply chain, between entities such as farmer and processor, and within an entity (i.e., from incoming loading dock bulk commodity to outgoing packaged pallets). Processes refer to practices in the production, handling, and marketing of grains or oilseeds that maintain the integrity and purity of the product. IPT programs can apply to crop varieties with unique product quality traits (e.g., non-GM soybeans, specific wheat varieties, and crops grown without pesticides).

---

[*] Excerpts modified from Sundstrom's *Identity Preservation of Agricultural Commodities*. Published by *Agricultural Biotechnology in California Series Publication 8077*, 2002, pp. 1-15. Found at http://anrcatalog.ucdavis.edu/pdf/8077.pdf.

Process	IP consideration

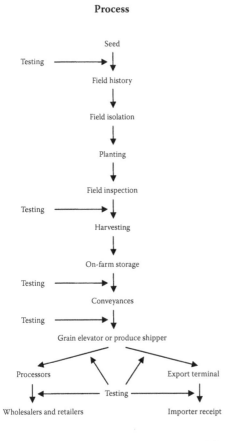

- Seed purity tested and confirmed
- Clean storage
- Previous crops
- Free of weeds and volunteers
- Retain records of field history

- Isolation standards met
- Borders and barriers present
- Time of planting and flowering

- All planting equipment cleaned and inspected

- Field inspected by certifying agency at proper times
- Value and purity items monitored

- Clean equipment and conveyances
- Preharvest inspection

- Clean storage facilities
- Multiple units for product segregation
- Maintain records and product identity

- All bins, trucks, etc., cleaned and inspected prior to transport

- Handling and processing facilies have documented IP protocols in place
- Facilities cleaned and inspected between lots
- Segregation maintained throughout product handling chain
- Maintain records and product identity
- Proper labeling

Regarding a farm IPT program, the following components apply (sample):

- *Seed:* Purity of seed is verified and seed is stored in clean bins.
- *Field history:* Field rotations ensure that the field is free of weeds and volunteers; field history records must be kept.
- *Field isolation:* Meets isolation standards; borders or barriers are used.
- *Planting:* Clean equipment is used during planting.
- *Harvesting:* Equipment and conveyers are cleaned before harvesting.
- *On-farm storage:* Storage bins are cleaned, product is segregated, records are kept.
- *Transportation:* Trucks are cleaned and inspected.
- *Primary elevators:* The quality management system is documented. This means that the entire process is documented and records are kept to ensure that processes are followed. Bins are cleaned and inspected. Product is kept segregated. Records are kept.
- *Transportation from the primary to terminal elevator:* Vehicles are cleaned and inspected.
- *Terminal elevators:* The quality management system is documented. This means that the entire process is documented and records are kept to ensure

that processes are followed. Bins are cleaned and inspected. Product is kept segregated. Records are kept.

- ***Transportation to market:*** Vehicles are cleaned and inspected.

If you are a farmer, an IPT contract usually indicates:

- You probably have to buy certified seed, either as a condition of your contract or to minimize your risk of not delivering on the contract specifications for varietal purity.
- You have to thoroughly clean equipment and machinery before using them for the IP crop.
- You are going to commit some on-farm storage to ensure that the IP crop is kept separate from other crops.
- You have to keep accurate records of crop rotations, seed use, chemical applications, harvest dates, and storage.
- You will likely be subject to some audits to ensure you are doing all of these things.
- When you deliver on the contract, a sample will likely be kept in case there is a problem with the shipment.
- All of this means higher prices for your product but at extra costs.

B.1.1 PLANTING SEED AND TOLERANCES

The purity of any commercial agricultural product propagated by seed begins with the purity of the seed planted. It is evident that the purity of the seed stock must equal or exceed the purity standards of the final product. However, it is virtually impossible to assure that all handling and conveyance equipment and storage facilities are completely free of contamination, so even foundation seed is seldom 100% pure. Currently, AOSCA purity standards (see Chapter 4) for certified seed average 98% across species. Consequently, IPT systems with product purity standards greater than 98% must begin with extraordinarily pure seed stocks. Different product tolerances are established in specialized IPT programs based on market-driven standards. It is not uncommon for a single commodity to have multiple quality tolerance thresholds based on diverse market needs.

B.1.2 FIELD HISTORY AND ELIGIBILITY

Fields eligible for IPT certification must not have grown a crop the previous year that could produce inseparable contaminating weeds or volunteer plants. In some cases, multiple-year rotations may be necessary between crops to achieve low contamination levels. Records and field maps must be maintained for up to 5 years to allow documentation of previous crop history.

B.1.3 FIELD ISOLATION

Crops must be isolated either spatially or temporally from potentially contaminating pollen sources. The degree of isolation depends on flower characteristics, sexual compatibility with neighboring crops, pollen quantity and viability, and mode of pollen dissemination. Self-pollinating crops such as wheat require relatively small isolation distances that are primarily intended to prevent mechanical mixtures during harvesting. Cross-pollinating crops require as much as 2 miles (3.2 kilometers) or more of isolation from plants of the same species to prevent outcrossing, depending on the flower structure and mode of pollen transfer. Insect- and wind-pollinated crops require various isolation distances depending upon the type of insect and the distance that pollen can be carried. Seed certification standards serve as a guide to minimum isolation distances. Isolation can also be achieved by planting crops at different times so that their flowering periods do not overlap. Border rows of the IPT crop are often left unharvested to intercept stray pollen and prevent contamination of the remainder of the field. Certifying agencies inspect fields and the surrounding areas to ensure that isolation standards are met.

B.1.4   EQUIPMENT AND FACILITIES

All equipment used in production, including planting, field maintenance, and harvest, must be cleaned and inspected before and after each use. All dryers, millers, storage facilities, and processing equipment must be cleaned and inspected between each product lot to ensure that segregation is maintained and no physical contamination occurs. Facilities' certification standards that handle IPT products must be established and published.

B.1.5   SAMPLING AND TESTING

In many cases, samples of a product must be tested at various stages to confirm product identification, purity, and quality. IPT programs must use statistically representative sampling and testing techniques to ensure reliable results. Test results are dependent on the sampling procedure, and a single sample at a single audit point is inadequate to evaluate an IPT system. Statistical procedures must be applied to accurately determine the number of samples and the numbers of seeds or grains required to generate a test result with an acceptable confidence level. The USDA's Grain Inspection, Packers and Stockyards Administration (GIPSA) guidelines on selecting a sampling protocol and on collecting bulk samples. The guiding principle is that the sample must be representative of the total quantity of material to be tested or test results are compromised. Significant differences in test results between laboratories may occur solely because of sampling differences. Analytical error in the testing laboratory can also result in test differences, but in many cases, sampling methods, rather than test sensitivity and accuracy, limit the ability to properly detect the presence or absence of specific crop traits.

In addition to using an appropriate sampling procedure, sampling must also be performed at meaningful audit points within the chain of product custody. Common sampling and testing points are at:

- The seed source for planting
- The field prior to harvest

- On-farm storage or local elevator receipt
- First processor receipt
- Export terminal receipt
- Overseas importer receipt
- Final processor receipt

B.1.6 RECORD MAINTENANCE AND LABELING

The party responsible for contracting IP services must maintain records of all field designations, harvest amounts, storage bin locations, and product transfers. IP products must be identified, segregated, and labeled at all times. Labeling standards depend on the product and market in which it is sold. Official auditing and labeling are available from various service providers to designate products meeting IP certification standards.

B.2 OUTLINE OF IDENTITY PRESERVATION PROCEDURES*

Seed Purchase—Seed Standard

Practice

- Grower should purchase certified seed (e.g., accredited to Association of Official Seed Certification Agencies standards or equivalent). "Bin run" seed not to be used.[†]

Documentation

- Grower must have sufficient documentation to prove that the seed purity and identity have been maintained, such as invoice or receipt of purchase for all quantities (for each bag of seed) and certified seed tag for each lot of seed purchased to produce the quantity of identity-preserved crop being contracted or delivered.

Planting

Practice

- Planter must be thoroughly cleaned and inspected prior to planting IP crop variety as detailed by equipment manufacturer (if available). This must be done regardless if grower uses his/her own equipment or uses a

---

[*] Used with permission from the Canadian Soybean Export Association Identity Preservation Procedures, Revision 2, February 21, 2003. Available at http://www.grainscanada.gc.ca/Prodser/ciprs/pdf/CSEA_procedures-e.pdf.

[†] Bin run grain is retained from a previous crop and is used as seed for next planting. Genetic purity and identity of "bin run" seed is uncertain. It is not produced under an AOSCA-approved pedigreed seed increase system and therefore it has not been field inspected by an accredited agency.

custom planter. Grower should attempt to use IP crop equipment before the equipment is used on other crops.

- Prior to planting, IP seed bags should be stored separately from other IP varieties and non-IP seed bags.
- Growers must ensure that the minimum isolation distance between IP crop and fields that do not require isolation is observed.[*]

<u>Documentation</u>

- Growers must detail cleaning procedure used and sign this document to authenticate that they have implemented the procedures described. In some cases auditors authenticate cleaning procedures.
- Proper isolation distance must be documented at the time of field inspection.
- Growers should keep detailed field maps, written history of previous crops grown on growers' and adjacent fields, and management practices of both fields that may affect IP crop attributes.

Field Season

<u>Practice</u>

- A second- or third-party field inspector must inspect the IP field during the growing season to confirm that isolation distances have been met and there is proper control of volunteer crops and weeds. The field inspector must also verify that the crop looks uniform as detailed in the variety description.
- If the IP crop is not being grown under contract (in which case the contracting party should conduct the field inspection), the grower should arrange for a qualified individual, at arms length from the operation of the farm, to conduct the field inspection.

<u>Documentation</u>

- The field inspection report must document that isolation distances have been met, there is proper control of weeds and volunteer crops, and that the IP crop variety appears to be characteristically uniform for the appropriate growth stage. The inspector and the grower must sign and date this report.
- Depending upon the contract, other factors such chemicals used, method of application, soil samples, and application weather conditions may need to be documented and certified.[†]

---

[*] Approved isolation distance for the IP crop must be used. The CSGA isolation standard for certified soybean seed is 3 meters between another soybean and another pulse crop (Bean, Fababean, Lentil, Lupin, Pea, or Peanut). There is no isolation distance necessary between soybeans and crops of Barley, Buckwheat, Canaryseed, Flax, Oat, Rye, Triticale, and wheat providing the crops do not overlap.

[†] Depending upon the contract/certification, ancillary documentation may need to be incorporated, such as Manure Management Plan, or other U.S. Department of Agriculture, U.S. Environmental Protection Agency, state, region, or association documentation.

Harvest

Practice

- Combine, equipment used to transfer, and conveyance vehicles/equipment used to transport IP crop must be thoroughly cleaned and inspected prior to harvesting, transfer, and transporting the IP crop. Grower should try to harvest, transfer, and transport IP crop before said equipment is used on other crops.

Documentation

- Grower must detail cleaning procedure used and sign this document to authenticate that they have implemented the procedures described.
- Grower must inspect truck and sign a document to authenticate that the truck/hopper was cleaned prior to loading.

On-Farm Storage

Practice

- Storage bin and equipment used to unload storage bin must each be thoroughly cleaned and inspected prior to loading and use.
- Storage bins used to store IP crops must be visually identified so that all persons working in the farm operation are aware that each bin should only be used for a particular IP crop.

Documentation

- Grower should keep full records of what was stored in their bin prior to filling with IP crop, records of crop type, and dates when bins were loaded, unloaded, and cleaned.
- Grower must sign a document indicating that their bin was thoroughly cleaned and inspected prior to filling and that the equipment used to load and unload the storage bin was thoroughly cleaned and inspected prior to usage.
- Grower must sign a document indicating that any storage bin used for an IP crop was visually identified.

Transportation

Practice

- Conveyance vehicles/equipment must be thoroughly cleaned and inspected prior to loading. This must be done regardless if grower uses his/her own equipment or uses custom trucking.

- Trucker must present documentation verifying the IP crop variety and name of the grower.[*]

Documentation

- Grower must inspect truck and sign a document to authenticate that the truck/hopper was cleaned prior to loading.
- Grower must fill out documentation for the trucker that identifies the IP crop variety being delivered and the grower name.

Elevator Receiving

Practice

- Elevator must have an IP manual that details their full IP procedures for receiving, storage, processing and loading.[†]
- Incoming loads must be identified and verified as an IP crop or a non-IP crop. The crop must be identified as IP, Special Quality White Hilum (SQWH), or crush. SQWH and crush soybeans are not qualified for IP certification. The crop is not unloaded as IP unless its identity is verified.
- Any non-IP loads received into the elevator must be tracked and accounted for. [‡]
- Elevator must take a sample from each load of IP crop received.
- Elevator pit/conveyor/legs must be thoroughly cleaned and inspected prior to receiving IP crops. Alternatively, they could also be dedicated to a specific IP crop.

Documentation

- Manual must be available for inspection by auditing authority.
- Scale tickets for incoming loads must indicate variety name and unloading/ storage details for all crops.
- Elevator must have detailed documentation for storage and tracking of non-IP loads that were received into the elevator.

---

[*] Trucker should be carrying a completed bill of lading. The producer, trucker, and receiver should sign the bill of lading. The trucker should also carry any additional documentation required by the receiving elevator.

[†] All relevant staff should be trained in IP procedures and should have access to the manual for reference. Receiving procedures should be detailed in IP manual.

[‡] Elevator should have detailed documentation showing which bins were used to store non-IP loads. Elevator should be able to show documentation demonstrating the end use for the non-IP crops.

- Elevator must retain documentation detailing variety name, moisture, weight, and grade details for each load.
- Elevator must have documentation to authenticate that pit/conveyor/legs have been thoroughly cleaned and inspected prior to receiving a specific IP crop. Records must include the date and the name of the employee who conducted the inspection.

Elevator Storage

Practice

- Elevator must keep detailed storage history.
- Storage bins/silos and equipment used to load/unload bins/silos must each be thoroughly cleaned and inspected prior to loading/unloading and used for IP crop.
- Elevator must identify all bins/silos that are used to store IP crop variety. Bins used to store SQWH and crush crops must also be identified. All elevator staff should be aware of and have access to bin/silo designation.

Documentation

- Elevator must have detailed storage history records. Records must indicate what crop or variety was stored in their silo/bins prior to it being used to store an IP crop. All tonnage loaded and unloaded should be recorded.
- Elevator must have records documenting that the silo/bin was thoroughly cleaned and inspected prior to loading with IP grain. Records must include the date and name of the employee who conducted the inspection.
- Elevator must have records documenting that all equipment used to load/ unload silos/bins with IP soybean crop were thoroughly cleaned and inspected prior to use. Again, records must include the date and the name of the employee who conducted the inspection.
- Elevator must have detailed bin and silo maps/schematics indicating which crop and variety is to be stored in each bin.*

Processing

Practice

- Conveyors/augers/legs and processing equipment must be cleaned prior to transporting and processing different IP varieties and different crops.

---

* Current elevator schematic should be available at pits and all other pertinent spots in elevator.

- Elevator must have documentation detailing the flow of IP grain through the entire processing system.

<u>Documentation</u>

- Elevator must have records showing that all transferring and processing equipment were each thoroughly cleaned and inspected prior to transferring and processing IP crop. Records must include the date(s) and the name(s) of employee(s) who conducted each of the inspections.
- Elevator must have written records detailing origin bin(s) used for unloading grain for processing and destination bin(s) used for storing the processed grain. Any bin movements prior to processing must be recorded. Elevator should record tonnage when grain is transferred to different bins and the tonnage that is transferred for processing.

Loading

<u>Practice</u>

- All containers/vessels/trucks must be inspected and cleaned as required prior to loading. The IP manual should detail procedures for rejection of container/vessels/trucks if they are not suitable for contract.
- Elevator must have documentation detailing the flow of IP grain handled through the elevator and should record tonnage when grain is transferred to different bins and the tonnage that is unloaded from the elevator.
- Elevator must document grain loading details for all crops (IP and non-IP) that exit the elevator system.

<u>Documentation</u>

- Elevator/exporter must have written records showing that containers/vessels/trucks have been inspected and cleaned as required prior to loading with IP grain. Records must have inspection date(s) and the name(s) of the employee(s) who conducted the inspection(s).
- Elevator must have written records detailing bins/silos used for storing IP grain that has not been processed but has been stored and unloaded from the elevator.
- Elevator must document and retain full records for all containers, trucks, and railcars loaded from the facility. Records must include container, truck, or railcar identification number, identification of the grain (IP variety, SQWH, or crush) and the quantity loaded. The bin that the grain has been loaded from must also be recorded.

Audit Standards

<u>Practice</u>

- The grower must retain grower documentation unless requested by the elevator. Documentation must be retained for a minimum period subject to the requirements of the HACCP Standard. Rule of thumb for HACCP records is three years.
- Elevator/exporter must have retained records to support an annual audit.

<u>Documentation</u>

- Elevator/exporter must declare on their sales contracts if they are selling crops under IP Standard.

Nonconforming Product

<u>Practice</u>

- The elevator/exporter shall ensure procedures exist to investigate the cause of potential and actual nonconformity.
- Nonconforming product includes any product that qualified as IP, but because of adventitious or intentional mixing no longer meets IP requirements.
- IP manual should detail how employees will inform the correct individual in the chain of command about nonconforming product.
- If the exporter has nonconforming product, they must show in their documentation that they have a procedure to address the situation. This must include either documentation for disposal, customer acceptance, or alternate non-IP sales arrangements.
- The elevator/exporter should have a corrective action procedure.

<u>Documentation</u>

- The elevator/exporter must have a written protocol detailing how they will address a situation where they have nonconforming product.
- The exporter must have documentation showing that nonconforming product has either been disposed of, that the customer has been informed and accepted the nonconformance, or that alternate non-IP sales arrangements were made.

B.3 EXAMPLE OF AN ON-FARM IPT PROGRAM CHECKLIST

See Table B.1 for an example of an on-farm IPT program checklist. (For other checklists, see Chapter 6 on Canada's Soybean Export Association Procedures and EurepGap.)

TABLE B.1
Program Checklist

General Checklist	Cleaned	Employee Inspected	Third-Party Inspected	Documented
Seed standards				
Planting				
Field season				
Harvest				
On-farm storage				
Transportation				
Elevator receiving				
Elevator storage				
Loading				
Audit standards				

	Action	Detailed Procedure	Employee Inspected	Third-Party Inspected	Documented
Seed standards					
Seed purity and ID certified/accredited					Invoice by bag and tag by lot
Pre-planting					
Soil sample					Yes
Field preparation		Yes			Yes
Minimum isolation distance	Yes		Yes		
Chemical application		Yes			Yes
Manure application		Yes			Yes
Planting					
Planter cleaned		Yes	Yes		
Storage separate	Yes				
Field map/history					Yes
Other management practice					Yes
Field season					
Inspection for weeds and volunteer crops		Yes	Yes	Yes	Yes
Proper growth inspected		Yes	Yes	Yes	Yes
Isolation distance			Yes	Yes	Yes
Mechanical weeding		Yes			Yes
Chemical application		Yes			Yes
Other management practice		Yes			Yes
Harvest					
Combine	Yes	Yes	Yes		Yes
Transfer equipment	Yes	Yes	Yes		Yes
Transport equipment	Yes	Yes	Yes		Yes
Other	Yes	Yes	Yes		Yes
On-farm storage					
Bin	Yes		Yes		Yes
Loading equipment	Yes		Yes		

TABLE B.1 *(Continued)*

	Action	Detailed Procedure	Employee Inspected	Third-Party Inspected	Documented
Bin ID	Yes	Yes			
Transportation					
Loading equipment	Yes		Yes		
Transfer equipment	Yes		Yes		
Truck	Yes		Yes		
Bill of lading					Yes
Elevator receiving					
Manual		Yes			Yes
Sample test(s)	Yes				Yes
Pit	Yes	Yes	Yes		
Conveyor/legs	Yes	Yes	Yes		
Conveyor/legs					
Elevator storage					
Storage history					Yes
Bin(s)/silo(s)	Yes	Yes	Yes		
Loading Equipment	Yes	Yes	Yes		
Unloading Equipment	Yes	Yes	Yes		
Bin/Silo ID	Yes				Yes
Site map	Yes				Yes
Processing					
Conveyors	Yes	Yes	Yes		Yes
Augers	Yes	Yes	Yes		Yes
Legs	Yes	Yes	Yes		Yes
Processing equipment	Yes	Yes	Yes		Yes
Map of flow					Yes
Storage history					Yes
Loading					
Containers	Yes	Yes	Yes		Yes
Vessels	Yes	Yes	Yes		Yes
Trucks	Yes	Yes	Yes		Yes
Railcars	Yes	Yes	Yes		Yes
IP manual					Yes
Storage history					Yes
Audit standards					
Grower					Yes
Elevator					Yes
Exporter					Yes
Nonconforming product					
Status change procedure		Yes			
IP manual		Yes			Yes
Exporter procedure		Yes			Yes
Custody change procedure		Yes			Yes

Source: Data adopted with permission from Canadian Soybean Export Association.

Appendix C

Official U.S. and Canadian Foundation Seed Agencies

Data used with permission from the Association of Official Seed Certifying Agency (AOSCA)'s website. Available at http://www.aosca.org/foundation%20seed%20agencies.htm.

TABLE C.1
Official U.S. and Canadian Foundation Seed Agencies

Ag Alumni Seed Improvement Association
Fayte Brewer, Manager
East Romney, IN 47981-0158
702 State Road 28
P.O. Box 158
Phone: 765-538-3145 Fax: 765-538-3600
E-mail: brewer@agalumniseed.com
Website: http://www.agalumniseed.com

Alabama Crop Improvement Association
Jim Bostick, Executive Vice President
P.O. Box 357
Headland, AL 36345-0357
Phone: 334-693-3988 Fax: 334-693-2212
E-mail: jpbostick@centurytel.net
Website: http://www.ag.auburn.edu/SSCA

Alaska Plant Materials Center
Kathi VanZant, Seed Analyst
HC04
P.O. Box 7440
Palmer, AK 99645
Phone: 907-745-4469 Fax: 907-746-1568
E-mail: kathi_vanzant@dnr.state.ak.us

Arizona Crop Improvement Association
Abed Anouti
2120 East Allen Road

Tucson, AZ 85719
Phone: 520-318-7271 Fax: 520-318-7272
E-mail: anouti@ag.arizona.edu

Arkansas, University of, Foundation Seed Program
Christopher Deren, Director
University of Arkansas
Rice Research & Extension Center
2900 Hwy. 130 East
Stuttgart, AR 72160
Phone: 870-673-2661 Fax: 870-673-4315
E-mail: cderen@uark.edu

California Foundation Seed Program
Larry R. Teuber, Director
University of California–Davis
Department of Plant Sciences Foundation Seed Program
One Shields Avenue
Davis, CA 95616-8780
Phone: 530-752-2461 Fax: 530-754-7283
E-mail: lrteuber@ucdavis.edu
Earl Booth, Seed Production Manager
Phone: 530-754-5184 Fax: 530-754-6122
E-mail: webooth@ucdavis.edu
Website: http://www.fsp.ucdavis.edu

(Continued)

509

TABLE C.1 *(Continued)*

Canadian Seed Growers' Association
Dale Adolphe, Executive Director
240 Catherine,
Box 8455
Ottawa K1G 3T1 Canada
Phone: 613-236-0497 Fax: 613-563-7855
E-mail: adolphed@seedgrowers.ca

Colorado Agronomy Foundation Seed
Aaron Brown, Manager
Department of Soil & Crop Sciences
Colorado State University
Fort Collins, CO 80523
Phone: 970-491-6202 Fax: 970-491-0565

Connecticut (No Agency)
Alton Van Dyke, Supervisor
Connecticut Department of Agriculture
765 Asylum Avenue
Hartford, CT 06115
Phone: 860-713-2565

Florida Foundation Seed Producers, Inc.
Tom Stadsklev, Manager Secretary
P.O. Box 309
Greenwood, FL 32443
Phone: 850-594-4721 Fax: 850-594-1068
E-mail: seed@digitalexp.com

Georgia Seed Development Commission
Mike Garland, Manager
2420 South Milledge Avenue
Athens, GA 30605
Phone: 706-542-5640 Fax: 706-227-7159
E-mail: mgarland@agr.state.ga.us
Website: http://www.gsdc.com

Idaho Foundation Seed Program
Kathy Stewart-Williams
3806 North 3600 East
Kimberly, ID 83341-5082
Phone: 208-423-6655 Fax: 208-423-6656
E-mail: williams@kimberly.uidaho.edu

Illinois Foundation Seeds, Inc.
Dale Cochran, Manager
P.O. Box 722
Champaign, IL 61824-0722
Phone: 217-485-6260 Fax: 217-485-3687
E-mail: dcochran@ifsi.com

Iowa Committee for Agricultural Development
Lynn E. Henn, Production Manager
4611 Mortensen Road, Suite 101
Ames, IA 50011-1010
Phone: 515-292-3497 Fax: 515-292-6272
E-mail: lhenn@iastate.edu
Website: http://www.ag.iastate.edu/centers/cad

Kansas State University Agronomy Department
Vernon A. Schaffer, Assistant Agronomist
Department of Agronomy Foundation Seed
  Kansas State University
2200 Kimball Avenue
Manhattan, KS 66502
Phone: 785-532-6115 Fax: 785-532-6094
E-mail: vas@ksu.edu

Kentucky Foundation Seed Project
Letha J. Drury, Manager
University of Kentucky
3250 Iron Works Pike
Lexington, KY 40511-8470
Phone: 859-281-1109 Fax: 859-253-3119
E-mail: ltomes@uky.edu

Maine Department of Agriculture
Bob Batteese, Acting Director
Division of Plant Industry
28 State House Station
Augusta, ME 04333-0028
Phone: 207-287-3891 Fax: 207-287-7548
E-mail: robert.batteese@maine.gov

Maryland Crop Improvement Association
William Kenworthy, Soybean Breeder
P.O. Box 169
Queenstown, MD 21658-0169
Phone: 301-405-1324 Fax: 301-314-9041
Bobbi Boyle, Secretary/Treasurer
Phone: 410-758-2007

Massachusetts State Seed Control Official
Department of Food & Agriculture
100 Cambridge
Boston, MA 02202
Phone: 617-727-3020 ext. 141 Fax: 617-727-7235

(Continued)

TABLE C.1 (Continued)

Michigan Crop Improvement Association
C. James Palmer, Foundation Seed Ops Manager
P.O. Box 21008
Lansing, MI 48909
Phone: 517-332-3546
Fax: 517-332-9301
E-mail: palmerj@michcrop.com

Minnesota Crop Improvement Association
  Foundation Seed Services
Roger Wippler, Manager
1900 Hendon Avenue
St. Paul, MN 55108
Phone: 612-625-7766, 800-510-6242
Fax: 612-625-3748
E-mail: wippl002@tc.umn.edu

Mississippi Foundation Seed Stocks
Randy Vaughan, Manager
Mississippi State University
Box 9811
Mississippi State, MS 39762
Phone: 662-325-2390 Fax: 662-325-8118
E-mail: rvaughan@pss.msstate.edu

Missouri Foundation Seed Stocks
Rick Hofen, University of Missouri
3600 New Haven Road
Columbia, MO 65201
Phone: 573-884-7333
Fax: 573-884-4880
E-mail: hofenrj@missouri.edu

Montana Foundation Seed Stocks
William E. Grey, Director
Department of Plant Sciences and Plant Pathology
214 Ag Biosciences Facility
P.O. Box 173150
Montana State University
Bozeman, MT 59717-3150
Phone: 406-994-5687
Fax: 406-994-7600
E-mail: wgrey@montana.edu

Nebraska Foundation Seed Division
Jeff Noel, Director
1071 County Road G, Room C
Ithaca, NE 68033-2234
Phone: 402-624-8038 or 402-624-8012
Fax: 402-624-8010
E-mail: jnoel2@unl.edu

Nevada Foundation Seed Stocks
P.O. Box 230
Lovelock, NV 89419
Phone: 702-273-2923 Fax: 702-273-7647

New Hampshire & New Jersey (No Agency)
See Northeast Foundation Seed Alliance

New Mexico Crop Improvement Association
Lonnie Mathews
USDA Building on West College Street
Las Cruces, NM 88003
Phone: 505-646-4125 Fax: 505-646-8137
E-mail: lomathew@nmsu.edu
Website: http://www.cahe.nmsu.edu/nmcia

New York Seed Improvement Project
Alan Westra, Manager
103C Leland Lab, Cornell University,
Ithaca, NY 14851-0218
Phone: 607-255-9869 Fax: 607-255-9048
E-mail: aaw4@cornell.edu

North Carolina Foundation Seed Producers
8220 Riley Hill Road
Zebulon, NC 27597
Phone: 919-269-5592 Fax: 919-269-5593

North Dakota Foundation Seed Stocks Project
Dale Williams, Director
270D Loftsgard Hall
P.O. Box 5051, North Dakota State University
Fargo, ND 58105-5051
Phone: 701-231-8140 Fax: 701-231-8474
E-mail: dale.williams@ndsu.nodak.edu
Website: http://www.ag.ndsu.nodak.edu/aginfo/
  seedstock/fss

Northeast Foundation Seed Alliance
Alan Westra, Manager
103C Leland Lab
Cornell University
Ithaca, NY 14853
Phone: 607-255-9869 Fax: 607-255-9048
E-mail: aaw4@cornell.edu

Ohio Foundation Seeds, Inc.
Jack D. Debolt, Manager
P.O. Box 6
Croton, OH 43013
11491 Foundation Road
Phone: 740-893-2501 Fax: 740-893-3183
E-mail: ofsi@earthlink.net

(Continued)

TABLE C.1 *(Continued)*

Oklahoma Foundation Seed Stocks, Inc.
D. L. (Doc) Jones, Coordinator
102 Small Grains Building
OSU Agronomy Research Station
Stillwater, OK 74078-6175
Phone: 405-624-7041 Fax: 405-624-6705
E-mail: doc@ofssinc.com

Oregon Foundation Seed
351B Crop Science Building
Oregon State University
Corvallis, OR 97331-3002
Phone: 541-737-5094
E-mail: Daniel.Curry@oregonstate.edu

Pennsylvania (No Agency)
See Northeast Foundation Seed Alliance

Rhode Island Department of Agriculture
Steve Volpe, Contact
22 Hayes Street
Providence, RI 01908
Phone: 401-277-2781 Fax: 401-277-6047

South Carolina Foundation Seed Association
G. Michael Watkins, Executive Vice President
1162 Cherry Road
Clemson University
Clemson, SC 29634-9952
Phone: 864-656-2520 Fax: 864-656-6879
E-mail: seedw@clemson.edu

South Dakota Foundation Seed
Jack Ingemansen, Manager
1200 North Campus Drive, P.O. Box 2207A
South Dakota State University
Brookings, SD 57007
Phone: 605-688-5418 Fax: 605-688-6633
E-mail: jack.ingemansen@sdstate.edu

Tennessee Foundation Seed
Jack R. Dunn, Manager
2640-C Nolensville Road
Nashville, TN 37211
Phone: 615-242-0467 Fax: 615-248-3461
E-mail: tfs@superiorseeds.org

Texas Foundation Seed Service
R. Steven Brown, Program Director
11914 Highway 70S
Vernon, TX 76384-8362
Phone: 940-552-6226 Fax: 940-552-5524
E-mail: rsbrown@ag.tamu.edu

University of Delaware Plant Science Department
Bob Uniatowski
Plant & Soil Sciences
152 Townsend Hall
Newark, DE 19716
Phone: 302-738-2531 Fax: 302-831-3651
E-mail: bobuni@udel.edu

Utah Crop Improvement Association
Stanford A. Young, Secretary Manager
4855 Old Main Hill
Utah State University
Logan, UT 84322-4855
Phone: 435-797-2082 Fax: 435-797-3376
E-mail: sayoung@mendel.usu.edu
Website: http://www.utahcrop.org

Vermont Department of Agriculture
Food & Marketing, Drawer 20
116 State Street
Montpelier, VT 05620-2901
Phone: 802-828-2431 Fax: 802-828-2361

Virginia Foundation Seed Division
Bruce Beahm, Manager
4200 Cople Highway
P.O. Box 78
Mount Holly, VA 22524
Phone: 804-472-3500 Fax: 804-472-4649
E-mail: bbeahm@rivnet.net

**Washington State Crop Improvement
Foundation Seed Service**
Darlene Hilkin,
WSU Seed House
Pullman WA 99164
Phone: 509-335-4365 Fax: 509-335-7007
E-mail: wscia@wsu.edu

TABLE C.1 *(Continued)*

West Virginia Associated Crop Growers
John A. Balasko, Secretary-Treasurer
1090 Agricultural Science Building
West Virginia University
P.O. Box 6108
Morgantown, WV 26506-6108
Phone: 304-293-6256

Wisconsin Foundation Seeds
Jim Albertson, Director
1575 Linden Circle
University of Wisconsin
Madison, WI 53706-1597
Phone: 608-262-9954 Fax: 608-262-0168
E-mail: jcalbert@facstaff.wisc.edu

Wyoming Seed Certification Service
Mike D. Moore, Manager
University of Wyoming Seed Certification Service
P.O. Box 983
Powell, WY 82435
Phone: 307-754-9815 Fax: 607-754-9820
E-mail: mdmoore@uwyo.edu
Website: http://www.wyseedcert.com

Source: Information used with permission from AOSCA.

Appendix D

GLOBALGAP (EurepGAP) Accreditation Bodies

Table D.1 (see below) is a listing of GLOBALGAP (EurepGap) accredited certification bodies (CBs), membership and certifying body fees, DAP German Accreditation System benchmarking fee schedule, and the Joint Accreditation System of Australia and New Zealand benchmarking fee schedule.[*]

TABLE D.1
Accreditation Bodies (2006)

Organization	Headquarters	Comments/Approved for
ABCERT GmbH	Germany	
Agrar-Control GmbH	Germany	Subscopes: Combinable crops, cattle and sheep, dairy, pigs, poultry
AGRIZERT GmbH Gesellschaft zur Qualitätsförderung	Germany	Subscope: Combinable crops
CERES—CERtification of Environmental Standards G	Germany	Subscopes: Cattle and sheep, dairy, pigs, poultry and combinable crops
Control Union Certifications B.V. (former Skal International)	Netherlands	Subscopes: Cattle and sheep, dairy, pigs, poultry, combinable crops
Efsis Ltd.	United Kingdom	Subscopes: Cattle and sheep, poultry
EUROCERT European Inspection and Certification	Greece	Subscopes: Poultry
FoodCert B.V.	Netherlands	Subscope: Combinable crops
Instituto Genesis	Brazil	Subscopes: Cattle and sheep, dairy, pigs, poultry and combinable crops
IRAM-Instituto Argentino de Normalizacion y Certificacion	Argentina	Subscopes: Cattle and sheep, dairy, poultry, combinable crops
Luxcontrol GmbH	Germany	Subscopes: Cattle and sheep, dairy, pigs, poultry and combinable crops
Organización Internacional Agropecuaria S.A.	Argentina	Subscopes: Cattle and sheep, dairy, combinable crops, poultry
Planejar Informatica e Certificacao Ltda.	Brazil	Subscopes: Cattle and sheep, dairy, Subscopes: Pigs and poultry, combinable crops

(Continued)

[*] Please see GLOBALGAP at http://globalgap.org for the most updated version of accreditation bodies.

TABLE D.1 *(Continued)*

Organization	Headquarters	Comments/Approved for
QAL GmbH	Germany	Subscopes: Cattle and sheep, dairy, pigs, poultry, combinable crops
Servico Brasileiro de Certificacoes Ltda	Brazil	Subscopes: Cattle and sheep, dairy, pigs, poultry and combinable crops
SGS BELGIUM NV	Belgium	Subscope: Combinable crops (restricted to Option 1 certification)
SGS Germany GmbH	Germany	Subscopes: Cattle and sheep, dairy, pigs, poultry and combinable crops
TVL—Thüringer Verband für Leistungs- u. Qualitätsprüfungen in der Tierzucht	Germany	Subscopes: Cattle and sheep, dairy, pigs and combinable crops
WQS Certificação de Produtos Ltda.	Brazil	Subscopes: Cattle and sheep, dairy, pigs, poultry and combinable crops

GLOBALGAP (EurepGAP)

Membership Fees

Fee Type	Applies to	Amount	Notes
Retail membership fee	Retailer and food-service membership	€ 3,600	Per calendar year
Group supplier membership	Produce group or producer organization	€ 2,500	Per calendar year; maximum 3,600 € per one organization covering 3 or more subscopes
Individual supplier membership	Each additional subscope	€ 1,550	Per calendar year; maximum 2,600 € per one organization covering 3 or more subscopes
Supplier membership extension	Each additional subscope	€ 520	Per calendar year up to maximum 1,050 €
Associate membership	CB consulting, plant-protection, or fertilizer industry, etc., and associations	€1,550– 3,600	Per calendar year; covers all scopes and subscopes

CB Fees

Fee Type	Applies to	Amount	Notes
Initial evaluation fee for applicant CB	Initial application	€ 300	For first application only
CB license fee "first scope"	Approval of first scope	€ 3,000	Per calendar year, includes one free participation for one person per year to a CB workshop of that scope and a 500 € voluntary association membership fee discount
CB license fee "each additional scope"	Approval of each additional scope	€ 500	Per calendar year, includes one free participation for one person per year to a CB workshop of each additional scope

(Continued)

TABLE D.1   *(Continued)*

Fee Type	Applies to	Amount	Notes
Certification license fee	Each audit and inspection based on the minimum frequency established in EurepGAP Gen. Reg. for Option 1 and 2	€ 20	For Option 1: One fee for each certification issued. For Option 2: The square root of the total number of producers + 1 for the group is multiplied by the certification license fee. Additional unannounced audits/ inspections (10% of all Option 1 and 2 Certificates per CB) are also charged at € 20 each.
Online training and examination fee	Each assigned auditor and/or inspection per scope and EurepGAP standard version	€ 150	Payable once for each auditor/inspector with a 3-year standard version validity period

See http://cb.eurep.org/GENERAL_EUREPGAP_FEE_TABLE_2007.pdf for more information.

DAP German Accreditation System for Testing

EurepGAP Benchmarking Fee Schedule

 DEUTSCHES AKKREDITIERUNGSSYSTEM PRÜFWESEN GMBH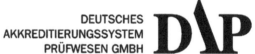

Process Step	Fee (€)	Notes
Standard owner application fee		
Fruit & Vegetable; Flowers & Ornamentals; and Integrated Aquaculture Assurance Standards	3,850	The application fee includes all associated administration costs, preliminary technical review, peer review facilitation, and independent technical review and report.
Integrated Farm Assurance (IFA) Standard		The following fees have been calculated based on the number of control points in each of the modules. If your standard includes a combination that is different from those below, please ask DAP for the correct fee applicable to your standard.
– Combinable crops	3,350	All farm base + combinable crops module
– Cattle and sheep, pig, or dairy	3,600	All farm base + livestock base + 1 species module
– Poultry	4,100	All farm base + livestock + poultry module
– Combinable crops + 1 species (except poultry)	4,100	All farm base + livestock base + 1 species module + combinable crops
– Combinable crops + poultry	4,550	All farm base + livestock base + poultry + combinable crops
Independent witness assessment (all standards)		
Scheme owner witnessing fee	1,400	Includes witness auditing (physical benchmarking), preparation, and reporting. The fee is the same for each of the above standards

(Continued)

TABLE D.1 *(Continued)*

Process Step	Fee (€)	Notes
Additional expenses		
Travel time	260	For travel in excess of 12 hours travel time, a flat fee will be charged each way.
Additional application processing	500/day	If applications are incomplete or where allocated time frames for processing are exceeded.

Joint Accreditation System of Australia and New Zealand (JAS-ANZ)

EurepGAP Benchmarking Fee Schedule

The application fee includes all associated administration costs, preliminary technical review, peer review facilitation, and independent technical review and report.

Process Step	Modules and Notes	Fee ($ AUD)
Fruit & Vegetables standard	NA	6,400
Flowers & Ornamentals standard	NA	6,400
Integrated Aquaculture Assurance standard	If your scheme or standard includes a combination that is different from those below, please contact JAS-ANZ for the correct fee applicable to your scheme or standard.	
	Base module (BM) + Salmonid module (SM)	6,600
	BM + chain of custody module (CCM)	6,700
	BM + CCM + SM	6,800
IFA standard	If your scheme or standard includes a combination that is different from those below, please contact JAS-ANZ for the correct fee applicable to your scheme or standard.	
	All farms module (AF) + crops-based module (CB) + Fruit & Vegetable (FV)	6,800
	AF + livestock-based module (LBM) + poultry module (PM)	6,800
	AF + CB + CCM (combined crops module)	6,400
	AF + LBM + pig module (PGM)	6,400
	AF + LBM + cattle & sheep module (CSM)	6,400
	AF + LBM + CSM + dairy module (DM)	6,500
Independent witness assessment (all standards) and additional expenses		
Scheme witnessing fee	Includes witness assessment (physical benchmarking) preparation and reporting. The fee is the same for each of the above standards and combinations.	2,400
Travel time	For travel in excess of 12 hours travel time is a flat fee and will be charged each way.	500

Source: Data adopted with permission from GLOBALGAP.

Appendix E

HACCP Training Providers

Table E.1 (see below) is a listing of HACCP national and international training providers.

TABLE E.1
HACCP Training Providers

National Sites

ABS Consulting
10301 Technology Drive
Knoxville, TN 37932
Phone: 865-676-2580 Fax: 865-671-5851
E-mail: kevans@absconsulting.com
Website: http://www.abs-jbfa.com/137.html

American Institute of Baking
P.O. Box 3999
Manhattan, KS 66505-3999
Phone: 800-242-2534 Fax: 785-537-1493
E-mail: sales@aibonline.org
Website: http://www.aibonline.org

ASI Food Safety Consultants, Inc.
7625 Page Boulevard
St. Louis, MO 63133
Phone: 800-477-0778 ext. 113
Fax: 314-727-4910
E-mail: jhuge@asifood.com
Website: http://www.asifood.com

Bizmanualz Inc./CVA International Ltd.
130 South Bemiston Avenue, Suite 101
Clayton, MO 63105
Phone: 304-863-5079 Fax: 314-863-6571
E-mail: chriz@bizmanualz.com
Website: http://www.Bizmanualz.com

BULLTEK Ltd.
4666 Wellesley Way, S101
Riverside, CA 92507
Phone: 888-BULLTEK Fax: 909-683-4013
E-mail: haccp@bulltek.com
Website: http://bulltek.com

Consulting Nutritional Services
26500 West Agoura Road, Suite 209
Calabasas, CA 91302
Phone: 818-874-9626 Fax: 818-874-9228
E-mail: cnsrd@aol.com or http://www.foodsafe.
com

D.L. Newslow & Associates, Inc.
8260 Cathy Ann Street
Orlando, FL 32818
Phone: 407-290-3156 Fax: 407-290-0252
E-mail: nancyemcdl@aol.com
Website: http://www.foodquality.com/newssems.
html

Environ Health Associates, Inc
2694 Magnolia Road
DeLand, FL 32720
Phone: 866-734-5187 Fax: 386-738-1465
E-mail: RCOSTA1@cfl.rr.com
Website: http://www.safefoods.tv

(Continued)

TABLE E.1 *(Continued)*

National Sites

Food Processors Institute/Food Products Association
1350 I Street, N.W. Suite 300
Washington, DC 20005-3305
Phone: 800-355-0983 Fax: 202-639-5932
E-mail: jepstein@nfpa-food.org
Website: http://www.fpi-food.org/
coursescheduke.cfm

Food Safe Services
P.O. BOX 5447
Pocatello, ID 83202
Phone: 877-770-8070
E-mail: kris@foodsafeservices.com
Website: http://www.haccpservice.com

Food Safety Specialists
1009 South Main Street
Fort Atkinson, WI 53538
Phone: 262-745-6087 Fax: 920-568-9270
E-mail: warreng175@compufort.com
Website: http://www.foodsafetyspecialists.com

Foodboss, LLC
P.O. Box 577455
Modesto, CA 95357
Phone: 209-869-5560
E-mail: haccp@foodboss.net
Website: http://www.foodboss.net

Hospitality Institute of Technology & Management
670 Transfer Road, Suite 21A
St. Paul, MN 55114
Phone: 651-646-7077 Fax: 651-646-5984
E-mail:osnyder@hi-tm.com
Website: http://www.hi-tm.com

Institute of Food Technologists
525 West Van Buren, Suite 1000
Chicago, IL 60607
Phone: 312-782-8424 Fax: 312-782-0045
E-mail: ajanguiano@ift.org
Website: http://www.ift.org/cms/?pid=1000408

National Marine Fisheries Service
11-15 Parker Street
Gloucester, MA 01930
Phone: 978-281-9124 Fax: 978-281-9125
E-mail: Karla.Ruzicka@noaa.gov
Website: http://seafood.nmfs.noaa.gov/
training.htm

National Meat Association
1970 Broadway, Suite 825
Oakland, CA 94612
Phone: 510-763-1533 Fax: 510-763-6186
E-mail: julie@nmaonline.org
Website: http://www.nmaonline.org

North Carolina University
Department of Food Science
Campus Box 7624
Raleigh, NC 27695
Phone: 919-513-2268
Fax: 919-515-7124
E-mail: foodsafety@ncsu.edu
Website: http://www.foodsafetytraining.info

Northern Sun Consulting
P.O. Box 2704
Baxter, MN 56425
Phone: 218-828-0214
E-mail: brookbob@uslink.net
Website: http://www.nscfoodsafety.com

NSF International World Headquarters
789 North Dixboro Road
Ann Arbor, MI 48105
Phone: 734-913-5703
Fax: 734-827-7795
E-mail:cphe@nsf.org
Website: http://www.nsf.org

Penn State University
Department of Food Science
University Park, PA 16802
Phone: 814-863-2298 Fax: 814-863-6132
E-mail: lfl5@psu.edu
Website: http://foodsafety.cas.psu.edu

(Continued)

TABLE E.1 *(Continued)*

National Sites

PhF Specialists
P.O. Box 7697
San Jose, CA 95160
Phone: 408-275-0161 Fax: 408-280-0979
E-mail: phfspec@pacbell.net
Website: http://www.phfspec.com

Professional Food Safety, Ltd.
11213 South Champlain
Chicago, IL 60628
Phone: 773-821-1943 Fax: 773-821-6910
E-mail: pfsltd@aol.com
Website: http://www.professionalfoodsafety.com

Silliker Laboratories
900 Maple Road
Homewood, IL 60430
Phone: 708-957-7878 Fax: 708-957-1483
E-mail: info@silliker.com
Website: http://www.silliker.com/courses.php

Southeastern Fisheries Association
1118-B Thomasville Road
Mount Vernon Square
Tallahassee, FL 32303
Phone: 850-224-0612 Fax: 850-222-3663
E-mail: bobfish@southeasternfish.org
Website: http://www.southeasternfish.org

SW Meat Associates & Texas A&M University
Southwest Meat Association
4103 South Texas Avenue, Suite 101
Bryan, TX 77802
Phone: 979-846-9011 Fax: 979-846-8198
E-mail: sma.jjh@tca.net
Website: http://www.southwestmeat.org

University of California–Davis
UC Davis Extension
1333 Research Park Drive
Davis, CA 95616
Phone: 800-752-0881 Fax: 530-757-8777
E-mail: questions@unexmail.ucdavis.edu
Website: http://extension.ucdavis.edu

University of Georgia
Food Science Extension
240 Food Science Building
Athens, GA 30602-7610
Phone: 706-542-2574 Fax: 706-583-0992
E-mail: ereynold@uga.edu
Website: http://fsext-outreach.ces.uga.edu/
  events/2005calendar.htm

Virginia Tech
Department of Food Science and Technology
Blacksburg, VA 24061
Phone: 540-231-3658 Fax: 540-231-9293
E-mail: jeifert@vt.edu

International Sites

Bizmanualz Inc./CVA International Ltd.
130 South Bemiston Avenue, Suite 101
Clayton, MO 63105
Phone: 304.863.5079 Fax: 314.863.6571
E-mail: chriz@bizmanualz.com
Website: http://www.Bizmanualz.com

Campden & Chorleywood Food Research Association
Station Road, Chipping Campden
Gloucestershire, GL55 6LD, United Kingdom
Phone: +44 (0)1386 842104 Fax: +44 (0) 1386
  842100
E-mail: training@campden.co.uk
Website: http://www.campden.co.uk/

Chartered Institute of Environmental Health
Chadwick Court 15 Hatfields
London SE1 8DJ United Kingdom
Phone: 020 7928 6006
Website: http://www.cieh.org

Food Industry Training/Reading, Science & Technology Centre
The University of Reading
Earley Gate, Whiteknights Road
Reading , RG6 6BZ , United Kingdom
Phone: +44 (0) 118 935 7346 Fax: +44 (0) 118
  935 7345
E-mail: info@fi t-r.com
Website: http://www.fi t-r.com

(Continued)

TABLE E.1 *(Continued)*

International Sites

Guelph Food Technology Center
88 McGilvray Street
Guelph, Ontario N1G 2W1 Canada
Phone: 519-821-1246 Fax: 519-836-1281
E-mail: gftc@gftc.ca
Website: http://www.gftc.ca/coursereg/list.cfm

International Flight Catering Association
Surrey Place, Mill Lane
Godalming, Surrey, GU7 1EY, United Kingdom
Phone: +44 (0) 1403 784363 Fax: +44 (0) 1483
 419780
E-mail: colin.banks3@btinternet.com
Website: http://www.ifcanet.com/teams/education/
 haccp/default.asp

International Inflight Food Service Association
5775 Peachtree-Dunwoody Road
Building G, Suite 500
Atlanta, GA 30342
Phone: 404-252-3663 Fax: 404-252-0774
E-mail: ifsa@kellencompany.com
Website: http://www.ifsanet.com

QMI Training/CSA Learning Center
Canadian Standards Association,
Learning Center
5060 Spectrum Way, Suite 100
Mississauga, Ontario, L4W 5N6 Canada
Phone: 800-463-6727 Fax: 416-747-2510
E-mail: learn@csa.ca
Website: http://www.csa.ca

Reading Scientific Services Ltd.
The University of Reading
Earley Gate, Whiteknights Road
Reading, RG6 6BZ, United Kingdom
Phone: +44 (0) 118 935 7346 Fax: +44 (0) 118
 935 7345
E-mail: info@fit-r.com
Website: http://www.rssl.com/OurServices/
 Training/Food

Source: Information used with permission from U.S. Department of Agriculture/U.S. Food and Drug
Administration HACCP training programs and resources database, 2006.

Appendix F

IFOAM-Accredited Certification Bodies

Figure F.1 (see below) is a listing of International Federation of Organic Agriculture Movements (IFOAM)-accredited certification bodies as of June 20, 2006.[*][†]

FIGURE F.1 IFOAM-accredited certification bodies. (Reprinted with permission from IFOAM.)

---

[*] Program(s) covered by the accreditation: A certification body may operate more than one certification program. However, the only program included in the scope of the IFOAM accreditation is listed here; for example, it may be certifying organic to a regulation or it may be certifying something other than organic such as "produced without genetically modified organisms."

[†] Categories included in accreditation scope: The certification body may certify various activities within its organic certification program. Accreditation is possible for certification of crop production,

| List of IFOAM Accredited Certification Bodies | October 20, 2008 |

Australian Certified Organic   Contract No: 39
PO Box 530, Ground level, 766 Gympie Rd (Bri ChermsideQueensland 4032

Australia   Phone +61 7 3350 5706 Fax +61 7 3350 5996 E-Mail info@aco.net.au

Programs included in accreditation scope:	ACO IFOAM Programme
Categories included in accreditation scope:	Crop production; Livestock; Wild products; Processing and handling; Retail; Certification transference; Grower groups
Not accredited organic certification programme(s):	ACO programme-Australian National Standard; Japan JAS; US NOP
Countries of operation:	Australia; Cook Islands; Fiji; Indonesia; Japan; Madagascar; Malaysia; New Zealand; Papua New Guinea; Singapore; Thailand.

Year first accredited: 2003   Date current contract: 4 Oct 06   Date of contract expiry: 31 Jul 10

Bioagricert srl   Contract No: 19
Via del Macabraccia 840033 Casalecchio di Ren (BO)

Italy   Phone +39 051 562 158 Fax +39 051 562 294 E-Mail giuseppe.tallarico@bioagricert.org

Programs included in accreditation scope:	Bioagricert private standards and seal programme designated as 'Bioagricert International'
Categories included in accreditation scope:	Crop production; Livestock; Processing and handling; Input manufacturing; Certification transference; Grower Groups
Not accredited organic certification programme(s):	European Regulation 2092/91; Japan JAS; US NOP
Countries of operation:	Italy; Mexico; Thailand

Year first accredited: 1996   Date current contract: 1 Aug 05   Date of contract expiry: 31 Jul 09

Bio-Gro New Zealand Ltd   Contract No: 23
PO Box 9693 Marion SquareWellington 6031

New Zealand   Phone + 64 4 801 9741 Fax +64 4 801 9742 E-Mail smason@bio-gro.co.nz

Programs included in accreditation scope:	BioGro private standards and seal programme
Categories included in accreditation scope:	Crop production; Livestock; Wild products; Processing and handling; Retail; Input manufacturing; Certification transference; Grower Groups; Aquaculture
Not accredited organic certification programme(s):	NZFSA Domestic Programme; US NOP
Countries of operation:	New Zealand; Cook Islands; Niue.

Year first accredited: 1996   Date current contract: 1 Jun 05   Date of contract expiry: 30 Jun 09

FIGURE F.1   *(Continued)*

(Continued)

livestock, wild products, processing, textile processing, aquaculture, input manufacturing, retailing, grower groups, and certification transference. Regarding nonaccredited organic certification program(s): If a certification body operates more than one certification program, this listing indicates any programs that are not included in the accreditation scope.

List of IFOAM Accredited Certification Bodies   October 20, 2008

Biokontroll Hungaria Nonprofit Kft.   Contract No: 47

H 1535 Budapest PF 800 PostalH 1027 BudapestMargit KRT 1

Hungary   Phone +36 1 336 1122Fax +36 1 315 1123 E-Mail info@biokontroll.hu

Programs included in accreditation scope:	Biokontroll Hungaria Nonprofit Kft. private standards and seal programme
Categories included in accreditation scope:	Crop production; Livestock; Wild products; Processing and handling; Retail; Certification transference; Aquaculture
Not accredited organic certification programme(s):	EU Regulation 2092/91
Countries of operation:	Hungary

Year first accredited: 2004   **Date current contract:**   1 May 08   **Date of contract expiry:**   30 Apr 12

Bioland e.V.   Contract No: 34

Kaiserstrasse 18D-55116 Mainz

Germany   Phone +49 613 123 9790Fax +49 613 123 979-27 E-Mail landbau@bioland.de

Programs included in accreditation scope:	Bioland e.V private standards and seal programme
Categories included in accreditation scope:	Crop production; Livestock; Wild products; Processing and handling; Certification transference; Aquaculture; Retailing
Not accredited organic certification programme(s):	None
Countries of operation:	Germany; Austria; Belgium; France; Italy; Netherlands

Year first accredited: 2002   **Date current contract:**   16 Feb 06   **Date of contract expiry:**   30 Apr 10

BIOPARKe.V.   Contract No: 42

Rövertannen 1318273 Güstrow

Germany   Phone +49 384 324 5030Fax +49 384 324 5032 E-Mail info@biopark.de

Programs included in accreditation scope:	Biopark e.V. private standards and seal programme
Categories included in accreditation scope:	Crop production; Livestock; Processing and handling; Retail; Certification transference
Not accredited organic certification programme(s):	None
Countries of operation:	Germany

Year first accredited: 2003   **Date current contract:**   23 Apr 07   **Date of contract expiry:**   30 Mar 11

FIGURE F.1   *(Continued)*

List of IFOAM Accredited Certification Bodies   October 20, 2008

BIOS S.r.l.   Contract No: 49

Via M. Grappa, 37/CMarosticaVI 36083

Italy   Phone +39 0424 471 125 Fax +39 0424 476 947 E-Mail info@certbios.it

Programs included in accreditation scope:	'BIOS International' programme - Italian Organic Standards
Categories included in accreditation scope:	Crop production; Livestock; Wild products; Processing and handling; Retail; Certification transference
Not accredited organic certification programme(s):	Biosuisse; Delinat; European Regulation 2092/91; Romania (code RO-ECO-009); US NOP
Countries of operation:	Italy

Year first accredited: 2004   Date current contract: 6 Sep 04   Date of contract expiry: 7 Feb 09

Bolicert   Contract No: 24

Calle 20 de Octubre Nº 1915, Edificio "MALAG Casilla 13030La Paz

Bolivia   Phone +591-2-2414972 Fax +591-2-2414972 E-Mail bolicert@mail.megalink.com

Programs included in accreditation scope:	Bolicert private standards and seal programme
Categories included in accreditation scope:	Crop production; Wild products; Processing and handling; Certification transference; Grower groups
Not accredited organic certification programme(s):	US NOP
Countries of operation:	Bolivia; Peru

Year first accredited: 1998   Date current contract: 17 Apr 06   Date of contract expiry: 30 Jun 10

CCOF Certification Services LLC   Contract No: 21

2155 Delaware Avenue, Suite 150Santa CruzCA 95060

USA   Phone +1 831 423 2263 Fax +1 831 423 4528 E-Mail ccof@ccof.org

Programs included in accreditation scope:	CCOF private standards and seal programme designated as 'Global Market Access'
Categories included in accreditation scope:	Crop production; Processing and handling; Certification transference.
Not accredited organic certification programme(s):	Private standards and logo - not designated as "International Program", US NOP
Countries of operation:	USA; Canada; Mexico

Year first accredited: 1997   Date current contract: 20 Sep 07   Date of contract expiry: 31 Mar 12

FIGURE F.1   *(Continued)*

List of IFOAM Accredited Certification Bodies October 20, 2008

CCPB SRL   Contract No: 28

Via Jacopo Barozzi N.840126 Bologna

Italy   Phone +39 0 51 6089811 Fax +39 0 51 254842 E-Mail ccpb@ccpb.it

Programs included in accreditation scope:	CCPB private standards and seal programme designated as 'CCPB Global Programme'
Categories included in accreditation scope:	Crop production; Livestock; Processing and handling; Certification transference
Not accredited organic certification programme(s):	European Regulation 2092/91; Japan JAS; US NOP
Countries of operation:	Italy

Year first accredited: 2000   Date current contract: 3 Aug 07   Date of contract expiry: 31 Oct 11

Debio   Contract No: 52

Bjorkelangen   N - 194

Norway   Phone +47 638 62 650 Fax +47 638 56 985 E-Mail morten@debio.no

Programs included in accreditation scope:	Debio IFOAM Programme (Debio-IFOAM standards; Debio Aquaculture standards & Debio Forestry standards
Categories included in accreditation scope:	Crop production; Livestock. Wild products; Processing and handling; Input manufacturing; Certification transference; Aquaculture
Not accredited organic certification programme(s):	European Regulation 2092/91; Demeter International Programme
Countries of operation:	Norway. Denmark. South Africa

Year first accredited: 2006   Date current contract: 17 Feb 06   Date of contract expiry: 31 Jul 10

Doalnara Certified Organic Korea, LLC   Contract No: 56

192-3 Jangyang-ri. Socho-myeonWonju-siGangwon-do 220833

South Korea   Phone +82 33 732 4234 Fax +82 33 732 4239 E-Mail doalnara@doalnara.or.kr

Programs included in accreditation scope:	Doalnara Certified Organic Korea, LLC International Standards and Seal Programme
Categories included in accreditation scope:	Crop production; Livestock: Wild products: Processing and handling; Certification transference
Not accredited organic certification programme(s):	Korean EFAPA
Countries of operation:	South Korea, China. South Africa

Year first accredited: 2006   Date current contract: 15 Nov 06   Date of contract expiry: 30 Apr 11

FIGURE F.1   *(Continued)*

List of IFOAM Accredited Certification Bodies   October 20, 2008

Ecoland e.V.   Contract No: 54

Raiffeisenstrasse 7Wolpertshausen   74549

Germany   Phone +49 791 93290 451 Fax +49 791 93290 459 E-Mail info@ecoland.de

Programs included in accreditation scope:	Ecoland Standards and Seal Programme
Categories included in accreditation scope:	Crop production; Livestock; Wild products; Processing & Handling; Certification transference
Not accredited organic certification programme(s):	None
Countries of operation:	Germany

Year first accredited: 2007   Date current contract: 16 Feb 07   Date of contract expiry: 30 Nov 10

Gäa e.V. Vereinigung okologischer Landbau Bundesverband   Contract No: 46

Gäa e.v BundesgeschaftsstelleArndtstrasse 11D-01099 Dresden

Germany   Phone +49 351 401 2389 Fax +49 351 401 5519 E-Mail Christian.Pein@gaea.de

Programs included in accreditation scope:	Gäa e.V. Vereinigung okologischer Landbau Bundesverband Private Standards and Seal Programme
Categories included in accreditation scope:	Crop production; Livestock, Wild products; Processing and handling; Inputs manufacturing, Aquaculture & Certification transference
Not accredited organic certification programme(s):	None
Countries of operation:	Germany

Year first accredited: 2003   Date current contract: 11 Jul 07   Date of contract expiry: 30 Jun 11

Global Organic Agriculturist Association   Contract No: 55

#492-4, Dongho-dong, BukkuTaegu City

South Korea   Phone + 82-53-326-9895 Fax +82-53-326-9896 E-Mail kkdeca@naver.com

Programs included in accreditation scope:	Global Organic Agriculturist Association International Standards and Seal Programme
Categories included in accreditation scope:	Crop production; Livestock; Wild products; Processing and handling; Retailing; Certification transference; Aquaculture.
Not accredited organic certification programme(s):	Korean EFAPA
Countries of operation:	South Korea, China

Year first accredited: 2006   Date current contract: 15 Nov 06   Date of contract expiry: 24 Apr 11

FIGURE F.1   *(Continued)*

List of IFOAM Accredited Certification Bodies   October 20, 2008

Instituto Biodinamico
Rua Prudente de Moraes, 53018.620-060Botucatu SP

Brazil   Phone +55 14 3882 5066 Fax +55 14 3815 9909 E-Mail ibd@ibd.com.br

Contract No: 14

Programs included in accreditation scope:	Orgânico Instituto Biodinâmico private standards and seal programme
Categories included in accreditation scope:	Crop production; Livestock; Wild products; Processing and handling; Input Manufacturing; Certification transference; Grower groups; Aquaculture; Fibre processing;
Not accredited organic certification programme(s):	IBD Organic; Demeter; European Regulation 2092/91; US NOP;
Countries of operation:	Brazil; Argentina; Bolivia; Mexico; Paraguay

Year first accredited: 1996   **Date current contract:** 29 Mar 07   **Date of contract expiry:** 28 Feb 11

International Certification Services Inc.
301 5th Ave. SEMedinaND 58467

USA   Phone +1 701 486 3578 Fax +1 701 486 3580 E-Mail Info@ics-intl.com

Contract No: 13

Programs included in accreditation scope:	Farm Verified Organic private standards and seal programme
Categories included in accreditation scope:	Crop production; Livestock; Wild products; Processing and handling; Retail; Input Manufacturing; Certification transference; Grower groups
Not accredited organic certification programme(s):	Quebec CAAQ; US NOP
Countries of operation:	USA; Brazil; Canada; China; Guatemala; Mexico; Paraguay; Tahiti

Year first accredited: 1995   **Date current contract:** 1 Sep 06   **Date of contract expiry:** 31 May 11

Istituto Mediterraneo Di Certificazione s.r.l.
Via Carlo Pisacane, 32 60019 SenigalliaAncona

Italy   Phone +39 071 792 8725 Fax +39 071 791 0043 E-Mail imcert@imcert.it

Contract No: 45

Programs included in accreditation scope:	Garanzia AMAB Programme
Categories included in accreditation scope:	Crop production; Livestock; Wild products; Processing and handling; Certification transference
Not accredited organic certification programme(s):	Catering and Farm Holiday Services; European Regulation 2092/91; Japan JAS; US NOP
Countries of operation:	Italy, Egypt, Tunisia, Lebanon, Turkey

Year first accredited: 2003   **Date current contract:** 28 Feb 08   **Date of contract expiry:** 28 Feb 12

FIGURE F.1   *(Continued)*

List of IFOAM Accredited Certification Bodies   October 20, 2008

Istituto per la Certificazione Etica e Ambientale

Contract No: 26

Strada Maggiore 2940125Bologna

Italy   Phone +39 051 272 986Fax +39 051 232 011E-Mail icea@icea.info

Programs included in accreditation scope:	Italian Organic Standard
Categories included in accreditation scope:	Crop production; Livestock; Processing & handling; Certification transference, Grower groups; Aquaculture; Wild harvest
Not accredited organic certification programme(s):	European Regulation 2092/91; Japan JAS; US NOP
Countries of operation:	Italy; Lebanon; Turkey

Year first accredited: 1999   Date current contract: 20 Nov 07   Date of contract expiry: 30 Nov 11

Japan Organic & Natural Foods Association

Contract No: 33

Takegashi Bldg. 3F, 3-5-3, KyobashiChuo-KuTokyo 104-003

Japan   Phone +81 33 538 1851Fax +81 33 538 1852E-Mail toshi@jona-japan.org

Programs included in accreditation scope:	JONA private standards and seal programme designated as 'International'
Categories included in accreditation scope:	Crop production; Wild products; Processing and handling; Certification transference.
Not accredited organic certification programme(s):	Japan JAS; JONA Original Programme; Input confirmation
Countries of operation:	Japan; China; Brazil (IFOAM) China; Brazil; Argentina; India; USA (ISO65)

Year first accredited: 2002   Date current contract: 2 May 03   Date of contract expiry: 30 Apr 09

LETIS S.A.

Contract No: 51

Urquiza 1564 (S2000 ANR)RosarioSanta Fe

Argentina   Phone +54 341 426 4244Fax +54 341 426 4244 E-Mail letis@letis.com.ar

Programs included in accreditation scope:	Letis IFOAM Programme
Categories included in accreditation scope:	Crop production; Livestock; Wild products; Processing and handling; Certification transference; Grower groups; Aquaculture
Not accredited organic certification programme(s):	European-Argentine Programme; US NOP
Countries of operation:	Argentina; Canada; Paraguay

Year first accredited: 2005   Date current contract: 16 Dec 05   Date of contract expiry: 30 Apr 10

FIGURE F.1   *(Continued)*

List of IFOAM Accredited Certification Bodies October 20, 2008

National Association Sustainable Agriculture Australia
Contract No: 11

PO Box 768StirlingS. Australia 5152

Australia Phone +61 8 8370 8455 Fax +61 8 8370 8381 E-Mail certification.manager@nasaa.com

Programs included in accreditation scope:	NASAA private standards and seal programme
Categories included in accreditation scope:	Crop production; Livestock; Wild products; Processing and handling; Input manufacturing; Aquaculture; Certification transference; Grower groups
Not accredited organic certification programme(s):	Japan JAS; US NOP; AQIS
Countries of operation:	Australia; East Timor; Indonesia; Nepal; New Zealand; Papua New Guinea; Samoa; Sri Lanka

Year first accredited: 1994   Date current contract: 23 Feb 07   Date of contract expiry: 30 Jun 11

Naturland - Verband für ökologischen Landbau e.V.
Contract No: 20

Kleinhaderner Weg 182168 Gräfelfing

Germany Phone +49 89 898 082-0 Fax +49 89 898 08290 E-Mail naturland@naturland.de

Programs included in accreditation scope:	Naturland Organic Worldwide Programme
Categories included in accreditation scope:	Crop production; Livestock; Wild products; Processing and handling; Certification transference; Grower groups.
Not accredited organic certification programme(s):	Naturland Private Standards Programme; US NOP
Countries of operation:	Germany; Australia; Austria; Bolivia; Brazil; China; Dominican Republic; Ecuador; Egypt; France; Greece; Guatemala; Hungary; India; Italy; Mexico; Netherlands; Nicaragua; Paraguay; Peru; Philippines; Saudi Arabia; South Africa; Sri Lanka; Switzerland; Taiwan; Tanzania; Uganda; United Kingdom; United States

Year first accredited: 1996   Date current contract: 9 Feb 05   Date of contract expiry: 30 Nov 08

Organic Agriculture Certification Thailand
Contract No: 32

619/43 Kiatngamwong Bldg, Ngamwongwan R Muang DistrictNonthaburi 11000

Thailand Phone +66 2 580 0934 Fax +66 2 580 0934 E-Mail info@actorganic-cert.or.th

Programs included in accreditation scope:	ACT private standards and seal programme
Categories included in accreditation scope:	Crop production; Wild products; Processing and handling; Input manufacturing; Certification transference; Grower groups; Aquaculture
Not accredited organic certification programme(s):	None
Countries of operation:	Thailand; Vietnam

Year first accredited: 2001   Date current contract: 18 Feb 05   Date of contract expiry: 30 Apr 09

FIGURE F.1 *(Continued)*

List of IFOAM Accredited Certification Bodies   October 20, 2008

Organic Certifiers Inc.
6500 Casitas Pass Road Ventura CA 93001

USA   Phone +1 805 684 6494 Fax +1 805 684 2767 E-Mail organic@west.net

Contract No: 50

Programs included in accreditation scope:	Organic Certifier's International Programme
Categories included in accreditation scope:	Crop production; Livestock; Wild products; Processing and handling; Input manufacturing; Certification transference; Grower Groups
Not accredited organic certification programme(s):	US NOP
Countries of operation:	USA, Mexico.

Year first accredited: 2005   Date current contract: 2 Sep 05   Date of contract expiry: 31 Aug 09

Organic Crop Improvement Association International
1340 N. Cotner Blvd. Lincoln NE 68505

USA   Phone +1 402 477 2323 Fax +1 402 477 4325 E-Mail info@ocia.org

Contract No: 30

Programs included in accreditation scope:	OCIA International (private standards and seal) Programme
Categories included in accreditation scope:	Crop production; Livestock; Wild products; Processing and handling; Retail; Certification transference; Grower groups.
Not accredited organic certification programme(s):	US NOP; CARTV (formerly CAAQ); JAS
Countries of operation:	USA; Brazil; Canada, China; Columbia, Costa Rica; Dominican Republic; Ecuador; El Salvador; Germany; Guatemala; Honduras; Japan; Mexico; Paraguay; Peru; Philippines; Timor Lorosa'e; Uganda

Year first accredited: 2000   Date current contract: 1 Jul 08   Date of contract expiry: 15 May 12

Organic Food Development & Certification Center of China
8 Jiangwangmiao Street, P.O. Box 4202 Nanjing   210042

P.R.China   Phone +86 25 8642 5370 Fax +86 25 8642 0606 E-Mail xiao@ofdc.org.cn

Contract No: 40

Programs included in accreditation scope:	OFDC private standards and seal programme
Categories included in accreditation scope:	Crop production; Livestock; Wild products; Processing and handling; Input manufacturing; Certification transference; Grower groups; Aquaculture
Not accredited organic certification programme(s):	Chinese National Standards
Countries of operation:	P.R.China

Year first accredited: 2003   Date current contract: 12 Dec 06   Date of contract expiry: 28 Feb 11

FIGURE F.1   *(Continued)*

List of IFOAM Accredited Certification Bodies   October 20, 2008

Organizacion Internacional Agropecuaria S.A.   Contract No: 31
Av. Santa Fe 930, B1641ABN AcassusoBuenos Aires

Argentina   Phone +54 11 4793 4340 Fax +54 11 4793 4340 E-Mail oia@oia.com.ar

Programs included in accreditation scope:	OIA IFOAM programme
Categories included in accreditation scope:	Crop production; Livestock; Wild products; Processing and handling; Certification transference.
Not accredited organic certification programme(s):	Argentine-European Programme; US NOP
Countries of operation:	Argentina; Uruguay

Year first accredited: 2001   **Date current contract:**   30 Nov 04   **Date of contract expiry:**   31 Dec 08

Organska Kontrola Sarajevo   Contract No: 57
Butmirska cesta 4071 000Sarajevo

Bosnia and Herz²hone +387 (0) 33 63 7301Fax +387 (0) 33 63 6768 E-Mail office@organskakontrola.ba

Programs included in accreditation scope:	Organska Kontrola private standards and seal programme
Categories included in accreditation scope:	Crop production; Livestock; Wild products; Processing and handling; Retail; Certification transference; Grower groups
Not accredited organic certification programme(s):	None
Countries of operation:	Bosnia & Herzogovina

Year first accredited: 2007   **Date current contract:**   18 Jul 07   **Date of contract expiry:**   31 Oct 11

Quality Assurance International   Contract No: 38
9191 Towne Centre Drive, Suite 510San DiegoCalifornia 92122

USA   Phone +1 858 792 3531 Fax +1 858 792 8665 E-Mail Maria@qai-inc.com

Programs included in accreditation scope:	QAI private standards and seal programme designated as 'QAI Global'
Categories included in accreditation scope:	Crop production; Wild harvest; Processing & handling; Retailing; Certification Transference; Grower groups
Not accredited organic certification programme(s):	European Regulation 2092/91; Japan JAS; QAI Fibre; Quebec CAAQ; US NOP
Countries of operation:	USA; Canada; Mexico; Paraguay

Year first accredited: 2003   **Date current contract:**   14 Mar 07   **Date of contract expiry:**   14 Mar 11

FIGURE F.1   *(Continued)*

List of IFOAM Accredited Certification Bodies   October 20, 2008

Tanzania Organic Certification Association   Contract No: 58

P. O. Box 70089Dar es Salaam

Tanzania   Phone +255 22 2124441Fax +255 22 2124441 E-Mail tancert@TanCert.org

Programs included in accreditation scope:	TanCert Private Standards and Seal Programme
Categories included in accreditation scope:	Crop production, Processing & handling, Livestock, Wild products, Input manufacturing, Grower groups and Certification transference
Not accredited organic certification programme(s):	None
Countries of operation:	Tanzania

Year first accredited: 2008   Date current contract: 31 Jul 08   Date of contract expiry: 30 Jun 12

Uganda Organic Certification Ltd   Contract No: 59

ACCORD Building, Gabba Road, P.O. Box: 33 Kampala

Uganda   Phone + 256 (0)41 269 416Fax + 256 (0)41 269 001 E-Mail info@ugocert.org

Programs included in accreditation scope:	Uganda Organic Certification Ltd. Private Standards and Seal Programme
Categories included in accreditation scope:	Crop production; Wild harvest; Processing & handling; Livestock; Certification Transference; Grower groups
Not accredited organic certification programme(s):	Utz Kapeh Sustainable Artisanal Wild Freshwater Fishery
Countries of operation:	Uganda

Year first accredited: 2008   Date current contract: 4 Aug 08   Date of contract expiry: 30 Jul 12

Washington State Dept. of Agriculture Organic Food Program   Contract No: 48

PO Box 42560, 1111 Washington StreetOlympiaWashington 98504-2

USA   Phone +1 360 902 1805Fax +1 360 902 2087E-Mail organic@agr.wa.gov

Programs included in accreditation scope:	Washington State Department of Agriculture Organic Program European Organic Verification Program
Categories included in accreditation scope:	Crop production; Wild products; Processing and handling
Not accredited organic certification programme(s):	US NOP
Countries of operation:	USA

Year first accredited: 2004   Date current contract: 29 Oct 04   Date of contract expiry: 31 Jan 09

FIGURE F.1   *(Continued)*

Appendix G

SQF Certification Bodies

Table G.1 (see below) is a listing of licensed SQF certification bodies.

TABLE G.1
Licensed SQF Certification Bodies

AIB International
1213 Bakers Way
Manhattan, KS 66505
Phone: 785-537-4750
Fax: 785-539-0106
E-mail: jkay@aibonline.org

Bureau Veritas Certification NA
515 West Fifth Street
Jamestown, NY 14701
Phone: 706-484-9002 ext. 104
Fax: 716-484-9004
E-mail: Ralph.mclouth@us.bureauveritas.com

NCS International Pty. Ltd. Inc.
 Food Operations
7 Leeds Street
Rhodes, New South Wales 2138 Australia
Phone: 61 (0) 7-3870-7556
Fax: 61 (0) 7-3870-4570
E-mail: Bill.McBride@ncsi.com.au

NSF International
789 North Dixboro Road
Ann Arbor, MI 48105
Phone: 734-827-6845
Fax: 734-827-7779
E-mail: gipple@nsf.org
Website: http://www.nsf.org/business/sqf/
 index.asp?program=SQF

SGS Societe Generale De Surveillance SA
Place des Alpes
Geneva, Switzerland
Phone: 41-22-739-91-11
Fax: 41-22-739-98-86
E-mail: Dick.Visser@sgs.com

TUV America, Inc.
5 Cherry Hill Drive
Danvers, MA 01923
Phone: 978-739-7021
Fax: 978-762-8414
E-mail: gminks@tuvam.com

Source: Data adopted with permission from SQF.

Appendix H

International Seed Federation

Table H.1 (see below) is a listing of the International Seed Federation's network of seed-trade and plant breeder associations.

TABLE H.1

International Seed Federation's Network of Seed-Trade and Plant Breeder Associations

Global Level	International Seed Federation (ISF)			
Regional Level	Africa & Middle East (AFSTA)	Asia/Pacific (APSA)	Europe (ESA) & (EESNET)	Americas (FELAS)

Seed-Related Organizations

AOSA (Association of Official Seed Analysts) ....http://www.aosaseed.com

AOSCA (Association of Official Seed Certifying Agencies) ....http://www.aosca.org

CPVO (Community Plant Variety Office)...http://www.cpvo.europa.eu

FAO (Food and Agricultural Organization of the United Nations) Seed-Plant Genetic Service..................
....http://www.fao.org/waicent/faoinfo/agricult/agp/agps/default.htm

ISHIs (International Seed Health Initiatives).......................http://www.worldseed.org/phytosanitary.htm

ISSS (International Society for Seed Science)...........................http://www.css.cornell.edu/ISSS/isss.htm

ISST (International Society of Seed Technologists) .... http://www.isstech.org

ISTA (International Seed Testing Association)..............................http://www.seedtest.org/en/home.html

OECD Organisation for Economic Co-operation and Development Seed ....
.... http://www.oecd.org/department/0,2688,en_2649_33905_1_1_1_1_1,00.html

UPOV (International Union for the Protection of New Varieties of Plants) ....http://www.upov.int

Plant Protection Organizations

FAO Plant Protection Service ....http://www.fao.org/waicent/faoinfo/agricult/agp/agpp/default.htm

ICPM (Interim Commission on Phytosanitary Measures)..............https://www.ippc.int/IPP/En/default.jsp

Regional Plant Protection Organizations

Comunidad Andina ....http://www.comunidadandina.org

EPPO (European and Mediterranean Plant Protection Organization) ....http://www.eppo.org

NAPPO (North American Plant Protection Organization)..http://www.nappo.org

PPPO (Pacific Plant Protection Organization) ....http://www.spc.org.nc/pps

Intergovernmental Organizations

United Nations .... http://www.un.org

(Continued)

TABLE H.1   *(Continued)*

FAO .... http://www.fao.org

CGRFA (Commission on Genetic Resources for Food Agriculture)............... http://www.fao.org/ag/cgrfa

Codex Alimentarius .... http://www.codexalimentarius.net/web/index_en.jsp

ICPM (Interim Commission on Phytosanitary Measures).............https://www.ippc.int/IPP/En/default.jsp

Plant Protection Service........................http://www.fao.org/waicent/faoinfo/agricult/agp/agpp/default.htm

Seed-Plant Genetic Research Service ....http://www.fao.org/waicent/faoinfo/agricult/agp/agps/default.htm

United Nations and Business .... http://www.un.org/partners/business/index.asp

UNCSD (U.N. Commission on Sustainable Development)....http://www.un.org/esa/sustdev/csd/policy.htm

Agenda 21 ....http://www.un.org/esa/sustdev/documents/agenda21/english/agenda21toc.htm

UNCTAD (U.N. Conference on Trade and Development)

UNDP (U.N. Development Program)

UNEP (U.N. Environment Program)

CBD (Convention on Biological Diversity)

CITES (Convention on International Trade in Endangered Species of Wild Fauna and Flora)

UNESCO (U.N. Educational, Scientific, and Cultural Organization)

UNIDO (U.N. Industrial Development Organization)

WIPO (World Intellectual Property Organization)

Other Intergovernmental Organizations

CGIAR (Consultative Group on International Agricultural Research)

CIMMYT (International Maize and Wheat Improvement Center)

IFPRI (International Food Policy Research Institute)

IPGRI (International Plant Genetic Resources Institute)

ECP/GR (European Cooperative Programme for Crop Genetic Resources Networks)

IRRI (International Rice Research Institute)

ISNAR (International Service for National Agricultural Research)

WTO (World Trade Organization)

SPS (Agreement on the Application of Sanitary and Phytosanitary Measures)

TBT (Agreement on Technical Barriers to Trade)

TRIPS (Agreement on Trade-Related Aspects of Intellectual Property Rights)

Other International Organizations

AFIC (Asian Food Information Centre)

AVRDC (Asian Vegetable Research & Development Center)

BASD (Business Action for Sustainable Development)

CAB International

EFB (European Federation of Biotechnology)

EUCARPIA (European Association for Research on Plant Breeding)

EUFIC (European Food Information Council)

GFAR (Global Forum on Agricultural Research)

IAMA (International Food and Agribusiness Management Association)

ICC (International Chamber of Commerce)

IFIC (International Food Information Council)

IFT (Institute of Food Technologists)

ICGEB (International Centre for Genetic Engineering and Biotechnology)

IISD (International Institute for Sustainable Development)

ISAAA (International Service for the Acquisition of Agri-Biotech Applications)

(Continued)

TABLE H.1 *(Continued)*

ISEB (International Society for Environmental Biotechnology)
ISHS (International Society for Horticultural Science)
WBCSD (World Business Council for Sustainable Development)

Seed Industry Associations

National Associations

ABRASEM (Brazil)
AIC (United Kingdom)
AIS (Italy)
AMSAC (Mexico)
AMSOL (France)
ANPROS (Chile)
ASA (Argentina)
ASF (Australia)
ASTA (United States)
BDP (Germany)
BRASPOV (Brazil)
BSPB (United Kingdom)
CFS (France)

CMSSA (Czech Republic)
CSBC (Argentina)
CSTA (Canada)
EEPES (Greece)
ESAS (Egypt)
JASTA (Japan)
KSA (Korea)
NZGSTA (New Zealand)
PIN (Poland)
Plantum (The Netherlands)
SANSOR (South Africa)
TURK-TED (Turkey)
YUSEA (Serbia and Montenegro)

Regional Associations

AFSTA (Africa)
APSA (Asia and Pacific)
EESNET (Eastern Europe)
ESA (Europe)

FELAS (Latin America)
WANA Seed Network (West Asia and North Africa)
WASNET (West Africa)

Crop-Specific Associations

Fleuroselect (flower seeds)

Other Industry Associations

Biotech Industry Associations

ABA (Australian Biotechnology Association)
AfricaBio
BIO (U.S. Biotechnology Industry Organization)
BIOTECanada
EuropaBio (European Association for Bioindustries)
FAB (Foro Argentino de Biotecnologie)
JBA (Japan Bioindustry Association)
NZBA (New Zealand Biotech Association)

Agri-Food Industry Associations

CropLife International
GAFTA (Grain and Feed Trade Association)
IAFN (International Agri-Food Network)
ICA (International Cooperative Alliance)
IFA (International Fertilizer Industry Association)
IFAP (International Federation of Agricultural Producers)
IFIF (International Feed Industry Federation)

(Continued)

TABLE H.1 *(Continued)*

International Conventions on Intellectual Property

Agreement on Trade-Related Aspects of Intellectual Property Rights (TRIPS)

International Convention for the Protection of New Varieties of Plants (UPOV)

Paris Convention for the Protection of Industrial Property

Patent Cooperation Treaty (PCT)

Patent Law Treaty (PLT)

International Conventions on Sanitary and Phytosanitary Matters

Agreement on the Application of Sanitary and Phytosanitary Measures (SPS)

International Plant Protection Convention (IPPC)

Rotterdam Convention on the Prior Informed Consent Procedure for Certain Hazardous Chemicals and
  Pesticides in International Trade

International Conventions on Biodiversity

Cartagena Protocol on Biosafety

Convention on Biological Diversity (CBD)

Convention on International Trade in Endangered Species of Wild Fauna and Flora (CITES)

International Treaty on Plant Genetic Resources for Food and Agriculture

International Undertaking on Plant Genetic Resources for Food and Agriculture

Other International Conventions

Agreement on Technical Barriers to Trade (TBT)

New York Convention on the Recognition and Enforcement of Foreign Arbitral Awards

Databases of Interest

AgNIC (Agriculture Network Information Center)

BINAS (UNIDO Biosafety Information Network and Advisory Service)

BIOBIN (OECD/UNIDO cooperative resource on safety in biotechnology)

Biotechfind

BioTrack (OECD database on biotechnology)

CBD Clearing-House Mechanism

EcoPort (plant and pest information system)

Essential Biosafety

FAOLEX (FAO Legislative Database)

FIS (Forage Information System)

IBM Patent Server

IP.com (Prior Art Database)

OrganicXseeds

PQR (EPPO Plant Quarantine Data Retrieval System)

SeedQuest

SINGER (CGIAR System-wide Information Network for Genetic Resources)

U.N. Statistics Division

WAICENT (FAO World Agricultural Information Center)

Web-agri

WIEWS (FAO World Information and Early Warning System on Plant Genetic Resources)

WTO Statistics

(Continued)

TABLE H.1 *(Continued)*

Discussion Forums of Interest

FAO Electronic Forum on Biotechnology in Food and Agriculture
UNEP Forum on Sustainable Agri-Food Production and Consumption

Seed Magazines

Asian Seed & Planting Material
Genesis (Argentina)
Germination
Seed & Crops
SEED News (Brazil)
The Seed News (Pakistan)
Seed Today
Seed World

Source: Excerpts used with permission from the International Seed Federation.

Appendix I

GS1 Methodology

Below is a summary of commonly used of data carriers for GS1 traceability and GS1 Methodology of Numbering and Identification Systems.

I.1 COMMONLY USED DATA CARRIERS FOR GS1 TRACEABILITY

To achieve traceability of foods within supply chains, it is essential to identify the food items concerned and to provide a seamless facility for maintaining identification of those items from source to consumer. Primary and secondary aspects of identification are necessary to ensure traceability. Primary identification is about identifying the source components whether they are crop-based, fish, or animals. Different techniques are available to determine identification at this level and are primarily deoxyribonucleic acid (DNA) or other molecular-based analytical methods that can be used to identify an individual entity to a reasonable degree of statistical confidence.

A range of data carrier technologies and an even wider range of products and systems are available to support identification at these various levels. The technologies that are considered particularly relevant to the needs for food traceability include:

- Linear bar codes
- Two-dimensional codes (multirow bar, matrix, and composite codes), including direct marking
- Contact memory devices
- Radiofrequency identification (RFID)
- Smart labels (passive and active devices)

For open systems usage, which is regularly the case for food supply chains, it is essential that the identification codes and any additional information relating to the item, such as batch number, weight, volume, other identification numbers, and use or sell-by date, adhere to a particular identification standard.[*]

---

[*] Through standardization, the hierarchy of packaging and item traceability can be better achieved. The GS1 standard provides the necessary coding structures for identification of items and other entities (e.g., location) and also specifies adopted data carriers, presently confined to bar and composite codes but currently pursuing standardization for RFID-based data carriers.

I.2   GS1 METHODOLOGY OF NUMBERING AND IDENTIFICATION SYSTEMS

Numbering for identification purposes essentially involves assigning strings of numbers to denote particular attributes such as company, location, product type, batch, lot, serial number, consignment, and individual item identification where considered appropriate. Other coding features may include quantities such as weight, volume, date, and so forth. The GS1 specifications account for the largest single usage of linear bar codes, not simply on consumer unit packaging but also on transit packaging and pallets at higher levels in supply chain applications. Because information requirements at different stages differ, the International Numbering Association (EAN)-128 standard uses the special data formats distinguished as "application identifiers" for defining the nature of application data, such as a batch number, a "best before" date, order number, and so on.

For traceability support and potentially process support (precision agriculture), location can play a significant role. The identification and coding of location can greatly assist traceability analyses in which rapid mapping of item movement and distribution histories can provide significant assistance in managing food hazards and associated problems. To accomplish this, global positioning systems (GPS) and RTLS are used for locating items. By using GPS-based coordinates for nodal location, mapping of supply chain structures and item movements can be readily achieved for analytical and planning purposes.

In recognizing the need for flexibility in defining traceability systems to satisfy different supply chain needs, it is necessary to identify a range of technologies and associated products to meet these needs that are tailored to data carrying capacity and capabilities to capture data. A sampling of technologies and products are listed below.

Automatic identification and data capture (AIDC) is a term that encapsulates the requirements for achieving traceability but is also a term that denotes an industry that serves a growing user community and a range of technologies and associated principles for automated item-level data acquisition.

Linear bar code data carriers are the most prominent and well established of the AIDC data carrier technologies, having been in widespread use since the early 1970s. They are a familiar sight on products in retail stores and on a wide variety of packaging and containers. They are used widely in manufacturing, asset management, document tracking, access control, warehouse management, and distribution.

Linear bar codes have been used extensively in retail and supply chain logistics for many years as an effective means of machine-readable identification and data transfer. The limitation on data capacity (typically 14–50 characters) and the various ways in which a bar code symbol can be formed determine the way in they can be used and the nature of the information handling systems required to fulfill process needs.[*]

---

[*] At first glance, one bar code may look much like any other. Although simple in concept, the way in which the bars and spaces are structured to carry data in digital form (the 1s and 0s, representing the data encoded) can be somewhat sophisticated. The rules by which they are structured determine to a large extent the type of bar code and the attributes they exhibit. The data-carrying part of the bar code symbol comprises several alternating dark (bar) and light (space) rectangular elements of variable width (some are based on narrow and wide bar/space elements). In addition to the bars and spaces that are used to represent the encoded data other structural features of the symbol can be distinguished that facilitate the reading or scanning process and enhance the security or integrity of the symbol.

- *Code 39 or Code 3 of 9* is one of the most widely used symbology for industrial and non-retail distribution applications.
- *EAN-13, EAN-8, and UPC:* Most retail items found in shops and supermarkets carry what are known as the EAN-13 (13-digit) or EAN-8 (8-digit) types of bar code symbols, structured in accordance with the GS1 symbology specifications. A similar form of EAN-13 and EAN-8 used in the United States and Canada encodes 12 digits and is known as the UPC-12 symbol.
- *EAN Interleaved Two of Five:* The EAN has adopted Interleaved 2 of 5 (ITF for short) as the symbology for coding transit packaging, partly because of the ease with which the symbols can be printed onto packaging materials such as corrugated cardboard. The EAN form of ITF accommodates 14 digits and is often referred to as ITF-14.
- EAN 128 symbology is a means of carrying GS1 numbers and specifying supplementary data (e.g., batch numbers) for product identification purposes.
- *Reduced Space Symbology (RSS)* is a family of three linear symbologies and variants (seven in total) specifically developed to accommodate the GS1 Global Trade Item Number (GTIN) on space-constrained items in which existing linear symbologies could not be used.
- *Composite symbologies* comprise a family of structures in which a GS1 linear symbol is linked to a two-dimensional symbol. The composite components (CCs) support supplementary application identifier data for the linear GS1 component.

I.3 IDENTIFICATION AND TRANSFORMATION IN INTERNAL PROCESSES

The very nature of food production and supply distinguishes transformations in the form of processing, combining, packaging, containerization, and so forth that require identification to facilitate a traceability structure. There is therefore a need for item and item-associated data management that can accommodate the wide-ranging transformation and transaction processes to be found in supply chains. In essence there are two primary categories of transformation that relate to both items and information for traceability purposes that for convenience are here referred to as "cascade and fusion."

A significant feature of any traceability system is the facility for communication and information exchange. Electronic data interchange (EDI) has for some time been applied as a fast and reliable means of achieving electronic, computer-to-computer exchange of information between trading partners with a supply chain legacy based on the use of the EANCOM® language [a subsystem of the EDIFACT (Electronic Data Interchange for Administration, Commerce and Transport)].

I.4 UNIVERSAL DATA CAPTURE PROTOCOL

The data capture appliance is the platform for transferring data from a data carrier or other item-attendant data source to a host repository or information management system. Ideally, the appliance is able to decode and recognize the source data that is captured and to communicate that source data, in an appropriate and possibly translated form, to a host device either in real time or at a later time by request or by operator action to transfer. The data capture appliance may be a linear bar code reader, a two-dimensional code scanner, an RFID interrogator, or any other of a range of data acquisition devices, which now includes sensory and locating devices. The data appliances are designed to read the appropriate data carrier by knowing the way in which the data is structured and conveyed, and what channel encoding data has to be stripped out to derive the source data. For application purposes, the host receiving the source data must know how it is structured in order to use it.

RFID is an important area of data carrier development, with new generation systems and products offering considerable potential for low-cost data carrier applications. RFID covers a range of data carrying technologies for which the transfer of data from the data carrier to host is achieved via a "radiofrequency" link. This contrasts with the touch memory type carriers in which the data transfer is via a conductive pathway.

Sensory devices in food production and distribution applications for data capture include measurements of temperature, pressure, humidity, vibration, biological and chemical agents, and so forth. Other systems include electronic eartags, injectable transponders, Bolus cylindrical tags (swallowed), Smart Labels, and Smart Active Labels (SALs) to shipping container identification identifiers.

Appendix J

U.S. Grains Council Office Locations

Table J.1 (see below) is a listing of U.S. Grains Council offices located worldwide. The U.S. Grains Council has offices in several markets. It is recommended that interested parties contact the council's headquarters in Washington, D.C. as a first point of contact for inquiries related to value-enhanced grains.

TABLE J.1
U.S. Grains Council Office Locations

Egypt
Dr. Hussein Soliman, Director
8 Abd El Rahman El Rafei Street
Floor No. 8, Flat 804, Mohandessin
Cairo, Egypt
Phone: 011-202-3-749-7078 or
011-202-3-761-3193
Fax: 011-202-3-760-7227
E-mail: cairo@grains.org

Japan
Tetsuo Hamamoto, Country Director
Fourth Floor, KY Tameike Building
1-6-19, Akasaka, Minato-ku
Tokyo 107-0052 Japan
Phone: 011-81-3-3505-0601
Fax: 011-81-3-3505-0670
E-mail: tokyo@grains.org

Korea
Byong Ryol Min, Director
303 Leema Building
146-1 Susong-dong, Chongro-ku
Seoul 110-755 Korea
Phone: 011-82-2-720-1891
Fax: 011-82-2-720-9008
E-mail: seoul@grains.org

Mediterranean and Africa
Kurt Shultz, Director
9 bis Avenue Louis Braille, A3
1002 Tunis-Belvedere
Tunis, Tunisia
Phone: 011-216-71-849-622 or
011-216-71-848-054
Fax: 011-216-71-847-165
E-mail: tunis@usgrains.net

Mexico and Central America
Julio Arturo Hernandez, Acting Director
Jaime Balmes No. 8-602 "C"
Col. Los Morales Polanco
Mexico, D.F., Mexico 11510
Phone: 011-52-55-5282-0244 or
011-52-55-5282-0973 or
011-52-55-5282-0977
Fax: 011-52-55-5282-0969
E-mail: mexico@grains.org

Middle East and Subcontinent
Joe O'Brien, Regional Director
Sweifeih, Al-Wakalat Street
Dream Center, Third Floor, Office 301
Amman 11831 Jordan
Phone: 011-962-6585-1254 or

(Continued)

TABLE J.1 *(Continued)*

011-962-6585-2064
Fax: 011-962-6585-4797
E-mail: usgc_jo@wanadoo.jo

People's Republic of China
Cary Sifferath, Senior Director
Unit 10C, 10th Floor China World Tower 1
No. 1 Jianguomenwai Avenue
Beijing 100004 China
Phone: 011-86-10-6505-1314 or
011-86-10-6505-2320
Fax: 011-86-10-6505-0236
E-mail: grainsbj@grains.org.cn

Southeast Asia
Adel Yusupov, Regional Director
Suite 3B-7-3A
Block 3B, Level 7

Plaza Sentral
Jalan Stesen Sentral 5
50470 Kuala Lumpur, Malaysia
Phone: 011-60-3-2273-6826
Fax: 011-60-3-2273-2052
E-mail: usgckl@usgc.com.my

Taiwan
Clover Chang, Director
Seventh Floor
157 Nanking East Road
Section 2
Taipei, Taiwan
Phone: 011-886-2-2508-0176 or
011-886-2-2507-5401
Fax: 011-886-2-2502-4851
E-mail: taipei@grains.org

Source: Information adopted with permission from U.S. Grains Council.

Appendix K

Examples of Novecta Corn Value-Enhanced Grains

The following is a list of Novecta's corn value-enhanced grains (VEGs) that add end-user value.

High-starch corn: Produces extractable starch yields greater than 69–70%. This characteristic improves wet milling plant economics.

High-oil corn: The most common type of nutritionally enhanced corn, typically offering an oil content of 6% or higher. Its primary use is as an ingredient in animal feed, but it is also used to produce corn oil for human consumption. High-oil corn yields are generally competitive with standard yellow dent hybrids.

High total fermentable starch corn: Defined as offering up to 4% greater ethanol yield compared with commodity corn.

Hard endosperm/food-grade corn: Produces hard endosperm for food-grade corn with higher levels of vitreous endosperm and with the pericarp nearly fully intact and easily removed. In dry milling, it yields higher levels of large grits. Also used in alkaline cooking processes to make tortilla chips and snack foods.

Low-phytate or high-available-phosphorous corn: Provides more available phosphorus than standard yellow corn. Its use can reduce the need to add supplemental phosphorus to livestock and poultry rations and reduce the level of phosphorus in livestock and poultry waste. Yields on these types of corn have been lower than conventional yellow dent hybrids.

Nutritionally enhanced corn: This enhanced corn refers to a group of hybrids with protein levels elevated to include more essential amino acids. It is best described as corn with modified feeding qualities developed for specific feed uses. Although primarily a livestock feed, it has some food applications.

High-amylose corn: Also known as amylomaize, high-amylose corn has a higher level of amylose (straight-chain) starch molecules than dent corn. Grown exclusively for the wet milling industry, its primary uses are in textiles, gum candies, biodegradable packaging materials, and adhesives for making corrugated cardboard.

High-lysine/opaque corn: Also known as opaque-2 corn, this corn has higher levels of lysine, an essential amino acid. It is a source of high-quality protein in nonruminant diets and can improve human nutrition in populations with diets high in corn. It is grown to a limited extent as a feed for poultry, swine, dairy cattle, and other livestock production needs.

High-oleic/high-oil corn: This corn offers a different mix of oleic and linoleic acids than standard dent corn. In swine rations, this compositional change will cause fat to be firmer and more stable, thereby extending freshness. Other quality attributes such as appearance and taste are still being evaluated.

Low-stress crack corn: This corn has a low percentage (typically less than 20%) of kernels with internal fissures, making the kernels less susceptible to mechanical damage during handling. Low-stress crack corn retains its grade better during storage and handling and increases processing quality.

Low-temperature dried corn (LTD): Defined by a handling characteristic versus any genetic or hybrid difference, LTD corn is typically field-dried or dried at temperatures less than 120°F. Like low-stress crack corn, it shows fewer cracks, is less susceptible to mechanical damage during handling, and processes better.

Non-GMO corn (GMO = genetically modified organism): Non-GMO corn is any corn hybrid that has not been genetically modified through biotechnology procedures to add a specific characteristic. All modification has occurred through natural breeding.

Nutritionally dense corn: Designed to contain a stacked trait set of genetics specific in nutrient density, quality, and consistency. This product is typically used in feeding animals to increase the efficiency of production.

Organic corn: This is corn grown without pesticides or chemical fertilizers, and the grain is not treated with pesticides in storage. Organic corn is grown for human consumption.

Post-harvest pesticide-free corn: This corn is not treated with pesticides after harvest. It is used as a livestock feed.

Waxy corn: Waxy corn is wet milled to produce specialized starch products for food and industrial uses. A small portion of waxy corn production is used for feed uses, primarily silage for dairy cattle. Waxy corn yields are typically 95–97% of standard yellow dent varieties.

White corn: White corn is primarily used for human consumption, for which its high kernel hardness makes it desirable for dry milling or alkaline processing into cereals, snacks, and Mexican food products. A small amount of white corn is wet milled to produce specialty products with very bright whiteness.

Appendix L

National Laws for Labeling Genetically Modified Foods

Table L.1 (see below) is a listing published by *The Organic & Non-GMO Report* of the national laws for labeling genetically modified (GM) foods. This table lists current or proposed labeling regulations around the world for GM foods.

TABLE L.1

National Laws for Labeling GM Foods

Regions/ Countries	Labels	Coverage and Tolerance	Effective Date
Asia			
China	NE	All foods containing GM content. No tolerance set. (not fully implemented)	2002
Hong Kong	V	All foods containing GM content; 5% tolerance.	NA
Indonesia	NE	Food law calls for labeling, but not yet implemented.	1996
India	NE	Labeling regulations under consideration since introduction of GM cotton in 2002.	NA
Japan	M	Regulations exempt additives, animal feed, and any ingredient representing less than 5% of content. Zero tolerance on unapproved varieties; 1% tolerance on unapproved varieties in animal feed.	2001
Malaysia	NE	Legislative proposal to require labeling, but not passed.	NA
Philippines	V	Labeling laws not passed. Now a voluntary system. GM imports must be declared and deemed safe.	NA
Russia	M	Labeling required on GM foods, similar to EU. 0.9% tolerance. Does not apply to animal feed or U.S. imports.	2004
South Korea	M	Processed foods with GM maize, soybean, or bean sprouts and potatoes; if one of top five ingredients; 3% tolerance. No GMO-free labels for processed products.	2001
Taiwan	M	Bulk and processed foods containing GM maize or soybeans; 5% tolerance.	2003 (bulk), 2004 (processed)
Sri Lanka	M	Proposed labeling regulations 4/2002.	NA

(Continued)

551

TABLE L.1 *(Continued)*

Thailand	M	Labeling if GM ingredient is one of top three ingredients..........................2003
		Applies to foods and raw products containing maize or soy;
		5% tolerance.
Vietnam	M	Transported GMOs must be labeled. NA
Africa		
Algeria	B	Prohibits import and commercialization of GM products. 2000
Angola	B	Ban on the imports and use of GM products except milled......................2004
		food-aid grains.
Benin	B	5-year moratorium on import, commercialization, and use of...................2002
		GM products.
Egypt	NE	Proposed labeling regulations similar to EU...NA
Ethiopia	B	No GM crops currently accepted. 2002
Mauritius	M	GMO bill requires labeling. 2004
Morocco	B	Prohibits imports of GM foods and products..2001
South Africa	M	Proposed labeling regulations June 2001 on the basis of.............................NA
		U.S. system of substantial equivalence.
Zambia	B	Ban on imports of GM products. 2002
Zimbabwe	B	Ban on imports of GM products, except milled maize. 2002

Middle East

Israel	NE	Proposed mandatory labeling in 2002 for maize and soy. ....NA
		products for human consumption; 1% tolerance.
		In 2004, government indicated it would not require labeling.
Saudi Arabia	M	Labeling of food imports, processed foods, GM fruit, ...2003
		vegetables, grains, planting seed; 1% tolerance.
United Arab	NE	Proposed labeling regulations. ....NA
Emirates		

Europe (National)

Albania	B	5-year ban on GM crops and foods.
Austria	B	Maintains ban on GM foods and crops..1997
		Follows EU regulations on labeling and traceability.
Bosnia and	B	Ban on imports of GM products. 2004
Herzegovina		
Bulgaria	M	Plans to adopt EU regulations ....NA
Croatia	M	Bans imports of GM products. Will adopt EU regulations. 2001
Norway	M	Labeling of GM foods. Reduced tolerance from 2% to 2003
		1%. No genetically engineered crops grown commercially.
Serbia/	M	Labeling required but not enforced. 2002
Montenegro		
Switzerland	M	Mandatory labeling of GM foods; 0.9% tolerance. 2005
		0.2% tolerance for non-GM foods.
Turkey	M	Requires GMO-free certification for imports. 2001
United	M	Follows EU regulations on labeling and traceability. 1999
Kingdom		Labeling also required for foods sold in restaurants,
		retail outlets, and caterers.

(Continued)

TABLE L.1 *(Continued)*

European Union[a]	M	Regulations 1829/2003 and 1830/2003 approved in 2004 ....2004 require labeling of all GM food and feed, including processed Derivatives; 0.9% tolerance. Also require complete traceability.

North and South America

Argentina	V	No labels required; voluntary labels allowed...NA
Brazil	M	Government decree requires mandatory labeling ....2004 of food and feed containing more than 1% GM content. Does not include highly processed products.
Chile	M	Labeling legislation proposed July 2000; 2% tolerance 2000
Canada	V	Voluntary labeling standards; 5% tolerance. 2004
Ecuador	NE	Labeling regulation not finalized. ....NA
Mexico	M	Agreement with United States to label foods with more than 5% GM .....2003 content for distributors but not for consumers.
United States	V	GM foods considered to be "substantially equivalent" ....1992 to conventional foods unless nutrition is affected. Voluntary guidelines to label GM and non-GM foods issued January 2001 never finalized.
Venezuela	B	Plantings of GM crops banned by president. 2004

Oceania

Australia and New Zealand	M	GM content in processed foods, fruits, vegetables; 1% tolerance..............2001

M, mandatory; V, voluntary; B, banned; NE, not established; NA, not applicable; EU, European Union; GMO, genetically modified organism.

[a] Includes Austria, Belgium, Cyprus, Czech Republic, Denmark, Estonia, France, Finland, Germany, Greece, Hungary, Ireland, Italy, Latvia, Lithuania, Luxembourg, Malta, The Netherlands, Poland, Portugal, Spain, Sweden, United Kingdom, Czech Republic, Cyprus, Estonia, Hungary, Latvia, Lithuania, Malta, Poland, Slovakia, and Slovenia (Phillips and McNeill, 2000).

Source: Data adopted with permission from *The Organic & Non-GMO Report.*

REFERENCES

Phillips, Peter W.B., and Heather McNeill. "Labeling for GM Foods: Theory and Practice." *AgBio Forum* 3, no. 4, 2000: 219–224.

Appendix M

Why Product Insurance Is Needed and What Is Offered

In an interview between Bernie Steves (vice president of brokerage house IBS and *Food Technology Source* (an Internet publication),* several aspects of product insurance were discussed. The first facet discussed was how closely do IBS consultants work with the insured customer? The use of insurance company consultants is not mandatory, but voluntary. Customers are suggested to take advantage of the resources that the insurance company offers. The costs, fees, and expenses of the consultants are covered under the policy. So consultants can go in at the client's request, walk them through, and offer advice.

Product insurance should lower or limit the risk. Generally speaking, with the globalization of the food industry, liability or recall is going to tend to increase risk with products being processed and manufactured throughout the world. Steves noted that more people want less processed food as opposed to more processed food, which increases their exposure from a contamination standpoint. A lot of times food that has been processed can lower exposure contaminants.

Regarding genetically modified organisms (GMOs), the public, especially in Europe, balks at genetically modified (GM) foodstuffs. An insurer looks at contamination as either accidental contamination or malicious contamination. The difference between the two from an insurer's standpoint is that with accidental contamination the product has to have resulted in, or likely will result in, bodily injury, sickness, disease, or death. So bearing this in mind, it has not been proven that GM products are dangerous to health. It is "just public perception," and given that they would not cause bodily injury between 90 and 125 days, which is the common period you would see on an insurance policy, there would be no recall.

So what exactly does a product recall policy cover? Steves explains that policies generally cover four separate areas. The first is recall expense, which would be the cost to inspect, withdraw, and destroy the bad product; the communications and

* Excerpts and modifications from an interview entitled Why You Need Product Recall Insurance, between Bernie Steves [Vice President, Insurance Brokers Services, Inc. (IBS)] and Food Technology Source. Interview available at http://www.foodtechsource.com/emag/011/trend.htm. IBS is an insurance brokerage firm and provider of product recall insurance. Copyright 2000 with permission from Cole-Parmer Instrument Company.

public relations; and transportation, staff overtime, and additional staff—whatever it takes to get the bad products off the shelf and good products back on the shelf. The second area covers replacement costs of the product that cannot be reused. Certain products can obviously be inspected, and, if not tampered with or contaminated, they can be put back on the shelf. Other products, especially fresh products or dairy products, would have a definite shelf life and would have to be destroyed, so the policy would cover the replacement costs. The third area that policies generally cover is rehabilitation expense and re-establishing the reputation and market share of the product line that was affected. This includes advertising costs, coupons, and so on. Finally, policies generally cover loss of business income for a period of 12 months following an incident. The last aspect of coverage is for the consultants: crisis and public relations consultants to assist the insured in the handling of an incident.

Appendix N

Sudan 1

An example of why product recall insurance is needed—Sudan 1, a short summary of "After Sudan 1: Can Food Firms Afford to Skip Product Recall Insurance?" by Lindsey Partos.[*]

In February of 2005, there was the discovery of Sudan 1, an illegal and potentially carcinogenic red coloring additive, in a consignment of Crosse and Blackwell Worcester sauce made by U.K. manufacturer Premier Foods, which triggered a mass recall in the U.K. food chain of more than 600 processed foods on the shelves that may have contained Sudan 1.

Early estimates of €143 million were figured for the cost of the recall, which included sales loss, destruction, management time, and consultants' fees plus the "softer" costs like brand damage.[†]

Surprisingly, in today's convoluted food chain with its daily exposure to risk, contaminated product insurance is not required by law, although this type of insurance could be an essential way to cover potential vulnerabilities. According to Marcos Garcia Norris, assistant vice president of the crisis management division at AIG Europe, approximately 70% of food and beverage firms do not have this type of insurance. Some policies' minimum premium is approximately £2,000 a year for £1 million of coverage. According to AIG Europe, the average premium ranges between £8,000 and £10,000 for smaller companies.

---

[*] From Partos, L. "After Sudan 1: Can Food Firms Afford to Skip Product Recall Insurance?" *FoodNavigator.Com*, March 2005. Available at http://www.foodqualitynews.com/Food-Alerts/After-Sudan-1-Can-Food-Firms-Afford-To-Skip-Production-Recall-Insurance.

[†] Against the backdrop of the massive Sudan 1 product recall in the United Kingdom, and the resultant cumbersome costs for key firms involved, Lindsey Partos analyzed whether food industry players could afford to opt out of product recall insurance.

Appendix O

OurFood.com Database

Table O.1 (see below) is the table of contents from OurFood's Database of Food and Related Sciences.

TABLE O.1

Database of Food and Related Sciences Table of Contents

OurFood
Database of Food and Related Sciences
Karl Heinz Wilm
E-mail: author@OurFood.com
October 15, 2008

Contents

(Continued)

TABLE O.1 *(Continued)*

Source: Data adopted with permission from Karl Heinz Wilm.

Appendix P

Cost-Benefit Spreadsheet—Complete

TABLE P.1
Cost-Benefit Spreadsheet—Complete

Background Information

Item	Measure Units	Standard	IPT 1	IPT 2	IPT 3	IPT 4
Personal information						
ID number		1	2	3	4	5
Name		Bill Smith				
Address		123 Main St. Ames IA 50014				
Phone number		515-123-4567				
E-mail		isu@iastate.edu				
Other						
Other						
General information						
Crop planted		Soybeans	UL Soybeans	UL Soybeans	UL Soybeans	UL Soybeans
Crop variety planted		DKB 2752				
Purity level (required)	%	N/A	5.0%	2.0%	1.0%	0.1%
Crop acres	acres	200	200	200	200	200
GIS acreage data	N/A					
Grain yield	bu/acre	55	55	55	55	55
Previously planted crop in field		Corn	Corn	Corn	Corn	Corn
Type of IP system		None	Non-GMO	Non-GMO	Non-GMO	Non-GMO
Trait(s) and/or attribute(s) of interest		None	Ultra-low linolenic	Ultra-low linolenic	Ultra-low linolenic	Ultra-low linolenic

(Continued)

TABLE P.1 *(Continued)*

Item	Measure Units	Standard	IPT 1	IPT 2	IPT 3	IPT 4
Other						
Other						
Hourly wage information						
Management	$/hr	$25.00	$25.00	$25.00	$25.00	$25.00
Labor	$/hr	$15.00	$15.00	$15.00	$15.00	$15.00
Meeting, offseason	$/hr	$40.00	$40.00	$40.00	$40.00	$40.00
Contract or hired professional	$/hr	$50.00	$50.00	$50.00	$50.00	$50.00
Other		$0.00	$0.00	$0.00	$0.00	$0.00
Other		$0.00	$0.00	$0.00	$0.00	$0.00
Operating assumptions						
Grain hauling, semi	$/mile	$0.250	$0.250	$0.250	$0.250	$0.250
Interest, carry-on operating money	%/yr	8.00	8.00	8.00	8.00	8.00
Capital interest	%/yr	6.00	6.00	6.00	6.00	6.00
Personal travel mileage	$/mile	$0.500	$0.500	$0.500	$0.500	$0.500
Personal travel meal expense	$/day	$50.00	$50.00	$50.00	$50.00	$50.00
Personal travel overnight expense	$/day	$100.00	$100.00	$100.00	$100.00	$100.00
Other		$0.00	$0.00	$0.00	$0.00	$0.00
Other		$0.00	$0.00	$0.00	$0.00	$0.00
		Revenues				
Production data/revenues						
Selling price-$/bu	$/bu	$8.53	$10.35	$15.00	$18.00	$20.00
Bushels sold	bu	11,000	11,000	11,000	11,000	11,000
Unit sale price	$	$93,830	$113,850	$165,000	$198,000	$220,000
		Costs				
Production data/costs						
Seed cost (total)	$/yr	$40,000	$35,000	$37,500	$40,000	$55,000
Volume purchased	lb	50,000	50,000	50,000	50,000	50,000
Seed $/bu @ 60 lb/bu	$/bu	$48.00	$42.00	$45.00	$48.00	$66.00
Pest management/fertilizer data/costs						
Management-applications	hr/yr	5.0	10.0	10.0	10.0	10.0
Labor-applications	hr/yr	8.0	20.0	20.0	20.0	20.0
Contracted expenses A	$/yr	$10,000	$28,000	$28,000	$28,000	$28,000
Contracted expenses B	$/yr	$0	$0	$0	$0	$0
Contracted expenses C	$/yr	$0	$0	$0	$0	$0
Total contracted expenses	$/yr	$10,000	$28,000	$28,000	$28,000	$28,000
Inputs-purchased products A	$/yr	$8,000	$15,000	$15,000	$15,000	$15,000
Inputs-purchased products B	$/yr	$0	$0	$0	$0	$0
Inputs-purchased products C	$/yr	$0	$0	$0	$0	$0
Total inputs-purchased products	$/yr	$8,000	$15,000	$15,000	$15,000	$15,000

(Continued)

TABLE P.1 *(Continued)*

Item	Measure Units	Standard	IPT 1	IPT 2	IPT 3	IPT 4
Crop value lost to cleanout	$/yr	$0	$500	$550	$650	$1,000
Other	$/yr	$0	$0	$0	$0	$0
Other	$/yr	$0	$0	$0	$0	$0
Total costs	$/yr	$18,245	$44,050	$44,100	$44,200	$44,550
Capital fixed costs						
Original capitalized equipment	$	$10,000	$10,000	$10,000	$10,000	$10,000
Sales value (10% orignial value)	$	$1,000	$1,000	$1,000	$1,000	$1,000
Payback period (yr)	yr	5	5	5	5	5
Interest on capitalized equipment costs	$/yr	$600	$600	$600	$600	$600
IP original capitalized equipment	$	$0	$1,000	$2,000	$5,000	$12,000
Sales value (10% orignial value)	$	$0	$100	$200	$500	$1,200
Payback period (yr)	yr	5	5	5	5	5
IP interest on capitalized costs	$/yr	$0	$60	$120	$300	$720
Other fixed costs	$/yr	$0	$0	$0	$0	$0
Other fixed costs	$/yr	$0	$0	$0	$0	$0
Total capital fixed costs	$/yr	$600	$660	$720	$900	$1,320
Working variable financial costs						
Labor interest costs	$/yr	$66	$101	$139	$203	$319
Purchased inputs interest costs	$/yr	$4,640	$6,240	$6,440	$6,640	$7,840
Storage/transport/marketing/training interest costs	$/yr	$8	$35	$23	$48	$48
Administrative interest costs	$/yr	$48	$48	$48	$48	$48
Certification/validation interest costs	$/yr	$0	$42	$42	$78	$118
Insurance/other interest costs	$/yr	$8	$28	$28	$32	$48
Total variable costs	$/yr	$4,769	$6,494	$6,720	$7,050	$8,421
Postharvest data/costs						
Management-planter cleanout	hr/yr	0.50	1.00	3.00	8.00	16.00
Management-combine cleanout	hr/yr	0.75	2.00	3.00	8.00	16.00
Management-storage, cleanout, maintenance upgrade	hr/yr	1.00	2.00	3.00	5.00	20.00
Labor-planter cleanout	hr/yr	1.50	1.00	3.00	5.00	3.00
Labor-combine cleanout	hr/yr	1.00	2.00	3.00	6.00	6.00
Labor-storage cleanout, maintenance upgrade	hr/yr	0.25	2.00	5.00	8.00	10.00
Other	$/yr	$0.00	$0.00	$0.00	$0.00	$0.00
Other	$/yr	$0.00	$0.00	$0.00	$0.00	$0.00
Total costs	$/yr	$97.50	$200.00	$390.00	$810.00	$1,585.00

(Continued)

TABLE P.1 *(Continued)*

Item	Measure Units	Standard	IPT 1	IPT 2	IPT 3	IPT 4
Postharvest costs						
Management-crop handling and separating	hr/yr	0.3	1.0	2.0	5.0	10.0
Labor-crop handling and separating	hr/yr	0.5	0.0	1.0	5.0	8.0
Other	$/yr	$0.00	$0.00	$0.00	$0.00	$0.00
Other	$/yr	$0.00	$0.00	$0.00	$0.00	$0.00
Total costs	$/yr	$13.75	$25.00	$65.00	$200.00	$370.00
Storage/transport data/costs						
On-farm storage cost	$/yr	$0.00	$10.00	$10.00	$10.00	$10.00
Transport distance to sale	miles	10	25	25	25	25
Off-farm storage cost	$/yr	$50.00	$0.00	$0.00	$0.00	$0.00
Transport distance to off-farm storage	miles	5	0	0	0	0
Transport distance to sale	miles	4	0	0	0	0
Shrinkage cost (estimate)	$/yr	$0.00	$0.00	$0.00	$0.00	$0.00
Other storage/transport costs	$/yr	$0.00	$0.00	$0.00	$0.00	$0.00
Other	$/yr	$0.00	$0.00	$0.00	$0.00	$0.00
Other	$/yr	$0.00	$0.00	$0.00	$0.00	$0.00
Total costs	$/yr	$54.75	$16.25	$16.25	$16.25	$16.25
Administrative data/costs						
Management-documentation hours (total)	hr/yr	1.3	4.0	5.0	6.0	8.0
Labor-documentation hours (total)	hr/yr	0.5	1.0	2.0	2.5	5.0
Office costs	$/yr	$100.00	$100.00	$100.00	$100.00	$100.00
Information technology costs	$/yr	$500.00	$500.00	$500.00	$500.00	$500.00
Other	$/yr	$0.00	$0.00	$0.00	$0.00	$0.00
Other	$/yr	$0.00	$0.00	$0.00	$0.00	$0.00
Total costs	$/yr	$638.75	$715.00	$755.00	$787.50	$875.00
Planning/preparation data/costs						
Management-field/bins inspection (pre-plant)	hr/yr	1.0	0.5	1.0	2.0	4.0
Management-product ID	hr/yr	0.3	0.0	1.0	2.0	3.0
Management-product isolation	hr/yr	0.0	0.0	1.0	2.0	2.0
Management-field/bins inspection	hr/yr	0.0	0.2	1.0	1.0	3.0
Management-documentation (total)	hr/yr	4.0	0.5	1.0	1.5	2.0
Labor-field/bins inspections (pre-plant)	hr/yr	1.0	0.0	1.0	2.0	4.0

(Continued)

TABLE P.1 *(Continued)*

Item	Measure Units	Standard	IPT 1	IPT 2	IPT 3	IPT 4
Labor-product ID	hr/yr	0.3	0.0	1.0	2.0	2.0
Labor-product isolation	hr/yr	0.0	0.0	1.0	2.0	4.0
Labor-field/bins inspection	hr/yr	0.0	0.0	1.0	1.0	4.0
Labor-documentation (total)	hr/yr	5.0	0.5	1.0	1.5	2.0
Other	$/yr	$0	$0	$0	$0	$0
Other	$/yr	$0	$0	$0	$0	$0
Total costs	$/yr	$225.00	$37.50	$200.00	$340.00	$590.00
Marketing/training						
Management-meeting/training	hr/yr	1.0	4.0	4.0	5.0	8.0
Management-meeting travel distance	miles/yr	25	100	100	100	100
Management-meeting meals	days/yr	0.0	2.5	2.5	2.5	2.5
Management-meeting overnights	days/yr	0	1	1	1	1
Management-meeting registration (costs)	$/yr	$25.00	$150.00	$0.00	$25.00	$25.00
Labor-meeting/training	hr/yr	4.0	0.0	1.0	2.0	3.0
Labor-meeting travel distance	miles/yr	10	0	0	100	100
Labor-meeting meals	days/yr	0.0	0.0	0.0	2.5	2.5
Labor-meeting overnights	days/yr	0	0	0	1	1
Labor-meeting registration (costs)	$/yr	$0.00	$0.00	$0.00	$10.00	$10.00
Other	$/yr	$0.00	$0.00	$0.00	$0.00	$0.00
Other	$/yr	$0.00	$0.00	$0.00	$0.00	$0.00
Total costs	$/yr	$242.50	$585.00	$475.00	$865.00	$1,025.00
Insurance costs						
IP insurance costs	$/yr	$100.00	$100.00	$100.00	$100.00	$100.00
Liability costs due to IP	$/yr	$0.00	$250.00	$250.00	$300.00	$500.00
Legal fees due to IP	$/yr	$0.00	$0.00	$0.00	$0.00	$0.00
Other risk mitigation costs	$/yr	$0.00	$0.00	$0.00	$0.00	$0.00
Other	$/yr	$0.00	$0.00	$0.00	$0.00	$0.00
Other	$/yr	$0.00	$0.00	$0.00	$0.00	$0.00
Total costs	$/yr	$100.00	$350.00	$350.00	$400.00	$600.00
Certification/validation data/costs						
Management-other IP tasks	hr/yr	0.0	7.0	7.0	7.0	7.0
Labor-other IP tasks	hr/yr	0.0	0.0	0.0	0.0	0.0
Annual fees	$/yr	$0.00	$25.00	$25.00	$25.00	$25.00
Audit costs	$/yr	$0.00	$500.00	$500.00	$750.00	$1,000.00
Inspection costs	$/yr	$0.00	$0.00	$0.00	$100.00	$200.00
Field test costs	$/yr	$0.00	$0.00	$0.00	$0.00	$0.00

(Continued)

TABLE P.1 *(Continued)*

Item	Measure Units	Standard	IPT 1	IPT 2	IPT 3	IPT 4
Laboratory test costs	$/yr	$0.00	$0.00	$0.00	$100.00	$250.00
Other-contracted hours	hr/yr	1.0	2.0	2.0	2.0	2.0
Other	$/yr	$0.00	$0.00	$0.00	$0.00	$0.00
Other	$/yr	$0.00	$0.00	$0.00	$0.00	$0.00
Total costs	$/yr	$50.00	$800.00	$800.00	$1,250.00	$1,750.00

Summary Results

		Standard	IPT 1	IPT 2	IPT 3	IPT 4
Total revenues		$93,830.00	$113,850.00	$165,000.00	$198,000.00	$220,000.00
Total costs		$65,036.71	$88,933.21	$92,091.51	$96,818.41	$116,102.31
Profit		$28,793.29	$24,916.79	$72,908.49	$101,181.59	$103,897.69
	Acres	200	200	200	200	200
	Yield/acre	55.0	55.0	55.0	55.0	55.0
	Production/bu	11,000	11,000	11,000	11,000	11,000

	Measure		Resource Units per Bushel				
Revenues	Units	Type	Standard	IPT 1	IPT 2	IPT 3	IPT 4
Production data/revenues							
Selling price - $/bu	$/bu	Var	$8.53	$10.35	$15.00	$18.00	$20.00

Cost Analysis

	Measure Units	Type	Standard	IPT 1	IPT 2	IPT 3	IPT 4
Production data/costs							
Seed costs-$/bu	$/bu	Var	$48.00	$42.00	$45.00	$48.00	$66.00
Pest management/fertilizer data/costs							
Management-applications	hr/yr	Var	$0.0114	$0.0227	$0.0227	$0.0227	$0.0227
Labor-applications	hr/yr	Var	$0.0109	$0.0273	$0.0273	$0.0273	$0.0273
Contracted expenses A	$/yr	Var	$0.9091	$2.5455	$2.5455	$2.5455	$2.5455
Contracted expenses B	$/yr	Var	$0.0000	$0.0000	$0.0000	$0.0000	$0.0000
Contracted expenses C	$/yr	Var	$0.0000	$0.0000	$0.0000	$0.0000	$0.0000
Total contracted expenses	$/yr	Var	$0.9091	$2.5455	$2.5455	$2.5455	$2.5455
Inputs-purchased products A	$/yr	Var	$0.7273	$1.3636	$1.3636	$1.3636	$1.3636
Inputs-purchased products B	$/yr	Var	$0.0000	$0.0000	$0.0000	$0.0000	$0.0000
Inputs-purchased products C	$/yr	Var	$0.0000	$0.0000	$0.0000	$0.0000	$0.0000
Total inputs-purchased products	$/yr	Var	$0.7273	$1.3636	$1.3636	$1.3636	$1.3636
Crop value lost to cleanout	$/yr	Var	$0.0000	$0.0455	$0.0500	$0.0591	$0.0909
Other	$/yr	Var	$0.0000	$0.0000	$0.0000	$0.0000	$0.0000
Other	$/yr	Var	$0.0000	$0.0000	$0.0000	$0.0000	$0.0000
Total costs	$/yr	Var	$1.6586	$4.0045	$4.0091	$4.0182	$4.0500

(Continued)

TABLE P.1 *(Continued)*

Revenues	Measure Units	Type	Resource Units per Bushel				
			Standard	IPT 1	IPT 2	IPT 3	IPT 4
Capital fixed costs							
Original capitalized equipment	$	Fix	$0.9091	$0.9091	$0.9091	$0.9091	$0.9091
Sales value (10% original value)	$	Fix	$0.0909	$0.0909	$0.0909	$0.0909	$0.0909
Payback period (yr)	yr	Fix					
Interest on capitalized equipment costs	$/yr	Fix	$0.0545	$0.0545	$0.0545	$0.0545	$0.0545
IP original capitalized equipment	$	Fix	$0.0000	$0.0909	$0.1818	$0.4545	$1.0909
Sales value (10% original value)	$	Fix	$0.0000	$0.0091	$0.0182	$0.0455	$0.1091
Payback period (yr)	yr	Fix					
IP interest on capitalized costs	$/yr	Fix	$0.0000	$0.0055	$0.0109	$0.0273	$0.0655
Other fixed costs	$/yr	Fix	$0.0000	$0.0000	$0.0000	$0.0000	$0.0000
Other fixed costs	$/yr	Fix	$0.0000	$0.0000	$0.0000	$0.0000	$0.0000
Total capital fixed costs	$/yr	Fix	$0.0545	$0.0600	$0.0655	$0.0818	$0.1200
Working variable financial costs							
Labor interest costs	$/yr	Var	$0.0060	$0.0092	$0.0126	$0.0185	$0.0290
Purchased inputs interest costs	$/yr	Var	$0.4218	$0.5673	$0.5855	$0.6036	$0.7127
Storage/transport/marketing/ training interest costs	$/yr	Var	$0.0007	$0.0032	$0.0021	$0.0044	$0.0044
Administrative interest costs	$/yr	Var	$0.0044	$0.0044	$0.0044	$0.0044	$0.0044
Certification/validation interst costs	$/yr	Var	$0.0000	$0.0038	$0.0038	$0.0071	$0.0107
Insurance/other interest costs	$/yr	Var	$0.0007	$0.0026	$0.0026	$0.0029	$0.0044
Total variable costs	$/yr	Var	$0.4336	$0.5904	$0.6109	$0.6409	$0.7656
Postharvest data/costs							
Management-planter cleanout	$/yr	Var	$0.0011	$0.0023	$0.0068	$0.0182	$0.0364
Management-combine cleanout	$/yr	Var	$0.0017	$0.0045	$0.0068	$0.0182	$0.0364
Management-storage, cleanout, maintenance upgrade	$/yr	Var	$0.0023	$0.0045	$0.0068	$0.0114	$0.0455
Labor-planter cleanout	$/yr	Var	$0.0020	$0.0014	$0.0041	$0.0068	$0.0041
Labor-combine cleanout	$/yr	Var	$0.0014	$0.0027	$0.0041	$0.0082	$0.0082
Labor-storage cleanout, maintenance, upgrade	$/yr	Var	$0.0003	$0.0027	$0.0068	$0.0109	$0.0136
Other	$/yr	Var	$0.0000	$0.0000	$0.0000	$0.0000	$0.0000
Other	$/yr	Var	$0.0000	$0.0000	$0.0000	$0.0000	$0.0000
Total costs	$/yr	Var	$0.0089	$0.0182	$0.0355	$0.0736	$0.1441
Postharvest costs							
Management-crop handling and separating	$/yr	Var	$0.0006	$0.0023	$0.0045	$0.0114	$0.0227

(Continued)

TABLE P.1 *(Continued)*

			Resource Units per Bushel				
Revenues	Measure Units	Type	Standard	IPT 1	IPT 2	IPT 3	IPT 4
Labor-crop handling and separating	$/yr	Var	$0.0007	$0.0000	$0.0014	$0.0068	$0.0109
Other	$/yr	Var	$0.0000	$0.0000	$0.0000	$0.0000	$0.0000
Other	$/yr	Var	$0.0000	$0.0000	$0.0000	$0.0000	$0.0000
Total costs	$/yr	Var	$0.0013	$0.0023	$0.0059	$0.0182	$0.0336
Storage/transport data/costs							
On-farm storage cost	$/yr	Var	$0.0000	$0.0009	$0.0009	$0.0009	$0.0009
Transport distance to sale	miles	Var	$0.0002	$0.0006	$0.0006	$0.0006	$0.0006
Off-farm storage cost	$/yr	Var	$0.0045	$0.0000	$0.0000	$0.0000	$0.0000
Transport distance to off-farm storage	miles	Var	$0.0001	$0.0000	$0.0000	$0.0000	$0.0000
Transport distance to sale	miles	Var	$0.0001	$0.0000	$0.0000	$0.0000	$0.0000
Shrinkage cost (estimate)	$/yr	Var	$0.0000	$0.0000	$0.0000	$0.0000	$0.0000
Other storage/transport costs	$/yr	Var	$0.0000	$0.0000	$0.0000	$0.0000	$0.0000
Other	$/yr	Var	$0.0000	$0.0000	$0.0000	$0.0000	$0.0000
Other	$/yr	Var	$0.0000	$0.0000	$0.0000	$0.0000	$0.0000
Total costs	$/yr	Var	$0.0050	$0.0015	$0.0015	$0.0015	$0.0015
Administrative data/costs							
Management-documentation hours (total)	hr/yr	Var	$0.0028	$0.0091	$0.0114	$0.0136	$0.0182
Labor-documentation hours (total)	hr/yr	Var	$0.0007	$0.0014	$0.0027	$0.0034	$0.0068
Office costs	$/yr	Var	$0.0091	$0.0091	$0.0091	$0.0091	$0.0091
IT costs	$/yr	Var	$0.0455	$0.0455	$0.0455	$0.0455	$0.0455
Other	$/yr	Var	$0.0000	$0.0000	$0.0000	$0.0000	$0.0000
Other	$/yr	Var	$0.0000	$0.0000	$0.0000	$0.0000	$0.0000
Total costs	$/yr	Var	$0.0581	$0.0650	$0.0686	$0.0716	$0.0795
Marketing/training							
Management-meeting/training	hr/yr	Var	$0.0036	$0.0091	$0.0091	$0.0114	$0.0182
Management-meeting travel distance	miles/yr	Var	$0.0011	$0.0045	$0.0045	$0.0045	$0.0045
Management-meeting meals	days/yr	Var	$0.0000	$0.0114	$0.0114	$0.0114	$0.0114
Management-meeting overnights	days/yr	Var	$0.0000	$0.0091	$0.0091	$0.0091	$0.0091
Management-meeting registration (costs)	$/yr	Var	$0.0023	$0.0136	$0.0000	$0.0023	$0.0023
Labor-meeting/training	hr/yr	Var	$0.0145	$0.0000	$0.0014	$0.0027	$0.0041
Labor-meeting travel distance	miles/yr	Var	$0.0005	$0.0000	$0.0000	$0.0045	$0.0045
Labor-meeting meals	days/yr	Var	$0.0000	$0.0000	$0.0000	$0.0114	$0.0114
Labor-meeting overnights	days/yr	Var	$0.0000	$0.0000	$0.0000	$0.0091	$0.0091

(Continued)

TABLE P.1 *(Continued)*

Revenues	Measure Units	Type	Standard	IPT 1	IPT 2	IPT 3	IPT 4
				Resource Units per Bushel			
Labor-meeting registration (costs)	$/yr	Var	$0.0000	$0.0000	$0.0000	$0.0009	$0.0009
Other	$/yr	Var	$0.0000	$0.0000	$0.0000	$0.0000	$0.0000
Other	$/yr	Var	$0.0000	$0.0000	$0.0000	$0.0000	$0.0000
Total costs	$/yr	Var	$0.0220	$0.0532	$0.0432	$0.0786	$0.0932
Insurance costs							
IP insurance costs	hr/yr	Fix	$0.0091	$0.0091	$0.0091	$0.0091	$0.0091
Liability costs due to IP	hr/yr	Var	$0.0000	$0.0227	$0.0227	$0.0273	$0.0455
Legal fees due to IP	hr/yr	Var	$0.0000	$0.0000	$0.0000	$0.0000	$0.0000
Other risk mitigation costs	hr/yr	Var	$0.0000	$0.0000	$0.0000	$0.0000	$0.0000
Other	hr/yr	Var	$0.0000	$0.0000	$0.0000	$0.0000	$0.0000
Other	hr/yr	Var	$0.0000	$0.0000	$0.0000	$0.0000	$0.0000
Total costs	hr/yr	Var	$0.0091	$0.0318	$0.0318	$0.0364	$0.0545
Certification/validation data/costs							
Management-other IP tasks	hr/yr	Var	$0.0000	$0.0159	$0.0159	$0.0159	$0.0159
Labor-other IP tasks	hr/yr	Var	$0.0000	$0.0000	$0.0000	$0.0000	$0.0000
Annual fees	$/yr	Fix	$0.0000	$0.0023	$0.0023	$0.0023	$0.0023
Audit costs	$/yr	Var	$0.0000	$0.0455	$0.0455	$0.0682	$0.0909
Inspection costs	$/yr	Var	$0.0000	$0.0000	$0.0000	$0.0091	$0.0182
Field test costs	$/yr	Var	$0.0000	$0.0000	$0.0000	$0.0000	$0.0000
Laboratory test costs	$/yr	Var	$0.0000	$0.0000	$0.0000	$0.0091	$0.0227
Other-contracted hours	hr/yr	Var	$0.0045	$0.0091	$0.0091	$0.0091	$0.0091
Other	$/yr	Var	$0.0000	$0.0000	$0.0000	$0.0000	$0.0000
Other	$/yr	Var	$0.0000	$0.0000	$0.0000	$0.0000	$0.0000
Total costs	$/yr	Var	$0.0045	$0.0727	$0.0727	$0.1136	$0.1591
Total revenues per bushel			$8.53	$10.35	$15.00	$18.00	$20.00
Total costs per bushel			$5.9124	$8.0848	$8.3720	$8.8017	$10.5548
Profit per bushel			$2.6176	$2.2652	$6.6280	$9.1983	$9.4452
IP insurance costs	hrs/yr	Fix	$0.0091	$0.0091	$0.0091	$0.0091	$0.0091
Liability Costs Due to IP	hrs/yr	Var	$0.0000	$0.0227	$0.0227	$0.0273	$0.0455
Legal Fees Due to IP	hrs/yr	Var	$0.0000	$0.0000	$0.0000	$0.0000	$0.0000
Other Risk Mitigation Costs	hrs/yr	Var	$0.0000	$0.0000	$0.0000	$0.0000	$0.0000
Other	hrs/yr	Var	$0.0000	$0.0000	$0.0000	$0.0000	$0.0000
Other	hrs/yr	Var	$0.0000	$0.0000	$0.0000	$0.0000	$0.0000
Total Costs	hrs/yr	Var	$0.0091	$0.0318	$0.0318	$0.0364	$0.0545

Source: Data developed through research.

Appendix Q

Questionnaire Spreadsheet Data

TABLE Q.1

Farm Survey Data

Participant	#1	#2	#3	#4	#5	#6	#7	#8
Years growing IP	7	15	7	2	10+	12	4	5
Wages ($)	55	30	30	25	20–25	23.50	20	10
Acres	None	400	199	100	600	300	287	340
Previous crop		Corn	Corn	Corn	Corn	Corn	Corn	Corn
Standard								
Seed cost ($)	26/acre	14,440	20.42	Roundup Ready 25		32.00	8,250	9.00
Pest cost ($)	14.90/acre	38,400	37.44	25		45.00	4,500	50
Total revenue ($)	439/acre	160,000	46/acre			350	105,000	293
Storage ($)	10.01/acre	2,000	0	0.20		0.35/bu	1,000	10
Transport ($)	0.14/bu	5,500	0.03			0.05/bu	2,000	3
Insurance ($)	100	6,000	Same			15/acre	1,800	7
IP								
Seed cost ($)	18.50/acre	6,800	31.50	17LL		37	5,500	9
Pest cost ($)	39.45/acre	34,800	37.44	40		55	11,800	50
Total revenue ($)	413/acre	140,000	61/acre			390/acre	112,500	306
Storage ($)	28.32/acre	1,700	0.12/bu	0.20		0.35/bu	1,500	10
Transport ($)	0.25/bu	5,000	0.06			0.25/bu	2,400	9
Insurance ($)	100	6,000	Same			15/acre	1,800	7

(Continued)

TABLE Q.1 *(Continued)*

Participant	#1	#2	#3	#4	#5	#6	#7	#8
Type	Ultra-low linolenic		Vistive NuTech Roundup Ready/ Pioneer seed beans		1% non-GMO/ seed	Seed beans		Non-GMO Soybean
ISO	Yes	No	No	No	Yes	No	Yes	No
Hours								
Preparation	49	10	1	2	4.00	5	4	2
Documentation	4	4	1	3	20	4	2	2
Planter	8	12	1 per variety	2	3	2	1.5	1
Management	7	10	4	-	2	21	4	1
Inspection	24	5	5	6	2	1	4	0
Combine	24	4	3 at $100	3	4	12	1	2
Handling	40	10	3	6	3	4	4	1
Managerial	5	5	2	-	10	1	6	1
Meeting hours	12	8	3	8		4	6	-
Miles	150	300	140	300		100	100	-
Overnights	0	1	0	-		-	0	-
Mileage rate ($)	0.33	0	0.35	0.40		0.35	0.40	-
Lodging/meals ($)	60	200	25			-	-	-
Other ($)	125	-	-			50.00	500	0

IP, identity preservation; GMO, genetically modified organism.

Source: Data developed through research.

Related Products, Services, and Organizations

Table 1 includes genetically modified organism (GMO) testing labs and test kit manufacturers, identity preservation (IP) firms, organic certification organizations, sustainable agriculture research organizations, associations and membership organizations, consultants, information management products, trade publications, and manufacturers of products related to producing sustainable non-GMO and organic crops (Non-GMO Sourcebook 2006).

TABLE 1
Related Products, Services, and Organizations

United States

Acres USA
P.O. Box 91299
Austin, TX 78709
• Phone: 512-892-4400 Fax: 512-892-4448
• E-mail: info@acresusa.com
• Website: http://www.acresusa.com
• Products/Services: Publications on eco-agriculture, distribution

Advanced Biological Marketing
P.O. Box 222
Van Wort, OH 45891
• Phone: 877-617-2461 Fax: 419-232-4664
• E-mail: abm@abm1st.com
• Website: http://www.abm1st.com
• Products/Services: Chemicals, seed treatments

AgraQuest
1530 Drew Avenue
Davis, CA 95616
• Phone: 800-962-8980 Fax: 530-750-0153
• E-mail: info@agraquest.com
• Website: http://www.agraquest.com
• Products/Services: Natural, and effective crop protection products. OMRI-listed

Agriculture Utilization Research Institute
P.O. Box 599
Crookston, MN 56716-0599
• Phone: 218-281-7600 Fax: 218-281-3759
• E-mail: lgjersvi@auri.org
• Website: http://www.auri.org
• Products/Services: Technical assistance for value-added products

AgriEnergy Resources
21417 1950 East Street
Princeton, IL 61356
• Phone: 815-872-1190 Fax: 815-872-1928
• E-mail: info@agrienergy.net
• Website: http//www.agrienergy.net
• Products/Services: Biological fertilizers

AGRIS Corporation
3820 Mansell Road, Suite 300
Alpharetta, GA 30022
• Phone: 800-795-7995 Fax: 770-238-5205
• E-mail: info@agris.com
• Website: http://www.agris.com
• Products/Services:Management and operations systems, including binSight for bin management, tracking, and traceability

(Continued)

573

TABLE 1 *(Continued)*

AgriSystems International
125 West Seventh Street
Wind Gap, PA 18091
• Phone: 610-863-6700 Fax: 610-863-4622
• E-mail: agrisys1@aol.com
• Website: http://www.agrisystemsinternational.
  com
• Products/Services: Consultants for certified
  organic products industry

AgVision Software for Agribusiness
1601 North Ankeny Boulevard
Ankeny, IA 50023
• Phone: 800-759-9492 Fax: 515-964-0473
• Website: http://www.agvisionsoftware.com
• Products/Services: Inventory management
  processing, conditioning, and accounting
  software for agricultural businesses

AIB International
1213 Bakers Way
Manhattan, KS 66505-3999
• Phone: 800-633-5137 ext. 193 Fax: 785-537-
  1493
• E-mail: molewnik@aibonline.org
• Website: http://www.aibonline.org
• Products/Services: IP services, audits and
  consulting, grain-related quality systems, and
  food safety audits

AOCS
2211 West Bradley Avenue
Champaign, IL 61821
• Phone: 217-359-2344 Fax: 217-351-8091
• E-mail: technical@aocs.org
• Website: http://www.aocs.org
• Products/Services: GMO testing, certification,
  reference materials, collaborative study organizers,
  international representatives, and publisher

Applied Genetics, Inc.
1900 Seventeenth Avenue S.
Brookings, SD 57006
• Phone: 605-691-9388 Fax: 605-697-7484
• E-mail: appliedgene@brookings.net
• Website: http://www.appliedgenetics.com
• Products/Services: GMO tests, PCR/DNA,
  ELISA, genetic profiling, and food safety testing

Arkansas State Plant Board–Seed Division
P.O. Box 1069
Little Rock, AR 72203
• Phone: 501-225-1598 Fax: 501-225-7213
• E-mail: mary.smith@aspb.ar.gov
• Website: http://www.plantboard.org/seed_cert8.
  html
• Products/Services: Identity preserved, quality
  assurance program for seed

BioDiagnostics, Inc.
507 Highland Drive
River Falls, WI 54022
• Phone: 715-426-0246 Fax: 715-426-0251
• E-mail: info@biodiagnostics.net
• Website: http://www.biodiagnostics.net
• Products/Services: Genetic purity, trait purity,
  germination, DNA, ELISA, analytical
  chemistry

Biogenetic Services, Inc.
801 32nd Avenue
Brookings, SD 57006
• Phone: 605-697-8500 Fax: 605-697-8507
• E-mail: biogene@brookings.net
• Website: http://www.biogeneticservices.com
• Products/Services: PCR and ELISA
  qualitative and quantitative methods to
  detect GMOs

Biogenic Enterprises
2545 Roanoke Avenue
Fredericksburg, IA 50630
• Phone: 563-237-5998 Fax: 563-237-5937
• E-mail: adhark@iowatelecom.net
• Products/Services: Certified organic live plant
  enzymes, pre-mixes, and soil products;
  consulting and distribution

BioProfile Testing Labs, LLC
2010 E. Hennepin Avenue, Suite 3-125
Minneapolis, MN 55413
• Phone: 651-428-8176 Fax: 612-378-1676
• E-mail: info@bioprofilelabs.com
• Website: http://www.bioprofilelabs.com
• Products/Services: GMO testing, ingredient
  testing

TABLE 1 *(Continued)*

California Crop Improvement Association
Parsons Seed Certification Center, UCD
1 Shields Way, Davis, CA 95616
• Phone: 530-752-0544 Fax: 530-752-4735
• E-mail: rfstewart@ucdavis.edu
• Website: http://www.ccia.ucdavis.edu
• Products/Services: IP for non-GMO, organic
crop & processing certification, AOSCA IP
program

California Seed & Plant Lab
7877 Pleasant Grove Road
Elverta, CA 95626
• Phone: 916-655-1581 Fax: 946-655-1582
• E-mail: randhawa@calspl.com
• Website: http://www.calspl.com
• Products/Services: Seed pathology, variety
identification, GMO testing

Canton Mills, Inc.
P.O. Box 97, 160 Mill Street
Minnesota City, MN 55959
• Phone: 800-328-5349, 507-689-2131
• Fax: 507-689-2400
• Products/Services: Shur-Gro natural organic and
sustainable fertilizers, diatomaceous earth, etc.

CCOF
1115 Mission Street
Santa Cruz, CA 95060
• Phone: 831-423-2263 Fax: 831-423-4528
• E-mail: ccof@ccof.org
• Website: http://www.ccof.org
• Products/Services: Third-party certification
agency for organic processors, growers,
retailers, & wholesalers

Cert ID
501 Dimick Drive
Fairfield, IA 52556
• Phone: 641-472-9979 Fax: 641-472-9198
• E-mail: info-na@certi-id.com
• Website: http://www.cert-id.com
• Products/Services: Cert-ID, non-GMO
certification

CII Laboratory Services
10835 Ambassador Drive
Kansas City, MO 64153
• Phone: 816-891-7337 Fax: 816-891-7450
• E-mail: ciisvc@ciilab.com
• Website: http://www.ciilab.com
• Products/Services: Grain testing & food
products

Critereon Company
21024 421st Avenue
Iroquois, SD 57353
• Phone: 605-546-2299 Fax: 605-546-2503
• E-mail: info@critereon.com
• Website: http://www.critereon.com
• Products/Services: System design, consulting
and administrative support; IP services; tracking
systems, and auditing

CropChoice
P.O. Box 33811, Washington, DC 20033
• Phone: 202-328-1209 Fax: 202-463-0862
• E-mail: editor@cropchoice.com
• Website: http://www.cropchoice.com
• Products/Services: Stories and briefs on GMO
issues, sustainable farming, corporate
agriculture, and trade policy

CropVerifeye, LLC
7311 W. Jefferson Boulevard
Ft. Wayne, IN 46804
• Phone: 866-432-3663 Fax: 260-459-7747
• E-mail: khockney@cropverifeye.com
• Website: http://www.cropverifeye.com
• Products/Services: Data management and
traceability, IP, and compliance for agri-food
markets

Custom Marketing Co.
1126 West Main Avenue
West Fargo, ND 58078-1311
• Phone: 800-359-1785
• Website: http://www.custommarketingco.com
• Products/Services: Pressure cure drying and
storage management for grain

(Continued)

TABLE 1 *(Continued)*

DePaul Industries
2738 North Hayden Island Drive
Portland, OR 97217
• Phone: 503-331-3822 Fax: 503-288-6514
• E-mail: lfletcher@depaulindustries.com
• Website: http://www.depaulindustries.com
• Products/Services: Organic co-packing services

Diversified Laboratory Testing, LLC
5205 Quincy Street
Mounds View, MN 55112
• Phone: 763-785-0484 Fax: 763-785-0584
• Website: http://www.dqcc.com
• Products/Services: GMO testing

DL Crank & Associates
707 Lake Street
Alexandria, MN 56308
• Phone: 320-763-2470
• E-mail: dcrank@rea-alp.com
• Contact: Don Crank
• Products/Services: Soy product and business
development

DRAMM Corporation
P.O. Box 1960,
Manitowoc, WI 54221
• Phone: 920-684-0227 Fax: 920-684-4499
• E-mail: fish@dramm.com
• Website: http://www.fishfertilizer.com
• Products/Services: OMRI listed organic fish
hydrolysate, six different blends; additives, etc.

EcoSmart Technologies, Inc.
318 Seaboard Lane, Suite 208
Franklin, TN 37067
• Phone: 800-723-3991 Fax: 615-261-7301
• E-mail: keden@ecosmart.com
• Website: http://www.ecosmart.com
• Products/Services: Crop protection products,
NOP compliant, OMRI listed

EnviroLogix Inc.
500 Riverside
Industrial Parkway
Portland, ME 04103
• Phone: 866-408-4597 Fax: 207-797-7533
• E-mail: info@envirologix.com
• Website: http://www.envirologix.com
• Products/Services: GMO and mycotoxin test kits

Eurofins Genescan, Inc.
2315 North Causeway Boulevard
Metairie, LA 70001
• Phone: 504-297-4330 Fax: 504-297-4335
• E-mail: gmo@gmotesting.com
• Website: http://www.gmotesting.com
• Products/Services: GMO testing, PCR and
ELISA methods, non-GM and IP certification

The Fertrell Company
P.O. Box 265,
Bainbridge, PA 17502
• Phone: 717-367-1566 Fax: 717-367-9319
• E-mail: theresia@fertrell.com
• Website: http://www.fertrell.com
• Products/Services: Organic fertilizer, soil
conditioners, natural supplements; sustainable
agricultural program and consulting

Genetic ID
P.O. Box 1810
Fairfield, IA 52556
• Phone: 641-472-9979 Fax: 641-472-9198
• E-mail: info@genetic-id.com
• Website: http://www.genetic-id.com
• Products/Services: GMO testing and consulting
services

Global Organic Alliance, Inc.
P.O. Box 530, 3185 Road 179
Bellefontaine, OH 43311
• Phone: 937-593-1232 Fax: 937-593-9507
• E-mail: kananen@logan.net
• Website: http://www.goa-online.org
• Products/Services: Organic certification
worldwide

Grain Journal
3065 Pershing Court
Decatur, IL 62526
• Phone: 217-877-9660 Fax: 217-877-6647
• E-mail: mark@grainnet.com
• Website: http://www.grainnet.com
• Products/Services: Trade magazine serving the
grain industry

TABLE 1 *(Continued)*

Illinois Crop Improvement Association, Inc.
P.O. Box 9013
Champaign, IL 61826
• Phone: 217-359-4053 Fax: 217-359-4075
• E-mail: dmiller@ilcrop.com
• Website: http://www.ilcrop.com
• Products/Services: Non- GMO testing,
  food-grade corn, NIR composition analysis, IP

Independent Organic Inspectors Association
P.O. Box 6
Broadus, MT 59317
• Phone: 406-436-2031 Fax: 406-436-2031
• E-mail: ioia@ioia.net
• Website: http://www.ioia.net
• Products/Services: Organic services and
  publications

Indiana Crop Improvement Association
7700 Stockwell Road
Lafayette, IN 47909
• Phone: 765-523-2535l Fax: 765-523-2536
• E-mail: icia@indianacrop.org
• Website: http://www.indianacrop.org
• Products/Services: Non-GMO testing, non-GMO
  certification and IP services for seed-food ISO
  9001-2000 regulations

Innovia Films, Inc.
1950 Lake Park Drive
Smyrna, GA 30080
• Phone: 770-970-8598 Fax: 770-970-8702
• E-mail: malcolm.cohn@innoviafilms.com
• Website: http://www.innoviafilms.com
• Products/Services: Nature-Flex film packaging
  sourced from renewable, non-GMO resources

Institute for Responsible Technology
P.O. Box 469
Fairfield, IA 52556
• Phone: 641-209-1765
• E-mail: info@responsibletechnology.org
• Website: http://www.responsibletechnology.org
• Products/Services: Educational materials on
  health and environmental risks of GMOs;
  books, videos, and CDs

Integrity Certified International
806 East Ohio Street
Lenox, IA 50851
• Phone: 800-815-7852 Fax: 641-333-2280
• E-mail: crayhon@ll.net
• Products/Services: Organic certification services

International Certification Services, Inc.
301 5th Avenue SE
Medina, ND 58467
• Phone: 701-486-3578 Fax: 701-486-3580
• E-mail: info@ics-intl.com
• Website: http://www.ics-intl.com
• Products/Services: Organic certification

Iowa Crop Improvement Association
4611 Mortensen Road Suite 101
Ames, IA 50014
• Phone: 515-294-6921 Fax: 515-294-1897
• E-mail: iowacrop@agron.iastate.edu
• Website: http://www.agron.iastate.edu/icia
• Products/Services: Seed certification and quality
  assurance, IP production inspection, and
  documentation services

**Iowa State University Extension–Value-Added
  Agriculture Program**
167 Heady Hall
Ames, IA 50011-1017
• Phone: 515-294-1938 Fax: 515-294-9496
• E-mail: ctordsen@iastate.edu
• Website: http://www.extension.iastate.edu/pages/
  valag
• Products/Services: Facilitate development of
  value-added agri-business, including feasibility
  studies, marketing and training

Iowa State University Seed Testing Lab
128 Seed Science Center
Ames, IA 50011
• Phone: 515-294-0117 Fax: 515-294-8303
• E-mail: curry@iastate.edu
• Website: http://www.seeds.iastate.edu/seedtest
• Products/Services: GMO testing with ELISA
  and PCR methods

(Continued)

TABLE 1 *(Continued)*

Iowa Testing Laboratories, Inc.
1101 North Iowa Avenue
P.O. Box 188
Eagle Grove, IA 50533
• Phone: 800-274-7645, 515-448-4741
• Fax: 515-448-3402
• E-mail: jack@iowatestinglabs.com
• Website: http://www.iowatestinglabs.com
• Products/Services: An independent lab providing
  analytical services to the agriculture industry for
  60 years

ISO-Ag
1421 Grand Avenue
Keokuk, IA 52632
• Phone: 319-524-3399 Fax: 319-524-3399
• E-mail: isoag@hotmail.com
• Website: http://www.iso-ag.com
• Products/Services: ISO 9000 implementation
  training for farmers

John Deere Ag Management Solutions
4140 Northwest 114th Street
Urbandale, IA 50322
• Phone: 515-331-4705
• E-mail: culpgordonj@johndeere.com
• Website:http://www.deere.com/en_us/ag/
  servicesupport/ams/index.html
• Products/Services: Automatic guidance,
  field documentation farm management software

Juneau Sales/AC Greenfix
17399 240th Street SE
Red Lake Falls, MN 56750
• Phone: 218-698-4222, 866-546-9297
• Fax: 218-698-4440
• E-mail: juneaufarms@gvtel.com
• Website: http://www.acgreenfix.com;
  http://www.calcium25.com
• Products/Services: AC Greenfix seed, organic
  remond salt, diatomaceous earth, foliar nutrient
  spray

Kamut Association
P.O. Box 6447
Great Falls, MT 59406
• Phone: 800-644-6450 Fax: 406-452-7175
• E-mail: debby@kamut.com
• Website: http://www.kamut.com
• Products/Services: Information service for
  Kamut brand products

Kansas Crop Improvement Association
2000 Kimball Avenue
Manhattan, KS 66502
• Phone: 785-532-6118
• Website: http://www.kscrop.org
• Products/Services: Seed certification, IP, quality
  assurance programs, organic inspections

Michael Fields Agricultural Institute
W2493 County Road SE
East Troy, WI 53120
• Phone: 262-642-3303 Fax: 262-642-4028
• E-mail: rdoetch@michaelfieldsaginst.org
• Website: http://www.michaelfieldsaginst.org
• Products/Services: Sustainable/organic outreach/
  education

Michigan Crop Improvement Association
P.O. Box 21008
Lansing, MI 48909
• Phone: 517-332-3546 Fax: 517-332-9301
• E-mail: info@michcrop.com
• Website: http://www.michcrop.com
• Products/Services: AOSCA IP/non-GMO
  field inspections, non-GM products bio-assay
  lab tests, and quality assurance

Midwest Organic Services Association
P.O. Box 344
Viroqua, WI 54665
• Phone: 608-637-2526 Fax: 608-637-7032
• E-mail: mosa@mosaorganic.org
• Website: http://www.mosaorganic.org
• Products/Services: Organic certification
  of producers, processors, and handlers
  throughout the United States NOP/ISO 65

Mid-West Seed Services
236 32nd Avenue
Brookings, SD 57006
• Phone: 605-692-7611 Fax: 605-692-7617
• E-mail: info@mwseed.com
• Website: http://www.mwseed.com
• Products/Services: Seed testing, AP/GMO and
  trait testing

TABLE 1 *(Continued)*

Midwest Shippers Association
400 South 4th Street, Suite 852
Minneapolis, MN 55415
• Phone: 612-252-1453 Fax: 612-339-5673
• E-mail: info@mnshippers.org
• Website: http://www.mnshippers.org
• Products/Services: Grower/processor specialty
  (IP) grain cooperative; variety of specialty grains

Minnesota Crop Improvement Association
1900 Hendon Avenue
St. Paul, MN 55108
• Phone: 612-625-7766 Fax: 612-625-3748
• E-mail: mncia@tc.umn.edu
• Website: http://www.mncia.org
• Products/Services: Certified seed, IP grain, and
  organic. Other services: field inspections, quality
  assurance and marketing services

Missouri Crop Improvement Association
3211 Lemone
Industrial Boulevard
Columbia, MO 65201
• Phone: 573-449-0586 Fax: 573-874-3193
• E-mail: moseed@aol.com
• Website: http://www.moseed.org
• Products/Services: IP, non-GMO, quality
  assurance and source-identified inspection,
  auditing, and testing programs

Missouri Enterprise Business Assistance Center
800 University Drive Suite 111
Rolla, MO 65401
• Phone: 573-364-8570 Fax: 573-364-6323
• E-mail: bthompson@missourienterprise.org
• Website: http://www.missourienterprise.org
• Products/Services: Technical assistance to
  manufacturers, and farmers for value-added
  enterprises

Natural Food Certifiers
119A South Main Street
Spring Valley, NY 10977
• Phone: 845-426-5098 Fax: 845-818-3598
• E-mail: natfcert@aol.com
• Website: http://www.nfccertification.com
• Products/Services: Organic, kosher, and vegan
  food certifier; USDA agent for NOP

Neogen Corporation
620 Lesher Place
Lansing, MI 48912
• Phone: 800-234-5333 Fax: 517-372-2006
• E-mail: mnichols@neogen.com
• Website: http://www.neogen.com
• Products/Services: GMO, mycotoxin/microbial
  test kits

New Mexico Organic Commodity Commission
4001 Indian School NE Suite 310
Albuquerque, NM 87110
• Phone: 505-841-9070 Fax: 505-841-9080
• E-mail: joan.quinn@state.nm.us
• Website: http://nmocc.state.nm.us
• Products/Services: Organic certification,
  education, and marketing assistance

North Dakota State Seed Department
P.O. Box 5257, 1313 18th Street N
Fargo, ND 58105
• Phone: 701-231-5400 Fax: 701-231-5401
• E-mail: ndseed@state-seed.ndsu.nodak.edu
• Website: http://www.ndseed.com
• Products/Services: Seed certification, IP,
  seed testing, seed quality, health, and
  GMO testing

Northern Plains Sustainable Agriculture Society
9824 79th Street SE
Fullerton, ND 58441
• Phone: 701-883-4304 Fax: 701-883-4304
• E-mail: tpnpsas@drtel.net
• Website: http://www.npsas.org
• Products/Services: Educational resources on
  organic systems

NSF International
789 North Dixboro Road
Ann Arbor, MI 48105
• Phone: 734-769-8010 Fax: 734-769-0109
• E-mail: donofrio@nsf.org
• Website: http://www.nsf.org
• Products/Services: Third-party certification and
  GMO testing

(Continued)

TABLE 1 *(Continued)*

Ohio Ecological Food & Farm Association
9665 Kline Road
West Salem, OH 44287
• Phone: 419-853-4060 Fax: 419-853-3022
• E-mail: organic@oeffa.com
• Website: http://www.oeffa.com
• Products/Services: Organic certification

Ohio Seed Improvement Association
6150 Avery Road, Box 477
Dublin, OH 43017
• Phone: 614-889-1136 Fax: 614-889-8979
• E-mail: osia@ohseed.org
• Website: http://www.ohseed.org
• Products/Services: Non- GMO field inspection,
lab testing, record keeping, and labeling

OMIC USA, Inc.
3344 North West Industrial Street
Portland, OR 97210
• Phone: 503-223-1497 Fax: 503-223-9436
• E-mail: labmgr@omicusa.com
• Website: http://www.omicusa.com
• Products/Services: GMO Testing, ISO certified

Oregon Department of Agriculture
635 Capital Street NE
Salem, OR 97301-2532
• Phone: 503-986-4620 Fax: 503-986-4737
• E-mail: jcramer@oda.state.or.us
• Website: http://www.oda.state.or.us/cid/identity
• Products/Services: Oregon IP Program for
producers/packers track products from "farm to
fork"

Oregon Tilth
470 Lancaster Drive NE
Salem, OR 97301
• Phone: 503-378-0690 Fax: 503-378-0809
• E-mail: pete@tilth.org
• Website: http://www.tilth.org
• Products/Services: Non-profit research for
sustainable agriculture and education.
Organic certification and membership
organizations

Organic Certifiers
6500 Casitas Pass Road
Ventura, CA 93001
• Phone: 805-684-6494 Fax: 805-684-2767
• E-mail: organic@west.net
• Website: http://www.organiccertifiers.com
• Products/Services: Organic certification,
accredited by NOP, IFOAM, ISO 65

Organic Consumers Association
6771 South Silver Hill Drive
Finland, MN 55603
• Phone: 218-353-7454 Fax: 218-353-7652
• Website: http://www.organicconsumers.org
• Products/Services: Advocacy for organic
integrity watchdogs

**Organic Crop Improvement Association
International, Inc.**
6400 Cornhusker Highway, Suite125
Lincoln, NE 68507
• Phone: 402-477-2323 Fax: 402-407-4325
• E-mail: info@ocia.org
• Website: http://www.ocia.org
• Products/Services: International organic
certification services, education and research
services

**Organic Crop Improvement Association
Northeast Wisconsin Organic Chapter, LLC**
N5364 Hemlock Lane
Kewauness, WI 54216
• Phone: 920-388-4369 Fax: 920-388-3408
• E-mail: kkinstetter@itol.com
• Products/Services: Offering organic certification
services to farmers throughout the Midwest

**Organic Crop Improvement Association
Wisconsin Chapter #1, Inc.**
5381 Norway Drive
Pulaski, WI 54162
• Phone: 920-822-2629 Fax: 920-822-4583
• E-mail: mnmsgang@netnet.net;
johnsonorganics@hotmail.com
• Products/Services: Organic certification

TABLE 1 *(Continued)*

Organic Farming Research Foundation
P.O. Box 440
Santa Cruz, CA 95061
• Phone: 831-426-6606 Fax: 831-426-6670
• E-mail: info@ofrf.org
• Website: http://www.ofrf.org
• Products/Services: Funds organic research and
advocates public support of organic research.

Organic Materials Review Institute
P.O. Box 11558
Eugene, OR 97440
• Phone: 541-343-7600 Fax: 541-343-8971
• E-mail: info@omri.org
• Website: http://www.omri.org
• Products/Services: Publishes information on
materials used in organic production. List of
commercial products

Organic Trade Association
P.O. Box 547
Greenfield, MA 01302
• Phone: 413-774-7511 Fax: 413-774-6432
• E-mail: info@ota.com
• Website: http://www.ota.com
• Products/Services: Membership organization
serving the organic industry

Origins, LLC
33 Lynwood Drive
Battle Creek, MI 49015-7911
• Phone: 259-441-7280 Fax: 419-844-1263
• E-mail: joe.colyn@juno.com
• Website: http://www.originz.net
• Products/Services: Consulting services on
strategies for food systems and a healthier world

Q Laboratories
1400 Harrison Avenue
Cincinnati, OH 45214-1606
• Phone: 513-471-1300 Fax: 513-471-5600
• E-mail: mgoins@qlaboratories.com
• Website: http://www.qlaboratories.com
• Products/Services: Full service, independent,
microbiology, and analytical chemistry lab
services.

Quality Assurance International
9191 Towne Center Drive, Suite 510
San Diego, CA 92122
• Phone: 858-792-3531 Fax: 858-797-8665
• E-mail: ellen@qai-inc.com
• Website: http://www.qai-inc.com
• Products/Services: Organic certification services
for growers, processors, traders, distributors, and
restaurants

Quality Certification Services
P.O. Box 12311
Gainesville, FL 32604
• Phone: 352-377-0133 Fax: 352-377-8363
• E-mail: qcs@qcsinfo.org
• Website: http://www.qcsinfo.org
• Products/Services: Certification services

The Rodale Institute
611 Siegfriedale Road
Kutztown, PA 19530-9320
• Phone: 610-683-1400 Fax: 610-683-8548
• E-mail: info@rodaleinst.org
• Website: http://www.rodalesinstitute.org
• Products/Services: Global; to achieve
regenerative food systems to improve
environmental/human health

**Richard E. Schell, Law Offices of Kurt A.
Wagner**
P.O. Box 3, 780 Lee Street, Suite 102
Des Plaines, IL 60016
• Phone: 847-759-9833 847-404-2950
• Fax: 847-635-0558
• E-mail: schell@wagneruslaw.com
• Website: http://www.wagneruslaw.com
• Products/Services: Legal counsel

Seed Savers Exchange
3094 N. Winn Road
Decorah, IA 52101
• Phone: 563-382-5990 Fax: 563-382-5872
• Website: http://www.seedsavers.org
• Products/Services: Membership organization
offering seeds of heirloom garden crops

(Continued)

TABLE 1   *(Continued)*

SGS North America
1019-1025 Harbor
Memphis, TN 38113
• Phone: 901-775-1660   Fax: 901-775-3308
• E-mail: sandy_holloway@sgs.com
• Website: http://www.sgs.com
• Products/Services: GMO testing, analytical
   testing services

Richard D. Siegel Law Offices
1400 16th Street NW
Washington, DC 20036
• Phone: 202-518-6364   Fax: 202-234-0399
• E-mail: rsiegel@ofwlaw.com
• Products/Services: Legal counsel and federal
   government representative for organic food and
   seed companies and certifiers

Silliker, Inc.
900 Maple Road
Homewood, IL 60430
• Phone: 708-957-7878   Fax: 708-957-1483
• E-mail: info@silliker.com
• Website: http://www.silliker.com
• Products/Services: GMO testing, technical
   consultation, plant/supplier audits

Soyatech, Inc.
1369 State Highway 102
Bar Harbor, ME 04609
• Phone: 207-288-4969 800-424-SOYA
• Fax: 207-288-5264
• E-mail: peter@soyatech.com
• Website: http://www.soyatech.com
• Products/Services: *Soy & Oilseed Bluebook*
   directory, Soyatech eNews Daily, market studies,
   consulting

Soy Works Corporation
3805 Vardon Court
Woodridge, IL 60517
• Phone: 630-853-4328
• E-mail: soyworks@msn.com
• Website: http://www.soyworkscorporation.com
• Products/Services: Marker pellets for bulk
   commodities

Star Dairy Resources
14035 Marsh Pike
Hagerstown, MD 21742
• Phone: 301-739-2025   Fax: 301-739-7029
• Products/Services: Consultant for dairy herd
   rations and soil consultant

Strategic Diagnostics, Inc.
111 Pencader Drive
Newark, DE 19702-3322
• Phone: 800-544-8881   Fax: 302-456-6782
• E-mail: sales@sdix.com
• Website: http://www.sdix.com
• Products/Services: Develops and sells a range
   of test kits for detection of GMO, mycotoxins,
   and pathogens

Dennis Strayer & Associates
302 Beverly Boulevard
Hudson, IA 50643
• Phone: 319-988-4187   Fax: 319-988-3922
• E-mail: dstrayer@prairieinet.net
• Products/Services: Consultant– IP analysis

The Synergy Company of Utah, LLC
2279 South Resource Boulevard
Moab, UT 84532
• Phone: 435-259-4787   Fax: 435-259-2328
• E-mail: "Contact Us" on website
• Website: http://www.thesynergycompany.com
• Products/Services: Certified organic, whole-
   food raw materials and contract
   manufacturing services. Kosher
   certified

Trade Acceptance Group, Ltd.
One Corporate Plaza, Suite 414
7400 Metro Boulevard
Edina, MN 55439
• Phone: 952-830-0064   Fax: 952-830-9054
• E-mail: curt@tradeacceptance.com
• Website: http://www.tradeacceptance.com
• Products/Services: Trade credit insurance and
   export finance

TABLE 1   *(Continued)*

University of Wisconsin—Center for Integrated Agricultural Systems
1450 Linden Drive, Madison, WI 53706
• Phone: 608-262-5200 Fax: 608-265-3020
• E-mail: mmmille6@wisc.edu
• Website: http://www.wisc.edu/cias
• Products/Services: Research, education, networking, and facilitation

Juliet A Zavon, Consulting
433012 McMillan Street
Cincinnati, OH 45219
• Phone: 513-333-0688
• E-mail: JulietZavon@fuse.net
• Products/Services: Consulting - supply chain management: analyzing and valuations, feasibility analysis

Canada

Annual Guelph Organic Conference
Box 116
Collingwood, Ontario L9Y 3Z4
• Phone: 705-444-0923 Fax: 705-444-0380
• E-mail: organix@georgian.net
• Website: http://www.guelphorganicconf.ca
• Products/Services: Annual (Jan. 27-29, 2006) international organic food and farming conference. Covers GMO issues

Greenpeace Canada
454 Laurier Est
Montréal, Quebec H2J 1E7
• Phone: 514-933-0021, ext. 15
• Fax: 514-933-1017
• E-mail: eric.darier@yto.greenpeace.org
• Website: http://www.greenpeace.ca
• Products/Services: Non-GMO foods shoppers' guide on the web

CFT Corporation
2020 Winston Park Drive, Suite 300
Oakville, Ontario L6H 6X7
• Phone: 800-561-8238 Fax: 905-829-5219
• E-mail: spocklington@cftcorp.com
• Website: http://www.cftcorp.com
• Products/Services: International freight forwarding company specializing in shipping commodities—IP grains

Grotek Manufacturing, Inc.
9850 201st Street
Langley, British Columbia V1M 4A3
• Phone: 604-882-7699 Fax: 604-882-7659
• E-mail: fonda@grotek.net
• Website: http://www.grotek.net
• Products/Services: Manufacturers of Earth Safe 100% organic and organic-based fertilizers

CSI (Centre for Systems Integration)
240 Catherine Street, Suite 200
Ottawa, Ontario K2P 2G8
• Phone: 613-236-6451 Fax: 613-236-7000
• E-mail: csi-east@storm.ca
• Website: http://www.csi-ics.com
• Products/Services: Organic certification of farms and processors (NOP, JAS, EEC Regulation No. 2092/91)

Manna International, Inc.
116 Industrial Park Crescent
Sault Ste., Marie, Ontario P6B 5P2
• Phone: 705-946-2662 Fax: 705-256-6540
• E-mail: intnl@sympatico.ca
• Website: http://www.mannainternationalinc.com
• Products/Services: IP/traceability consulting; certification, auditing, and sourcing service

Garantie Bio-Ecocert
71 St-Onésine
Levis, Quebec G6V 5Z4
• Phone: 418-838-6941 Fax: 418-838-9823
• E-mail: info@garantiebio-ecocert.qc.ca
• Website: http://www.garantiebio-ecocert.com
• Products/Services: Organic certification

Norseman, Inc.
21 Keppler Crescent
Ottawa, Ontario K2H 5Y1
• Phone: 613-829-4378 Fax: 613-721-2168
• E-mail: wid@norseman.ca
• Website: http://www.norseman.ca
• Products/Services: Protector liners for food-grade shipping containers

(Continued)

TABLE 1 *(Continued)*

OCPP/Pro-Cert Canada, Inc.
1099 Monarch Road
Lindsay, Ontario K9V 4R1
• Phone: 877-867-4264, 705-374-5602
• Fax: 705-374-5604
• E-mail: ocpp@lindsaycomp.on.ca
• Website: http://www.ocpro-certcanada.com
• Products/Services: Certification for organic
  producers and processors

**Organic Producers Association of Manitoba
  Cooperative, Inc.**
101-247 Wellington Street, W
P.O. Box 940
Virden, Manitoba R0M 2C0
• Phone: 204-748-1315 Fax: 204-748-6881
• E-mail: info@opam.mb.ca
• Website: http://www.opam.mb.ca
• Products/Services: Organic certification,
  accredited to SCC, USDA

QMI Organic, Inc.
4167-97 Street, 2nd Floor
Edmonton, Alberta T6E 6E9
• Phone: 800-268-7321 Fax: 780-496-2464
• E-mail: clawrence@qmi.com
• Website: http://www.qmiorganic.com
• Products/Services: Organic certification services

Québec Vrai
390 Principale
Ste-Monique, Quebec J0G 1N0
• Phone: 819-289-2666 Fax: 819-289-2999
• E-mail: info@quebecvrai.org
• Website: http://www.quebecvrai.org
• Products/Services: Organic certification

Asia/Australia

AgriQuality Pty. Ltd.
3-5 Lillee Crescent
Tullamarine, Victoria 3043 Australia
• Phone: 61-3-8318-9018 Fax: 61-3-8318-9001
• Website: http://www.agriquality.com
• Products/Services: GMO testing for
  the Australasian food and agriculture
  industries

China Certification and Inspection Group
Tower B Number 9 East Madian Road Haidian
  District, Beijing 100088 China
• Phone: 86-10-8226-2829
• E-mail: inspect@ccic.com
• Website: http://www.ccic.com
• Products/Services: Certification services
  of non-GMO/IP, non-GMO control, and
  GMO testing services

Hong Kong DNA Chips
1/F, Cosmos Center
108 Soy Street, Mongkok
Kowloon, Hong Kong SAR, China
• Phone: 852-2111-2123 Fax: 852-2111-9762
• E-mail: info@dnachip.com.hk
• Website: http://www.dnachip.com.hk
• Products/Services: GMO testing, non-GMO
  certification, and non-GMO supply chain
  consultation

NASAA
P.O. Box 768
Stirling, SA 5152 Australia
• Phone: 61-8-8370-8455 Fax: 61-8-8370-8381
• E-mail: enquiries@nasaa.com.au
• Website: http://www.nasaa.com.au
• Products: Organic certification (USDA national
  organic program, JAS)

TABLE 1 *(Continued)*

Europe

Cert ID, Ltd.
Vesey House High Street, Sutton Coldfield
West Midlands, B72 1XH United Kingdom
• Phone: 44-121-321-1777
• Fax: 44-121-321-2999
• E-mail: info-uk@cert-id.com
• Website: http://www.certi-id.com
• Products/Svcs: Non-GMO certification, IP
services

Consumers International
24 Highbury Crescent
London N5 1RX United Kingdom
• Phone: 44-20-7226-6663 Fax: 44-20-7354-0607
• E-mail: dcuming@consint.org
• Website: http://www.consumersinternational.
org/gm
• Products/Services: Federation of consumer
organizations dedicated to consumers' rights
worldwide

DB Information Systems
9 Station Road
Adwick-le-Street
Doncaster, DN6 7DB United Kingdom
• Phone: 44-1302-330837
• Fax: 44-1302-724731
• E-mail: david.trueman@dbis.biz
• Website: http://www.dbis.biz
• Products: CommTrac software provides
traceability, IP procedures and compliance with
quality-assurance requirements

Genetic ID (Europe) AG
Am Mittleren Moos 48
Augsburg D-86167 Germany
• Phone: 49-821-747-7630 Fax: 49-821-747-7639
• E-mail: info-europe@genetic-id.com
• Website: http://www.genetic-id.com
• Products/Services: GMO testing

IdentiGEN, Ltd.
Unit 9 Trinity Enterprise, Centre, Pearse Street
Dublin 2, Ireland
• Phone: 353-1-677-0221 Fax: 353-1-677-0220
• E-mail: gmtesting@identigen.com
• Website: http://www.identigen.com
• Products/Services: DNA meat traceability,
GMO testing, and DNA food
diagnostics

SGS Netherlands
Malledijk 18 - P.O. Box 200
NL-3200 Spijkenisse, Netherlands
• Phone: 31-181-693297 Fax: 31-181-693572
• E-mail: sgs.nl.agro@sgs.com
• Website: http://www.sgs.nl
• Products/Services: Non- GMO/IP auditing, lab
and certification

Soil Association
Bristol House, 40-56 Victoria Street
Bristol B51 6BY United Kingdom
• Phone: 44-117-314-5000 Fax: 44-117-314-5001
• E-mail: info@soilassociation.org
• Website: http://www.soilassociation.org
• Products/Services: Independent, nonprofit
organization that sets organic standards,
supports organic United Kingdom farmers

(Continued)

TABLE 1   *(Continued)*

South America

Argencert SRL
B. de Irigoyen 972 - piso 4 - Of, "B"
Buenos Aires, C1072ATT Argentina
- Phone: 54-11-4363-0033 Fax: 54-11-4363-0202
- E-mail: argencert@argencert.com.ar
- Website: http://www.argencert.com.ar
- Products/Services: Organic certification, NOP certification, GMO-free certification

GeneScan do Brasil Ltda
Av Antonio Gazzola, 1001
3º Andar
Itu, SP 13 301 245 Brazil
- Phone: 55-11-4023-0522 Fax: 55-11-4023-0625
- E-mail: info@genescan.com.br
- Website: http://www.genescan.com.br
- Contact: Pablo Molloy
- Products/Services: GMO analysis and IP services

Source: Data adopted with permission from *The Organic & Non-GMO Report*

Glossary of Terms

The terms below are a consolidation and refinement of like-terms originating from various standards and organizational sources. Each term will generally have a number or set of numbers following it. These numbers represent the source(s) document from which it came. Terms without numbers after them originate from various dictionaries or articles.

TERMS

Accreditation—a process of vouching for the fulfillment of requirements, to certify as meeting requirements, usually by a third party. Also, a determination made by a sovereign authority (often the "Secretary" position) that authorizes a private, foreign, or state entity to conduct certification activities as a certifying agent under sanction or jurisdiction. 10, 11

Act—The Organic Foods Production Act of 1990, as amended (7 U.S.C. 6501 et seq.). 11

Action level—the limit at or above which the Food and Drug Administration will take legal action against a product to remove it from the market. Action levels are based on unavoidability of the poisonous or deleterious substances and do not represent permissible levels of contamination where it is avoidable. 11

Active ingredient—in any pesticide product, the component that kills, or otherwise controls, target pests. Pesticides are regulated primarily on the basis of active ingredients. 5

Advance Shipment Notice (ASN)—also referred to as a Ship Notice/Manifest, the ASN is a communication (normally via electronic means and known as the EDI transaction sets 856 or 857) of the contents, ship date, and time of an expected shipment. The ship notice/manifest enables the receiver or retailer to identify short shipments before receipt and plan warehouse receiving more efficiently. 9

Adventitious pollen—in this usage pollen, which is not inherent, but accidental; is acquired; it is pollen out of place, coming from an outside source. *Adventitious pollen intrusion* describes pollen coming from surrounding, undesirable sources. *Adventitious presence*, in the case of non-GMO production, this refers to the accidental or unintended introduction or presence of genetically modified (GM) material, or foreign genetic material from another variety, crop, or weed, in a seed or grain shipment or ingredients into a non-GMO product line. This can happen, for example, during processing, shipping, the mislabeling of lot numbers, and improper cleaning of equipment. 1, 4, 11

Aflatoxin—a highly carcinogenic natural toxin produced by a fungus (*Aspergillus flavus*), which occurs when crops are grown, but more often stored under warm, humid conditions. Most commonly associated with corn, peanuts, and soybeans. Shipments of grain containing high levels of aflatoxin are generally rejected. 11

Agricultural inputs—all substances or materials used in the production or handling of agricultural products. 10

Agricultural product/product of agricultural origin—any agricultural commodity or product, whether raw or processed, including any commodity or product derived from livestock, that is marketed for human consumption (excluding water, salt, and additives) or livestock consumption. 10

Allowed synthetic—a substance that is included on the National List of synthetic substances allowed for use in organic production or handling (USDA NOP). 10

Animal drug—any drug as defined in section 201 of the Federal Food, Drug, and Cosmetic Act, as amended (21 U.S.C. 321), that is intended for use in livestock, including any drug intended for use in livestock feed but not including such livestock feed. 10

Antibodies—proteins that are produced by an organism in response to exposure to foreign substances called antigens, or neutralizing proteins generated in reaction to foreign proteins in the blood and that produce immunity against certain microorganisms or their toxins. Antigens are most often foreign proteins. Antibodies to an antigen are very specific for that antigen and usually bind very tightly to the antigen. The ability of antibodies to bind strongly and specifically to antigens can be used as the basis for qualitative and quantitative assays (*Enzyme-linked immunoabsorbent assay* ELISA). 1, 11

Application identifier (AI)—the field of two or more characters at the beginning of an element string that uniquely defines its format and meaning. They are predefined numbers enclosed by parentheses used in the EAN.UCC-128 barcode symbol to delineate additional information about the item. 6

Application—a group of software programs that provides functionality for the business (examples are General Ledger, Order Entry, Inventory, Quality Control, etc.).

Applicator—a person applying potentially harmful chemicals, such as pesticides, herbicides, fungicides, fertilizers, and certain industrial chemicals. 8

Area of operation—the types of operations: crops, livestock, wild-crop harvesting or handling, or any combination thereof that a certifying agent may be accredited to certify under this part. 10

Assay—qualitative, quantitative, or semiquantitative analysis of a substance to determine its components. 8

Attribute—a measurable characteristic, or trait that differentiates from similar products (percentage of oil or starch content); or a piece of information reflecting a characteristic related to an identification number (e.g., Global Trade Item Number [GTIN], SSCC). 9, 11

Audit or audit trail—a process of certifying a process by a third party; a systematic and functionally independent examination to determine whether activities and related results comply with planned objectives; see ISO 9000:2000, a systematic and functionally independent examination to determine whether quality and food safety activities and results comply with planned procedures and whether these procedures are implemented effectively and are suitable to achieve objectives, such as for documentation that is sufficient to determine the source, transfer of ownership, and transportation of any agricultural product labeled as "100 percent organic," the organic ingredients of any agricultural product labeled as "organic" or "made with organic (specified ingredients)" or the organic ingredients of any agricultural product containing less than 70 percent organic ingredients identified as organic in an ingredients statement. 5, 7, 10, 11

Authorized inspector—inspectors used to carry out on-site audits and inspections. Individuals assigned to these tasks are usually either employees or specially selected and trained to conduct this type of work. They meet certain criteria laid down in specified standard to receive written standard approval as authorized inspectors. All authorized certification inspectors working for the agent should be demonstrably impartial and independent evaluators of client compliance within specified standards. 3

Bacillus thuringiensis (Bt)—a naturally produced soil bacterium that produces toxins that are deadly to some insects; a type of bacteria commonly sprayed by organic farmers as a natural insecticide. When ingested by certain insects, the bacterium secretes an endotoxin that ruptures the insect's mid gut, causing it to die. Different forms of *Bt* are effective against insects of the orders Lepidoptera (a group of certain caterpillars, moths, and butterflies), Coleoptera (beetles, e.g., Colorado potato beetles), and Diptera (flies and mosquitoes). By the nature of its action, *Bt* is believed to be harmless to mammals, birds, fish, and certain beneficial insects. Through genomic research, seed breeders are able to insert the gene sequence, giving rise to the endotoxin into the DNA of certain plants, such as corn and cotton, producing a natural insecticide; a group of rod-shaped soil bacteria found all over the earth that produce "cry" proteins, which are indigestible by, yet still "bind" to, specific insects' gut lining receptors, so those "cry" proteins are toxic to certain classes of insects (i.e., corn borers). 1, 8, 11

Barcode—the array of bars and spaces representing data. The combination of symbol characters and features required by a particular symbology, including quiet zones, start and stop characters, data characters, check characters, and other auxiliary patterns that together form a complete scannable entity. Also known as the barcode symbol.

Batch—a batch unites trade items that have undergone the same transformation processes. 6

Benchmark—a measurable set of variables used as a baseline or reference in evaluating the performance of Quality Schemes. 5

Bill of Lading (BOL)—a document that establishes the terms of a contract between a shipper and a transportation company. It serves as a document of title, a contract of carriage, and a receipt for goods. 9

Biodiversity—assemblage of living organisms from all sources including terrestrial, marine, and other aquatic ecosystems and the ecological complexes of which they are part; the number and types of organisms in a region or environment. 1, 5

Bioengineering—the technique of removing, modifying, or adding genes to a chromosome to change the information it contains. By changing this information, genetic engineering changes the type or the amount of proteins an organism is capable of making. 11

Biosafety protocol (Convention on Biological Diversity)—the international treaty governing the conservation and use of biological resources around the world that was signed by more than 150 countries at the 1992 United Nations Conference on Environment and Development. 11

Biotechnology—a set of biological techniques developed through basic research and now applied to research and product development. In particular, the use of recombinant DNA techniques; the science of using living things, such as plants or animals, either to develop new products or to make modifications to existing products. Current methods include the transfer of a gene from one organism to another. The application of the techniques of molecular biology and/or recombinant DNA technology, or in vitro gene transfer, to either develop products or impart specific capabilities to organism. 1, 11

Blending—the process of drawing measured amounts of different lots of cultivars from bins and mixing these parts into a uniform blend by grain assemblers and millers. 11

Breeder seed—a class of certified seed that is produced and directly controlled by the originating or sponsoring plant breeding institution, firm, or individual and is the source for the production of seed of the other classes of certified seed. The seed may occur from natural selection or through systematic plant breeding programs and shall be grown/handled to maintain its original genetic purity and identity. 11, 13

Buffer zone—an area located between a certified production operation, or portion of a production operation, and an adjacent land area that is not maintained under organic management. A buffer zone must be sufficient in size or other features (e.g., windbreaks or a diversion ditch) to prevent the possibility of unintended contact by prohibited substances applied to adjacent land areas with an area that is part of a certified operation; the region near the border of a protected area; a transition zone between areas managed for different objectives. 5, 10

Calibration—a measurement of the uncertainty degree of the machinery used to apply any product. Set of operations that establish, under specified conditions, the relationship between values of quantities indicated by measuring instrument and the corresponding values realized by standards. 5

Canola, "Canada Oil"—a strain of the rapeseed plant with a low level of toxic erucic acid (a monounsaturated omega-9 fatty acid, denoted 22:1 ω-9) used by Canadian breeders to produce oil used for cooking. 11

Carrier—party that provides freight transportation services. A physical or electronic mechanism that carries data. 9

Cereals—members of the grass family in which the seed is the most important part used for food and feed. 11

Certification body—a body that is responsible for verifying that a product sold or labelled as "organic" is produced, processed, prepared, handled, and imported according to these guidelines. 7

Certification or certified—a determination made by a certifying agent that a production or handling operation is in compliance with the Act and the regulations in this part, which is documented by a certificate of organic operation; all those actions leading to the issuing of a certificate in terms EN45011/ ISO Guide 65 Product Certification; the procedure by which official certification bodies or officially recognized certification bodies provide written or equivalent assurance that foods or food control systems conform to requirements. Certification of food may be, as appropriate, based on a range of inspection activities that may include continuous on-line inspection, auditing of quality assurance systems, and examination of finished products; to certify something means to ensure or to confirm that something is true. Certification means that, as a certification body, it is assured that the systems or products certified comply with the criteria that are laid down in a written standard. 4, 5, 7, 10

Certified operation—a crop or livestock production, wild-crop harvesting or handling operation, or portion of such operation that is certified by an accredited certifying agent as using a system of organic production or handling as described by the Act and the regulations in this part. 10

Certified seed—the progeny of breeder, foundation, and registered seed. It shall be handled to maintain its genetic purity and identity. 11, 13

Certifying agent—any entity accredited by the Secretary or recognized authority as a certifying agent for the purpose of certifying a production or handling operation as a certified production or handling operation. 10

Chain of custody—an unbroken trail of acceptability that ensures the physical security of data, records, and/or samples. Also, a process used to maintain and document the chronological history of the evidence. 5

Channeling—a recently coined term used to describe the process of maintaining commodities in separate market channels. For producers, channeling means having a contract for specialized grain, especially bioengineered varieties, before the seed is planted. For elevators, it means taking responsibility to keep that grain separate; to monitor its movement and make sure it goes only to approved markets. Channeling systems may not have the strict traceability that is part of identity-preserved systems. 11

Chromatography—a technique for separating complex mixtures of chemicals or proteins into their various constituents; any of various techniques for the separation and measurement of complex mixtures that rely on the differential affinities of substances for certain gases or liquid mobile media and for a stationary adsorbing medium through which they pass, such as paper, gelatin, magnesia, or a form of silica. 1, 8

Claim—any representation that states, suggests, or implies that a food has particular qualities relating to its origin, nutritional properties, nature, processing, composition, or any other quality; oral, written, implied, or symbolic representations, statements, or advertising or other forms of communication presented to the public or buyers of agricultural products that relate to, for example, the organic certification process or the term "100 percent organic," "organic," or "made with organic (specified ingredients or food group[s])," or, in the case of agricultural products containing less than 70 percent organic ingredients, the term, "organic," on the ingredients panel. 7, 10

Client—used in description of a business that is subject to certification. In contrast to this, a certified client usually supplies products to his or her industrial customers. Retail or supermarket chains typically do not need certification for products or as producers because even their own private label products are manufactured by established food manufacturers. These, however, may wish to obtain certification for the benefit of their private consumer customers. 3

Commingle—the act of bringing together or mixing of more than one lot of grain; physical contact between unpackaged organically produced and non-organically produced agricultural products during production, processing, transportation, storage, or handling, other than during the manufacture of a multi-ingredient product containing both types of ingredients. 10, 11

Commoditize—a word formed from commodity, which describes the movement of an item with low volume of trade that moves into higher volume and becomes like a commodity. 11

Commodity—something bought and sold, in agriculture usually a common item of grains and oilseeds as opposed to specialty grains and oilseeds, crops used for general uses rather than special uses. 11

Commodity mindset—the frame of mind that deals with commodities, that bases all trade off of commodities; as opposed to specialty items, commodities tend to have a continuously fluctuating price that is affected by many worldwide factors of weather, supply, demand, economics, and politics, whereas specialty items may place a more stable price based on end-use values. 11

Company number—a component of the GS1 Company Prefix. GS1 and GS1 Member Organizations assign GS1 Company Prefixes to entities that administer the allocation of EAN.UCC system identification numbers. 9

Competitive assay—a form of immunoassay in which residues in the sample compete with known amounts of the test analyte for a limited number of antibody binding sites on the test media. The outcome of the competition is visualized with a color development reaction. In all competitive immunoassays, the sample concentration is inversely proportional to color development: darker color = lower concentration of the target analyte; lighter color = higher concentration of the target analyte. 8

Composite sample—a sample assembled from several subsamples of equal size; a composite bin sample might be made up of equal size samples taken from each truckload going into that bin. 11

Conditioning—the act of cleaning, which removes impurities or foreign matter, damaged or diseased seeds from a lot of seed or grain; this procedure may also include sizing the seeds into like or uniform size groupings. 11

Consumer—an individual who buys products or services for personal use and not for manufacture or resale; persons and families purchasing and receiving food to meet their personal needs. 5, 7

Container—packaging of food for delivery as a single item, whether by completely or partially enclosing the food and includes wrappers. A container may enclose several units or types of packages when such is offered to the consumer. 7

Contamination—the possibility of "making something impure or unclean by contact or mixture" is something that is central to the risk assessment within certification. The question that is always present is this: How high is the risk of adventitious contamination at a given point in the production or handling chain? Besides this type of accidental contamination, intentional contamination by way of mixing is also possible. Certain jurisdictions with labeling regulations, such as the EU, do permit a certain level of adventitious contamination, but intentional contamination is ruled out, even if it stays below the labeling threshold. This is also recognized, for instance, by the national regulations of many countries, including all EU member states, which require that the so-called precautionary principle be met. This means that a production system must meet "all reasonable precautions." It is only reasonable to expect of a food manufacturer to do as much as possible to come close to perfection. Legislation in all countries recognizes that as long as human or technical error is possible, true perfection itself is something unattainable. 3

Content guarantee—for example, the "non-GMO" assurance given by Cert ID is not a content guarantee but a process guarantee. This means that Cert ID ensures that the systems the company's inspectors have audited comply with the Cert ID Standard and will thus produce only product that complies with the Cert ID Standard as well, with a GMO content well below 0.1 percent. No certifier can guarantee that every grain, kernel, or bean meets this requirement at all times. 3

Conventional breeding—those plant-breeding procedures that do not involve transgenic methods. 11

Corn—an American terminology, *Zea mays* is often called maize in most countries and is the primary crop or grain produced from this member of the grass family. 11

Cover product—a close-growing product grown to protect and improve soils between periods of regular products or between trees and vines in orchards and vineyards. 5

Critical control point (CCP)—a point, step, or procedure at which control can be applied and a safety hazard can be prevented, eliminated, or reduced to acceptable levels. 5

Critical defect—a deviation at a CCP, which may result in a hazard. 5

Critical limits—the maximum or minimum value to which a physical, biological, or chemical hazard must be controlled at a critical control point to prevent, eliminate, or reduce to an acceptable level the occurrence of the identified food safety hazard (adopted from Corlett, 1998 as the 1996 FSIS-USDA/1997 NACMCF definition). 5

Crop protection product risk analysis—covers the following risks, exceeding MRLs, legal registration issues, residue analysis decision taking, and reasons behind decision taking for residue analysis. 5

Crop residues—the plant parts remaining in a field after the harvest of a crop, which include stalks, stems, leaves, roots, and weeds. 10

Crop—a plant or part of a plant intended to be marketed as an agricultural product or fed to livestock. 10

Cross-pollination—to apply pollen of one flower to the stigma of another; commonly refers to the pollinating of the flowers of one plant by pollen from another plant; referring to pollination by another plant, as opposed to "self" pollination (pollen from the flower pollinates the stigma of the same plant). 11

Cry proteins—a class of proteins produced by *Bacillus thuringiensis* (Bt) bacteria. Cry proteins are toxic to certain categories of insects but harmless to mammals and beneficial insects. Examples are Cry1Ab, cryIII, Cry9C protein. 1, 8, 11

Cultivar—a horticultural race or variety of a plant that has originated and persisted only under cultivation; synonymous with variety. 2

Cultivation—digging up or cutting the soil to prepare a seed bed; control weeds; aerate the soil; or work organic matter, crop residues, or fertilizers into the soil. 10, 11

Cultural methods—methods used to enhance crop health and prevent weed, pest, or disease problems without the use of substances; examples include the selection of appropriate varieties and planting sites; proper timing and density of plantings; irrigation; and extending a growing season by manipulating the microclimate with green houses, cold frames, or windbreaks. 10

Custom operator—an equipment owner who uses the equipment for hire in production activities for other parties, party doing custom planting or harvesting of someone else's crop. 11

Customer—anyone (party) who receives, consumes, or purchases products or services from a supplier. 5, 9

Date of packaging—date on which the food is placed in the immediate container in which it will be sold. 7

Demeter or Demeter brand (Demeter-International e. V.)—a nonprofit organization, produces products derived from Biodynamic Agriculture, and represents a worldwide biodynamic certification system used to verify production in more than 60 countries. Only strictly controlled and contractually bound partners are permitted to use the brand and labeling. Originated by Rudolf Steiner in his "Agriculture Course" given in Koberwitz in 1924. A comprehensive verification process ensures strict compliance with the International Demeter Production and Processing Standards, as well as applicable organic regulations in the various countries. Biodynamic agriculture goes beyond the standard demands of organics by incorporating three additional principles: (1) the farm is a sustainable ecosystem in itself; (2) use of biodynamic preparations enhances the activities in the compost and the soil; and (3) the notion that dynamic forces of the sun, moon, planets, and constellations on plants ultimately nourish humans physically, spiritually, and emotionally. Biodynamic is the oldest non-chemical agricultural movement, predating the organic agriculture movement by some twenty years and has now spread throughout the world. It is used, without a gap, through every step, from agricultural production to processing and final product packaging. The holistic Demeter requirements exceed government-mandated regulations. They exclude the use of synthetic fertilizers and chemical plant protection agents in agricultural crop production, or artificial additives during processing, but also require very specific measures to strengthen the life processes in soil and foodstuffs. Demeter farmers and processors actively contribute toward the shaping of a future worth living for, creating healthy foods of distinctive tastes, truly "Foods with Character." 13

Deoxyribonucleic acid (DNA)—a nucleic acid that carries the genetic information in the cell and is capable of self-replication; the substance within cells that carries the "recipe" for the organism and is inherited by offspring from parents, transmitted from parents to offspring. DNA consists of two long chains of nucleotides twisted into a double helix and joined by hydrogen bonds between the complementary bases adenine (A) and thymine (T) or cytosine (C) and guanine (G). The sequence of nucleotides determines individual hereditary characteristics. In a plant or animal, it possesses the individual hereditary characteristics in the DNA that are modified, and a product derived from them is called a genetically modified organism (GMO). 1, 3, 11

Detectable residue—the amount or presence of chemical residue or sample component that can be reliably observed or found in the sample matrix by currently approved analytical methodology. 10

Dockage—a factor in the grading of grains and oilseeds, which includes waste and foreign material, which can be readily removed by the use of screens, sieves, and other cleaning devices. Dockage is always determined and reported on the inspection certificate. The term is also used to describe the amount of money deducted because of a deficiency in quality. 11

Document—the certificate, paperwork, or electronic record conveying authoritative information, which might be trade, legal, or testing information related to identity-preserved (IP) trade. 11

Documentation audit—a review by an auditing panel of the company's quality and food safety management system manual. 5

Drift—the physical movement of prohibited substances from the intended target site onto an organic operation or portion thereof. 10

EAN.UCC system—(European Article Number/Uniform Code Council) now known as GS1. It is a global standard numbering system to identify services and products. It comprises those standards endorsed by the EAN Member Organizations (including UCC and ECCC in North America). The system includes specifications, standards, and guidelines to identify services and products. Examples include EAN/UCC 128, now referred to as a GS1-128, EAN/UCC-13, EAN/UCC-8, and EANCOM©. 9

Electronic data interchange (EDI)—a form of electronic commerce in which the computer-to-computer exchange of business data is in a standardized, structured format. It is a voluntary public standard. 9

Electronic product code (EPC)—an electronically coded 96-bit tag, which may contain a Global Trade Identification Number (GTIN). Unlike a UPC number, which only provides information specific to product, the EPC gives each product its own serialization number, giving greater accuracy in tracking. The EPC was the creation of the MIT AutoID Center, a consortium of more than 120 global corporations and university labs. The EPC system is currently managed by EPCglobal Inc., a subsidiary of the Electronic Article Numbering International group (now known as GS1) and the Uniform Code Council (UCC) (now known as GS1 US), creators of the UPC barcode. The EPC is used utilizing radio frequency identification, or RFID. 9

End user—the ultimate user of a product; at the user end of a supply or value chain; sometimes this may be the last manufacturer in a chain, or sometimes the ultimate consumer is referred to as the end user. 11

Endotoxin—a poisonous substance found within a cell, usually in the outer membrane. Originally contained in bacteria, gene research has discovered how to implant certain endotoxins in the genetic makeup of other organisms such as plants. For example, the endotoxin secreted by the bacteria *Bacillus thuringiensis* is now included in the genome of certain varieties of corn and cotton plants to provide a natural defense against the European corn borer and the cotton bollworm. 8

Enhanced value—in IP, a product having a value higher than a commodity, which usually contains a special trait or attribute that increases the value over similar products. 11

Enzyme immunoassay (EIA)—an immunoassay using a color-changing enzyme-substrate system for indicating results.

Enzyme-linked immunoabsorbent assay (ELISA)—immunological assay techniques that can be used to measure qualitatively and quantitatively a specific protein. 8, 11

Escherichia coli (E. coli)—a bacterium found in the intestine of animals and humans used extensively in genetic engineering. *E. coli* can be fatal to humans if undercooked meat is digested. 1

Excluded methods—a variety of methods used to genetically modify organisms or influence their growth and development by means that are not possible under natural conditions or processes and are not considered compatible with organic production. Such methods include cell fusion, microencapsulation and macroencapsulation, and recombinant DNA technology (including gene deletion, gene doubling, introducing a foreign gene, and changing the positions of genes when achieved by recombinant DNA technology). Such methods do not include the use of traditional breeding, conjugation, fermentation, hybridization, in vitro fertilization, or tissue culture. 11

Extensible markup language (XML)—a computer language used to exchange data. XML is a form of electronic commerce used similarly to EDI.

Farm—an agricultural production unit or group of agricultural production units, covered by the same operational procedures, farm management, and decision-making activities. 5

Farmer—person or business representing the farm (horticultural, agricultural, or livestock, according to the relevant scope) who has legal responsibility for the products sold by that farming business. 5

Feed—edible materials that are consumed by livestock for their nutritional value. Feed may be concentrates (grains) or roughages (hay, silage, fodder). The term "feed" encompasses all agricultural commodities, including pasture ingested by livestock for nutritional purposes. 10

Feed additive—a substance added to feed in micro quantities to fulfill a specific nutritional need, i.e., essential nutrients in the form of amino acids, vitamins, and minerals. 10

Feed grains—also known as coarse grains. This category includes corn, sorghum, barley, oats, rye, and millet. 11

Fertilizer—a single or blended substance containing one or more recognized plant nutrient(s) that is used primarily for its plant nutrient content and is designed for use or claimed to have value in promoting plant growth. 10

Food—any substance, whether processed, semiprocessed, or raw, which is intended for human consumption, and includes drinks, chewing gum, and any substance that has been used in the manufacture, preparation, or treatment of "food" but does not include cosmetics or tobacco or substances used only as drugs. 7

Food additive—any substance not normally consumed as a food by itself and not normally used as a typical ingredient of the food, whether it has nutritive value, the intentional addition of which to the food for a technological purpose in the manufacture, processing, preparation, treatment, packing, packaging, transport, or holding of such food results, or may be reasonably expected to result (directly or indirectly), in it or its byproducts becoming a component of or otherwise affecting the characteristics of such foods. The term does not include "contaminants" or substances added to food for maintaining or improving nutritional qualities. 7

Food safety—for consumers, safety is the most important ingredient of their food. Past crises have undermined public confidence in the capacity of the food industry and of public authorities to ensure that food is safe. Governing bodies, such as the European Commission, have identified food safety as one of their top priorities. Food safety today usually means modernizing legislation and industry practice according to a coherent and transparent set of rules, reinforcing controls from the farm to the table and increasing the capability of the scientific advice system, to guarantee a high level of human health and consumer protection. Food safety usually means the assurance that food will not cause harm to the consumer when it is prepared and consumed according to its intended use. 3, 5

Forage—vegetative material in a fresh, dried, or ensiled state (pasture, hay, or silage), which is fed to livestock. 10

Foundation seed—a class of certified seed, which is the progeny of breeder or foundation seed and is produced and handled under procedures established by the certifying agency for producing the foundation class of seed, for maintaining genetic purity and identity. 11, 13

Fresh Produce Traceability (EAN FPT) guidelines—aimed at providing a common approach to tracking and tracing fresh produce by mean of an internationally accepted numbering and barcoding system: the EAN.UCC system. See http://www.ean-int.org/Doc/TRA_0402.pdf

Full Traceability Database—as a certification body Cert ID stores all of its data in a decentralized database called the *Full Traceability Database.* Be it inspector audit reports, decisions of the Certification Committee, port facility photographs, or a laboratory's Analysis Reports, they are all stored in this database in electronic form. Those data pertaining to the supply chain of a given Cert ID client are available to this client, again in electronic form. The client company is then able to use these data in the event of a challenge from government authorities, or from any other side, to demonstrate that its production system complies with the Cert ID Standard and has thus met the so-called precautionary principle as required by food legislation in many countries. 3

Gas chromatography (GC)—an analytical method in which a sample is vaporized and injected into a carrier gas (called the mobile phase; usually helium) moving through a column. The quantity of a particular compound in the mixture is determined by comparing detector response with the response to known standards. Identification of unknown compounds is only possible if the detector used is a mass spectrometer. The technique can require extensive cleanup and preparation of the sample, the use of costly equipment, and operation by a highly trained technician. 8

Gene—the fundamental physical and functional unit of heredity; made up of a particular sequence of nucleotides found on a particular chromosome. Regardless of the source, a specific gene is composed of a sequence of DNA that usually represents the coded description or blueprint for a specific protein. 1, 11

Gene flow—the concept that in natural ecosystems genes can move within and among plant species (often by cross-pollination), transfer of genetic material by interbreeding from one plant population to another that changes the composition of the gene pool of the receiving population. 1, 2, 11

Gene stacking—involves combining traits (e.g., herbicide tolerance and insect resistance) in seed. 1, 11

Genetic engineering—the selective, deliberate alteration of genes (genetic material) by humans. This term has come to have a very broad meaning, including the manipulation and alteration of the genetic material of an organism in such a way as to allow it to produce endogenous proteins with properties different from those of the normal, or to produce entirely different proteins altogether. 11

Genetic modified organism (GMO)—sometimes also referred to as GM (genetically modified) or GMF (genetically modified food) or GE (genetically engineered); similar abbreviations exist in other languages are produced through techniques in which the genetic material has been altered in a way that does not occur naturally by mating and/or natural recombination; also, food ingredients in which the host DNA has been altered by insertion of gene sequences from another organism. These modifications are made to reduce the cost or improve the effectiveness of agricultural chemical applications (input traits) or to enhance the quality, appearance, or value of resulting food products (output traits). According to the Gentechnikgesetz (GenTG) from 20.06.90 (Genetic Technique Law) in Germany, they are organisms whose genetic materials were modified in a way that is not found in nature under natural conditions of crossbreed or natural recombination. The GMO must be a biological unit, which is able to multiply itself or to transmit genetic material. Examples of modifications are techniques by which genetic material prepared outside of the cell is introduced directly in the organism. These techniques include microinjection, macroinjection, microencapsulation, cell fusion, as well as hybridization procedures by which living cells are formed with a new combination of genetic material using methods. 3, 7, 8, 11

Genome—the basic set of chromosomes of an organism; the entire DNA "recipe" for an organism, found in every cell of that organism. 1, 10

Genotype—the genetic makeup of an individual; the hereditary constitution of an organism. 1, 11

Germplasm—hereditary material, in crop breeding, the totality of genes and genetic materials available for the improvement of a crop. 1, 11

Global EAN Party Information Register (GEPIR)—GEPIR is an International Catalogue of EAN.UCC numbers including Global Trade Identification Numbers (GTIN) and Global Location Numbers (GLN). 6

Global Individual Asset Identifier (GIAI)—the Global Individual Asset Number identifies serial identification numbers for objects and containers, bins, boxes etc. that do not require categorization. Its function is limited to the container itself, not its content. 6

Global Location Number (GLN)—a 13-digit number used to identify a location. The GLN consists of two parts: a company prefix and a four-digit location number assigned by the owner of the GLN.

Global Returnable Asset Identifier (GRAI)—the Global Returnable Asset Identifier identifies all reusable entities owned by a company, used for transport and storage of goods. 6

Global Trade Item Number (GTIN)—the umbrella term for several kinds of item numbers and a shorthand term for the EAN.UCC Global Trade Item Number. A GTIN may use the EAN.UCC-8, UCC-12, UCC-13, or UCC-14 Data Structure. This data structure comprises a 14-digit number that has four components: (1) an indicator, (2) a manufacturer prefix, (3) a unique number to that manufacturer, and (4) a check digit. The GTIN has gained a lot of traction in the consumer packaged goods (CPG) marketplace and has largely been the accepted standard for the packaged goods side of the business. The recommendation in this book is to use the GTIN (EAN.UCC-14 Data Structure) at the case level. 9

GMO testing—analyses of samples of food or agricultural products for the presence of GMOs, often also for the quantity of GMOs and types of GMO events. 3

GMO traceability—general traceability, as a special characteristic of the basic idea of the precautionary principle, in force in the EU since April 18, 2004, stipulates the requirement for operators to label products containing or made from GMOs, but also the need for traceability of products containing or made from GMOs (EU Regulation [EC] No. 1830/2003). However, some food ingredients are processed so deeply that even the most delicate PCR testing is sometimes unable to distinguish whether they were produced from genetically modified or from conventional plants. The reason for this is a total absence of DNA molecules in such cases. If, for example, soy oil is refined so well that all DNA molecules have been filtered out, even the best PCR analysis is unable to make a statement. Therefore, the new EU regulation mentioned above reaches far beyond the old rules presented up to this point by not focusing exclusively on the detectability by way of PCR analysis. From now on, food, feed, ingredients, and additives must be labeled if they are or consist of a GMO or if they have been made from GMOs, regardless of whether these can be detected in the food/feed. Labeling independent of detection is possible only if the information about the application of GMOs is handed down the entire production chain, from the producer to the retailer. The EU regulation requires that the food industry and at least the suppliers of raw materials for the feed industry set up appropriate traceability systems. Based on the documentation to be kept by food and feed operators on raw materials purchased, there will be inspections from now on as to whether they originate entirely or partly from genetically modified plants and must thus be labeled. 3

Good Agricultural Practices (GAP)—program addresses site selection, adjacent land use, fertilizer usage, water sourcing and usage, pest control and pesticide monitoring, harvesting practices (including worker hygiene, packaging storage, field sanitation, and product transportation), and cooler operations. Standard operating procedures are developed and incorporated into the GAP program, providing guidance with respect to potential points for contamination and preventative or corrective measures to mitigate their effects.

Governmental entity—any domestic government, tribal government, or foreign governmental subdivision providing certification services. 10

Grain—crops or seeds produced from the cereal crop species. 11

GS1—the entire GS1 organization consisting of GS1 Head Office and the worldwide network of GS1 Member Organizations. GS1 is a voluntary standards organization charged by the GS1 board with the management of the EAN.UCC system and the Global Standard Management Process (GSMP). The EAN.UCC system standardizes barcodes, EDI transactions sets, XML schemas, and other supply chain solutions for more efficient business. 9

Handle—to sell, process, or package agricultural products, except such term shall not include the sale, transportation, or delivery of crops or livestock by the producer thereof to a handler.

Handler—any person engaged in the business of handling agricultural products, including producers who handle crops or livestock of their own production, except such term shall not include final retailers of agricultural products who do not process agricultural products; usually referring to people or companies that move, transfer, or store products but are not involved in the growing, conditioning, or processing of that product; handling refers to moving, transferring, and storing activities and may be performed by almost anyone in a supply or value chain. 10, 11

Handling operation—any operation or portion of an operation (except final retailers of agricultural products who do not process agricultural products) that receives or otherwise acquires agricultural products and processes, packages, or stores such products. 10

Hazard Analysis and Critical Control Point (HACCP)—a food safety program for preventing hazards that could cause food-borne illnesses by applying science-based controls, from raw material to finished products, which includes analyze hazards, identify critical control points, establish preventive measures with critical limits for each control point, establish procedures to monitor the critical control points, establish corrective actions to be taken when monitoring shows that a critical limit has not been met, establish procedures to verify that the system is working properly, and establish effective recordkeeping to document the HACCP system.

Herbicide—a chemical that controls or destroys undesirable plants. Some herbicides (such as synthetic triazines) selectively kill broad-leaved plants while leaving grass-leafed plants (i.e., cereal crops) unharmed. Other herbicides, such as paraquat, kill all plants. 5, 8

High-performance liquid chromatography (HPLC)—an analytical method in which a sample is injected into a stream of liquid (called the mobile phase; usually a mixture of water and an organic solvent) moving through a column. The quantity of a particular compound in the mixture is determined by comparing detector response with the response to known standards. Identification of unknown compounds is only possible if the detector used is a mass spectrometer. The technique can require extensive cleanup and preparation of the sample, the use of costly equipment, and operation by a highly trained technician. 8

Identity preservation (IP)—legislators of several countries have satisfied consumer and industry demands by enacting mandatory labeling laws for foods containing ingredients derived from genetically modified (GM) crops. To comply with these labeling laws, food manufacturers must be able to document the genetic purity of both GM and non-GM ingredients. This can be accomplished by preserving the identity of a crop from seed to final product (identity preservation or IP) and by thus enabling the various players in a supply chain to document traceability. This means being able to trace back from the final product to the crops from which ingredients were manufactured. Traceability is not possible without IP systems. It requires that manufacturers have a complete understanding of the supply chain for primary and secondary ingredients and blends. New specifications are being developed with well-defined expectations regarding purity and handling. Audit systems ensure compliance by farmers, grain elevators, processors, ingredient suppliers, and food manufacturers. IP is a system of maintaining the seg-regation of a grain or oilseed crop from planting the seed to delivery to the final end user by using a carefully controlled production and distribution system that maintains integrity of the crop being delivered. IP can involve any system of raw material management that segregates or preserves the identity of the source or nature of the materials; a stringent handling pro-cess that separates GM crops and their derived products and provides docu-mentation at each transfer point in the food chain. 2, 3, 11

Immunoassay—an analytical test to measure or detect a substance using antibody-antigen reactions. A technique that makes use of the specific binding between an antigen and its related antibody to identify or quantify a sub-stance in a sample. 8

Industry Product Database (IPD)—an initiative in the produce sector to help address product identification. It enables a retailer's SKU# to be mapped to a supplier's product code (i.e., GTIN or other number). This helps facili-tate data synchronization between trading partners. See www.pma.com/IPDFactSheet.

Inspection—the act of examining (visual observation) and evaluating of food, the production, handling operation, or system of procedural or product quali-ties of an applicant in a production or delivery system; for control of food, raw materials, processing, and distribution, including in-process and fin-ished product testing, to verify compliance to requirements; such as ISO series or for certification or certify operations to determine compliance to a standard or Act. 6, 7, 10, 11

Intellectual property (IP)—the legal rights associated with inventions, artistic expressions, and other products of the imagination (e.g., patent, copyright, and trademark law.). 1

International Organization for Standardization (ISO)—promotes the develop-ment of standardization and related activities to facilitate the international exchange of goods and services, and to developing cooperation in economic activity. For example, ISO 9000 comprises eight quality management principles that can be used as a framework to guide organizations toward improved performance. 6

IP systems—production or handling systems where IP (identity preservation) has been implemented. 3, 11

Isolation—in planting field isolation refers to the distance required from other fields of the same crop to minimize cross-pollination, sometimes referred to as a "buffer" strip. The isolation distance can sometimes be modified (reduced) by planting additional IP crop along the field edges and harvesting as non-IP product. 11

Isolation standards—the standards that dictate distances and modifications by crop, for the production of seed or identity-preserved crop. Isolation standards may be set by a third-party certifying body or by a production company. 11

Label—a display of written, printed, or graphic material on the immediate container of an agricultural product or any such material affixed to any agricultural product or affixed to a bulk container containing an agricultural product, except for package liners or a display of written, printed, or graphic material that contains only information about the weight of the product. 7, 10

Labeling threshold—the level of GMO content of consumer products, as defined by some governments, above which a label on the packaging must indicate that the product inside contains GMOs. The best-known examples by now are probably the two thresholds stipulated in the two EU regulations that must be implemented fully since April 18, 2004 (EU Regulation [EC] No. 1829/2003 and No. 1830/2003, both of September 22, 2003). 4

Lateral flow membrane assay—a form of immunoassay. Results of the test are indicated by the presence or absence of one or more additional "test lines" that are expected between point of sample application and the control line. May be in either cassette or dipstick format. Usually requires no additional reagents. 8

Livestock—any domestic or domesticated including bovine (including buffalo and bison), ovine, porcine, caprine, equine, poultry, and bees raised in the production of food, fiber, feed, or other agricultural-based consumer products; wild or domesticated game; or other non-plant life, except such term shall not include aquatic animals. The products of hunting or fishing of wild animals shall not be considered part of this definition. 7, 10

Lot—a definitive quantity of a commodity produced essentially under the same conditions; any number of containers that contain an agricultural product of the same kind located in the same conveyance, warehouse, or packing house and that are available for inspection at the same time. 7, 10

Maize—the common name for *Zea mays*; in most countries maize is the primary crop or grain produced from a member of the grass family. In America, the word "corn" is more common. 11

Mandatory requirements—the data that must be exchanged between trading partners to accomplish traceability. 9

Marker—a genetic flag or trait used to verify successful transformation and to indirectly measure expression of inserted genes. For example, a gene used as a marker in Bt11 confers tolerance to the herbicide Liberty. 11

Mass spectrometry—a technique for determining the composition of a molecule and its fragments. 1

National List—a list of allowed and prohibited substances as provided for in the Act of the USDA's NOP. 10

National Organic Program (NOP)—the program authorized by the Act for the purpose of implementing its provisions. 10

National Organic Standards Board (NOSB)—a board established by the Secretary under 7 U.S.C. 6518 to assist in the development of standards for substances to be used in organic production and to advise the Secretary on any other aspects of the implementation of the National Organic Program. 10

Niche market—markets of specialty items, usually having higher value than commodity items; special markets set up around specialty products. 11

Nonagricultural substance—a substance that is not a product of agriculture, such as a mineral or a bacterial culture, which is used as an ingredient in an agricultural product. For the purposes of this part, a nonagricultural ingredient also includes any substance, such as gums, citric acid, or pectin, that is extracted from, isolated from, or a fraction of an agricultural product so that the identity of the agricultural product is unrecognizable in the extract, isolate, or fraction. 10

Non-GMO—an organism that has not been modified by transgenic breeding techniques, as opposed to GMO (genetically modified organism). Many organizations refrain entirely from using terms such as "GMO-free," "GE-free," etc., terms that would imply a 100 percent absence of GMOs. Certification to a 0.0 percent GMO content threshold is impossible for two reasons, each one being sufficient on its own: (1) GMO testing can only be conducted with representative samples, never with an entire lot (otherwise, nothing would be left for consumption.); and (2) the PCR testing method for GMOs is able to test to detection limits as low as between 0.1 percent and 0.01 percent, depending on the tested material. It does not "reach" as low as 0.0 percent. At the same time, organizations endeavor to enable their clients to attain a production output that is, in fact, as "free" of GMOs as possible. This is accomplished by rigid input testing for GMOs and by certifying to a rigorous standard that ensures a minimization of contamination risks throughout the entire IP chain. 3, 11

Nonsynthetic (natural)—a substance that is derived from mineral, plant, or animal matter and does not undergo a synthetic process as defined in section 6502(21) of the Act [7 U.S.C. 6502(21)]. For the purposes of this part, nonsynthetic is used as a synonym for natural as the term is used in the Act. 10

Nutraceuticals—either a food or a portion of food that possesses medical or health benefits. 11

Official accreditation—the procedure by which a government agency having jurisdiction formally recognizes the competence of an inspection and/or certification body to provide inspection and certification services. For organic production, the competent authority may delegate the accreditation function to a private body. 7

Officially recognized inspection systems/officially recognized certification systems—systems that have been formally approved or recognized by a government agency having jurisdiction. 7

One-up/one-down traceability—under a one-up/one-down system each participant within the food continuum is responsible for maintaining records about the products he or she receives and where they were shipped to, or sold. 9

Organic—a labeling term that refers to an agricultural product produced in accordance with the Act and the regulations in this part. 10

Organic fertilizer—organic fertilizers mean materials of animal origin used to maintain or improve plant nutrition and the physical and chemical properties and biological activity of soils, either separately or together; they may include manure, compost, and digestion residues. The use of compost, originated from enhanced treated sewage sludge, can be seen as organic fertilizer. 5

Organic matter—the remains, residues, or waste products of any organism. 10

Organic production—a production system that is managed in accordance with the Act and regulations in this part to respond to site-specific conditions by integrating cultural, biological, and mechanical practices that foster cycling of resources, promote ecological balance, and conserve biodiversity. 10

Organic system plan—a plan of management of an organic production or handling operation that has been agreed to by the producer or handler and the certifying agent and that includes written plans concerning all aspects of agricultural production or handling described in the Act and the regulations in subpart C of this part. 10

Pallet—a platform with or without sides, on which a number of packages/pieces may be loaded to facilitate handling by a lift truck. 9

Paper trail—the documents that provide assurance in every step of a transaction that traces the production from its very beginning to point of reference at present; traces all origins and procedures of handling the item. 11

Paper transaction—the documentation that refers to a sale, the contracting of a process, or other work with a product or process; the physical movement of the product may or may not happen at the same time. 1

Pasture—land used for livestock grazing that is managed to provide feed value and maintain or improve soil, water, and vegetative resources. 1

Pathogen—an agent that causes disease, especially a living microorganism such as a bacterium, virus, or fungus. 8

Pesticide—any substance which alone, or in any formulation with one or more substances, is defined as a pesticide in section 2(u) of the Federal Insecticide, Fungicide, and Rodenticide Act [7 U.S.C. 136(u) et seq] is used to destroy pests, which commonly includes insecticides, herbicides, fungicides, and nematicides in chemical combination. 8, 10

Phytosanitary Certificate—a certificate issued by authorities to satisfy import regulations of foreign countries; indicates that a shipment has been inspected and found free from harmful pests and plant diseases.

Plant protection product—any active substances containing one or more active substances and preparations intended to: protect plants or plant products against all harmful organisms or prevent the action of such organisms. Such as for preventing, destroying, attracting, repelling, or controlling any pest or disease, including unwanted species of plants or animals during the production, storage, transport, distribution, and processing of food, agricultural commodities, or animal feeds. Influence the life processes of plants, other than as a nutrient (e.g., growth regulators); or destroy parts of plants, check or prevent undesired growth of plants. 7, 5

Plants—live plants and live parts of plants, including fresh fruit and seeds. 5

Pollen flow—the normal flow or path of pollen, in a cross-pollinated crop, carried by wind or other means, from the male sex organ to the female sex organ. 11

Pollination—the transfer of pollen between the male germ cell of a plant (anther) and the female reproductive system (stigma) in seed plants. 2, 11

Polymerase chain reaction (PCR)—a method for creating millions of copies of a particular segment of DNA. If a scientist needs to detect the presence of a very small amount of a particular DNA sequence, PCR can be used to amplify the amount of that sequence until there are enough copies available to be detected. A very sensitive, rapid biochemical assay system for detection of specific sequences of DNA that is often used to indicate the presence or absence of specific genes. PCR can be used to determine whether an organism contains specific DNA sequences. The presence of specific sequences might be an indicator that a plant has been modified through biotechnology. A transgenic trait is made up of a promoter that controls the expression of the gene, the gene, and a terminator that assists in the insertion of a gene. PCR detects the transgenic trait by replicating a particular portion (promoter, gene, or terminator) of the trait that is present. 3, 4, 8, 11

Practice standard—the guidelines and requirements through which a production or handling operation implements a required component of its production or handling organic system plan. A practice standard includes a series of allowed and prohibited actions, materials, and conditions to establish a minimum level performance for planning, conducting, and maintaining a function, such as livestock health care or facility pest management, essential to an organic operation. 10

Precautionary principle—an approach used to the management of risk, when scientific knowledge is incomplete. 1

Prepackaged—made up in advance in a container, ready for offer to the consumer, or for catering purposes. 7

Preparation—the operations of slaughtering, processing, preserving, and packaging of agricultural products and alterations made to the labeling concerning the presentation of the organic production method. 7

Process guarantee—the "non-GMO" assurance given by Cert ID is a process guarantee, as opposed to a content guarantee. This means that Cert ID ensures that the systems the company's inspectors have audited comply with the Cert ID standard and will thus produce only product that complies with the Cert ID standard as well, with a GMO content well below 0.1 percent. No certifier can guarantee that every grain, kernel, or bean meets this requirement at all times. 3

Processing—cooking, baking, curing, heating, drying, mixing, grinding, churning, separating, extracting, slaughtering, cutting, fermenting, distilling, eviscerating, preserving, dehydrating, freezing, chilling, or otherwise manufacturing and includes the packaging, canning, jarring, or otherwise enclosing food in a container. 10

Produce—the harvested product of the product after it has been harvested, before it is sold. 5

Produce handling—low risk post-harvest activities carried out on the produce that is still owned by the certified farmer/group of farmers, on or off farm, i.e., packing, storage, and transport ex farm, but excluding harvesting and on-farm transport from point of harvest to first point of storage/packing. Processing of produce is not covered by produce handling. Packing carried out at point of harvest is considered produce handling. In addition, any storage, chemical treatments, trimming, washing, or any other handling where the product may have physical contact with other materials or substances. 5

Producer—a person who engages in the business of growing or producing food, fiber, feed, and other agricultural-based consumer products. 10

Product code—a number issued internally by the supplier to distinguish it from other products. Used by itself, the product code has no value to anyone other than the supplier. Product traceability describes the qualitative followup of products. It essentially relies on correct recordkeeping and the thoroughness of information concerning the product. A manufacturer uses it to find the causes of a quality fault either upstream, if the incident could have occurred at his or her supplier's premises, or downstream, if the incident could have occurred during shipping, for example.

Product tracing—the capability to identify the origin of a particular unit and/or batch of product located within the supply chain by reference to records held upstream in the supply chain. Products are traced for purposes such as product recall and investigating complaints. Within the context of EurepGAP Integrated Farm Assurance, this means tracing product from the farmer's immediate customer back to the farmer and certified farm. 5

Product tracking—the capability to follow the path of a specified unit of a product through the supply chain as it moves between organizations. Products are tracked routinely for obsolescence, inventory management, and logistical purposes. In the context of this document, the focus is on tracking produce from the grower to retail point of sale. Within the context of EurepGAP Integrated Farm Assurance, this means tracking product from the farmer to his or her immediate customer. 5, 9

Production—the operations undertaken to supply agricultural products in the state in which they occur on the farm, including initial packaging and labeling of the product. 7

Production lot number/identifier—identification of a product based on the production sequence of the product showing the date, time, and place of production used for quality control purposes. 10

Production output—the products/trade units that have been produced and/or shipped from a trading partner in the food supply chain and may include animals (including fish), plants, and their products, as well as foods produced from these products/trade units. 9

Prohibited substance—a substance, the use of which in any aspect of organic production or handling is prohibited or not provided for in the Act or regulations. 10

Protein—a complex biological molecule composed of a chain of amino acids that are assembled in the linear order specified by the gene that encodes the protein. Proteins are usually biologically active only when the chain of amino acids is folded into a specific three-dimensional conformation. Proteins have many different biological functions; for example, enzymes, antibodies, and hair are proteins. 11

Protocol—the rules or process describing a procedure. 11

Radio frequency identification (RFID)—RFID tags are small integrated circuits connected to an antenna, which can respond to an interrogating RF signal. The tag is affixed to or incorporated into a product to track its movement and attributes of the product. They offer wireless electronic communication using radio frequency allowing electronic memory to be read and written. 9, 6

Random sample—a limited sample of product or observation, so assembled from the total array as to be truly representative of its characteristics or properties; taken without personal bias of the sampler or observer. 11

Recombinant DNA (rDNA)—DNA molecules created by splicing together two or more different pieces of DNA. 1

Record—a record is a document that contains objective evidence that shows how well activities are being performed or what kind of results are being achieved; any information in written, visual, or electronic form that documents the activities undertaken by a producer, handler, or certifying agent to comply with standards and regulations. 5, 10

Refugium or refuge—an area planted with non-transgenic plants (e.g., non-*Bt* corn or alternative host for European corn borer), where susceptible pests can survive and produce a local population capable of mating with any possible resistant survivors from *Bt* corn. 11

Registered seed—a class of certified seed, which is the progeny of breeder or foundation seed and is produced and handled under procedures established by the certifying agency for producing the registered class of seed, for the purpose of maintaining genetic purity and identity. 11, 13

Requirements—the criteria set down by the competent authorities relating to trade in foodstuffs covering the protection of public health, the protection of consumers, and conditions of fair trading. 7

Residue methods—analytical techniques involving an abstractive chemical or physical process such as evaporation, distillation, filtration, chromatography, or immunoassay. 9

Residue testing—an official or validated analytical procedure that detects, identifies, and measures the presence of chemical substances, their metabolites, or degradation products in or on raw or processed agricultural products. 10

Retail food establishment—a restaurant; delicatessen; bakery; grocery store; or any retail outlet with an in-store restaurant, delicatessen, bakery, salad bar, or other eat-in or carry-out service of processed or prepared raw and ready-to-eat-food. 10

Risk analysis/assessment—an estimate of the probability of the occurrence of a hazard or other non-conformity with regard to quality and food safety; the evaluation of the likelihood and severity of adverse effects on public health arising, for example, from the presence in foodstuffs of additives, contaminants, residues, toxins, or disease-causing organisms. 5, 7

Sample—a part or piece taken randomly as representative of a whole, in agricultural products a sample of grain or oilseeds that is taken to observe or test as representative of a larger lot.

Sampling protocol—most certification plans contain a description of where, when, and how samples of product to be certified are to be drawn and treated subsequently. This procedural code is called sampling protocol. 3, 11

Seed certifying agency—(a) an agency authorized under the laws of a state, territory, or possession to officially certify seed and that has standards and procedures approved by a higher authority to ensure the genetic purity and identity of the seed certified, or (b) an agency or a foreign country determined to adhere to procedures and standards for seed certification comparable with those adhered to generally by seed certifying agencies. 11

Seed purity—determined by observation or testing that gives the percentage of pure seed that is of the described variety/hybrid and not other materials such as inert mater, weeds seeds, or other crop seeds. 11

Segregation—the process of keeping separate; keeping crops separate by variety or type. 11

Self-inspection—internal inspection of the registered product carried out by the farmer on his or her farm using a checklist based on the EurepGAP checklist. 5

Serial number—links to the supplier's produce description attributes. The combination of supplier ID and serial number uniquely identifies the pallet globally.

Serial Shipping Container Code (SSCC)—an 18-digit number that identifies the nature of the container, the company prefix identifying the owner, and a serialized number. There is no relationship between the SSCC and the GTINs on the shipment. This number (often represented in a barcode) is also known as the "license plate" used on variable content containers, pallets, and shipments.

Shipping advice—notice sent to a local or foreign buyer advising that shipment has gone forward and contains details of packing, routing, etc.

Shipping Container Code (SCC)—the Shipping Container Symbol is the 14-digit number applied to intermediate packs and shipping containers containing UCC-12, EAN/UCC-13, or EAN/UCC-8 marked items. 9

Soybean(s)—a legume, the botanical name of which is *Glycine max* (L.); a summer annual varying in height from less than a foot to more than 6 feet and in habit of growth from erect to prostrate. The seeds (soybeans) are borne in pods that grow in clusters of three to five with each pod usually containing two, three, or more seeds. The oil content varies from 13 to 25 percent and from 38 to 45 percent protein (on a moisture-free basis). Both the oil and protein components are used extensively for food, feed, and industrial uses. 11

Split operation—an operation that produces or handles both organic and non-organic agricultural products. 10

Spot market—a market based on an immediate, momentary response based on the conditions at the time as opposed to a contracted market. 11

StarLink incident—this event in June 2000 involved a genetically engineered corn (maize) variety approved in the United States for animal consumption only. Lab tests showed that it was found in a brand of taco shells, a type of Mexican food, offered in retail stores. It was soon apparent, through sampling and testing in many locations throughout the United States, that StarLink corn was present all over the country. The result was a recall project that, at one point, brought the American corn logistics to a complete standstill for a day or two. The price tag on all of this for the companies involved soon grew into billions of dollars. It is thought that more IP systems and traceability could have reduced these costs considerably. 3

State organic program (SOP)—a state program that meets the requirements of section 6506 of the Act, is approved by the Secretary, and is designed to ensure that a product that is sold or labeled as organically produced under the Act is produced and handled using organic methods. 10

Subcontractor—specific farm operations performed under contract between the farmer and the contractor. The contractor furnishes labor, equipment, and materials to perform the operation. Custom harvesting of grain, spraying, and picking of fruit, and sheep shearing are examples of custom work. Within the EurepGAP context, subcontractors are those organizations/individuals contracted by the farmer/farmer group to carry out specific tasks that are covered in the EurepGAP Control Points and Compliance Criteria. 5

Supplier ID—assigned by EAN member organizations (including ECCC and UCC in North America); also known as the company's prefix.

Supply chain—a series of linked stages that provide goods or services; the layers of processes involved in the manufacture of goods or provision of services. All business activities needed to satisfy the demand for trade items or services from the initial requirements for raw material or data to final delivery to the end user. 6, 9

Synthetic—a substance that is formulated or manufactured by a chemical process or by a process that chemically changes a substance extracted from naturally occurring plant, animal, or mineral sources, except that such term shall not apply to substances created by naturally occurring biological processes. 10

System certification—this general certification bears in mind the principles of EU Regulation (EC) No 178/2002 of the EU Parliament and of the Council about General Principles of Food Law of January 28, 2002. It serves as a platform on which process certification can be established. One of its main aspects is the presence of traceable IP. 3

System check—audit of the Internal Quality Management and Control System. 5

Systematic sample—a sampling method used to provide a reasonable substitute for a random sample; designed to remove some of the fallacies of a completely random sample by using a system to define the method. 11

Techniques of genetic engineering—modification includes, but is not limited to, recombinant DNA, cell fusion, microinjection and macroinjection, encapsulation, and gene deletion and doubling. Genetically engineered organisms will not include hybridization. 7

Third party—a party not involved directly in services or business; an outside party. 11

Third-party certification—certification of any product does not physically change the product in any way. However, if certified according to a publicly available standard that ensures that certain quality-improving measures are taken in processing a product, certification can add value to the certified product. This will then reflect in added consumer confidence in the product and, consequently, in added value to the supplier of the product. Such certification makes sense only if provided by a so-called third party, i.e., an organization that is neither the manufacturer nor supplier of the product nor a consumer advocacy group. 3

Threshold—regarding GMOs, allowable level of GM crop or derived food ingredient that does not trigger a legal requirement for labeling. 2, 10

Tolerance—the maximum legal level of a pesticide chemical residue in or on a raw or processed agricultural commodity or processed food. The permissible variation from the standard for a product; in IP would usually refer to the allowable limit for mixture of other varieties or types. 10, 11

Toxin—a poisonous substance that is produced by living cells or organisms and is capable of causing disease or other measurable pathological effect. 8

Traceable resource unit (TRU)—unique identification and traceability in any system hinges on the definition of what is the batch size or the traceable resource unit (TRU). For batch processes, a TRU is a unique unit, meaning that no other unit can have exactly the same, or comparable, characteristics from the point of view of traceability. When dealing with continuous processing, the definition of a TRU can be difficult. It may depend on the raw material TRU or on a change in processing conditions, as different activities according to the definition give different TRUs. A consistent definition must be maintained but what constitutes a TRU is decided by the system designer. The identification of a TRU may change during the product route when, for example, TRUs are pooled. This results in a new TRU, which must be given a new identification different from that of any of the original TRUs. The size of a TRU may also change, for example, when one batch is split into several batches. However, the individual TRUs can only keep the identification of the original TRU as long as the activities occurring to the individual TRUs are identical (Moe, 1998).

Traceability—the ability to retrace the history, use, or location of a product (that is the origin of materials and parts, the history of processes applied to the product, or the distribution and placement of the product after delivery) by the means of recorded identification. The ability, within an identity-preserved (IP) system, to trace both the crop product and the system of product segregation from the beginning of the production process (the seed source) to the end use of the product. From an information management point of view, implementing a traceability system within a supply chain involves systematically associating a flow of information with a physical flow. The objective is to be able to obtain predefined information concerning batches or groups of products (also predefined) at any given moment, using one or more key identifiers.

Traceability is not possible without an existing IP system. Traceability requires that manufacturers have a complete understanding of the supply chain for primary and secondary ingredients and blends. From the point of view of the user, traceability may be defined as following-up products in both a qualitative and quantitative manner within space and time. Since the coming into force of Regulation (EC) No. 1830/2003 in the EU, the term of "adventitious or technically unavoidable" GMO contamination has assumed a special relevance. This regulation is about the traceability and the labeling of products containing GMOs or made from them. Products with GMO content in excess of a certain threshold (labeling threshold) must be labeled as containing GMOs since April 18, 2004. Even if this threshold is not exceeded, a product must be labeled if the GMO contamination was not "adventitious or technically unavoidable." 2, 3, 5, 9, 12

Tracing—the capability to identify the origin of a particular unit and/or batch of product located within the supply chain by reference to records held upstream in the supply chain. Products are traced for purposes such as product recall and investigating complaints. In the context of this document, the focus is on tracing produce from retail to grower. The ability to reconstruct the historical flow of a product from records. 9

Tracking—the capability to follow the path of a specified unit of a product through the supply chain as it moves between organizations. Products are tracked routinely for obsolescence, inventory management, and logistical purposes. In the context of this document, the focus is on tracking produce from the grower to retail point of sale. The ability to follow products through the supply chain. 9

Transgenic—a plant or animal modified by genetic engineering to contain DNA from an external source is called transgenic. An organism whose cells contain genetic material derived from a source in addition to or other than the parents. Containing genes transferred from species to another. Having altered genetic makeup, often resulting in different physical and developmental characteristics. 8, 11

Transparency—a term applied to a process; transparency means that nothing has been hidden from view. Meetings have been announced in advance; hearings have been open to the public; public comments have been collected; and once decisions have been made, the rationale for the policy adopted is explained clearly. 11

Uniform Code Council (UCC)/GS1 US—a United States-based membership organization that jointly manages the EAN.UCC System with EAN International, and administers the EAN.UCC System in the United States and Canada. UCN Number is a Unique Component Identification Number.

United States Food and Drug Administration (FDA)—a federal agency that has developed voluntary guidelines for Good Agricultural Practices (GAP) for reducing the potential for microbial contamination of produce. GAP guidelines are established to ensure a clean and safe working environment for all employees while eliminating the potential for contamination of food products.

Universal Product Code (UPC) number—the standard barcode symbol for retail food packages in the United States and Canada. 10

Validation audit—a comprehensive evaluation of the entire Quality and Food Safety Management system to ensure that the procedures as documented in the company's Quality and Food Safety Management System manual are implemented and are effective. 5

Value chain—a descriptive term for a supply chain where product values are increased along the movement from initial product to the final product. 11

Variety—an assemblage of cultivated individuals that are distinguished by any characters (morphological, physiological, cytological, chemical, or others) significant for the purposes of agriculture, and which retain their distinguishing features when reproduced or reconstituted. A category within a species of crop plants. Plants of a variety are related by descent and are characterized by morphological, physiological, and adaptation traits. In seed certification terms, variety means a subdivision of a kind that is distinct, uniform, and stable. "Distinct" in the sense that the variety can be differentiated by one or more identifiable morphological, physiological, or other characteristics from all other varieties of public knowledge. "Uniform" in the sense that variations in essential and distinctive characteristics are describable. "Stable" in the sense that the variety will remain unchanged to a reasonable degree of reliability in its essential and distinctive characteristics and its uniformity when reproduced or reconstituted as required by the different categories of varieties. 11, 13

Verification—confirmation by examination and provision of evidence that specified requirements have been met, providing a means for checking that the deviation between values indicated by a measured instrument and corresponding known values of a measured quantity are consistently smaller than the maximum allowable error defined in a standard or specification peculiar to the management of the measuring equipment. 5, 11

Verification audit—a routine unannounced audit of the Quality and Food Safety Management System after approval to ensure that the Quality and Food Safety Management System in place is adequately maintained. 5

Volunteer plant—are crop plants that persist for a few seasons without deliberate cultivation. Plants that are produced from seeds of the previous cropping cycle, seeds that have fallen to the ground during harvesting activities and then germinate and grow in the following crop. 1, 9

Weed—any plant growing where it is not wanted. In agriculture a plant that has good colonizing capability in a disturbed environment, and can usually compete with a cultivated species therein. Weeds are typically considered as unwanted, economically useless, or pest species. 5

Yield drag—a slang term indicating a yield reduction of a specialty crop variety compared with similar commodity-type varieties. 11

GLOSSARY SOURCES

1. Pew Initiative on Food and Biotechnology, Resources, Glossary, http://pewagbiotech.org/resources/glossary.

2. From tracking genes from seed to supermarket: techniques and trends, by Carol A. Auer. *TRENDS in Plant Science* 8:591–97. http://www.canr.uconn.edu/plsci/auer/TIPS-tracking-genes.pdf.

3. Cert-ID, http://www.cert-id.com and http://www.cert-id.com/industry_industry_glossary.htm.

4. Mid-West Seed Services, Inc., http://www.mwseed.com.

5. EurepGAP_GR_IFA_V2-0Mar05_update_08Jun06, EurepGAP c/o Food-PLUS GmbH, http://www.eurepgap.org.

6. Trace-I Guideline, Glossary, 2003, p. 102. This glossary provides a list of important definitions and abbreviations that are used in these guidelines, http://www.gsluk.org/uploaded/doc_library/Traceability%20 guidelines340.pdf.

7. The Codex General Standard for the Labeling of Prepackaged Foods, http://www.fao.org/docrep/005/Y2770E/y2770e02.htm.

8. EnviroLogix Glossary, http://www.envirologix.com/artman/publish/article_5.shtml and http://www.envirologix.com.

9. Can-Trace Technology Guidelines, 2006, Glossary of Terms, pp. 21–22, http://www.can-trace.org/portals/0/docs/Can-Trace%20Technology%20 Guidelines%20Mar%202006%20-%20mjf.pdf.

10. USDA NOP, Subpart A—Definitions, § 205.2 Terms defined, http://www.ams.usda.gov/NOP/NOP/standards/DefineReg.html.

11. Dennis Strayer. 2002. *Identity-Preserved Systems—A Reference Handbook*. Boca Raton, FL: CRC Press LLC.

12. Demeter-International, http://www.demeter.net.

13. State Indiana Crop Improvement Association, Information Site and Seed Technology Center, http://www.indianacrop.org.

Directory of Resources

AgBioForum Journal (Chapter 9): http://www.agbioforum.org/v3n1/v3n1a08-caswell.pdf

AgMotion International Trade Company (Chapter 5): http://www.agmotion.com

AgVision/DMI Computer Technologies (Chapter 10): http://www.dmi-agvision.com

AIB International (American Institute of Baking—International) (Chapter 5): http://www.aibonline.org

Alpro Soya (Chapter 5): http://www.alprosoya.co.uk/homepage/_en-UK/index.html

American Soybean Association (ASA) (Chapter 9): http://www.soygrowers.com/step/darmstadt.htm

AmericanERP, LLC (Enterprise Resource Planning) (Chapter 10): http://www.AmericanERP.com

ANZFA Food Industry Recall Protocol (Chapter 12): http://www.foodstandards.gov.au

Association of Official Seed Certifying Agencies (AOSCA) (Chapter 10): http://www.aosca.org

ATTRA - National Sustainable Agriculture Information Service (Chapter 9): http://attra.ncat.org

Batch Processing (Chapter 3) (Dupuy): http://www.sciencedirect.com

Biogenetics Services, Inc. (Chapter 8): http://www.biogeneticservices.com

BRS, Ltd. (Chapter 7): http://www.brsltd.org

BSI Global (Chapter 7): http://www.bsi-global.com/index.xalter

California Seed and Plant Laboratory, Inc. (Chapter 8): http://www.calspl.com

Caliso Consulting, LLC (Chapter 7): http://www.caliso9000.com/index.shtml

Canadian Grain Commission (Chapter 6): http://www.grainscanada.gc.ca/main-e.htm

Canadian Grain Commission Laboratory (Chapter 8): http://www.grainscanada.gc.ca/grl/grl-e.htm

Canadian Seed Institute (CSI) and CSI Centre for Systems Integration (Chapter 4): http://www.csi-ics.com

Canadian Soybean Export Association (Chapter 6): http://www.canadiansoybeans.com

Can-Trace (Chapter 6): http://www.can-trace.org

Cargill - InnovaSure IdP (Chapter 5): http://www.cargilldci.com/innovasure/index.shtm

Cert ID (Chapter 7): http://www.cert-id.com

CII Laboratory Services (Chapter 8): http://www.ciilab.com

Clarkson Grain Company, Inc. (Chapter 5): http://www.clarksongrain.com/2002CGweb.htmm

Codex Alimentarius (Codex) and FAO/WHO Food Standards (Chapter 6): http://www.codexalimentarius.net/web/index_en.jsp

Co-Extra (Co-Existence and Traceability) (Chapter 11): http://www.coextra.org/ default.html

Critereon Company, LLC (Chapter 11): http://www.critereon.com/index.html

CSB-System (Chapter 10): http://www.csb-system.com

EnviroLogix Inc. (Chapter 8): http://www.envirologix.com

Eurofins GeneScan, Inc. (Chapter 8): http://www.eurofins.com

European Union (EU) Standards (Chapter 6): http://europa.eu

Farm Foundation (Chapter 9): http://www.farmfoundation.org

Farm Identity Preservation and Traceability Program and its Components, an Example of a (General) Program: Appendix B (Sundstrom and Williams)

Food Consulting Company (Chapter 11): http://www.foodlabels.com/index.htm

Food Standards Agency (United Kingdom) (Chapter 9): http://www.foodstandards. gov.uk

FoodTrace (Chapter 11): http://www.eufoodtrace.org/index.php

FoodTrack Inc. and FoodTrack Incident Report (Chapter 12): http://www.food track.net

FoodTrust Certification (Chapter 7): http://www.foodtrustcert.com

Genetic ID, Inc. (Chapter 8): http://www.genetic-id.com

GLOBALGAP (Chapter 6): http://www.eurepgap.org/Languages/English/index_ html

Grain Elevator and Processing Society (GEAPS)/Purdue Distance Learning Program (Chapter 5): http://www.geaps.com

GS1 and EAN.UCC (Chapter 10): http://www.gs1.org and http://www.eanucc.org

HACCP standards, HACCP Web.com, and HACCP training providers (Chapter 6)

Halal standard (Chapter 6)

IdentityPreserved.com (Chapter 10): http://www.identitypreserved.com

Indiana Crop Improvement Association (Chapter 4): http://www.indianacrop.org

International Federation of Organic Agriculture Movements (IFOAM) and International Organic Accreditation Services (IOAS) (Chapter 6): http://www. ioas.org

International Food and Agribusiness Management Association (IAMA) (Chapter 9): http://www.ifama.org

International Food Policy Research Institute (IFPRI) (Chapter 9): http://www.ifpri.org

International Organization for Standards (ISO)—22000 (Chapter 6): http://www.iso. org/iso/en/ISOOnline.frontpage

International Seed Federation (ISF) (Chapter 9): http://www.worldseed.org

Introduction to 3rd Party Certification/Validation by Laboratories (Chapter 8) (Lees, Schwägele, and Auer)

Iowa Crop Improvement Association (Chapter 4): http://www.agron.iastate.edu/icia

Japanese Agricultural Standard (JAS) (includes organic standards) (Chapter 6): http://www.maff.go.jp/eindex.html

John Deere - Food Origins (Chapter 5): http://www.deere.com/en_US/deerecom/ agriservices/inc.html

Kosher Standard (Chapter 6)

Linnet: Croplands—The System (CTS) (Chapter 10): http://www.croplandthe system.com

MapShots, Inc. (Chapter 10): http://www.mapshots.com

MicroSoy Corporation (Chapter 5): http://www.microsoyflakes.com/index.htm

Mid-west Seed Services, Inc. (Chapter 8): http://www.mwseed.com

Minnesota Crop Improvement Association (Chapter 4): http://www.mncia.org

National Starch–TrueTrace (Chapter 5): http://www.nationalstarch.com

Neogen Corporation (Chapter 8): http://www.neogen.com

Northern Great Plains, Inc. (Chapter 9): http://www.ngplains.org

Northland Grain & Seed, Northland Organic, and Pacifica Research (Chapter 5): http://www.northlandorganic.com/index.html and http://www.pacificaresearch.com

Novecta, LLC (Chapter 11): http://www.novecta.com/index.html

Organic Crop Improvement Association (OCIA) International (Chapter 6): http://www.ocia.org

Organic Farming (Overview) (Chapter 6) (Delate)

Other Regional and Religious Systems (Chapter 6)

OurFood.com—Database of Food & Related Sciences (Chapter 12): http://www.OurFood.Com

Pacifica Software (Chapter 10): http://www.pacificaresearch.com/Company.html

PathTracer (Chapter 10): http://www.pathtracer.net

Pioneer Hi-Bred International, Inc.—MarketPoint (Chapter 5): http://www.pioneer.com and http://www.pioneer.com/marketpoint/traceability/default.htm

Product Recall: Disasters waiting to happen (Chapter 12) (Weaver)

Product Recall Insurance (Chapter 12): http://www.productrecallins.com

Protein Technologies International, Ltd. (Chapter 8): http://www.emeraldinsight.com/Insight/ViewContentServlet?Filename=/published/emeraldfulltextarticle/pdf/0170990603.pdf

QMI—Management Systems Registration (Chapter 7): http://www.qmi.com

Seedsmen Professional Liability Insurance (Chapter 12): http://www.rattnermackenzie.com

SGS SA (Société Générale de Surveillance) (Chapter 7): http://www.sgs.com

Strategic Diagnostics, Inc. (Chapter 8): http://www.sdix.com

Sudan 1, an example of why product recall insurance is needed: Appendix O (Partos)

The Organic & Non-GMO Report (and *2006 Non-GMO Sourcebook*) (Chapter 11): http://non-gmoreport.com

The SQF Institution (Chapter 6): http://www.sqfi.com

TRACE (TRAcing Food Commodities in Europe) (Chapter 11): http://www.trace.eu.org/index.php

TraceFish (Chapter 5): http://www.tracefish.org

TÜV America, Inc. (Chapter 7): http://www.tuvamerica.com/home.cfm

Understanding the Recall Concept in the Food Industry (Chapter 12) (Kaletunç): http://ohioline.osu.edu/AEX-Fact/0251.htm

U.S. Department of Agriculture (USDA)—General (Chapter 6): http://www.usda.gov/wps/portal/usdahome

U.S. Food and Drug Administration (FDA) Standards (Chapter 6): http://www.fda.gov

U.S. Department of Agriculture—Grain Inspection, Packers & Stockyards Administration (GIPSA) (Chapter 6): http://www.gipsa.usda.gov/GIPSA/webapp?area=home&subject=landing&topic=landing

U.S. Department of Agriculture—National Organic Program (NOP) (Chapter 6): http://www.ams.usda.gov/nop/indexIE.htm

Value-Enhanced Grains (VEG) Solutions (website) (Chapter 11): http://www.veg-rains.org

Vertical Software, Inc. (Chapter 10): http://home.verticalsoftware.net

Why Food Recall Insurance Is Needed—A Short History of Food Recall Insurance (Chapter 12)

Why Product Insurance Is Needed and What is Offered: Appendix N (Steves)

Bibliography

Adolphe, Dale. 1999. Genetically modified organisms—what can they mean for you? *Proceedings of GEAPS Exchange '99*, 71–77.

Anonymous. 2005. Consumers drive changes in the food system. *Feed & Grain* 44:39–40.

Auer, Carol. 2003. Tracking genes from seed to supermarket: techniques and trends. *TRENDS in Plant Science* 8:591–97.

Bantham, Amy, and Jean-Louis Duval. 2004. Connecting food chain information for food safety/security and new value—Area III: Food safety and food security. http://www. ifama.org/conferences/2004conference/papers/bantham1113.pdf.

Bender, Karen. 2003. Product differentiation and identity preservation: implications for market development in U.S. corn and soybeans. Presented at the Symposium of Product Differentiation and Market Segmentation in Grains and Oilseeds: Implications for Industry in Transition, Washington, DC.

Bhalgat, Mahesh K., William P. Ridley, Allan S. Felsot, and James N. Seiber, eds. 2004. *Agricultural Biotechnology: Challenges and Prospects*. 2004. New York: American Chemical Society (ACS) (symposium series 866), Oxford University Press.

Bodria, Luigi. 2002. System integration and certification. The market demand for clarity and transparency—Part 2. Presented at the Club of Bologna, Italy Meeting, Nov. 16, 2002. *Agricultural Engineering International: the CIGR Journal of Scientific Research and Development*. Invited Overview Paper. vol. V, February 2003.

Boyle, Robert, Bill Jorgenson, William Pape, and Jerred Pauwels. 2004. How do you measure traceability performance? *Food Traceability Report* March:14–15.

Boyle, Robert D. 2004. Full-product traceability: intelligent tracking is key to supply chain integrity. *APICS—The Performance Advantage* 5:26–27.

Brown, Richard. 1999. Producing non-GM soy protein through identity preservation. *Nutrition & Food Science* 6:281–84.

Bryar, Peter J. 1999. From paddock to plate—the first step. Proceedings of the Third International and Sixth National Research Conference on Quality Management, RMIT Storey Hall, Melbourne, Australia.

Bullock, David, Marion Desquilbet, and Elisavet Nitsi. 2000. The economics of non-GMO segregation and identity preservation. Paper presented at the AAEA annual meetings, Tampa, FL.

Caswell, Julie A. 2000. Labeling policy for GMOs: to each his own? *AgBioForum* 3:53–57.

Clapp, Stephen. 2002. Traceability: a global perspective. *Food Traceability Report* 1:16–19.

Delate, Kathleen. 2003. *Fundamentals of Organic Agriculture*. Ames, IA: Iowa State University Extension, Bulletin PM 1880.

Demand for product recall insurance is set to spiral in the new year. 2004. *Food Manufacture—The Information Resource for Food and Drink Processing*. Crawley, UK: William Reed Publishing Ltd, pp. 22.

Dupuy, Clement, Valerie Botta-Genoulaz, and Alain Guinet. 2005. Batch dispersion model to optimise traceability in food industry. *Journal of Food Engineering* 70:333–39.

Economic impacts of genetically modified crops on the agri-food sector: a first review. 2000. Commission of the European Communities Directorate-General for Agriculture.

Enhancing Brand Value—Tracking and Tracing Applications in the Food Industry. 2004. White Paper by Rockwell Automation.

Fearne, Andrew. 1998. The evolution of partnerships in the meat supply chain: insights from the British beef industry. *Supply Chain Management: An International Journal* 3:214–31.

Fonsah, Greg. 2005. Tackling traceability: having a traceback plan in place allows growers to monitor produce from the farm to the packing shed to the consumer. *American/Western Fruit Grower* 14:14.

Food traceability and assurance in the global food system. 2004. *Farm Foundation's Traceability and Assurance Panel Report July 2004*. Oak Brook, IL: Farm Foundation.

From GMO to nano: a familiar debate over a new technology. 2006. *The Non-GMO Report* 6:1–2.

Garrett, Arthur S. 2005. *Food Recall Insurance: Why Your Company Should Take a Look*. Washington, DC: Keller and Heckman LLP.

Germans find Italian organic standards wanting. 2006. *Agra Europe Weekly* no. 2218:N/6.

Glassheim, Eliot, Jerry Nagel, and Cees D. Roele. 2005. The new marketplace in European agriculture: environmental and social values within the food chain. *Plains Speaking* 8:5–12.

Golan, Elise, Barry Krissoff, and Fred Kuchler. 2004. Food traceability one ingredient in a safe and efficient food supply. *Amber Waves* 2:14–21.

Golan, Elise, Barry Krissoff, Fred Kuchler, Linda Clavin, Kenneth Nelson, and Gregory Price. 2004. *Traceability in the U.S. Food Supply: Economic Theory and Industry Studies*. Washington, DC: USDA Economic Research Service, Agricultural Report Number 830 (AER-830).

Grande, Bill. 2003. Servicing IP production and marketing: a third party role. Presented at the Product Differentiation and Market Segmentation in Grains and Oilseeds: Implications for Industry in Transition symposium, sponsored by Economic Research Service, USDA, and The Farm Foundation, Washington, DC.

Gruère, Guillaume P. 2006. An analysis of trade related international regulations of genetically modified food and their effects on developing countries. International Food Policy Research Institute (IFPRI) EPT Discussion Paper 147, Environment and Production Technology Division.

Hobbs, Jill E. 2002. Traceability and country of origin labelling. Prepared for the Theme Day "Consumer Driven Agricultural Trade" at the International Agricultural Trade Research Consortium Annual Meeting and presented at the Policy Dispute Information Consortium 9th Agricultural and Food Policy Information Workshop, April 25, 2003, Montreal, Canada.

Huffman, Wallace E. 2004. Production, identity preservation, and labeling in a marketplace with genetically modified and non-genetically modified foods. *Plant Physiology* 134:3–10.

Identity preservation: "a processors perspective." Riverland Oilseed Processors online PowerPoint presentation. Available at http://australianoilseeds.com/__data/page/44/Identity_Preservation-Jon_Slee,_Riverland_Oilseed_Processors.pdf.

Jenkins, C. 2003. HACCP further up the food chain. *Food Quality* 10:55–56.

Jones, Michael K., and Bruce R. Rich. 2004. The food industry from feeling full to fulfillment. Paper presented to the Social Change in the 21st Century Conference Centre for Social Change Research, Queensland University of Technology.

Jorgenson, Bill. 2004. Technology for the supply chain: benefits, costs, perceptions. Presented at the International Food and Agribusiness Management Association (IAMA) meeting, June 14, 2004, Montreux, Switzerland.

Kalaitzandonakes, Nicholas, Richard Maltsbarger, and James Barnes. 2001. Global identity preservation costs in agricultural supply chains. *Canadian Journal of Agricultural Economics* 49:605–15.

Kaletunç, Gönül, and Ferhan Özadali. 2002. Understanding the recall concept in the food industry. *Ohio State University Extension Fact Sheet*.

Kim, Henry M., Mark S. Fox, and Michael Gruninger. 1995. An ontology of quality for enterprise modelling. Paper presented at the 4th Workshop of Enabling Technologies, Infrastructure for Collaborative Enterprises (WET-ICE '95), Berkeley Springs, WV.

Latimer, Jack. 1946. Friendship among equals. Interview with Willy Kuert; Swiss delegate to the London Conference. The conference of national standardizing organizations, which established ISO, took place in London October 14–26, 1946.

Lees, Michèle, ed. 2003. *Food Authenticity and Traceability*. Cambridge, UK: Woodhead Publishing Ltd, CRC Press.

Lipton, Gigi. 1998. Product traceability: a guide for locating recalled manufactured goods. Quality Congress. ASQ's 52nd Annual Quality Congress Proceedings. ABI/INFORM Global.

Maltsbarger, Richard, and Nicholas Kalaitzandonakes. 2000. *Study Reveals Hidden Costs in IP Supply Chain*. Economics & Management of Agrobiotechnology Center (EMAC), University of Missouri-Columbia.

Mayer, Cheryl. 2003. Tracking & traceability. *Germination—The Magazine of the Canadian Seed Industry* 7:10–12.

Miskin, Michael. 2006. Connecting the supply chain: traceability management systems can do more than track problems. *Food in Canada* 3:29.

Moe, Tina. 1998. Perspectives on traceability in food manufacture. *Trends in Food Science & Technology* 9:211–14.

Noonan, John, and Graham McAlpine. 2003. Management of value chains through SQF: a global standard for food safety and quality. Presented at the Muresk 75th Anniversary Conference: From farm to fork: linking producers to consumers through value chains.

Nunn, Janet. 2000. What lies behind the GM label on UK foods? *AgBioForum* 3:250–54.

Pape, Will. 2006. Connecting the dots for food recalls. *Food Traceability Report* 1:12–13.

Parcell, Joe. 2001. An initial look at the Tokyo grain exchange non-GMO soybean contract. *Journal of Agribusiness* 19:85–92.

Partos, Lindsey. 2005. After Sudan 1: can food firms afford to skip product recall insurance? http://www.foodnavigator.com/content/view/print/83829.

Peres, Bruno, Nicolas Barlet, Gérard Loiseau, and Didier Montet. 2007. Review of the current methods of analytical traceability allowing determination of the origin of foodstuffs. *Food Control* 18:228–35.

Phillips, Peter W.B., and Heather McNeill. 2000. Labeling for GM foods: theory and practice. *AgBioForum* 3:219–24.

Pinto, D.B., I. Castro, and A.A. Vicente. 2006. The use of TIC's as a managing tool for traceability in the food industry. *Food Research International* 39:772–81.

Pointon, David. 1995. Wintulichs Pty. Ltd. and marketing environmental disturbance "Garibaldi effect." *Agribusiness Review* 3:37–50.

Proposal for a regulation of the European Parliament and the Council concerning traceability and labelling of genetically modified organisms and traceability of food and feed products produced from genetically modified organisms and amending Directive 2001/18/EC, July 2001. Commission of the European Communities. Available at http://europa.eu.int/comm/food/fs/biotech/biotech09_en.pdf.

Rethinking the role of the producer. 2003. From the European Study Tour, Northern Great Plains, Annual Report. Available at http://www.ngplains.org/documents/2003%20Annual%20Report%20Final.pdf.

Roseboro, Ken, ed. 2006. *The 2006 Non-GMO Sourcebook*. Fairfield, IA: Writing Solutions, Inc.

Schellhorn, Eric. 2005. Kosher in the mainstream. *Food Processing* May. http://www.foodprocessing.com/articles/2005/397.html?page=print.

Schwägele, Fredi. 2005. Traceability from a European perspective. *Meat Science* 71:164–73.

Send in the clones: FDA set to approve food from cloned animals. 2007. *The Organic & Non-GMO Report* 7:1, 3.

Smith, Gary C., John D. Tatum, Keith E. Belk, et al. 2005. Traceability from a US perspective. *Meat Science* 71:174–93.

Smith Ian, and Anthony Furness. 2006. *Improving Traceability in Food Processing and Distribution.* Cambridge, UK: Woodhead Publishing Ltd, and Boca Raton, FL: CRC Press LLC.

Smith, Marilyn. 1998. Overview of food safety in oilseeds and oilseed products. Proceedings of the Australian Postharvest Technical Conference, Canberra, Australia.

Smyth, Stuart, and Peter W. B. Phillips. 2002. Product differentiation alternatives: identity preservation, segregation, and traceability. *AgBioForum* 5:30–42.

Smyth, Stuart, and Peter W. B. Phillips, et al. 2004. *Regulating the Liabilities of Agricultural Biotechnologies.* Cambridge, MA: CABI Publishing.

Steves, Bernie. 2000. An interview between IBS vice-president and Food Technology Source (FTS) forum. May/June. http://www.foodtechsource.com/emag/011/trend.htm.

Strayer, Dennis. 2002. *Identity-Preserved Systems: A Reference Handbook.* Boca Raton, FL: CRC Press LLC.

Sundstrom, Frederick J., Jack Williams, Allen Van Deynze, and Kent J. Bradford. 2002. Identity preservation of agricultural commodities. *Agricultural Biotechnology in California Series Publication 8077*, pp. 1–15.

Tanner, Steven N. 2001. Testing for genetically modified grain. *Proceedings of GEAPS Exchange '01.* Washington, DC: USDA, GIPSA.

Wagner, Gary L., and Eliot Glassheim. 2003. Traceability of agricultural products. Available at http://www.ngplains.org/documents/traceability_report.pdf.

Weaver, Patrick. 2003. Product recalls: disasters waiting to happen. *Australian Business Report* June–July:2–5.

Index